Microbial Biofilms

Microbial Biofilms

Properties and Applications in the Environment, Agriculture, and Medicine

Edited by

Bakrudeen Ali Ahmed Abdul

CRC Press
Taylor & Francis Group
Boca Raton London New York

CRC Press is an imprint of the
Taylor & Francis Group, an **informa** business

First edition published 2021
by CRC Press
6000 Broken Sound Parkway NW, Suite 300, Boca Raton, FL 33487-2742

and by CRC Press
2 Park Square, Milton Park, Abingdon, Oxon, OX14 4RN

CRC Press is an imprint of Taylor & Francis Group, LLC

Library of Congress Cataloging-in-Publication Data
Names: Abdul, Bakrudeen Ali Ahmed, editor.
Title: Microbial biofilms : properties and applications in the environment, agriculture, and medicine / Bakrudeen Ali Ahmed Abdul.
Description: Boca Raton : CRC Press, 2020. | Includes bibliographical references and index. | Summary: "This book provides a broad range of applications and recent advances in the search of biofilm materials in nature. It also explains the future implications for biofilms in the areas of advanced molecular genetics, pharmaceuticals, pharmacology and toxicology"— Provided by publisher.
Identifiers: LCCN 2020024162 (print) | LCCN 2020024163 (ebook) | ISBN 9780367415068 (hardback) | ISBN 9780367415075 (ebook)
Subjects: LCSH: Biofilms.
Classification: LCC QR100.8.B55 M55 2020 (print) | LCC QR100.8.B55(ebook) | DDC 579/.17—dc23
LC record available at https://lccn.loc.gov/2020024162
LC ebook record available at https://lccn.loc.gov/2020024163

ISBN: 978-0-367-41506-8 (hbk)
ISBN: 978-0-367-41507-5 (ebk)

Typeset in Times
by codeMantra

Contents

PART III Microbial Biofilms: Chemical Sciences, Natural Products, and Biotechnological Approaches

PART IV Microbial Biofilms: Biomass, Plant Growth, Soil Nutrient, and Wastewater Management

Preface

Biofilms are polymicrobials that associate with abiotic and biotic (natural and artificial) surfaces. These adherent cells are embedded in an extracellular matrix that contains extracellular polymeric substances (lipopolysaccharides, proteins, etc.). Biofilm-forming microorganisms include archaea, fungi, protozoa, algae, and bacteria. The emergent advanced strategies such as quorum quenching, bacteriophages, nanotechnology, bacteriocin, numerous enzymes, and natural substances/extracts are hopeful that might aid to find suitable antibiofilm approaches which could be superior to the conventional ones. The biofilm structure depends on the availability of light, grazing capability, nutrient availability, hydrodynamics, and nature of substratum. It plays various roles in agriculture, environment, biotechnology, industries, health, and wastewater management.

This book aims to collect the material for biofilm conservation, management, bioenergy system, and biomaterials used in environmental systems. Moreover, biofilms have several prospective bioactive compounds, and these compounds have been increased through biotechnological approaches. Biofilms are used to improve biomass, plant growth, soil nutrient, and wastewater management and are also applied in the treatment of human diseases.

This book is divided into five parts:

- Chapter 1–2 provide the detailed information about biofilms quality, quantity improvement, and conservation in biotechnological practices through fermentation.
- Chapter 3–5 deals with biofilms derivatives impact on agriculture, uses in bioenergy conservation, and application in biomaterial preparation from environmental and aquaculture system are described.
- Chapters 6–9 provide the detailed information about biofilm materials affecting secondary metabolites (natural products) production and their application in agriculture.
- Chapters 10–14 deal with biofilms' applications in wastewater treatment, bioelectricity production, and bioremediation used in spirulina production.
- Chapters 15–20 describe the usage of microbial biofilms and their derivatives of essential oil application in chronic diseases, human pathogenic fungi, and vaginal infections.

This book provides cumulative information about microbial biofilm preparation, improvement methods and application in agriculture, aquaculture, and environment as well application in the treatment of human chronic diseases. Hence, this book will be an important reference for the marine science, biotechnology, and natural product scientists, and whoever working in the microbial biofilm field.

Preface

Editor

Dr. Bakrudeen Ali Ahmed Abdul is an Associate Professor, Head of the Department of Biochemistry, and Dean of the School of Life Sciences PRIST Deemed University, Vallam, Thanjavur, Tamil Nadu, India. His research areas include the application of plant biochemistry, bioactive compound production, biotechnological methods, development of pharmaceutical products, and pharmacological studies. He is a member of the Indian Association of Biomedical Scientists (IABMS) and the Malaysian Society of Plant Physiology (MSPP). He received the Agathiyar Award from the IABMS in 2007 for his work. He received his bachelor's degree (chemistry), master's degree (biochemistry), and Ph.D. (plant science-biochemistry) from Bharathidasan University, Tiruchirappalli, Tamil Nadu.

Contributors

Sidra Abbas
Department of Biotechnology
Fatima Jinnah Women University
Rawalpindi, Pakistan

Bakrudeen Ali Ahmed Abdul
Department of Biochemistry
Centre for Research and Development
PRIST Deemed University
Thanjavur, Tamil Nadu, India

Ayaz Ahmed
Dr. Panjwani Center for Molecular Medicine and Drug Research, International Center for Chemical and Biological Sciences
University of Karachi
Karachi, Pakistan

Noor-ul-ain Ali
Department of Microbiology
Abdul Wali Khan University
Mardan, Pakistan

Mohd. Musheer Altaf
Department of Life Science
Institute of Information Management and Technology
Aligarh, Uttar Pradesh, India

S. Ambiga
Department of Biochemistry
Centre for Research and Development
PRIST Deemed University
Thanjavur, Tamil Nadu, India

Saadia Andleeb
Atta-ur-Rahman School of Applied Biosciences, Industrial Biotechnology
National University of Sciences and Technology
Islamabad, Pakistan

Jacqueline Cosmo Andrade Pinheiro
Laboratory Bioassay—LABIO
Federal University of Cariri—UFCA,
Brejo Santo, CE, Brazil

Fozia Anjum
Department of Chemistry
Government College University
Faisalabad, Pakistan

A. Thaminum Ansari
Department of Chemistry
Muthurangam Government Arts College
Vellore, Tamil Nadu, India

N. Arunkumar
Department of Microbiology, School of Basic and Applied Sciences
Central University of Tamil Nadu
Thiruvarur, Tamil Nadu, India

K. Ashokkumar
Department of Crop Improvement, Cardamom Research Station
Kerala Agricultural University
Idukki, Kerala, India

Bárbara Buhl
Department of Chemistry Biotechnology graduate program
Vale do Taquari University—Univates
Lajeado, RS, Brazil

Sayed Muhammad Ata Ullah Shah Bukhari
Department of Microbiology
Abdul Wali Khan University
Mardan, Pakistan

Nuno Cerca
Laboratory of Research in Biofilms Rosário Oliveira, Centre of Biological Engineering
University of Minho
Braga, Portugal

Sabrina Grando Cordeiro
Department of Chemistry Biotechnology graduate program
Vale do Taquari University—Univates
Lajeado, RS, Brazil

Bruna Costa
Department of Chemistry Biotechnology graduate program
Vale do Taquari University—Univates
Lajeado, RS, Brazil

Henrique Douglas Melo Coutinho
Laboratory of Microbiology and Molecular Biology—LMBM
Regional University of Cariri—URCA
Crato, CE, Brazil

Bárbara de Azevedo Ramos
Center of Biological Sciences and Department of Biochemistry
Federal University of Pernambuco–UFPE
Recife, PE, Brazil

Camila Roberta de Castro
Department of Chemistry Biotechnology graduate program
Vale do Taquari University—Univates
Lajeado, RS, Brazil

Elisete Maria de Freitas
Department of Chemistry Biotechnology graduate program
Vale do Taquari University—Univates
Lajeado, RS, Brazil

Luciene Ferreira de Lima
Laboratory of Microbiology and Molecular Biology—LMBM
Regional University of Cariri—URCA
Crato, CE, Brazil

Ali Demirci
Department of Agricultural and Biological Engineering
Pennsylvania State University
University Park, Pennsylvania

Eduardo Miranda Ethur
Department of Chemistry Biotechnology graduate program
Vale do Taquari University—Univates
Lajeado, RS, Brazil

Anila Fariq
Department of Biotechnology
Fatima Jinnah Women University
Rawalpindi, Pakistan

Hina Fatima
Department of Biochemistry
University of Agriculture
Faisalabad, Pakistan

Mahroze Fatima
Department of Fisheries and Aquaculture
University of Veterinary and Animal Sciences
Lahore, Pakistan

Maria Audilene Freitas
Laboratory of Medical Mycology Sylvio Campos
Federal University of Pernambuco—UFPE
Recife, PE, Brazil

Vijay Kumar Garlapati
Department of Biotechnology & Bioinformatics
Jaypee University of Information Technology
Waknaghat, Himachal Pradesh, India

Pranab Goswami
Department of Biosciences and Bioengineering
Indian Institute of Technology
Guwahati, Assam, India

Peterson Haas
Department of Chemistry
Biotechnology graduate program
Vale do Taquari University—Univates
Lajeado, RS, Brazil

Ruhil Hayati Hamdan
Faculty of Veterinary Medicine
Universiti Malaysia Kelantan
Kota Bharu, Kelantan, Malaysia

Lucélia Hoehne
Department of Chemistry
Biotechnology graduate program
Vale do Taquari University—Univates
Lajeado, RS, Brazil

Hafiz M. N. Iqbal
School of Engineering and Sciences
Tecnologico de Monterrey
Monterrey, NL, Mexico

Muhammad Irfan
National Institute for Biotechnology and Genetic Engineering
Faisalabad, Pakistan

Muhsin Jamal
Department of Microbiology
Abdul Wali Khan University
Mardan, Pakistan

Nor Fadhilah Kamaruzzaman
Faculty of Veterinary Medicine
Universiti Malaysia Kelantan
Kota Bharu, Kelantan, Malaysia

M. Karthik
Department of Agricultural Microbiology
Tamil Nadu Agricultural University
Coimbatore, Tamil Nadu, India

Carla Kauffmann
Department of Chemistry
Biotechnology graduate program
Vale do Taquari University—Univates
Lajeado, RS, Brazil

Anum Khalid Khan
Dr. Panjwani Center for Molecular Medicine and Drug Research, International Center for Chemical and Biological Sciences
University of Karachi
Karachi, Pakistan

Mohd Sajjad Ahmad Khan
Department of Basic Sciences
Imam Abdulrahman Bin Faisal University
Dammam, Kingdom of Saudi Arabia

Sher Jamal Khan
Institute of Environmental Sciences and Engineering (IESE), School of Civil and Environmental Engineering (SCEE)
National University of Sciences and Technology (NUST)
Islamabad, Pakistan

Waheed S. Khan
National Institute for Biotechnology and Genetic Engineering
Faisalabad, Pakistan

R. Krishnamoorthy
Department of Crop Management (Agricultural Microbiology)
Vanavarayar Institute of Agriculture
Pollachi, Tamil Nadu, India

Daniel Kuhn
Department of Chemistry Biotechnology graduate program
Vale do Taquari University—Univates
Lajeado, RS, Brazil

Ehsan Mahdinia
Center for Biopharmaceutical Education and Training
Albany College of Pharmacy and Health Sciences
Albany, New York

Hira Munir
Department of Biochemistry and Biotechnology
University of Gujrat
Gujrat, Pakistan

Sunandan Naha
Department of Biosciences and Bioengineering
Indian Institute of Technology
Guwahati, Assam, India

Leandra Andressa Pacheco
Department of Chemistry Biotechnology graduate program
Vale do Taquari University—Univates
Lajeado, RS, Brazil

Tan Li Peng
Faculty of Veterinary Medicine
Universiti Malaysia Kelantan
Kota Bharu, Kelantan, Malaysia

Aluisie Picolotto
Department of Chemistry Biotechnology graduate program
Vale do Taquari University—Univates
Lajeado, RS, Brazil

Lutfur Rahman
National Institute for Biotechnology and Genetic Engineering
Faisalabad, Pakistan

Sana Raza
Department of Microbiology
Abdul Wali Khan University
Mardan, Pakistan

Redaina
Department of Microbiology
Abdul Wali Khan University
Mardan, Pakistan

Asma Rehman
National Institute for Biotechnology and Genetic Engineering
Faisalabad, Pakistan

Muhammad Tjammal Rehman
Department of Biochemistry
University of Agriculture
Faisalabad, Pakistan

Rengasamy Sathya
Department of Microbiology
Centre for Research and Development
PRIST Deemed University
Thanjavur, Tamil Nadu, India

Talita Scheibel
Department of Chemistry Biotechnology graduate program
Vale do Taquari University—Univates
Lajeado, RS, Brazil

Ytan Andreine Schweizer
Department of Chemistry
Biotechnology graduate program
Vale do Taquari University—Univates
Lajeado, RS, Brazil

Surajbhan Sevda
Department of Biosciences and Bioengineering
Indian Institute of Technology
Guwahati, Assam, India
National Institute of Technology
Warangal, Warangal, Telangana, India

Liloma Shah
Department of Microbiology
Abdul Wali Khan University
Mardan, Pakistan

Syed Zakir Hussain Shah
Department of Zoology
University of Gujrat
Gujrat, Pakistan

Muhammad Shahid
Department of Biochemistry
University of Agriculture
Faisalabad, Pakistan

Hafiz Muhammad Aamir Shahzad
Institute of Environmental Sciences and Engineering (IESE), School of Civil and Environmental Engineering (SCEE)
National University of Sciences and Technology (NUST)
Islamabad, Pakistan

Swati Sharma
Department of Biotechnology & Bioinformatics
Jaypee University of Information Technology
Waknaghat, Himachal Pradesh, India

Ata Ullah
National Institute for Biotechnology and Genetic Engineering
Faisalabad, Pakistan

Thangaprakasam Ushadevi
Department of Microbiology
Centre for Research and Development
PRIST Deemed University
Thanjavur, Tamil Nadu, India

Azra Yasmin
Department of Biotechnology
Fatima Jinnah Women University
Rawalpindi, Pakistan

Muhammad Bilal Yazdani
National Institute for Biotechnology and Genetic Engineering
Faisalabad, Pakistan

Part I

Microbial Biofilms: Properties, Biodiversity, Conservation, and Management

1 Microbial Biofilms

Properties, Biodiversity, Conservation and Management

Muhsin Jamal, Sayed Muhammad Ata Ullah Shah Bukhari, Sana Raza, Liloma Shah, Redaina, and Noor-ul-ain Ali
Abdul Wali Khan University

Saadia Andleeb
National University of Sciences and Technology

CONTENTS

1.1 INTRODUCTION

Biofilm is defined as a microbial community adhered to abiotic or biotic surfaces surrounded by self-produced EPS. The genetic diversity of biofilm forming organisms verifies that biofilm is an olden form of life of a microorganism (Chandra et al. 2001). For the first time biofilm was noticed by Van Leeuwenhoek on tooth surface (Heukelekian and Heller 1940). Biofilms are thought to be an aggregation of microorganisms adhered to a surface enclosed within a self produced EPS (Hall-Stoodley and Stoodley 2009). Bacterial cells are protected from adverse ecological conditions by biofilm and thus show extreme resistance to antibiotics (Brown et al. 1988). Biofilms are extremely harmful in industrial, natural, and medical settings. For example, development of biofilm at the surface of medical devices like implants or catheters frequently causes infections which are then hard to treat (Hatt and Rather 2008). However, infections have been related to the development of biofilm on surfaces of human on urinary tract, skin, and teeth (Hatt and Rather 2008). However, biofilm formation on human surfaces is not at all times harmful. For instance, biofilms of dental plaque contain lots of species, and the disease-causing abilities are determined commonly by the composition of the community (Kreth et al. 2008). Biofilms are formed universally. For instance, biofilms made on ships hulls and within pipes can cause serious complications (De Carvalho 2007). In natural surroundings, the formation of biofilm frequently permits symbioses (mutualistic). For example, *Actinobacteria* frequently rise on ants, permitting ants to sustain microbe-free fungal gardens (Danhorn and Fuqua 2007). There are several profits that a community of bacteria may obtain from biofilm formation. Biofilms protect microbial cells by different ways such as providing antimicrobial resistance, defense from grazing of protozoan, and also evading host defense system (Anderson and O'Toole 2008). The basic structural unit of biofilms is microcolony, which is a discrete group of cells of bacteria surrounded by the matrix of EPS. Microcolonies comprise 10%–25% of cells and 79%–90% of the matrix of EPS (Costerton 1999). The extrapolymeric substance comprises mostly water that helps to transport the nutrients inside biofilm, and furthermore, enzymes, DNA, protein, and RNA are also present (Table 1.1). The biofilm structure comprises two important constituents, i.e., compactly closed cells and water (Jamal et al. 2015). The development of biofilm is

TABLE 1.1
Composition of Biofilm

S. No	Different Components	Percentage of Matrix
1	Protein	<1%–2% (including enzymes)
2	Polysaccharide	1%–2%
3	DNA/RNA	<1%–2%
4	Water	Up to 97%
5	Microbial cell	2%–5%

Source: Adapted from Jamal et al. (2015).

regulated via different environmental and genetic factors. Genetic researches have revealed that cell membrane proteins, extracellular polysaccharides, bacterial mobility, and signaling molecules show significant parts in biofilm development (Pratt and Kolter 1998). Extrapolymeric substance has a major part in the development of biofilms. Genetic investigations on *P. aeruginosa* displayed that gene stimulation is essential for EPS synthesis (Davies and Geesey 1995). Nutrient availability, pH, temperature, and oxygen presence play a major part in biofilm development (Kim and Frank 1995). Biofilms have a significant diversity of microbes. Algae, fungi, archaea, protozoa, and bacteria make significant parts of the matrix of biofilm and contribute to biodiversity and aquatic ecosystem processes (Jackson and Jackson 2008).

Biofilms have adverse effects on human activity, and numerous strategies are used for the prevention and removal of biofilm. Traditionally, chemical and physical approaches like chlorination, ultraviolet disinfection, and flushing are used for controlling and removing biofilms. However, due to the lack of both effectiveness and safety of these stratifies, the concerns still persist (Srinivasan et al. 2008). In the late twentieth century, researchers tried to prove through experimentation that several naturally occurring compounds have antimicrobial characteristics (Cowan 1999). However, antibiofilm activities have been accredited to numerous natural substances like diverse essential oil (EO), plant extracts, and honey, and such properties are investigated expansively. Phages are supposed to be the largest group of microbes in the environment having antibacterial activity irrespective of their resistance (Pires et al. 2011). Quorum sensing (QS) plays a crucial role in the formation of biofilm, and QS inhibition is a hopeful approach for preventing biofilms (Ueda and Wood 2009). Nanotechnology is also a promising tool for control and prevention of biofilm, e.g., silver nanoparticles (AgNPs) (Kalishwaralal et al. 2010). Bacteriocins are antimicrobial peptides produced by prokaryotes which are commonly used against closely related bacterial species having a role in controlling bacterial biofilms (Hetrick et al. 2009). In this chapter, we have discussed biofilms in detail, such as biofilm formation, properties, biodiversity, and several current managements for the control of biofilms.

1.2 STEPS OF MICROBIAL BIOFILM FORMATION

The development of biofilm is a complex mechanism that relies on the medium, genetic control, signaling, etc. (Okada et al. 2005). The biofilm formation process happens in a series of events that leads to the adaptation underneath diverse nutritional and ecological circumstances (Hentzer et al. 2005; (Flemming 2007). In the first step of development of biofilm, free-floating microbial cells attach to a living or nonliving surface irreversibly via different forces like van der Waals' forces (Palmer et al. 2007). In the early stages, the bacterial attachment to the surface is uneven and could effortlessly be detached via detergents, disinfectants, etc. In the second step, bacteria attach to the surface irreversibly and proliferate easily (Donlan 2009). The matrix produced by bacteria contains different molecules like protein and enzymes, and water is the major component that supports the biofilm till its maturation (Palmer et al. 2007). In the third step, bacteria form a biofilm that is not fully mature. In this stage, bacteria produce EPS, and among bacteria, a strong quardination system (QS) is developed (Otto 2013). Among bacteria the only communication way is QS, which

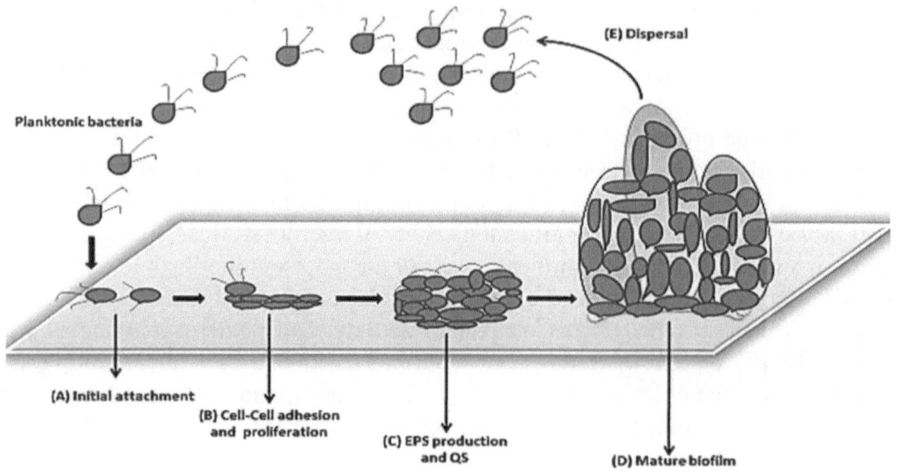

FIGURE 1.1 Different phases (steps) of biofilm formation on an abiotic solid surface. These steps are (a) initial attachment, (b) cell attachment, and (c) EPS production followed by (d) maturation of biofilm and dispersion. (From Jamal et al., 2015, with permission.)

chiefly helps in the regulation of biofilms (Hammer and Bassler 2003). In the fourth step, a thick mature biofilm is formed, leading to the disruption of the matrix, and bacteria are dispersed and colonized on the new surface (Solano et al. 2014). The mechanism of biofilm formation is shown in Figure 1.1.

1.3 PROPERTIES OF MICROBIAL BIOFILMS

Biofilms are microbial communities. The capability of microbes to form biofilm plays an important role in the pathogenicity and also in making the microbes resistant to antimicrobials. Fungal and bacterial species that have the ability to form biofilms produce surface adhesion molecules which make an extracellular matrix (ECM) (Visick et al. 2016). Environment controls the biofilm dynamics by affecting mechanical, structural, chemical and physical properties, and microbial interactions (Billings et al. 2015). Biofilms are exposed to numerous challenges, and in order to understand in a good way their mechanical characteristics during various steps of biofilms development (Wang et al. 2018), Lieleg and colleagues studied numerous biofilms of *Pseudomonas* and noticed their prominent influence of antimicrobials on the properties of biofilms (Lieleg et al. 2011). The viscoelastic characteristics of biofilms of *P. aeruginosa* showed resistance to strong forces which are applied and also to chemical treatment, allowing them to proficiently improve from mechanical harm. The mechanical properties of matured biofilms are affected by bacterial sensing (Kovach et al. 2017). During the formation of biofilms, the shear stress increased as an outcome of biofilms having high cell densities, higher dry mass, and lower thickness. However, biofilms which were below lower shear stresses showed much resistance to elimination via treatment with chemicals (Lemos et al. 2015).

Polysaccharides secreted by bacteria play a significant role in the ecology of biofilm, playing a part in QS and signaling (Rinaudi et al. 2010).

1.3.1 Quorum Sensing

QS is a gentle cellular mechanism by means of which bacterial species generate and identify signaling molecules through which coordination is done among cells (Waters and Bassler 2005). Autoinducers are extracellular signaling molecules which provide the intercellular communication inside the cells of bacteria. When the signaling molecules are accumulated in the medium, it makes each bacterial cell to estimate the overall number of bacteria (cellular density) by a process called quorum sensing (Fuqua et al. 1994). Signaling molecules within bacteria, i.e., gram negative, are noncrucial amino acids termed as acyl-homoserine lactones (acyl-HSL) (Fuqua and Greenberg 1998). Formation of these acyl-HSL molecules are followed by diffusion through the cell membrane and graduall accumulation in the medium. When inside the medium the levels of signaling molecules increase, they arrive into the cell and bind to the HSL receptor. A complex containing a receptor and signaling molecule then attaches to an appropriate gene of target and stimulates transcription of the target gene. Oligopeptides are used by the gram-positive bacterial species (Bassler 1999). Pathogenicity of numerous human microbes amongst *P. aeruginosa* and *Staphylococcus aureus* is controlled through QS (Novick and Muir 1999).

1.3.2 General Stress Response

It is supposed that slower development of several subpopulations of cells inside a biofilm is an outcome of general stress response. It is a regulatory process that makes the bacteria capable to survive when conditions are not favorable in the environment. General stress response comprises several physiological alterations in the cells of bacteria and passes them into a stationary stage. This consequence results in the resistance of bacterial cell to numerous unfavorable conditions within environment, such as nutrient deficiency, pH changes, and the action of numerous chemical substances (Mah and O'Toole 2001). On the molecular stage, general stress response regulation is done through an RNA polymerase sigma S (RpoS) protein that performs just as a sigma subunit of RNA polymerase. RpoS regulates an intricate gene network accountable for the passage of cells of bacteria into a stationary stage (Foley et al. 1999).

1.4 BIODIVERSITY OF MICROBIAL BIOFILMS

The diversity of aquatic microbes is noticed to be an important part of the functioning of an aquatic ecosystem (Cotner and Biddanda 2002) Within rivers and streams, the life of microbes is dominated by means of benthic-type biofilms that regulate vital processes of ecosystem (Battin et al. 2008). Biofilms have a significant diversity of microbes. Different microorganisms such as archaea, bacteria, fungi, algae, viruses, and protozoa can be involved in biofilms and make essential components of the matrix of biofilm which further contribute to biodiversity (Jackson and

Jackson 2008). Mainly, *Beta-Proteobacteria* frequently dominates biofilms within rivers, streams, and lakes (Besemer et al. 2012). Biofilms can be formed on decaying or living plants and even on aggregates of diatoms within lakes (Buesing et al. 2009). Additional taxonomic classes particularly found within biofilms consist of *Acidobacteria* (mainly at lower pH), *Verrucomicrobia*, *Firmicutes*, *Actinobacteria* and *Gamma and Delta-Proteobacteria*, *Planctomycetes*, *Deinococcus-Thermus*, and *Gemmatimonadete* (Romaní i Cornet et al. 2014). Algae, more frequently *Chlorophyta* and *Bacillariophyta*, offer substrates via lysis products and exudates and serve as a main source of carbon for microbes of heterotrophic biofilm (Romaní i Cornet et al. 2004). Fungi, particularly *Ascomycota*, show a significant part in the organic matter decomposition which is found submerged (Heino et al. 2014). Finally, protists (involving amoebae, ciliates, and flagellates) and viruses could regulate the growth of biofilm and cause changes in function, diversity, and architecture of biofilms (Böhme et al. 2009). The collective impacts of dispersal limitations and niche-based processes have a role in diversity and community composition of biofilms found in epilithic streams in spite of the impact of strong ecological factors having obvious impacts (Griebler et al. 2014). Some other studies have also justified that microbial community composition are chiefly governed by ecological processes, but dispersion likewise shows a role (Wang et al. 2013). Diversity and community composition of biofilms are driven by a wide variety of ecological factors. Temperature of water, for example, is associated with the community structure and diversity of hyporheic and benthic stream biofilms (Freimann et al. 2014). Additionally, pH is associated with the composition of communities of bacterial and fungal species on rotten leaves, and the diversity of bacteria of such communities decreased with the rising pH (Heino et al. 2014). Nutrients and organic carbon, as a requirement for growth of biofilm, could change the biodiversity, composition of community, and architecture of biofilms (Battin et al. 2007).

1.5 CONSERVATION OF MICROBIAL BIOFILMS

Multiple species of biofilms are found on the surfaces of plants and in the rhizosphere of the plants, adhering to transport vessels of plants, leaves, roots, and stems and serving to sustain biodiversity mutually on plants as well as on microbial level. One way they help in conserve biodiversity at the microbial level is to ensure the survival of bacteria in the harsh environments (Angus and Hirsch 2013). In circumstances, bacterial species that makes the biofilm have an extrapolymeric substance that provides them a safeguard, and in planktonic form they do not have such safeguard. Extrapolymeric substance provides them protection against environmental stress factors, permitting them to face in a good way the changing environmental circumstances like alterations in temperature or pH, desiccation, or rare nutrients availability, which might arise both from human activities and from casualties, for example as a result of change in climate or deforestation. Biofilms likewise have influences on the health of plants and productivity (obviously, dependent on biofilms types, such effects can be advantageous or harmful for host plant). For example, numerous bacteria inside the matrix of EPS not just advance biocontrol, i.e., the plants' defense from phytomicrobes, but correspondingly promote growth of plants by hormone production and through further processes, thus signifying maintainable replacements to agrochemicals (Bogino et al. 2013). In brief, biofilm not only improves the fitness of specific bacteria but also improves the productivity and health of plants (Bogino et al. 2013). Biofilms that are photosynthetic make the basis of tropic chains within numerous aquatic habitats; the same is case with diatoms (which is main microalgae class); and on estuaries, it dominates the intertidal microphytobenthos (MPB) (Macintyre et al. 1996). The extracellular polymeric substance made by diatoms of MPB encourages the sediment particle aggregation and hence critically influences the estuarine intertidal mudflat stabilization (Paterson and Black 1999). The same environmental part within terrestrial habitats is performed through extracellular polymeric substance which is made via the cyanobacteria found in biological soil crusts (BSCs) (Adessi et al. 2018). MPBs are communities which are dominated by diatoms, and these cover the sandbanks (epipsamic MPB) and estuarine mudflats (epipelic MPB). Epipelic MPB is predominantly appropriate for the functioning of ecosystem, as it is mainly accountable for higher yield of intertidal mudflats, which are significant constituents of the productivity of estuaries (Underwood and Kromkamp 1999). Biologists praised the unseen beauty of communities of MPB (Marques da Silva 2015) and called it as the "secret gardens," because under the lens of microscope, they show unexpected beauty (Macintyre et al. 1996). Moreover, in addition to the ecological role of such communities, additional role to biodiversity could be highlighted for MPB, specifically, esthetic importance of them (Costanza et al. 1997). The species centrality defines a relatively conservative strategy for the conservation of biodiversity, which includes the conservation of viable populaces of several microbial species, like indicator species, which are the species that are assumed to show a biodiversity state of several areas or endangered species. Numerous criticisms are made to such strategy. Certainly, a difference between practice and theory is highlighted: many conservationists argue to defend the species, which is a conservation unit indeed, and for conservation of biodiversity, each individual population should be managed (Casacci et al. 2014). Preferably, the whole individual populations of entire species must need conservation and must be conserved; and the criteria for prioritizing are required. In order to continue this several targets should be selected for conservation related actions. It has been noticed that placing collectively conservation of biodiversity and evolution improbably creates a type of challenge: evolution means change, and conservation means that all the things should be kept as it is or it means to bring them into their previous state. However, conservation must likewise be understood in terms of conserving evolutionary and ecological processes required to encourage natural dynamics (Smith et al. 1993), and it is documented that decisions related to conservation must be based on evolutionary thoughts (Höglund 2009). Specifically, it is claimed that species-based strategy must be integrated via keeping in the mind that how such entities must be maintained that possess evolutionary potential (Casetta and Marques da Silva 2015). Here in this chapter, it is emphasized that collectives of multiple species like biofilms may hold a specific tendency to change in comeback to ecological fluctuations. In short, evolutionary potential is a central aspect to be considered in conservation-related actions. If the evolutionary capability is considered, it will broaden the focus of conservation-related actions from species to a greater amount of individuals (i.e., several populations and species, communities of multispecies, and perhaps several ecosystems also), and it may happen the essential of prioritization that economic limitations dictate to biodiversity conservation exercise while going further than a simple population-based or species-based strategy.

1.6 MANAGEMENT OF MICROBIAL BIOFILMS

The pathogenicity of microorganisms increases day by day due to biofilm formation. Numerous issues are linked with the formation of biofilms. Therefore, scientists are trying to develop new strategies to control biofilm (Cowan 1999). However, antimicrobial properties are found in different natural materials, quorum quenching (QQ), bacteriophages, bacteriocin, biosurfactants (BSs), nanotechnology, and numerous enzymes. Such approaches are helpful in the management of biofilm. Some of the important management strategies are mentioned here.

1.6.1 Plants Extracts in the Management of Microbial Biofilms

Numerous extracts of plant and active substances have expansively examined to eliminate the biofilm of *Propionibacterium acne* (Coenye et al. 2012). Such investigation confirmed that among119 extracts of plants, 5 (*Polygonum cuspidatum*, *Malus pumila*, *Rhodiola crenulate*, *Dolichos lablab*, and *Epimedium brevicornum*) presented an effective antibiofilm action. Investigators have described that *P. cuspidatum* and *E. brevicornum* extracts along with active constituents (resveratrol and icartin) show a significant antibiofilm action (Ravichandiran et al. 2012). *Lagerstroemia speciosa* is found in South Asia which is a therapeutic plant. The potential of fruit extracts of *L. speciosa* was associated with formation of the biofilm of *P. aeruginosa* PAO1 (Singh et al. 2012). Investigators established that extracts of *L. speciosa* show an inhibition to biofilm at the level of 10 mg/mL. A potential antibiofilm action separately has been shown by Dandasa and Green tea (Faraz et al. 2012). The later investigation exhibited that mutually Green tea and Dandasa at the levels of 12.5 and 6.2 mg/mL possess potential antibiofilm against *Streptococcus mutans* and at levels of 3.1 and 12.5 mg/mL against *E. coli*, respectively. The probable strong inhibitory impact counter to biofilm was analyzed for fresh *Allium sativum* extract, i.e., fresh garlic extract (FGE) (Harjai et al. 2010). The researchers studied four plants (*Vaccinium oxycoccus*, *Hippophae rhamnoides*, *Juglans regia*, and *Azadirachta indica*). The extracts of neem (*A. indica*) showed a potential antibiofilm activity against *M. smegmatis* (Carneiro et al. 2011).

1.6.1.1 Honey

Honey is a natural product produced by honey bees. Honey is extensively common and has antibacterial, anti-inflammatory, antioxidant, and wound curing characteristics. It owns antimicrobial activities against 60 fungal and a large number of bacterial species (Molan 2002). Currently, honey has been described as a potential substance/agent for inhibiting biofilm formation (Maddocks et al. 2012). It has been confirmed that honey prevents the biofilm formation of *Enterococcus* spp. (Ng et al. 2014). Honey could likewise cause a reduction in biofilm of *E. coli* - O157:H7 (Lee 2011). Honey contains defensin-1 as an antibacterial peptide, which inhibits the growth of bacteria and in turn inhibits the formation of biofilms (Santangelo 2013).

1.6.1.2 Essential Oil

EOs are aromatic hydrophobic liquids obtained from plant materials (leaves, herbs, buds, twigs, flowers, seeds, wood, fruits, roots, and bark). There are about 3,000 EOs, which are known for their antibacterial activities. Due to the use of EOs, the cell wall and cell membrane of microbes are damaged, morphology is altered, and the cytoplasmic materials are coagulated (Hammer et al. 1999). Furthermore, it is likewise described that EOs cause in-activation of bacterial species without creating resistance to antimicrobial agents (Ali et al. 2005). In particular, availability of many EOs, low mammalian toxicity, and quick degradation in environment make them as safe antibiofilm agents (Isman 2000). Cinnamon oil is a common example of EO, which is extensively applied in food industries due to its specific aroma and is effective against several species of bacteria (Filoche et al. 2005).

1.6.1.3 Vegetable Oil

Vegetable oil (Brazil nut oil) has antibiofilm properties to control dental biofilm (Filogônio et al. 2011). However, investigators noticed that the addition of vegetable oil to dentifrice (which is commercially accessible) enhanced the management of dental biofilm. This signifies that such oil might help in inhibition and/or management of periodontal illnesses and caries.

1.6.2 Quorum Quenching

QS is implemented through secretion of smaller extracellular signaling molecules which acts like an autoinducer to initiate genomic plans. Three chief systems of QS are notable: autoinducing peptide (AIP) QS system in (gram-positive) bacterial specie, acetyl homoserine lactone (AHL) QS system in (gram-negative) bacterial species and autoinducer 2 (AI-2) QS system within both bacterial species (gram-positive and gram-negative). Numerous reports have detailed the significance of QS in the formation of biofilm (Coenye 2010). QS is associated to control swarming of bacteria and maturing of architecture of biofilms (Ueda and Wood 2009). A wide variety of molecules are identified to disrupt the QS system. Over the past decades, an excessive amount of QQ enzymes have been identified from numerous bacterial species (gram-positive and gram-negative). Such enzymes are classified into three classes: (i) oxidoreductases, (ii) AHL lactonases, and (iii) AHL acylase. The QQ strategy results in dissociating the architecture of the biofilm but not kill the microorganisms of the biofilm. However, Quorum sensing inhibitors (QSIs) possess the capability to make the bacteria (which do formation of biofilms) more sensitive to drugs. Consequently, a mixture of antibiotics and QSI complexes are suggested to manage the biofilm (Hentzer et al. 2003). A number of QS inhibitors are shown in Table 1.2.

1.6.3 Bacteriophages

Bacteriophages or bacterial viruses are supposed to be the most abundant entities in the biosphere. Because of the development of frequent resistance to antibiotics, phages (as natural killer of bacteria) can offer a better option for the eradication of

TABLE 1.2
Natural Compounds as QS Inhibitors

S. No	Natural Compounds	Source	QS Activity	Year of Publication
1	Brominated furanone	Marine algae	Anti-QS brominated furanone was able to inhibit biofilm formation in *Salmonella typhimurium*	2006
2	Taxifolin/distylin (dihydroquercetin)	Malagasy plant extract (*Combretum albiflorum*)	Reduce production of pyocyanin and elastase in *P. aeruginosa* PAO1	2011
3	*Cuminum cyminum*	*Cuminum cyminum* extract	Reduce LuxR-dependent biofilm formation and swarming motility of *P. aeruginosa*	2012
4	Iberin (1-isothiocyanato-3(methylsulfinyl)propane)	Horseradish extract (*Armoracia rusticana*)	Inhibit expression of QS-regulated lasB-gfp and rhlA-gfp genes responsible for virulence factor in *P. aeruginosa*	2012
5	4′5-diOH-flavone-7-rhgluc)	Citrus extract	Decrease the QS-mediated biofilm formation and swimming motility in *Y. enterocolitica*	2012
6	Morin (2′,3,4′,5,7-pentahydroxyflavone)	Grapefruit (*Artocarpus heterophyllus*)	Inhibit LasR- and RhlR-dependent protease, elastase and hemolysin in *P. aeruginosa* PAO1	2012
7	Sulforaphane (1-isothiocyanato-4-(methylsulfinyl)butane)	Broccoli	Reduce the expression of lasI-luxCDABE reporter in *P. aeruginosa*	2013
8	Erucin (4-methylthiobutyl isothiocyanate)	Broccoli	Reduce the expression of lasI-luxCDABE reporter in *P. aeruginosa*	2013

Source: Data adapted from Lade, H., et al., *Int. J. Biol. Sci.*, 10, 550, 2014, Sadekuzzaman, M., et al., *Compr. Rev. Food Sci. Food Safe.*, 14, 491–509, 2015, Ali, S., "The Role of Quorum Sensing in Survival, Biofilm Formation and Gene Expression of *Listeria monocytogenes*." Doctoral dissertation, Department of Food Science, Griffiths, Mansel, 2011, Biradar, B., and Devi, P., *J. Contemp. Dent. Pract.*, 12, 479–85, 2011. With permission.

bacterial species. Bacteriophages are presently well thought as a probable substitute to antibacterials for infections of bacteria, particularly for disrupting or inhibiting biofilm. Isolation of bacteriophage is simpler and fast, and manufacture is comparatively cheap. Phages are specific to their host and do not disturb the usual microflora. In addition, bacteriophages are also ecofriendly and their replication occurs at the site of target as longer as the bacterial host cells persist over there. Until now no adversative complications have been being noticed using phages (Pires et al. 2011). Phages have been confirmed as antibiofilm substances in many studies. For instance, "T4 phage" could efficiently infect and replicate in biofilms of *E. coli* and cause interruption of the matrix of biofilm via eliminating the bacteria cells (Meng et al. 2011). Phages could penetrate in the EPS by means of phage derivative enzymes (polysaccharide depolymerase) or by diffusion. Phage enzymes have the capability to disturb the structure of biofilm (Hughes et al. 1998). There exists considerable indication that bacteriophage depolymerases affect the biofilm (Donlan 2009). A list of different studies has been provided in Table 1.3 showing the applications of bacteriophages against biofilms.

1.6.4 Enzymes

The extracellular enzymes specifically act on the numerous morphological constituents of extracellular polymeric substances such as exopolysaccharides, proteins, and extracellular DNA (eDNA), and such enzymes help in the dispersion of biofilms. Enzymes enable the cell detachment from the colony of biofilm, and their planktonic cells discharge into the atmosphere (Xavier et al. 2005). By purifying such enzymes, clinicians could hypothetically add them exogenously to preformed biofilms at raised levels so as to accomplish interventional dispersal, enabling biofilm-related microorganisms much vulnerable to the antimicrobials/antibiotics and host immune system (Kaplan 2010). Here, we have mentioned several enzymes that have been examined for the dispersion of biofilms such as proteases, glycoside hydrolases, lyase, and deoxyribonucleases (Table 1.4.).

1.6.5 Nanotechnology

It is thought that nanotechnology-dependent strategies would offer positive progressions to avoid antibiotic-resistant infections of biofilm of medical instruments and biomaterials. A minor quantity of investigations has described the usage of surface-coated nanoparticle (NP) as biofilm preventing agents (Taylor and Webster 2011). Additionally, NPs are slightly sufficient to do penetration in cell walls of microbes and even layers of biofilm that could produce irretrievable impairment to DNA and cell membranes (Suci et al. 2007).

1.6.5.1 Nanoparticles in Antibiofilm Therapy

Antibacterial metals including silver, gold, copper, zinc, and titanium are recognized to possess antibiofilm and antibacterial actions, which are used as alternatives to antibiotics without the development of antimicrobial resistance. It is being recognized that metal-based NPs possess more improved antibiofilm actions compared to (microsized)

TABLE 1.3
Different Phages Have Been Used to Infect a Variety of Bacterial Biofilms Reduction/Removal

S. No.	Bacteria	Phages	Year of Publication
1	*S. epidermidis*	456	2006
2	*S. aureus*	φl1, φl2	2007
3	*S. epidermidis*	K	2007
4	*E. coli TG1*	T7	2007
5	*S. maltophilia*	C2	2008
6	*P. fluorescens*	φS1	2008
7	Treponema	φtd1	2010
8	*K. pneumonia*	KPO1K2	2010
9	*L. monocytogenes*	P100	2010
10	*Actinomycetemcomitans*	Aabφ01-1	2011
11	*Aggregatibacter*	Aabφ01	2011
12	*S. aureus*	SAP-26	2011
13	*P. aeruginosa*	PAO1 and ATCC-10145	2011
14	*P. aeruginosa*	φMR299-2 and φNH-4	2012
15	*A. baumannii*	AB7-IBB1	2012
16	*A. baumannii*	AB7-IBB2	2012
17	*Streptococcus mutans*	φAPCM01	2015
18	*Klebsiella pneumonia*	Z	2015
19	*E. faecalis*	EFDG1	2015
20	*S. aureus*	phiIPLA-RODI	2015
21	*S. mutans*	ϕAPCM01	2015
22	*Enterococcus faecalis*	Ef11/φFL1C(Δ36)PnisA	2016
23	*P. aeruginosa-2995*	AZ1	2017
24	*E. cloacae*	MJ-2	2018
25	*P. aeruginosa-2949*	MA-1	2019

Source: Adapted from Sillankorva, S., et al., *Bmc Biotechnol.*, 8, 79, 2008, Jamal, M., et al., *J. Med. Microbiol.*, 64, 454–62, 2015, Elbreki, M., et al., *J. Viruses*, 2014, 20, 2014, Szafrański, S.P., et al., *J. Biotechnol.*, 250, 29–44, 2017, Fernández, L., et al., *Antibiotics*, 8, 126, 2019, Adnan, M., et al., *Biologicals*, 63, 89–96, 2020, Al-Zubidi, M., et al., *Infect. Immun.*, 87, e00512–19, 2019, Jamal, M., et al., *Folia Microbiol.*, 64, 9101–11, 2019, Jamal, M., et al., *Life Sci.*, 190, 21–28, 2017. With permission.

complements (Jones et al. 2008). The AgNPs are effective against *S. epidermidis* and *P. aeruginosa* biofilms (Kalishwaralal et al. 2010). Recently, titanium dioxide (TiO_2) and ethylene diamine tetra acetic acid (EDTA) were used against *C. albicans* biofilms (Haghighi et al. 2013). Nitric oxide (NO) containing silica NPs have been described for their antibiofilm potential counter to *P. aeruginosa*, *C. albicans*, *E. coli*, *S. epidermidis*, and *S. aureus* biofilms (Hetrick et al. 2009). A list of nanoantimicrobials with inherent antimicrobial potentials is shown in Table 1.5.

TABLE 1.4
Bacterial Enzymes Implicated in Active Biofilm Dispersal

S. No	Enzymes	Substrate	Origin	Year of Publication
1	Hemagglutinin protease (HAP)	Bacterial receptors on human intestinal cells	*Vibrio cholera*	1992
2	Exopolysaccharide lyase	Unknown	*Pseudomonas fluorescens*	1998
3	Chitinase	Chitin	Pseudoaltermon as sp. S91	2000
4	Dispersin B	Poly-β(1,6)-N-acetyl-D-glucosamine (PNAG) *Aggregatibacter*	*Aggregatibacter actinomycetemcomitans*	2003
5	Endo-β-1,4-mannanase	Unknown	*Xanthomonas campestris*	2003
6	Dispersin B (or DspB)	Poly-b-1,6-GlcNAc implicated as an adhesion factor for biofilms of several bacterial species	*Actinobacillus actinomycetemcomitans*	2003
7	Spl protease	Cleaving the cell wall-associated protein, EbpS	*Staphylococcus aureus*	2008
8	Hyaluronidase	Hyaluronan	*Streptococcus intermedius*	2008
9	Aureolysin	Unknown	*Staphylococcus aureus*	2008
10	Thermonuclease	eDNA	*Staphylococcus aureus*	2009
11	LapG protease	LapA exopolysaccharide-binding protein	*Pseudomonas putida*	2010

Data from Kaplan, J. á. B., *J. Dent. Res.*, 89, 205–18, 2010, Xavier, J.B., et al., *Microbiology*, 151, 3817–32, 2005, Thallinger, B., et al., *Biotechnol. J.*, 8, 97–109, 2013, Jamal, M., et al., *J. Med. Microbiol.*, 64, 454–62, 2015. With permission.

1.6.6 Bacteriocins

BSs or microbial surfactants belong to a heterogenous class of surface-active substances which are amphiphilic with small molecular weights. A huge variety of bacteriocins that are being considered consequently to target the microbial biofilms are lantibiotics. Numerous bacteriocins are used in the preservation of food, counting health benefits (Chikindas et al. 2018). Possibly the utmost carefully investigated bacteriocin is the lantibiotic nisin. Such a lantibiotic was previously being stated to be effective at pervading biofilms, and its derivatives possess an improved capability to diffuse in a complex biofilm (Field et al. 2015). Murinda et al. (2003) have reported that bacteriocins pediocin, colicin E6, and nisin have shown a very slight or no cytotoxicity towards kidney cells of vero monkey (Murinda et al. 2003). A number of bacteriocins that have been reported against bacterial biofilms for their antibiofilm potential are mentioned in Table 1.6.

TABLE 1.5
Categorization of Nanoantimicrobials with Inherent Antimicrobial Potential

S. No	Antimicrobial Activity against Multidrug Resistant Pathogens	Mode of Action	Nanosystem	Year of Publication
1	*E. coli*	Disrupt cell walls through Reactive Oxygen Species (ROS)	Aluminum (Al) NPs	2009
2	Methicillin-resistant *S. aureus*	Generate holes in the cell wall. Bind to the DNA and inhibit the transcription process	Gold (Au) NPs	2010
3	*S. aureus*, *S. epidermidis*, and *E. coli*	Through ROS-generated oxidative stress. ROS, superoxide radicals (O_2'), singlet oxygen ($1O_2$), hydroxyl radicals (OH′), and hydrogen peroxide (H_2O_2)	Iron-containing NPs	2012
4	Colloidal bismuth NPs completely eradicated the biofilm formation caused by *Streptococcus mutans*	Bismuth NPs usually are used in conjugation with X-ray treatment to cure multidrug-resistant (MDR) pathogens	Bismuth NPs	2013
5	Multiple-antibiotic resistant *Helicobacter pylori*	Alter the Krebs cycle, and amino acid and nucleotide metabolism	Bismuth (Bi) NPs	2013
6	Mesoporous silica NPs were effective against MDR *E. coli*	Si NPs disrupt microbial biofilms by releasing nitric oxide. It also causes lipid peroxidation of cell membranes	Silica NPs	2014
7	*B. subtilis*, *S. aureus*, and *E. coli*	Reduce bacteria at the cell wall. Disrupt the biochemical processes inside bacterial cells	Copper oxide (CuO) NPs	2014

Source: Data from Jamal, M., et al., *Life Sci.*, 190, 21–28, 2017; Rudramurthy, G.R., et al., *Molecules*, 21, 836, 2016; Sousa, C., et al., *Science against Microbial Pathogens: Communicating Current Research and Technological Advances*, Formatex, Badajoz, 2011; Natan, M., and Banin, E., *FEMS Microbiol. Rev.*, 41, 302–22, 2017; Ramasamy, M., and Lee, J., *BioMed Res. Int.*, 2016, 17, 2016. With permission.

1.7 CONCLUSIONS

Biofilms are the leading lifestyle of microbes in nearly all habitats and is a serious issue in marine, healthcare settings, laboratories, and food industries. Microbial biofilms are associated with a large number of human infectious diseases, and these biofilms are posing major public health issues. Efforts are needed for the conservation of biodiversity of microbial biofilms. Practical ways are recommended through which advancements might be done within the conservation of microbial biofilm. Expansion of operative approaches for fighting against biofilms is a stimulating job. Several advanced antibiofilm strategies are available. The emergent advanced strategies such

TABLE 1.6
Activity of the Lantibiotic Group of Bacteriocins against Biofilms

S No.	Biofilm former	Bacteriocins	Effects	Year of Publication
1	*S. epidermidis*	Nisin	Loss of green fluorescence from biofilm, loss of viability and membrane integrity	2010
2	*L. monocytogenes* 4032	Nisin	Mature biofilms on stainless steel and polypropylene recalcitrant to nisin	2011
3	*S. aureus*, *S. epidermidis*, *P. acnes*	Gallidermin	Prevention of biofilm formation. Persister cells survived	2012
4	MRSA	Nisin in combination with antibiotics	Nisin-antibiotic combinations prevented biofilm formation	2012
5	*S. aureus* strains	Nisin and lysozyme	1× MIC nisin prevented biofilm formation	2012
6	MRSA	Nisin, lacticin Q, nukacin ISK-1	Nisin and lacticin Q potent against biofilm, causing pore formation, efflux of ATP from biofilm. No antibiofilm activity for nukacin ISK-1	2013
7	MRSA	Nisin with ciprofloxacin/ daptomycin	Synergy between nisin and ciprofloxacin/daptomycin against biofilm	2013
8	*S. aureus*	Nisin, bovicin HC5	Reduced adhesion to polystyrene. Reduced expression of genes involved in biofilm formation	2014
9	*S. pseudintermedius*	Nisin I4V	I4V inhibited formation and reduced biomass of biofilms	2015
10	*S. pseudintermedius*	Nisin I4V	I4V potent against DSM21284 biofilms	2016
11	*L. monocytogenes* F2635	Nisin M21A with citric acid or cinnamaldehyde	Nisin combined with EOs effective against biofilm	2016
12	*C. albicans*	EntV	Prevention of *C. albicans* biofilm formation	2017
13	*E. faecalis*	Nisin ZP and sodium hypochlorite	Thickness and biovolume of biofilm decreased	2017

Source: Data from Mathur, H., et al., *NPJ Biofilms Microbi.*, 4, 1–13, 2018, Chopra, L., et al., *Sci. Rep.*, 5, 13412, 2015, Santos, V.L., et al., *Curr. Develop. Biotechnol. Bioeng. Hum. Anim. Health Appl.*, 403, 2016. With permission.

as bacteriophages, bacteriocins, NPs (e.g., Ag, Au, Cu, Ni, Ti, Zn), natural materials, nanotechnology, QQ, BSs, and numerous enzymes (proteases, glycoside hydrolases, lyase, deoxyribonucleases, etc.) are hopeful and might aid to search antibiofilm approaches that are better than conservative ones. Furthermore, advanced practices in combination with conventional approaches (disinfectants, physical methods, and antibiotics) would be helpful in solving the issue of biofilm in the near future.

REFERENCES

Adessi, Alessandra, Ricardo Cruz de Carvalho, Roberto De Philippis, Cristina Branquinho, and Jorge Marques da Silva. "Microbial Extracellular Polymeric Substances Improve Water Retention in Dryland Biological Soil Crusts." *Soil Biology and Biochemistry* 116 (2018): 67–69.

Adnan, Muhammad, Muhammad Rahman Ali Shah, Muhsin Jamal, Fazal Jalil, Saadia Andleeb, Muhammad Asif Nawaz, Sidra Pervez, Tahir Hussain, Ismail Shah, Muhammad Imran, and Atif Kamil. "Isolation and Characterization of Bacteriophage to Control Multidrug-Resistant *Pseudomonas aeruginosa* Planktonic Cells and Biofilm." *Biologicals* (2020). doi: 10.1016/j.biologicals.2019.10.003.

Ali, Saleh. "The Role of Quorum Sensing in Survival, Biofilm Formation, and Gene Expression of *Listeria monocytogenes*." Doctoral dissertation, Department of Food Science, Griffiths, Mansel, 2011.

Ali, Shaik Mahaboob, Aleem A. Khan, Irshad Ahmed, Muhammad Musaddiq, Khaja S. Ahmed, Hanmanlu Polasa, L. Venkateswar Rao, Chittoor M. Habibullah, Leonardo A. Sechi, and Niyaz Ahmed. "Antimicrobial Activities of Eugenol and Cinnamaldehyde against the Human Gastric Pathogen *Helicobacter pylori*." *Annals of Clinical Microbiology and Antimicrobials* 4, no. 1 (2005): 20.

Al-Zubidi, M., M. Widziolek, A. F. Gains, R. E. Smith, K. Ansbro, A. Alrafaie, C. Evans, C. Murdoch, S. Mesnage, C. W. I. Douglas, and A. Rawlinson. "Identification of Novel Bacteriophages with Therapeutic Potential that Target *Enterococcus faecalis*." *Infection and Immunity* 87, no. 11 (2019): e00512–19.

Anderson, G. G, and G. A. O'Toole. "Innate and Induced Resistance Mechanisms of Bacterial Biofilms." In Romeo, Tony. (eds) *Bacterial Biofilms. Current Topics in Microbiology and Immunology*, vol. 322, pp. 85–105. Springer, Berlin, Heidelberg, 2008.

Angus, Annette A., and Ann M. Hirsch. "Biofilm Formation in the Rhizosphere: Multispecies Interactions and Implications for Plant Growth." *Molecular Microbial Ecology of the Rhizosphere* 1 (2013): 701–12.

Bassler, Bonnie L. "How Bacteria Talk to Each Other: Regulation of Gene Expression by Quorum Sensing." *Current Opinion in Microbiology* 2, no. 6 (1999): 582–87.

Battin, Tom J., Louis A. Kaplan, Stuart Findlay, Charles S. Hopkinson, Eugenia Marti, Aaron I. Packman, J. Denis Newbold, and Francesc Sabater. "Biophysical Controls on Organic Carbon Fluxes in Fluvial Networks." *Nature Geoscience* 1, no. 2 (2008): 95.

Battin, Tom J., William T. Sloan, Staffan Kjelleberg, Holger Daims, Ian M. Head, Tom P. Curtis, and Leo Eberl. "Opinion: Microbial Landscapes: New Paths to Biofilm Research." *Nature Reviews Microbiology* 5, no. 1 (2007): 76.

Besemer, Katharina, Hannes Peter, Jürg B. Logue, Silke Langenheder, Eva S. Lindström, Lars J. Tranvik, and Tom J. Battin. "Unraveling assembly of stream biofilm communities." *The ISME Journal* 6, no. 8 (2012): 1459–1468.

Billings, Nicole, Alona Birjiniuk, Tahoura S. Samad, Patrick S. Doyle, and Katharina Ribbeck. "Material Properties of Biofilms—A Review of Methods for Understanding Permeability and Mechanics." *Reports on Progress in Physics* 78, no. 3 (2015): 036601.

Biradar, Baswaraj, and Prapulla Devi. "Quorum Sensing in Plaque Biofilms: Challenges and Future Prospects." *Journal of Contemporary Dental Practice* 12, no. 6 (2011): 479–85.

Bogino, Pablo C., María De Las Mercedes Oliva, Fernando G. Sorroche, and Walter Giordano. "The Role of Bacterial Biofilms and Surface Components in Plant-Bacterial Associations." *International Journal of Molecular Sciences* 14, no. 8 (2013): 15838–59.

Böhme, Anne, Ute Risse-Buhl, and Kirsten Küsel. "Protists with Different Feeding Modes Change Biofilm Morphology." *FEMS Microbiology Ecology* 69, no. 2 (2009): 158–69.

Brown, Michael R. W., David G. Allison, and Peter Gilbert. "Resistance of Bacterial Biofilms to Antibiotics a Growth-Rate Related Effect?" *Journal of Antimicrobial Chemotherapy* 22, no. 6 (1988): 777–80.

Buesing, Nanna, Manuela Filippini, Helmut Bürgmann, and Mark O. Gessner. "Microbial Communities in Contrasting Freshwater Marsh Microhabitats." *FEMS Microbiology Ecology* 69, no. 1 (2009): 84–97.

Carneiro, Victor Alves, Hélcio Silva dos Santos, Francisco Vassiliepe Sousa Arruda, Paulo Nogueira Bandeira, Maria Rose Jane Ribeiro Albuquerque, Maria Olívia Pereira, Mariana Henriques, Benildo Sousa Cavada, and Edson Holanda Teixeira. "Casbane Diterpene as a Promising Natural Antimicrobial Agent against Biofilm-Associated Infections." *Molecules* 16, no. 1 (2011): 190–201.

Casacci, Luca Pietro, Francesca Barbero, and Emilio Balletto. "The 'Evolutionarily Significant Unit' Concept and Its Applicability in Biological Conservation." *Italian Journal of Zoology* 81, no. 2 (2014): 182–93.

Casetta, Elena, and Jorge Marques da Silva. "Biodiversity Surgery: Some Epistemological Challenges in Facing Extinction." *Axiomathes* 25, no. 3 (2015): 239–51.

Chandra, Jyotsna, Duncan M. Kuhn, Pranab K. Mukherjee, Lois L. Hoyer, Thomas McCormick, and Mahmoud A. Ghannoum. "Biofilm Formation by the Fungal *Pathogencandida albicans*: Development, Architecture, and Drug Resistance." *Journal of Bacteriology* 183, no. 18 (2001): 5385–94.

Chikindas, Michael L., Richard Weeks, Djamel Drider, Vladimir A. Chistyakov, and Leon M. T. Dicks. "Functions and Emerging Applications of Bacteriocins." *Current Opinion in Biotechnology* 49 (2018): 23–28.

Chopra, Lipsy, Gurdeep Singh, Kautilya Kumar Jena, and Debendra K. Sahoo. "Sonorensin: A New Bacteriocin with Potential of an Anti-Biofilm Agent and a Food Biopreservative." *Scientific Reports* 5 (2015): 13412.

Coenye, Tom, Gilles Brackman, Petra Rigole, Evy De Witte, Kris Honraet, Bart Rossel, and Hans J. Nelis. "Eradication of Propionibacterium Acnes Biofilms by Plant Extracts and Putative Identification of Icariin, Resveratrol, and Salidroside as Active Compounds." *Phytomedicine* 19, no. 5 (2012): 409–12.

Coenye, Tom. "Social Interactions in the *Burkholderia cepacia* Complex: Biofilms and Quorum Sensing." *Future Microbiology* 5, no. 7 (2010): 1087–99.

Costanza, Robert d'arge, R. Degroot, S. Farber, M. Grasso, B. Hannon, K. Limburg, S. Naeem, R. V. Oneill, J. Paruelo, R. G. Raskin, P. Sutton, and M. VandenBelt. "The Value of the World's Ecosystem Services and Natural Capital." *Nature* 387 (1997): 253–60.

Costerton, J. William. "Introduction to Biofilm." *International Journal of Antimicrobial Agents* 11, no. 3–4 (1999): 217–21.

Cotner, James B., and Bopaiah A. Biddanda. "Small Players, Large Role: Microbial Influence on Biogeochemical Processes in Pelagic Aquatic Ecosystems." *Ecosystems* 5, no. 2 (2002): 105–21.

Cowan, Marjorie Murphy. "Plant Products as Antimicrobial Agents." *Clinical Microbiology Reviews* 12, no. 4 (1999): 564–82.

Danhorn, Thomas, and Clay Fuqua. "Biofilm Formation by Plant-Associated Bacteria." *Annual Review of Microbiology* 61 (2007): 401–22.

Davies, David Gwilym, and Gill G. Geesey. "Regulation of the Alginate Biosynthesis Gene algC in *Pseudomonas aeruginosa* During Biofilm Development in Continuous Culture." *Applied Environtal Microbiology* 61, no. 3 (1995): 860–67.
De Carvalho, Carla C. C. R. "Biofilms: Recent Developments on an Old Battle." *Recent Patents on Biotechnology* 1, no. 1 (2007): 49–57.
Donlan, Rodney M. "Preventing Biofilms of Clinically Relevant Organisms Using Bacteriophage." *Trends in Microbiology* 17, no. 2 (2009): 66–72.
Elbreki, Mohamed, R. Paul Ross, Colin Hill, Jim O'Mahony, Olivia McAuliffe, and Aidan Coffey. "Bacteriophages and Their Derivatives as Biotherapeutic Agents in Disease Prevention and Treatment." *Journal of Viruses* 2014 (2014): 20.
Faraz, Naveed, Saba Sehrish, and Rehana Rehman Zia-Ul-Islam. "Antibiofilm Forming Activity of Naturally Occurring Compound." *Biomedica* 28, no. 2 (2012): 171–75.
Fernández, Lucía, Diana Gutiérrez, Pilar García, and Ana Rodríguez. "The Perfect Bacteriophage for Therapeutic Applications—A Quick Guide." *Antibiotics* 8, no. 3 (2019): 126.
Field, Des, Paul D. Cotter, R. Paul Ross, and Colin Hill. "Bioengineering of the Model Lantibiotic Nisin." *Bioengineered* 6, no. 4 (2015): 187–92.
Filoche, S. K., K. Soma, and C. H. Sissons. "Antimicrobial Effects of Essential Oils in Combination with Chlorhexidine Digluconate." *Oral Microbiology and Immunology* 20, no. 4 (2005): 221–25.
Filogônio, Cíntia de Fátima Buldrini, Rodrigo Villamarim Soares, Martinho Campolina Rebello Horta, Cláudia Valéria de Sousa Resende Penido, and Roberval de Almeida Cruz. "Effect of Vegetable Oil (Brazil Nut Oil) and Mineral Oil (Liquid Petrolatum) on Dental Biofilm Control." *Brazilian Oral Research* 25, no. 6 (2011): 556–61.
Flemming, Hans-Curt, Thomas R. Neu, and Daniel J. Wozniak. "The EPS Matrix: The 'House of Biofilm Cells.'" *Journal of Bacteriology* 189, no. 22 (2007): 7945–47.
Foley, I., P. Marsh, E. M. H. Wellington, A. W. Smith, and M. R. W. Brown. "General Stress Response Master Regulator rpoS Is Expressed in Human Infection: A Possible Role in Chronicity." *Journal of Antimicrobial Chemotherapy* 43, no. 1 (1999): 164–65.
Freimann, Remo, Helmut Bürgmann, Stuart E. G. Findlay, and Christopher T. Robinson. "Spatio-Temporal Patterns of Major Bacterial Groups in Alpine Waters." *PLoS One* 9, no. 11 (2014): e113524.
Fuqua, Clay, and E. Peter Greenberg. "Self-Perception in Bacteria: Quorum Sensing with Acylated Homoserine Lactones." *Current Opinion in Microbiology* 1, no. 2 (1998): 183–89.
Fuqua, W. Claiborne, Stephen C. Winans, and E. Peter Greenberg. "Quorum Sensing in Bacteria: The Luxr-Luxi Family of Cell Density-Responsive Transcriptional Regulators." *Journal of Bacteriology* 176, no. 2 (1994): 269.
Griebler, Christian, Florian Malard, and Tristan Lefébure. "Current Developments in Groundwater Ecology—From Biodiversity to Ecosystem Function and Services." *Current Opinion in Biotechnology* 27 (2014): 159–67.
Haghighi, Farnoosh, Shahla Roudbar Mohammadi, Parisa Mohammadi, Saman Hosseinkhani, and Reza Shipour. "Antifungal Activity of TiO_2 Nanoparticles and EDTA on *Candida albicans* Biofilms." *Infection, Epidemiology and Microbiology* 1, no. 1 (2013): 33–38.
Hall-Stoodley, Luanne, and Paul Stoodley. "Evolving Concepts in Biofilm Infections." *Cellular Microbiology* 11, no. 7 (2009): 1034–43.
Hammer, Brian K., and Bonnie L. Bassler. "Quorum Sensing Controls Biofilm Formation in *Vibrio cholerae*." *Molecular Microbiology* 50, no. 1 (2003): 101–04.
Hammer, Katherine A., Christine F. Carson, and Thomas V. Riley. "Antimicrobial Activity of Essential Oils and Other Plant Extracts." *Journal of Applied Microbiology* 86, no. 6 (1999): 985–90.

Harjai, Kusum, Ravi Kumar, and Sukhvinder Singh. "Garlic Blocks Quorum Sensing and Attenuates the Virulence of *Pseudomonas aeruginosa*." *FEMS Immunology & Medical Microbiology* 58, no. 2 (2010): 161–68.

Hatt, J. K., and P. N. Rather. "Role of Bacterial Biofilms in Urinary Tract Infections." In Romeo, Tony. (eds) *Bacterial Biofilms*, pp. 163–92. Springer, Berlin, Hiedelberg, 2008.

Heino, Jani, Mikko Tolkkinen, Anna Maria Pirttilä, Heidi Aisala, and Heikki Mykrä. "Microbial Diversity and Community–Environment Relationships in Boreal Streams." *Journal of Biogeography* 41, no. 12 (2014): 2234–44.

Hentzer, á. M., á. L. Eberl, and á. M. Givskov. "Transcriptome Analysis of *Pseudomonas aeruginosa* Biofilm Development: Anaerobic Respiration and Iron Limitation." *Biofilms* 2, no. 1 (2005): 37–61.

Hentzer, Morten, Hong Wu, Jens Bo Andersen, Kathrin Riedel, Thomas B. Rasmussen, Niels Bagge, Naresh Kumar, Mark A. Schembri, Zhijun Song, Peter Kristoffersen, Mike Manefield, John W. Costerton, Søren Molin, Leo Eberl, Peter Steinberg, Staffan Kjelleberg, Niels Høiby, and Michael Givskov. "Attenuation of *Pseudomonas aeruginosa* Virulence by Quorum Sensing Inhibitors." *The EMBO Journal* 22, no. 15 (2003): 3803–15.

Hetrick, Evan M., Jae Ho Shin, Heather S. Paul, and Mark H. Schoenfisch. "Anti-Biofilm Efficacy of Nitric Oxide-Releasing Silica Nanoparticles." *Biomaterials* 30, no. 14 (2009): 2782–89.

Heukelekian, Henry, and Agnes Heller. "Relation between Food Concentration and Surface for Bacterial Growth." *Journal of Bacteriology* 40, no. 4 (1940): 547.

Höglund, Jacob. *Evolutionary Conservation Genetics*. Oxford University Press, Oxford, 2009.

Hughes, K. A., I. W. Sutherland, J. Clark, and M. V. Jones. "Bacteriophage and Associated Polysaccharide Depolymerases—Novel Tools for Study of Bacterial Biofilms." *Journal of Applied Microbiology* 85, no. 3 (1998): 583–90.

Isman, Murray B. "Plant Essential Oils for Pest and Disease Management." *Crop Protection* 19, no. 8–10 (2000): 603–08.

Jackson, Evelyn F., and Colin R. Jackson. "Viruses in Wetland Ecosystems." *Freshwater Biology* 53, no. 6 (2008): 1214–27.

Jamal, Muhsin, Saadia Andleeb, Fazal Jalil, Muhammad Imran, Muhammad Asif Nawaz, Tahir Hussain, Muhammad Ali, Sadeeq ur Rahman, and Chythanya Rajanna Das. "Isolation, Characterization and Efficacy of Phage MJ2 against Biofilm Forming Multi-Drug Resistant *Enterobacter cloacae*." *Folia Microbiologica* 64, no. 1 (2019): 101–11.

Jamal, Muhsin, Saadia Andleeb, Fazal Jalil, Muhammad Imran, Muhammad Asif Nawaz, Tahir Hussain, Muhammad Ali, Chythanya Rajanna Das. "Isolation and Characterization of a Bacteriophage and Its Utilization against Multi-Drug Resistant *Pseudomonas aeruginosa*-2995." *Life Sciences* 190 (2017): 21–28.

Jamal, Muhsin, Tahir Hussain, Chythanya Rajanna Das, and Saadia Andleeb. "Characterization of Siphoviridae Phage Z and Studying Its Efficacy against Multidrug-Resistant *Klebsiella pneumoniae* Planktonic Cells and Biofilm." *Journal of Medical Microbiology* 64, no. 4 (2015): 454–62.

Jamal, Muhsin, Ufaq Tasneem, Tahir Hussain and Saadia Andleeb. "Bacterial biofilm: its composition, formation and role in human infections." *Journal of Microbiology and Biotechnology* 4 (2015): 1–14.

Jamil, Bushra, Habib Bokhari, and Mohammad Imran. "Mechanism of Action: How Nano-Antimicrobials Act?" *Current Drug Targets* 18, no. 3 (2017): 363–73.

Jones, Nicole, Binata Ray, Koodali T. Ranjit, and Adhar C. Manna. "Antibacterial Activity of ZnO Nanoparticle Suspensions on a Broad Spectrum of Microorganisms." *FEMS Microbiology Letters* 279, no. 1 (2008): 71–76.

Kalishwaralal, Kalimuthu, Selvaraj BarathManiKanth, Sureshbabu Ram Kumar Pandian, Venkataraman Deepak, and Sangiliyandi Gurunathan. "Silver Nanoparticles Impede the Biofilm Formation by *Pseudomonas aeruginosa* and *Staphylococcus epidermidis*." *Colloids and Surfaces B: Biointerfaces* 79, no. 2 (2010): 340–44.

Kaplan, J. á. B. "Biofilm Dispersal: Mechanisms, Clinical Implications, and Potential Therapeutic Uses." *Journal of Dental Research* 89, no. 3 (2010): 205–18.

Kim, Kwang Y., and Joseph F. Frank. "Effect of Nutrients on Biofilm Formation by Listeria Monocytogenes on Stainless Steel." *Journal of Food Protection* 58, no. 1 (1995): 24–28.

Kovach, Kristin, Megan Davis-Fields, Yasuhiko Irie, Kanishk Jain, Shashvat Doorwar, Katherine Vuong, Numa Dhamani, Kishore Mohanty, Ahmed Touhami, and Vernita D. Gordon. "Evolutionary Adaptations of Biofilms Infecting Cystic Fibrosis Lungs Promote Mechanical Toughness by Adjusting Polysaccharide Production." *NPJ Biofilms and Microbiomes* 3, no. 1 (2017): 1.

Kreth, Jens, Yongshu Zhang, and Mark C. Herzberg. "Streptococcal Antagonism in Oral Biofilms: *Streptococcus sanguinis* and *Streptococcus gordonii* Interference with *Streptococcus mutans*." *Journal of Bacteriology* 190, no. 13 (2008): 4632–40.

Lade, Harshad, Diby Paul, and Ji Hyang Kweon. "Quorum Quenching Mediated Approaches for Control of Membrane Biofouling." *International Journal of Biological Sciences* 10, no. 5 (2014): 550.

Lee, Jin-Hyung, Joo-Hyeon Park, Jung-Ae Kim, Ganesh Prasad Neupane, Moo Hwan Cho, Chang-Soo Lee, and Jintae Lee. "Low Concentrations of Honey Reduce Biofilm Formation, Quorum Sensing, and Virulence in *Escherichia coli* O157: H7." *Biofouling* 27, no. 10 (2011): 1095–104.

Lemos, Madalena, Filipe Mergulhão, Luís Melo, and Manuel Simões. "The Effect of Shear Stress on the Formation and Removal of *Bacillus cereus* Biofilms." *Food and Bioproducts Processing* 93 (2015): 242–48.

Lieleg, Oliver, Marina Caldara, Regina Baumgärtel, and Katharina Ribbeck. "Mechanical Robustness of *Pseudomonas aeruginosa* Biofilms." *Soft Matter* 7, no. 7 (2011): 3307–14.

MacIntyre, Hugh L., Richard J. Geider, and Douglas C. Miller. "Microphytobenthos: The Ecological Role of the 'Secret Garden' of Unvegetated, Shallow-Water Marine Habitats. I. Distribution, Abundance, and Primary Production." *Estuaries* 19, no. 2 (1996): 186–201.

Maddocks, Sarah E., Marta Salinas Lopez, Richard S. Rowlands, and Rose A. Cooper. "Manuka Honey Inhibits the Development of *Streptococcus pyogenes* Biofilms and Causes Reduced Expression of Two Fibronectin Binding Proteins." *Microbiology* 158, no. 3 (2012): 781–90.

Mah, Thien-Fah C., and George A. O'Toole. "Mechanisms of Biofilm Resistance to Antimicrobial Agents." *Trends in Microbiology* 9, no. 1 (2001): 34–39.

Marques da Silva, Jorge. "Reconciling Science and Nature by Means of the Esthetical Contemplation of Natural Diversity." *Rivista di estetica* 59 (2015): 93–113.

Mathur, Harsh, Des Field, Mary C. Rea, Paul D. Cotter, Colin Hill, and R. Paul Ross. "Fighting Biofilms with Lantibiotics and Other Groups of Bacteriocins." *npj Biofilms and Microbiomes* 4, no. 1 (2018): 1–13.

Meng, Xiangpeng, Yibo Shi, Wenhui Ji, Xueling Meng, Jing Zhang, Hengan Wang, Chengping Lu, Jianhe Sun, and Yaxian Yan. "Application of a Bacteriophage Lysin to Disrupt Biofilms Formed by the Animal Pathogen *Streptococcus suis*." *Applied and Environmental Microbiology* 77, no. 23 (2011): 8272–79.

Molan, Peter C. "Re-Introducing Honey in the Management of Wounds and Ulcers—Theory and Practice." *Ostomy Wound Manage* 48 (2002): 28–40.

Murinda, S. E., K. A. Rashid, and Robert F. Roberts. "In Vitro Assessment of the Cytotoxicity of Nisin, Pediocin, and Selected Colicins on Simian Virus 40–Transfected Human Colon and Vero Monkey Kidney Cells with Trypan Blue Staining Viability Assays." *Journal of Food Protection* 66, no. 5 (2003): 847–53.

Natan, Michal, and Ehud Banin. "From Nano to Micro: Using Nanotechnology to Combat Microorganisms and Their Multidrug Resistance." *FEMS Microbiology Reviews* 41, no. 3 (2017): 302–22.

Ng, Wen-Jie, Kit-Yin Lim, Ju-Yee Chong, and Ka-Lok Low. "In Vitro Screening of Honey against Enterococcus Spp. Biofilm." *Journal of Medical and Bioengineering* 3, no. 1 (2014): 23–28.

Novick, Richard P., and Tom W. Muir. "Virulence Gene Regulation by Peptides in Staphylococci and Other Gram-Positive Bacteria." *Current Opinion in Microbiology* 2, no. 1 (1999): 40–45.

Okada, Masahiro, Isao Sato, Soo Jeong Cho, Hidehisa Iwata, Toshihiko Nishio, David Dubnau, and Youji Sakagami. "Structure of the *Bacillus subtilis* Quorum-Sensing Peptide Pheromone ComX." *Nature Chemical Biology* 1, no. 1 (2005): 23.

Otto, Michael. "Staphylococcal Infections: Mechanisms of Biofilm Maturation and Detachment as Critical Determinants of Pathogenicity." *Annual Review of Medicine* 64 (2013): 175–88.

Palmer, Jon, Steve Flint, and John Brooks. "Bacterial Cell Attachment, the Beginning of a Biofilm." *Journal of Industrial Microbiology & Biotechnology* 34, no. 9 (2007): 577–88.

Paterson, David M., and Kirby S. Black. "Water Flow, Sediment Dynamics, and Benthic Biology." *Advances in Ecological Research* 29 (1999): 155–93.

Pires, Diana, Sanna Sillankorva, Alberta Faustino, and Joana Azeredo. "Use of Newly Isolated Phages for Control of *Pseudomonas aeruginosa* PAO1 and ATCC 10145 Biofilms." *Research in Microbiology* 162, no. 8 (2011): 798–806.

Pratt, Leslie A, and Roberto Kolter. "Genetic Analysis of *Escherichia coli* Biofilm Formation: Roles of Flagella, Motility, Chemotaxis, and Type I Pili." *Molecular Microbiology* 30, no. 2 (1998): 285–93.

Ramasamy, Mohankandhasamy, and Jintae Lee. "Recent Nanotechnology Approaches for Prevention and Treatment of Biofilm-Associated Infections on Medical Devices." *BioMed Research International* 2016 (2016): 17.

Ravichandiran, Vinothkannan, Karthi Shanmugam, Kornepati Anupama, Sabu Thomas, and Adline Princy. "Structure-Based Virtual Screening for Plant-Derived SdiA-Selective Ligands as Potential Antivirulent Agents against Uropathogenic *Escherichia coli*." *European Journal of Medicinal Chemistry* 48 (2012): 200–05.

Rinaudi, Luciana V., Fernando Sorroche, Ángeles Zorreguieta, and Walter Giordano. "Analysis of the mucR Gene Regulating Biosynthesis of Exopolysaccharides: Implications for Biofilm Formation in *Sinorhizobium meliloti* Rm1021." *FEMS Microbiology Letters* 302, no. 1 (2010): 15–21.

Romaní i Cornet, Anna M., Carles Borrego i Moré, Verónica Díaz Villanueva, Anna Freixa Casals, Frederic Gich Batlle, and Irene Ylla i Monfort. "Shifts in Microbial Community Structure and Function in Light-and Dark-Grown Biofilms Driven by Warming." *Environmental Microbiology* 16, no. 8 (2014): 2550–67.

Rudramurthy, Gudepalya Renukaiah, Mallappa Kumara Swamy, Uma Rani Sinniah, and Ali Ghasemzadeh. "Nanoparticles: Alternatives against Drug-Resistant Pathogenic Microbes." *Molecules* 21, no. 7 (2016): 836.

Sadekuzzaman, M., S. Yang, M. F. R. Mizan, and S. D. Ha. "Current and Recent Advanced Strategies for Combating Biofilms." *Comprehensive Reviews in Food Science and Food Safety* 14, no. 4 (2015): 491–509.

Santangelo, E. F. "Honey." 2013. http://flipper.diff.org/app/items/info/4617.

Santos, V. L., R. M. Nardi Drummond, and M. V. Dias-Souza. "Bacteriocins as Antimicrobial and Antibiofilm Agents." *Current Developments in Biotechnology and Bioengineering: Human and Animal Health Applications* (2016): 403.

Sillankorva, Sanna, Peter Neubauer, and Joana Azeredo. "*Pseudomonas fluorescens* Biofilms Subjected to Phage phiIBB-PF7A." *BMC Biotechnology* 8, no. 1 (2008): 79.

Singh, Brahma N., Hardeep B. Singh, Akanksha Singh, Braj R. Singh, Aradhana Mishra, and C. Shekhar Nautiyal. "*Lagerstroemia speciosa* Fruit Extract Modulates Quorum

Sensing-Controlled Virulence Factor Production and Biofilm Formation in *Pseudomonas aeruginosa*." *Microbiology* 158, no. 2 (2012): 529–38.
Smith, Timothy B., Michael William Bruford, and Robert K. Wayne. "The Preservation of Process: The Missing Element of Conservation Programs." In Smith, Timothy B., Michael William Bruford, and Robert K. Wayne (eds) *Ecosystem Management*, pp. 71–75. Springer, New York, 1993.
Solano, Cristina, Maite Echeverz, and Iñigo Lasa. "Biofilm Dispersion and Quorum Sensing." *Current Opinion in Microbiology* 18 (2014): 96–104.
Sousa, Cláudia, Claudia M. Botelho, and Rosário Oliveira. "Nanotechnology Applied to Medical Biofilms Control." In Méndez-Vilas, Antonio (ed) *Science against Microbial Pathogens: Communicating Current Research and Technological Advances*, pp. 878–88. Formatex, Badajoz, 2011.
Srinivasan, Soumya, Gregory W. Harrington, Irene Xagoraraki, and Ramesh Goel. "Factors Affecting Bulk to Total Bacteria Ratio in Drinking Water Distribution Systems." *Water Research* 42, no. 13 (2008): 3393–404.
Suci, Peter A., Deborah L. Berglund, Lars Liepold, Susan Brumfield, Betsey Pitts, Willy Davison, Luke Oltrogge, Kevin O. Hoyt, Sharon Codd, Philip S. Stewart, Marian Young. "High-Density Targeting of a Viral Multifunctional Nanoplatform to a Pathogenic, Biofilm-Forming Bacterium." *Chemistry & Biology* 14, no. 4 (2007): 387–98.
Szafrański, Szymon P., Andreas Winkel, and Meike Stiesch. "The Use of Bacteriophages to Biocontrol Oral Biofilms." *Journal of Biotechnology* 250 (2017): 29–44.
Taylor, Erik, and Thomas J. Webster. "Reducing Infections through Nanotechnology and Nanoparticles." *International Journal of Nanomedicine* 6 (2011): 1463.
Thallinger, Barbara, Endry N. Prasetyo, Gibson S. Nyanhongo, and Georg M. Guebitz. "Antimicrobial Enzymes: An Emerging Strategy to Fight Microbes and Microbial Biofilms." *Biotechnology Journal* 8, no. 1 (2013): 97–109.
Ueda, Akihiro, and Thomas K. Wood. "Connecting Quorum Sensing, c-di-GMP, Pel Polysaccharide, and Biofilm Formation in *Pseudomonas aeruginosa* through Tyrosine Phosphatase TpbA (PA3885)." *PLoS Pathogens* 5, no. 6 (2009): e1000483.
Underwood, Graham J. C., and Jacco Kromkamp. "Primary Production by Phytoplankton and Microphytobenthos in Estuaries." *Advances in Ecological Research* 29 (1999): 93–153.
Visick, Karen L., Mark A. Schembri, Fitnat Yildiz, and Jean-Marc Ghigo. "Biofilms 2015: Multidisciplinary Approaches Shed Light into Microbial Life on Surfaces." *Journal of Bacteriology* 198, no. 19 (2016): 2553–63.
Wang, Jianjun, Ji Shen, Yucheng Wu, Chen Tu, Janne Soininen, James C. Stegen, Jizheng He, Xingqi Liu, Lu Zhang, and Enlou Zhang. "Phylogenetic Beta Diversity in Bacterial Assemblages across Ecosystems: Deterministic Versus Stochastic Processes." *The ISME Journal* 7, no. 7 (2013): 1310.
Wang, Yang, Yuxin Wang, Liyun Sun, Daniel Grenier, and Li Yi. "*Streptococcus suis* Biofilm: Regulation, Drug-Resistance Mechanisms, and Disinfection Strategies." *Applied Microbiology and Biotechnology* 102, no. 21 (2018): 9121–29.
Waters, Christopher M., and Bonnie L. Bassler. "Quorum Sensing: Cell-to-Cell Communication in Bacteria." *Annual Review of Cell and Developmental Biology* 21 (2005): 319–46.
Xavier, Joao B., Cristian Picioreanu, Suriani Abdul Rani, Mark C. M. van Loosdrecht, and Philip S. Stewart. "Biofilm-Control Strategies Based on Enzymic Disruption of the Extracellular Polymeric Substance Matrix—A Modeling Study." *Microbiology* 151, no. 12 (2005): 3817–32.

2 Microbial Biofilm in Clinical Bioremediation Practices in Human Health

Rengasamy Sathya, Thangaprakasam Ushadevi, S. Ambiga, and Bakrudeen Ali Ahmed Abdul
PRIST Deemed University

CONTENTS

2.1 INTRODUCTION

Biofilm is a natural system of bacterial evolution all over in the environmental spaces. The consequence of biofilm development is improved resistance to harmful environmental effects and resistance to antibiotics and antimicrobial mediator. Based on the quorum sensing (QS) mechanism, which plays a vital role in the biofilm enlargement and equilibrium the environment once the bacteria mass grows into high (Lu et al. 2019). Biofilm formation covers single groups or multiple groups of species and acts as a complex community (bacteria, fungi, algae, cyanobacteria, etc.) with different environmental factors linked to nutritional requirements. The biofilm enclosed with slimy extracellular polymeric substances (EPS) produced specific biofilm forming microbes. Biofilms are formed on the surface of biotic and abiotic locations that can be dominant in natural resources and industrial and hospital infections. Biofilm, measured as a hydrogel, is coated with a complex polymer and is easily attached on the surface of tooth, glass lenses, clothes, hospital materials, etc.

2.2 BIOFILM TOPOGRAPHIES

Biofilms are comprised of unicellular and multicellular groups that coexist via symbiotic relationship with counterparts to survive the suitable environment. The main features of the biofilms are self-bearing EPS consisting of polysaccharide, proteins, and extracellular DNA, which help to protect the biofilm from external threats (Percival and Kite 2007).The biofilm structure has incorporated with water networks, nutrient and electron transport, and other reduced complexes may present. The living and nonliving cells and the biofilm cell wall are composed of teichoic acid, host products, DNA, and N-acetyl glucosamine. The capability of bacteria is to observe and settle in contradiction of the external surface, and spreading the mode of infection.

2.3 SPECIES INTERACTIONS

The mixed species communication on biofilm is practically a great challenging one. Biofilm formation covers several biological, physical, and chemical routes depending on environmental and hydrodynamic settings. The formation of biofilm includes several stages: (i) attachment (reversible and irreversible), (ii) colonization, (iii) maturation, and (iv) dispersion (Figure 2.1). Hence, several topographies of the microbe cell surface encourage the attachment to the cell surface (flagella, pili, fimbriae, and glycocalyx) (Donlan and Costerton 2002).The aggregation of biofilm microbes in sessile or free-floating nature has the major benefit of improved tolerance to environmental factors such as temperature, pH, nutrients, predation, toxic substances, and antibiotics.

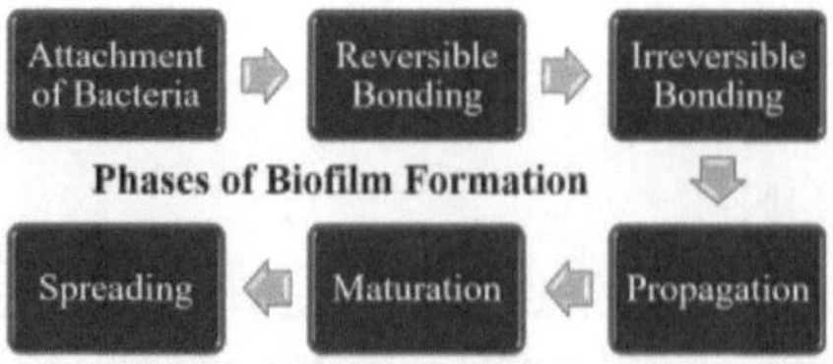

FIGURE 2.1 Biofilm developmental stages. The biofilm developments are consist of six vital stages: (i) Attachment of bacteria stick to the surface of biomaterial, (ii) During reversible bonding of bacteria can move around the material, (iii) Irreversible bonding of bacteria can enrich the maturation process, (iv) Bacteria enables to spread and colonize on new surface areas (spreading), (v) Maturation is the critical life cycle of biofilm, (vi) propagation is the detachment of single to multi-group of cells and release of seeding dispersal.

2.4 HOSPITAL INFECTION

Nosocomial infections are seen in around 2 million cases of people, and 90,000 deaths increased in the United States alone (Darouiche 2004; Guggenbichler et al. 2011). Biofilm infections occur in 60%–70% of nosocomial infections (hospital acquired) associated with certain kinds of fixed medical devices. Nosocomial infections are the leading the cause of death in the United States with 2 million patients affected per annum. In American unaided hospitals, 1.7 million infections and 99,000 deaths occurred every year, and 3 million people affected in European countries catch infectious diseases and cause around 50,000 deaths (ECDC 2007). The record broadly spreading hospital "superbug" methicillin resistant *Staphylococcus aureus* (MRSA) which are regular root of septicaemia or bacteremia in clinical surroundings (HPA 2012a). Mostly 80% of the pathogens are linked with persistent infections (Donlan 2008; NHLBI 2002). Hence, there is a crucial need to reduce the illness and death rates arising from acute and chronic infections worldwide. More than 80% of biofilm infections have been occurred in human, and animal, which created a major health problem. Together commensal and pathogenic microbes cause biofilm formation. Biofilm contaminations create many clinical challenges, counting diseases linked with unproductive species, chronic inflammation, and antibiotic resistance (Romling and Balsalobre, 2012). These infections arise on the body and reduce wound healing, further causing lung diseases, urinary tract infections, cystic fibrosis, etc.

The hospital infections are virtually accumulated through all medical devices (contact lenses, prostheses, pacemakers, catheters, etc.). Entirely medical devices or tissue engineering is vulnerable to microbial growth and contamination. The primary infection of medical devices most probably occurs from tiny number of microbes, which are moved to the device over the patients or health care products, contaminated water, worker's skin, and other springs. In the hospital, the medical device related infection bears a big economic problem on health service and is associated with increased patient illness and death (Donlan 2008).

The report has been comprehensively studied in microorganisms, and most frequently health care associated infections caused by *S. aureus, S. epidermidis, P. aeruginosa* and *Enterobacter* are the most dominant gram-negative bacteria biofilm producing pathogen associated with medical devices (Figure 2.2). Around 90% of the biofilm is encapsulated as an extracellular polysaccharide (EPS). EPS initiates external bonding, and acts as a framework for cells, enzymes, and antibiotic near assign (Stewart and Costerton 2001; Vega and Gore 2014). *Pseudomonas aeruginosa* is linked with cystic fibrosis, and *S. aureus* is responsible for wound infection (Hoiby, Ciofu, and Bjarnsholt 2010).

2.4.1 DEVICE LINKED BIOFILM CONTAMINATIONS

Microorganisms are attached towards both types of soft and hard medical devices. A variety of microorganisms adhere to contact lenses, notably *E. coli*, *S. aureus*, *S. epidermidis,* and *P. aeruginosa*, and the point of adherence depends on several factors such as water content, substrate nature, electrolyte mass, type of bacteria, and

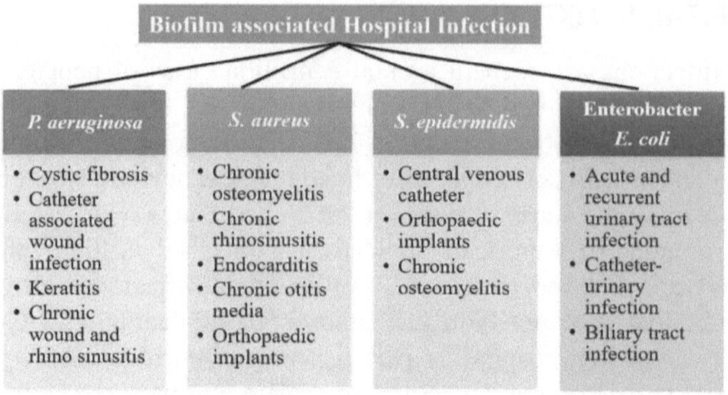

FIGURE 2.2 Pathogens causing Biofilm infections. Biofilm forming bacteria are usually pathogenic and lead to the cause of hospital-acquired infection. Most of the acute and chronic infections are linked with biofilm bacteria. Biofilm producers cause a variety of health issues and more harmful for excreting toxins and greatly resistant to the number of antibiotics. *P. aeruginosa, S. aureus, S. epidermidis, Enterobacter,* and *E. coli* are the major bacteria which causes severe hospital infections.

a combination of polymers. The infection on contact lenses of a patient diagnosed with keratitis is caused by *P. aeruginosa.* Biofilm can be easily attached on contact lenses that are often in the lens storage cases. The growth of biofilm on a central venous catheter infection depends on the nature of fluid. Notably gram-positive bacteria *S. epidermidis* and *S. aureus* cannot grow well in intravenous fluids, while gram-negative bacteria *P. aeruginosa, Enterobacter,* and *Klebsiella* species withstand growth development in such fluids (Percival and Kite 2007).

The attachment of biofilm on mechanical heart valves and surrounding tissues causes endocarditis. The group of bacteria *Enterococcus, S. epidermidis,* and *S. aureus* gets accumulated and causes an unpleasant condition. At the time of surgical implantation of prosthetic heart valves, tissue damage may appear as a result of aggregation of platelets and fibrin at the point of location, and microbes have good ability to colonize the particular part. In the popular open catheter method, contaminations highly flowed causing the urinary tract infections to spread very quickly. In the closed method urine gathers in a plastic bag, and during this condition *E. coli, Enterococcus faecalis, K. pneumonia,* and other bacteria pollute the device (Stickler 1996).

2.4.2 Nondevice Linked Biofilm Contaminations

The periodontitis infections occurred in the gums, injuries of soft tissues, and teeth. The nondevice infection is caused by poor oral hygiene. The microbial infection occurred on the surface of mucosa. The microbial mass surrounding the teeth may initiate penetration of epithelial cells and release harmful toxic substances, and plaque formation appeared within three weeks. The plaque can mineralize with calcium and phosphate and develop calculus. The group of bacteria enters the bones via the blood stream, causing trauma and Osteomyelitis (Kokare et al. 2007). The microbes that pass through the blood stream affect the metaphysis of bone, and the infection leads to the accumulation of White Blood Cell (WBC). The WBC shot the phagocytosis using secretory enzymes. The responsible enzyme caused different symptoms such as lysis of the bone, quick formation of the pus, spread via blood vessels, slowdown of the proper blood flow causing tissue injury, and deterioration of the affected bone parts.

2.5 QUORUM SENSING

QS bacteria can harvest and discharge the chemical indicator molecule (cell to cell communication by autoinducer). The molecule enhanced the absorption and density of the biofilm. Generally, gram-positive and gram-negative bacteria use QS communication circuit to regulate a variety of physiological actions. The target gene is made exactly by the bacterial QS system to regulate the gene expression. The bacterial cells provide the extracellular signaling molecules. These molecules initiated intercellular communication of groups of biofilm microbes. A current report on the inhibition of QS and biofilm growth has been comprehensively studied by numerous sources.

The potent natural secondary metabolite is usnic acid derived from lichen, the combination of RNAIII-inhibiting peptide plus usnic acid linked with QS, which activated the inhibition of *S. aureus* biofilm growth and altered the morphological features of *P. aeruginosa* (Francolini et al. 2004). Several natural compounds (Penicillic acid, solenopsin-A, ellagic acid, catechin and curcumin derivatives) effects on the inhibition of QS were studied. (Rasamiravaka et al. 2015). Plant derived Quorum Sensing Inhibitors (QSIs) habitually prevent the biofilm growth specifically shared with antibiotics (Brackman et al. 2011).The most traditional Chinese treatment baicalein proteolytically consumed the signal receptor (TraR gene) in *P. aeruginosa*, which are expected to create antibiofilm efficacy. The fruit extract of *Lagerstroemia speciosa* initiated the downregulation of QS responsible genes (las and rhl) and N-acyl-homoserine lactones (HSL) in *P. aeruginosa* (Jakobsen et al. 2012).

2.6 HOST IMMUNE REACTION

The group of microbial biofilm is generally apparent in situ, and the interspecies connection within a biofilm is significant in the control of biofilms (Martin et al. 2017). The host immune response has evolved to protect the host from infection based on the innate and adaptive immunity. The binding of microbial substances to the innate immune receptor activates signaling cascades inside the cell and initiates phagocytic process for the production of antibiofilm substances. The host response to biomaterials follows a cascade of events initial planting causes a disconcertion of the homeostatic device, location of the process of healing. The classic view of host response to biomaterials is separated into various overlapping stages (Figure 2.3). The numerous cell phenotypes involve the immune and inflammatory host reactions of biofilm (Anderson 2001).

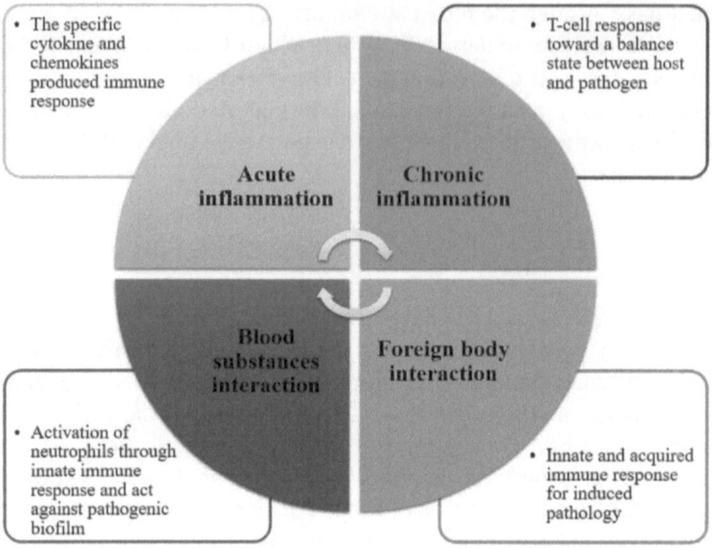

FIGURE 2.3 Different stages of host immune response during infection. The four vital steps involved during host immune response to the pathogens: (i) Blood substances interaction, (ii) Foreign body interaction, (iii) Acute inflammation, (iv) chronic inflammation.

2.7 CONTROL OF BIOFILM

Biofilm formation is an existing challenge in medical, industrial, and environmental needs and dignified as a world safety problem of the humankind. Biofilms are of great significance in control of health care related and other infections. This chapter focuses on the current status of hospital infections created by harmful biofilm forming bacteria. The primary aim is to degrade the harmful biofilm colonizing bacteria by beneficial microbes and other natural sources through bioremediation aspects.

Generally, biofilm creating bacteria are highly resistant to different antimicrobial agents and resistant towards toxic, desiccation, antibiotics, and chemical compounds. Since biofilms can tolerate very tough form and endure the host immune system, there is a need for novel treatment routes to cure biofilm-related infections. Most of the recent researches have been focusing on the formation of anti-biofilm agents that are nontoxic, such molecule don't lead the future drug resistant. This chapter, we dignified the current research of new antibiofilm agents derived from various natural sources, the derivatives of imidazole, indole, phenols, triazole, sulfide, furanone, peptides, silver nanoparticles etc., take ability to diffuse bacterial biofilm in *in-vivo* that reduce biofilm formation (Figure 2.4).

The current status of the research concentrated more upon classy methods for cleansing and the alteration of medical devices to inhibit the biofilm formation. The complex community of microbial biofilm can cause tissue damage and gets attached to the surface of medical devices. The prevention of a biofilm infection represents that serious awareness is required to minimize the existence of biofilm forming pathogens. This chapter reviews the most expressive to reduce the effect of biofilm

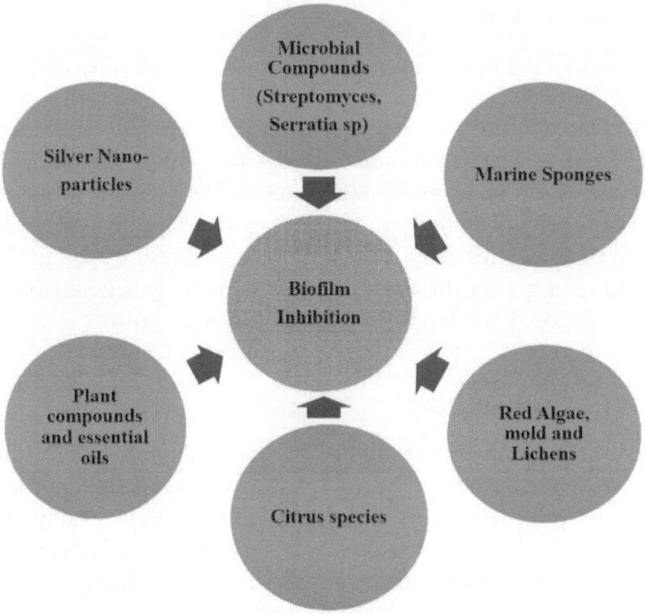

FIGURE 2.4 Biofilm inhibitions by natural sources. Natural sources are a vital role in reducing the growth of biofilm-forming bacteria. Natural anti-biofilm compounds are capable of producing therapeutic substances. The anti-biofilm inhibitors (microbial compounds, marine sponges, red algae, citrus species, plant compounds and essential oils, nanoparticles) act as an essential role to reduce the biofilm-based infections.

through beneficial microorganisms. The need of special care is made to reduce the formation of biofilm by antibiofilm process.

2.7.1 Natural Biofilm Inhibitors

Billion years of sensitive pressures take certain increase to many plans in bacterial survival; adapt multiple organisms into the environment. The marine sponges are the best example for the reduction of biofilm formation on clinical bioremediation aspects. The *Agelas conifer* species is the best example for the decrease of biofilm. The natural products and compounds are extracted from sponges, such as 2-aminoinidazole, bromoageleferin, and oroidin. These natural compounds can easily disperse and inhibit biofilm formation. Recent reports proved to inhibit the activity of MRSA (Reyes et al. 2011).

The emodin compound has been extracted from mold, lichens, and various plants. Emoidin significantly inhibited biofilm formation and hooked on the surface and affected the QS capacity of *P. aeruginosa* (Ding et al. 2011). Antibiofilm agents that can together diffuse and kill the biofilm bacteria might have certain valuable applications that remain sporadic (Figure 2.4). Essential oils (EOs) are natural antibiofilm agents that inhibit virulent bacteria, fungi, and viruses (Hammer, Carson, and Riley 1999).

2.7.2 Electrogenic Biofilm

The current advancement in biofilm research has been applied to control the energy crisis system. The recent approach is used in microbial fuel cells (MFCs). MFCs generate electricity by using the chemical energy present in organic and inorganic substances. The electrogenetic and nonelectrogenetic microbes are involved in the part of synergistic electrogenic biofilm (Zhou et al. 2013). There are various antimicrobial agents and antibiotics that are applied against biofilm prevention. Recently many possible techniques applied for the inhibition of biofilm, specifically inhibition of QS via breaking of matrix through F-actin. A similar genetic analysis is exposed in prokaryotic and eukaryotic biofilms (Buommino et al. 2014).

2.7.3 Synthetic QS Inhibitors

The synthetic inhibitors have been highly suggested as a desirable solution to the eradication of unwanted biofilms. The combination of natural QSI, garlic extract, and antibiotic tobramycin leads to killing of all the biofilms. Remarkably, the inhibition of biofilm cell growth has been measured and the inhibition of QS can be forced on one species by another one in the group of biofilm (Han et al. 2017).

2.7.4 Nanoparticles

The significant nanomaterials are used to control biofilm formation (silver, titanium oxide, copper oxide, zinc oxide, silica nanoparticles). The nanomaterials are applied on the surface of the tooth and act as a carrier with specific binding properties. The remarkable nanomaterials such as silver nitrate and silver nanoparticles are effective against oral biofilm pathogen (Allaker and Memarzadeh 2014). Hence, the nanoparticles are considered as a promising source for the inhibition of dental caries of biofilm. AgNPs display the antibiofilm activity against *E. faecalis*, which is identified as the major cause of secondary and persistent endodontic injuries. The advantage mode of AgNPs initiates the antibiofilm ability (Besinis, De Peralta, and Handy 2014).

Nanoparticles are commonly stated as taking an inhibitory effect against biofilm cells, e.g., *P. aeruginosa* and *S. aureus*. The activity is connected with ATP linked breakdown process, absorbency of the outer membrane, and the group of hydroxyl radicals that are encouraged by bactericidal substances (Hwang, Hwang and Choi 2012). Potential transpeptidase is appropriate as the mark for novel antivirulence agent since it is difficult in the bacterial connections (Cascioferro et al. 2015). The currently available antibiofilm inhibitors are given in Table 2.1.

2.7.5 In Vitro and In Vivo Approach

The new methods are applied for the prevention of biofilm formation. The antiadhesive treatment involves the physical, chemical, and structural behaviors that reduce the adhesion of biofilm formation. The in vitro studies of biofilm prevention include photodynamic therapy, matrix degradation, regaining of signal blockers, interpolation through biofilm ruling, and initiation of biofilm dispassion and progress of

TABLE 2.1
Antibiofilm Inhibitors Derived from Natural Sources

Antibiofilm Inhibitors	Sources	Biofilm Producers
Skyllamycins B and C	*Streptomyces* sp.	*P. aeruginosa*
Imidazole derivatives,	Sponge	*S. aureus (MRSA)*
2-aminoimidazole, bromoageleferin and oroidin	Agelas conifer	*P. aeruginosa*
Marine derived compounds, natural furanone	*Red algae, Delisea pulchra*	*P. aeruginosa*
Ageloxime D	*Agelas nakamurai* sponge	*S. epidermidis*
Cembranoid	*Pseudoplexaura, flagellosa, Eunicea knightii*	*S. aureus,P. aeruginosa*
Ageloxime D	*Agelas nakamurai* sponge	*S. epidermidis*
Bromopyrrole alkaloids	Marine sponges	*S. aureus*
Cathelicidin peptide LL-37	Human	*P. aeruginosa, S. aureus*
Glycolipid biosurfactant	*Serratia marcescens*	*P. aeruginosa*
Cathelicidin peptide LL-37	Human	*P. aeruginosa*
Carolacton	*Sorangium cellulosum* strain Soce	*S. mutans*
Emodin, anthraquinone	Mold, lichens	*P. aeruginosa*
Benzimidazole	Plant source	*P. aeruginosa,(CF-145), Klebsiella pneumonia*
Indole derivatives, 5-hydroxyindole carboxyaldehyde pyrroloindoline trizoleamide	Plant source	*E. coli* O157: H7, *S. aureus* (MRSA) *P. aeruginosa*
Flavonoids, phloretin	Apple	*E. coli* O157: H7, *E. coli* K12
Hyperforin	*Hypericum perforatum*	*S. epidermidis* *S. aureus*, *Enterococcus faecalis*
7-Epiclusianone	*Rheedia brasiliensis*	*S. mutans*
Isolimonic acid, triterpenoid	Citrus species	*E. coli* O157: H7
Chelerythrine	*Chelidonium majus*	*S. aureus* ATCC 6538P *S. epidermidis*
Proanthocyanidin A2-Phosphatidylcholine	*Krameria lappacea*	*Staphylococcus* strains
Polyphenolic compound, tannic acid	Tea plant	*S. aureus*
Ginkgolic acid C15:1	Plant	*E. coli* O157: H7, *E. coli* K12 , *S. aureus*
EO derivatives, carvacrol, oregano oil, and thymol	Plant source	*S. aureus, S. epidermidis*
Bgugaine	*Arisarum vulgare*	*P. aeruginosa*
Casbane diterpene	*Croton nepetaefolius*	*K. pneumonia, S. aureus* *S. epidermidis, P. aeruginosa*
Resveratrol oligomer, ϵ-viniferin	Carex plants	*P. aeruginosa* PAO1, PA14
Sulfur derivatives, S-phenyl-1-cystein sulfoxide, diphenyl disulfide	*Petiveria alliacea*	*P. aeruginosa*
Carolacton	*Sorangium cellulosum* Soce 960	*S. mutans*

cytotoxic approaches to treat biofilm making bacteria. Therefore vast work has studied antibiofilm treatment by in vitro methods, whereas very little evidence has proved in vivo treatment of biofilm prevention. Notably, the in vivo treatment of *P. aeruginosa* using HSL QS signaling combination of antibiotics has antibiofilm properties (Christensen, Van Gennip and Jakobsen, 2012). For example, furanone products, mannitol pilicide, EspA protease, Biophage-PA cocktail with different antibiotics, and lytic bacteriophage. Generally, the combination therapy is used for biofilm treatment by in vivo (Jakobsen et al. 2012). Hence, numerous tasks to be encountered in the progress of novel antibiofilm therapies but in vitro improvement of biofilm prevention have made leading advancement than for in vivo approaches. Therefore novel approaches are needed to disclose the prevention of biofilm.

2.8 FUTURE RESOLUTION

In this chapter, we are looking forward to the research on natural sources to reduce the biofilm formation based on the remediation aspects. The current updates of biofilm eradication through the process of *in-vitro* and *in-vivo* studies for instructive the mechanism of action and for evaluating the clinical prospective of the novel antibiofilm compounds.

In the decision, biofilm development is on medical devices and nondevice related human health problems. Hence, it is essential to reduce biofilm formation to avoid hospital infections. Moral aseptic conditions and practices are required to shirk biofilm formation. The best way is to treat and control biofilm formation through novel antibiofilm compounds. They are derived from natural sources to eliminate biofilm associated infections. The valuable features of the biological products are applied to control the growth of biofilms. The capability of some novel natural products to work effectively to reduce the biofilm formation on hospital-acquired infection is recently identified. Despite the rising number, new possible antibiofilm compounds are a suitable source for the elimination of biofilm.

REFERENCES

Allaker, R. P., and K. Memarzadeh. 2014. Nanoparticles and the control of oral infections. *International Journal of Antimicrobial Agents* 43 (2):95–104.

Anderson, J. M. 2001. Biological responses to materials. *Annual Review of Materials Research* 31 (1):81–110.

Besinis, A., T. De Peralta, and R. D. Handy. 2014. Inhibition of biofilm formation and antibacterial properties of a silver nanocoating on human dentine. *Nanotoxicology* 8 (7):745–754.

Brackman, G., P. Cos, L. Maes, H. J. Nelis, and T. Coenye. 2011. Quorum sensing inhibitors increase the susceptibility of bacterial biofilms to antibiotics in vitro and in vivo. *Antimicrobial Agents Chemotherpy* 55 (6):2655–2661.

Buommino, E., M. Scognamiglio, G. Donnarumma, A. Fiorentino, and B. D'Abrosca. 2014. Recent advances in natural product-based antibiofilm approaches to control infections. *Mini Reviews Medicinal Chemistry* 14 (14):1169–1182.

Cascioferro, S., D. Raffa, B. Maggio, M. V. Raimondi, D. Schillaci, and G. Daidone. 2015. Sortase A inhibitors: recent advances and future perspectives. *Journal of Medicinal Chemistry* 58 (23):9108–9123.

Christensen, L. D., Van Gennip, M., and T. H. Jakobsen. 2012. Synergistic antibacterial efficiency of early combination treatment with tobramycin and quorum sensing inhibitors against *P. aeruginosa* in an intraperitoneal foreign body infection mouse model. *Journal of Antimicrobial Chemotherapy* 67: 1198–1206.

Darouiche, R. O. 2004. Treatment of infections associated with surgical implants. *The New England Journal of Medicine* 350 (14):1422–1429.

Ding, X., B. Yin, L. Qian, Z. Zeng, Z. Yang, H. Li, Y. Lu, and S. Zhou. 2011. Screening for novel quorum-sensing inhibitors to interfere with the formation of *Pseudomonas aeruginosa* biofilm. *Journal of Medical Microbiology* 60 (Pt 12):1827–1834.

Donlan, R. M. 2008. Biofilms on central venous catheters: is eradication possible? *Current Topics in Microbiology Immunology* 322:133–161.

Donlan, R. M., and J. W. Costerton. 2002. Biofilms: survival mechanisms of clinically relevant microorganisms. *Clinical Microbiological Rev*iews 15 (2):167–193.

European Centre for Disease and Control (ECDC). 2007. *Annual Epidemiological Report on Communicable Diseases in Europe.* Stockholm, Sweden. http://www.ecdc.eu.int/intex.html.

Francolini, I., P. Norris, A. Piozzi, G. Donelli, and P. Stoodley. 2004. Usnic acid, a natural antimicrobial agent able to inhibit bacterial biofilm formation on polymer surfaces. *Antimicrobial Agents Chemotherapy* 48 (11):4360–4365.

Guggenbichler, J. P., O. Assadian, M. Boeswald, and A. Kramer. 2011. Incidence and clinical implication of nosocomial infections associated with implantable biomaterials—catheters, ventilator-associated pneumonia, and urinary tract infections. *GMS Krankenhhyg Interdiszip* 6 (1):Doc18.

Hammer, K. A., C. F. Carson, and T. V. Riley. 1999. Antimicrobial activity of essential oils and other plant extracts. *Journal of Applied Microbiology* 86 (6):985–990.

Han, Q., B. Li, X. Zhou, Y. Ge, S. Wang, M. Li, B. Ren, H. Wang, K. Zhang, H. H. Xu, and X. Peng. 2017. Anti-caries effects of dental adhesives containing quaternary ammonium methacrylates with different chain lengths. *Materials (Basel)* 10 (6):643.

Hoiby, N., O. Ciofu, and T. Bjarnsholt. 2010. *Pseudomonas aeruginosa* biofilms in cystic fibrosis. *Future Microbiology* 5 (11):1663–1674.

Jakobsen, T. H., M. van Gennip, R. K. Phipps, M. S. Shanmugham, L. D. Christensen, M. Alhede, M. E. Skindersoe, T. B. Rasmussen, K. Friedrich, F. Uthe, and P. Ø. Jensen. 2012. Ajoene, a sulfur-rich molecule from garlic, inhibits genes controlled by quorum sensing. *Antimicrobial Agents and Chemotherapy* 56 (5):2314–2325.

Kokare, C. R., S. Kadam, K. Mahadik, and Prof. B. Chopade. 2007. Studies on bioemulsifier production from marine Streptomyces sp. S1. *Indian Journal of Biotechnology* 6:78–84.

Lu, L., W. Hu, Z. Tian, D. Yuan, G. Yi, Y. Zhou, Q. Cheng, J. Zhu, and M. Li. 2019. Developing natural products as potential anti-biofilm agents. *Chinese Medicine* 14:11.

Martin, B., Z. Tamanai-Shacoori, J. Bronsard, F. Ginguené, V. Meuric, F. Mahé, and M. Bonnaure-Mallet. 2017. A new mathematical model of bacterial interactions in two-species oral biofilms. *PLoS One* 12 (3):e0173153.

National Heart Lung and Blood Institute (NHLBI). 2002. *Research on Microbial Biofilms (PA-03-047).* National Institute of Health, Bethesda, MD.

Percival, S. L., and P. Kite. 2007. Intravascular catheters and biofilm control. *Journal of Vascular Access* 8 (2):69–80.

Rasamiravaka, T., Q. Labtani, P. Duez, and M. El Jaziri. 2015. The formation of biofilms by *Pseudomonas aeruginosa*: a review of the natural and synthetic compounds interfering with control mechanisms. *BioMed Research International* 2015:759348.

Reyes, S., R. W. Huigens, 3rd, Z. Su, M. L. Simon, and C. Melander. 2011. Synthesis and biological activity of 2-aminoimidazole triazoles accessed by Suzuki-Miyaura cross-coupling. *Organic and Biomolecular Chemistry* 9 (8):3041–3049.

Romling, U., and C. Balsalobre. 2012. Biofilm infections, their resilience to therapy and innovative treatment strategies. *Journal of Internal Medicine.* 541–561.

Stewart, P. S., and J. W. Costerton. 2001. Antibiotic resistance of bacteria in biofilms. *Lancet* 358 (9276):135–138.
Stickler, D. J. 1996. Bacterial biofilms and the encrustation of urethral catheters. *Biofouling* 9 (4):293–305.
Vega, N. M., and J. Gore. 2014. Collective antibiotic resistance: mechanisms and implications. *Current Opinion in Microbiology* 21:28–34.
Zhou, M., H. Wang, D. J. Hassett, and T. Gu. 2013. Recent advances in microbial fuel cells (MFCs) and microbial electrolysis cells (MECs) for wastewater treatment, bioenergy, and bioproducts. *Journal of Chemical Technology & Biotechnology* 88 (4):508–518.

Part II

Microbial Biofilms: Bioenergy and Biomaterials Used in Environment Systems

3 Role of Microbial Biofilm in Agriculture and Their Impact on Environment

Current Status and Future Prospects

Asma Rehman, Lutfur Rahman, Ata Ullah, Muhammad Bilal Yazdani, Muhammad Irfan, and Waheed S. Khan
National Institute for Biotechnology and Genetic Engineering

CONTENTS

3.1 INTRODUCTION

Conventional agricultural practices have gained worldwide attention about its adverse effects on environment (Hazell and Wood 2007), leading to new avenues for improved crop yield and productivity (Singh, Pandey, and Singh 2011). Biofertilizers have opened the new gateway for the use of beneficial microorganisms to improve plant growth and yield through better nutrient supply, thus maintaining environmental health and soil productivity (O'Connell 1992). The soil is a favorable habitat for a diverse population of microbes that strongly interact with soil (Seneviratne et al. 2010). These soil microbes interact with plants not only to promote plant growth but also to act as biocontrol, alleviate biotic/abiotic stresses, and greatly influence physical, chemical, and biological properties of soil through biogeochemical cycles, thus facilitating soil nutrient balance. A plethora of microbes were tested, for the biofertilizers have shown better results than the conventional agrochemicals. However, there was a limitation in the use of such microbial inocula because they cannot compete with indigenous microbial flora of the soil, leading to the development of biofilms. Microbial biofilm is the assembly of integrated, surface mounted communities of cells which are encased in a polymeric matrix comprising polysaccharides (Costerton et al. 1995) and are associated with various surfaces like metals, ceramics, wood, glass, above and below the ground, and on medical implants, devices, living tissues, etc. The concept of biofilm arises from the statement "viscous intermediary substance" given by Burton-Sanderson in 1870 (Vlamakis et al. 2013). Biofilm formation is a ubiquitous characteristic exhibited by bacteria/fungi. During the formation of biofilms, the microorganisms adhere to the natural/artificial surface with the help of structure (flagella, pilus) found on their surface or other molecules (proteins, carbohydrates, lipids) or their conjugated products (lipopolysaccharides) (Stewart and Franklin 2008).

Microbial growth and biofilm formation depends on various factors such as organism genetic makeup and physiological conditions (Davey and O'Toole 2000). Biofilm formation is also a strategy of microorganisms to resist various stresses like nutrient deflation, medium pH, oxidative stress, and other toxic compounds present in the environment (Karatan and Watnick 2009). It is also a strategy to remain in a convenient position through efficient colonization (Jefferson 2004). In comparison to the planktonic cells the rate of genetic material transfer among microorganisms in biofilm is fast due to nearer cell contact (Hausner and Wuertz 1999). The transformational frequencies in biofilms are 10–600 times higher when compared with planktonic cells. As a result of such genetic transformations, alteration in gene expression and regulation in microbes are present in biofilms as compared to individual microbial strains (Li et al. 2001).

3.2 BIOFILM DEVELOPMENT, MECHANISMS, AND FACTORS AFFECTING THEIR FORMATION

The development of biofilm is a stepwise process (Monds and O'Toole 2009) and involves various steps from microbial cell attachment to the formation of mature biofilms, and the general mechanism is shown in Figure 3.1, as reported in the literature (Yin et al. 2019). This section describes biofilm development, its mechanisms, and the factors affecting their formation.

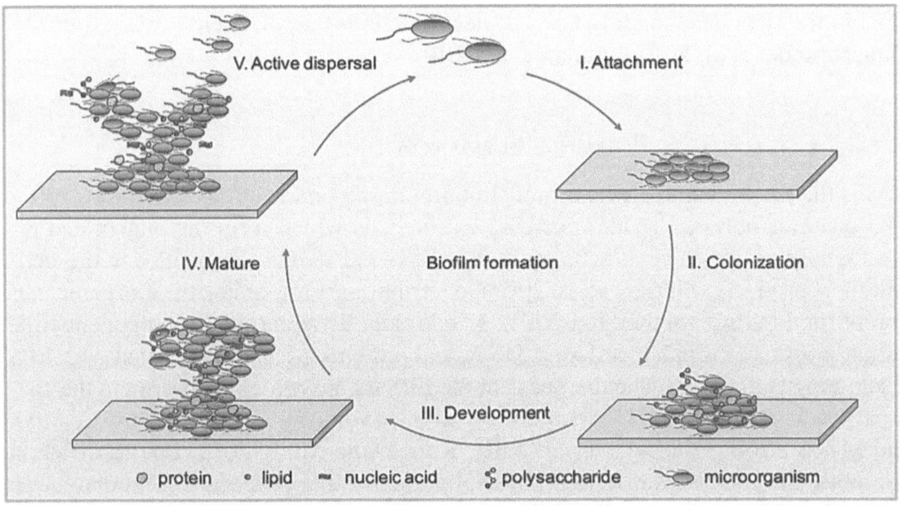

FIGURE 3.1 The various steps and mechanism of microbial biofilm formation. Steps involved in biofilm formation are (1) Attachment: it is the irreversible adhesion of microorganisms to the surface. (2) Colonization: this step involves the attachment of microbes to the surface and surrounding cells to form colonies through flagella, pili, and other substances. (3) Biofilm development: accumulation of multi microbial layers and secretions of EPS). (4) Maturation: in this step microbial communities arrange themselves in three-dimensional form. (5) Active dispersal: at this stage microbes get dispersed to its planktonic state from the aggregated biofilm. (This is an open access article distributed under the Creative Commons Attribution License which permits unrestricted use, distribution, and reproduction in any medium, provided the original work is properly cited, Yin, W., et al., Int. J. Mol. Sci., 20, 3423, 2019. With Permission.)

3.2.1 Attachment and Aggregation

Attachment of microbial cell to biotic or abiotic surface is the initial step towards biofilm development. The attachment of microbial cells to the abiotic surface is referred as adhesion, while the term cohesion is used for microbial cell to cell interaction and attachment (Boland, Latour, and Stutzenberger 2000). The attachment depends on many factors like nature of the surface, quantity of extracellular polymeric substance (EPS) released by bacteria, and the presence or absence of flagella and fimbriae (Donlan 2002). The cell attaches rapidly to nonpolar, hydrophobic surfaces (Flemming and Wingender 2001). Biofilm formation is influenced by different cell surface adhesins (Agladze, Jackson, and Romeo 2003) and cell aggregation (Prigent-Combaret et al. 2001). Aggregation helps to expand the biofilm and it depends on the type of interaction between microorganisms. There are two types of aggregation on the basis of interaction: auto- and coaggregations. Autoaggregation is the attachment between bacteria of similar strains, whereas the attachment between two or more species is referred as coaggregation (Rickard et al. 2003, Simões, Simoes, and Vieira 2008, Yang et al. 2011). Aggregation is influenced by different factors like bacterial surface-attached molecules including proteins, cell wall, and different appendages (fimbriae, flagella) (Karched, Bhardwaj, and Asikainen 2015). Coaggregation plays an important role in the enhancement of biofilm because the multispecies attachment in the same surrounding leads to new functions and different genetic transfers (Kolenbrander et al. 2010, Hannan et al. 2010).

3.2.2 Extracellular Polymeric Substances

EPS is the fundamental element in the biofilm development (Bianciotto et al. 2001). EPS contains polysaccharides excreted by the cells which form the matrix and lay the foundation for their attachment and considered as the "cement" for the cells and their products (Starkey et al. 2004), contributing for a majority of organic carbon of the biofilms (Sutherland 2001). In addition, different other components like humic substances, lipid, metal ions, DNA, and proteins are also associated with EPS (Flemming et al. 2016). The thickness of the EPS matrix varies according to the bacterial species (Sleytr 1997). The microbial EPSs have various compositions (Czaczyk and Myszka 2007, Naessens et al. 2005, Kumar and Mody 2009) due to different microbes and thus have different physical and chemical properties (Stoodley et al. 2002). The EPS matrix has the ability to sustain the survival rate of microbial cells in the soil by tolerating the drought stress, which is a major breakthrough for the areas which have limited water resources (Roberson and Firestone 1992). EPS is commonly constituted of different carbohydrates like acetate esters, pyruvates, formates, and succinates (Czaczyk and Myszka 2007). Gram-positive bacteria's EPS have a special feature in the form of the presence of polypeptides (Sleytr 1997). The bacteria involved in biofilm formation have wide varieties of EPS synthesis such as cellulose (Jonas and Farah 1998), dextran (Naessens et al. 2005), alginate (Ertesvåg and Valla 1998), and xanthan (Sutherland 2001). EPS also has the feature of facilitating the transition from reversible to irreversible adhesion of single planktonic cells (Stoodley et al. 2002) and helps further in the adhesion of biofilm forming bacteria on the desired surface (Berk et al. 2012). The soluble EPS during the period of biofilm maturation stage forms a filamentous matrix which accelerates cell aggregation and helps to maximize nutrient distribution in the cells (Janissen et al. 2015). The nature of EPS, their quality, and the secreted amount have a great influence on biofilm development (Matthysse et al. 2005).

3.2.3 Nutritional Factors

There is a great association between available nutrients and the inoculated biofilms. Any kind of fluctuation in the pattern of nutrients, i.e., release or uptake in each part of plant, significantly affects the formation of biofilm as well as its morphology (Ramey et al. 2004). Furthermore, moisture content and nutrient availability are the major components required for microbial cell aggregation and biofilm development (Lindow and Leveau 2002). Carbon source in the medium has a great influence on EPS synthesis and is considered as a key component involved in the regulation and maintenance of biofilms (Chai et al. 2012, Zhang et al. 2014, Assaf, Steinberg, and Shemesh 2015). The high level of nutrients affects gene expression and is responsible for biofilm formation (Musk, Banko, and Hergenrother 2005, Glick et al. 2010). It has been reported that a low level of nutrients may cause damage of biofilm by affecting their component synthesis. Amino acids function as a dual source for carbon and nitrogen for the rhizosphere bacteria (Valle et al. 2008). In rhizospheric bacteria biofilms formation assembly, amino acids play a significant role as reported previously (Kolodkin-Gal et al. 2010). It has been reported that a high level of amino acid synthesis occurs in biofilms, formed by gram-negative bacterial strains as compared to their planktonic cellular growth (Valle et al. 2008).

3.2.4 Genetic Factors

The formation of biofilms from cellular adhesion to the substrate and to the mature biofilm development is a well-regulated established process (Prüß et al. 2006). Gene expression in biofilm formation is influenced by highly regulated and complex pathways (Prigent-Combaret et al. 2001). A cyclic diguanylate /cyclic di- Guanosine monophosphate (GMP) (c-di-GMP) has a critical role in the determination of bacterial cell's position as to remain free in the environment or in biofilm (Römling, Gomelsky, and Galperin 2005, Wolfe and Visick 2008). It has the capability to regulate transcription by the attachment to the effector proteins (Weber et al. 2006, Hickman and Harwood 2008) and involve in adhesin secretion and localization (Monds et al. 2007). Moreover function and motility of flagella and EPS synthesis can also depend on c-di-GMP activity (Weinhouse et al. 1997, Merighi et al. 2007). It has been reported that enzyme histidine kinase (Chen et al. 2012) and sRNA (small noncoding RNA) also have a significant effect on biofilm formation (Chambers and Sauer 2013). Moreover, gene expression regulation is very much important in microbial cells due to the that their occur synthesis and secretion of cellular polymeric substances, which play significant role in the attachment of cells to the surfaces. However, it is also notable that genes which are involved in the synthesis of chemotaxis proteins and proteins for the development of flagella are also regulated to ensure the movement of microbes (Barraud et al. 2006, Rollet, Gal, and Guzzo 2009).

3.2.5 Environmental Factors

The formation of biofilms is significantly governed by various environmental factors such as pH of media, availability of oxygen, temperature, etc., and these physical and chemical environmental factors have been reposted for their controlling capability over gene expression, amino acid synthesis, and other cellular processing (Olson 1993). pH on media is one of the important parameters that influence microbial cell attachment to the surface. However, some microbes have a specific mechanism to synthesize extracellular substances to the surrounding, resulting in medium pH alteration and facilitating cellular adhesion (Oliveira et al. 1994). As EPS is produced by the microbial cells and biofilm formed is the accumulation of such substances surrounding the cells, the secretion, quality, and quantity of these extracellular substances are very much important (Ohashi and Harada 1994). Different EPSs such as alginate, gellan, and xanthan have been reported, and a special type of jelly-like structure can form hydrogen bonding easily. Further, the viscosity of such polysaccharides also depends on temperature fluctuations (Villain-Simonnet, Milas, and Rinaudo 2000). In the formation of *Pseudomonas aeruginosa* biofilm, the hydrophilicity of alginate increases as the acetylated uronic acid is secreted in high quantity (Kumar and Anand 1998). As temperature fluctuation has significant effects on biofilm development (Kaplan and Fine 2002), enzymes need an optimal temperature for their catalytic activity (Stepanović et al. 2003) for nutrient metabolism. When fluctuation occurs in the environmental temperature, it affects bacterial metabolic activities which ultimately influence cellular secretion of polymers required for the adhesion to the surface (Herald and Zottola 1988). Each microorganism need an optimal temperature for biofilm development, some microbes form biofilms more efficiently at lower temperature while other at higher. (Nisbet et al. 1984). For some microbes to form biofilms, an insufficient quantity of oxygen and water may lead to a disturbance in the process of biofilm development (Thormann et al. 2005).

3.3 SOIL NUTRIENTS: THEIR DEPLETION AND IMPACT ON PLANTS AND RECOVERY THROUGH BIOFILMS

3.3.1 Essential Nutrients: Their Effects on Plant Growth and Deficiency Symptoms

Soil is the foundation of agriculture, and agriculture is the main source of food and nutrition. Plant growth and development basically depend upon soil composition. To sustain the productivity, nutrient availability and balance are very much important in the agriculture land. The depletion of such essential nutrients in the soil may cause damage to the crops, productivity and alternatively to economy (Johnston, Poulton, and Coleman 2009).

Major elements required by the plants are carbon, hydrogen, oxygen, phosphorous, nitrogen, calcium, potassium, magnesium, iron, sulfur, zinc, manganese, copper, molybdenum, boron, and chlorine. The uptake of these nutrients occurs either directly from the environment or indirectly in the integrated form with other nutrients/elements or compounds through a special mechanism. Plants fulfill their requirements of carbon, hydrogen, and oxygen from atmosphere and soil water, while the uptake of remaining essential elements occurs either in the form of soil minerals, organic matter, or through organic and inorganic fertilizers. Each plant in a particular environment needs a specific level of nutrients known as optimal nutritional rang. Plants show nutritional deficiency symptoms when the availability of nutrients become less as compared to the optimum required value. Besides these, plants also respond to excessive amount of nutrients as it may affect their growth due to toxicity. Thus, adequate amounts of nutrients are required in the cultivated soil, but if needed nutrients should be supplemented in appropriate quantity to the plants to avoid adverse effects.

Plants get nutrients from the surrounding environment for their growth and development. Plants cannot uptake nutrients directly in their existing state for metabolism, but need them in some modified forms such as ions. Nitrogen in the form of nitrate (NO_3^-) and ammonium NH_4^+ ions are available for plant's uptake. Plants metabolize these ions as a source of nitrogen with a combination of C, H, O, and S for the synthesis of amino acids and proteins, which the plants use for further growth and development. Moreover, N is also the key component of plant enzymes, chlorophyll molecule, and several vitamins, thus improving the quality and quantity of vegetables, fruits, and grains. Its deficiency leads to growth retardation due to reduction in cell division and chlorosis (leaf color from green to yellow) because of the effect on chlorophyll. Further, depletion of N also causes a reduction in protein contents of fruits, seeds, and flowering as well as cause a significant reduction in plant yields and quality due to early maturity.

Phosphorous (P) is another nutritionally important element available to plants in orthophosphate ions form (HPO_4^{2-} and $H_2PO_4^-$). It plays a significant rule in photosynthesis and respiration as well as forms a part of ATP, ADP, etc. involved in energy storage and transfer. The components of nucleic acids (DNA and RNA) play an important role in genetic information storage. It has been reported that P is needed in enough concentration to increase the metabolism rate in young plant cells for enhancing cell division and alternatively plant growth. Also involved in the development of roots, seeds and fruits play a significant role in crop quality improvement and reduction of plant disease prevalence. Deficiency of P causes fragility of plants, growth retardation, and delay maturity which may lead to poor seed and fruit development.

Plants obtain potassium (K) in the ionic (K^+) form and is not involved in the synthesis of vital organic molecules just like N and P. However, it is very much important for enzymatic reactions, acting as enzyme activator in metabolic activities. Its presence is important for stomata opening and closing to control water and gracious transport and maintenance of electrical charges for ATP synthesis, and is thus indirectly involved in protein synthesis. Also it plays a role in disease resistance and quality and quantity improvement of grains, seeds, fruits, and vegetables. Deficiency symptoms occur due to potassium in plants and include chlorosis, stunt growth, and reduced quality and quantity of productivity.

The available form of calcium (Ca) to plants is Ca^{2+} ion, and functionally it is involved in the formation, plasticity, and permeability of cell wall and plasma membrane to maintain integrity exchange of materials. Ca acts as an activator for several enzymes, having a significant role in protein synthesis and carbohydrate transport, and plays a vital role as detoxifying and acid neutralizing agent in combination with other compounds. As a neutralizing agent, it indirectly improves productivity by reducing soil acidity and is essential for the production of seed in peanuts. As a result of calcium deficiency, leaves and root tips of plants become brown in color and finally dies. Further, its shortage may cause secondary effects like soil acidity, plant fragility, and premature fall of buds and blossoms.

Another important nutritional component is magnesium (Mg), which occurs in Mg^{2+} form. As a major component of chlorophyll molecule, it is actively involved in the process of photosynthesis. As a cofactor for enzymes, it also plays a vital role in reactions such as phosphorylation. Besides these Mg is the central element for stabilizing of ribosome subunits and is also responsible for the sugar molecules inside the plants. As a major component of chlorophyll, its deficiency will significantly disturb chlorophyll quantity and will cause chlorosis and premature leaf dropping.

Sulfate ion SO_4^{2+} is the available form of sulfur (S) to plants. As a component of certain amino acids, it involves in protein synthesis and stabilization. It plays a significant role in chlorophyll formation and metabolism of thiamine, coenzyme A, and B vitamin biotin. In plants, S deficiency leads to growth retardation, delayed maturity, and chlorosis. Deficiency symptoms are in someway similar to that of N; thus, it is necessary for a plant scientist to diagnose properly.

Boron (B) is available to plants in the form of borate, H_3BO_3. Functionally it is involved in cellular activation, root growth promotion, pollen germination, and pollen tube growth. Further it is associated with the activity of certain enzymes, lignin and cell wall synthesis, as well as sugar transport. Deficiency symptoms include growth inhibition, thickening, and curling of leaves, as a result of which they become brittle.

Copper (Cu) is an important component among plant's nutrients, available in its ionic form, i.e., Cu^{++} to plants. This nutrient is essential for photosynthetic enzyme systems, and it is the component of some plant pigments including chlorophyll. Growth retardation, apical meristem necrosis, and chlorotic effects are the main deficiency symptoms of Cu.

Chlorine (Cl) occurs in its ionic (Cl^-) form, as a result of which plants easily uptake it from its environment. Cl plays a vital role in oxygen evaluation, water content regulation in tissues, and photosynthesis. Further, it increases osmotic pressure and reduces certain fungal disease, but its deficiency may cause chlorosis and plant wilting.

Plants uptake Iron (Fe) both in its ionic forms Fe^{2+} and Fe^{3+}. Biologically it is important in heme enzyme system, and is involved in photosynthesis and respiration like peroxidase, catalase, and cytochrome oxidase, etc. It is also the component of ferredoxin protein, needed for the reduction of sulfate and nitrate, and is associated with protein metabolism and chlorophyll maintenance. Just like Magnesium, Fe deficiency causes interveinal chlorosis because it is involved in chlorophyll formation.

The primary functions of Manganese (Mn) in plants are, as a part of several enzymes (pyruvate carboxylase, indole acetic acid oxidase, etc.), involved in different pathways: activation, oxidation, and reduction processes occur during photosynthesis, i.e., photolysis in photosystem-II. Plants obtain it from the environment in its two ionic forms (Mn^{2+} and Mn^{3+}). Moreover, its deficiency in plants shows symptoms like chlorosis, yellow spots on leaves, as well as marsh spots, i.e., the necrotic area developed on cotyledons of seeds.

The available form of Molybdenum (Mo) which facilitates its uptake for plants is MoO_4. It is the component of enzymes (nitrogenase and nitrate reductase) that is significantly important for normal assimilation of N. On association with N assimilation, its deficiency symptoms resemble with that of nitrogen, although necrotic spots appear at leaf margins due to nitrogen accumulation. Its inadequacy also causes growth retardation and delays flower formation.

Zinc is symbolized as "Zn," and the available form is Zn^{++} ion. It is involved in several different functions of metabolism, such as in the synthesis of tryptophan for indole acetic acid production, part of metalloenzymes like dehydrogenases and carbonic anhydrase, and also play a vital role in RNA and protein synthesis. Its deficiency leads to chlorosis in young leaves just like Fe, but the difference is that in case of iron chlorosis occurs interveinal, while in Zn, it appears at the basal part of leaves (Uchida 2000).

3.3.2 Microbes and Microbial Biofilms as Nutrients Providing Factories for Agriculture

Mostly nitrogen fertilizers are used to provide nitrogen to the plants, but they may have adverse effects on the environment and other organisms. As a microbial community present in the soil, it has the ability to convert free nitrogen into absorbable form for plants. Hence such bioagents can be used as nitrogen fixers, and with their use environmental pollution can be reduced. Seneviratne et al. (2011) have developed biofilms used for the restoration of soil nutrients, especially for nitrogen fixation. Further he reported that such bioagents, used for nitrogen fixation, can reduce the use of chemical fertilizers up to 50% (Seneviratne et al. 2011). For the restoration of soil nitrogen colonization of microbial community of a single strain has been shown in Figure 3.2. The scanning electron microscopy and green fluorescent protein (GFP) tagged stains have been cultured (Ramey et al. 2004). It has been reported that *Azotobacter* is globally accepted as a major inoculum in biofertilizers and is used for biofilm formation. Further it was concluded that it may enhance and promote plant growth with other PGPRs, when inoculated in combination (Gauri, Mandal, and Pati 2012). Fungal biofilm known as bradyrhizobial-fungal biofilm has been reported as a nitrogen fixating bioagent in greenhouse condition. It was further reported that when single strain of fungi, which was active for nitrogenase activity introduced so it was unable to perform nitrogen fixation. Although when biofilm was developed, the same fungi was found active for nitrogen fixation. (Jayasinghearachchi and Seneviratne 2004). Another study conducted and reported that cyanobacteria fix nitrogen in cyanolichens for their nutrition (Seneviratne and Indrasena 2006). Moreover, Seneviratne and Jayasinghearachchi reported a rhizobial biofilm for nitrogen fixation and enhanced nitrogenase activity (Seneviratne and Jayasinghearachchi 2005).

It has been reported that *Bradyrhizobium japonicum* SEMIA 5019-*Penicillium* spp. when living free in soil have no significant effects on nutrients fixation and fertility of soil. But their biofilms can enhance the availability of phosphorous in

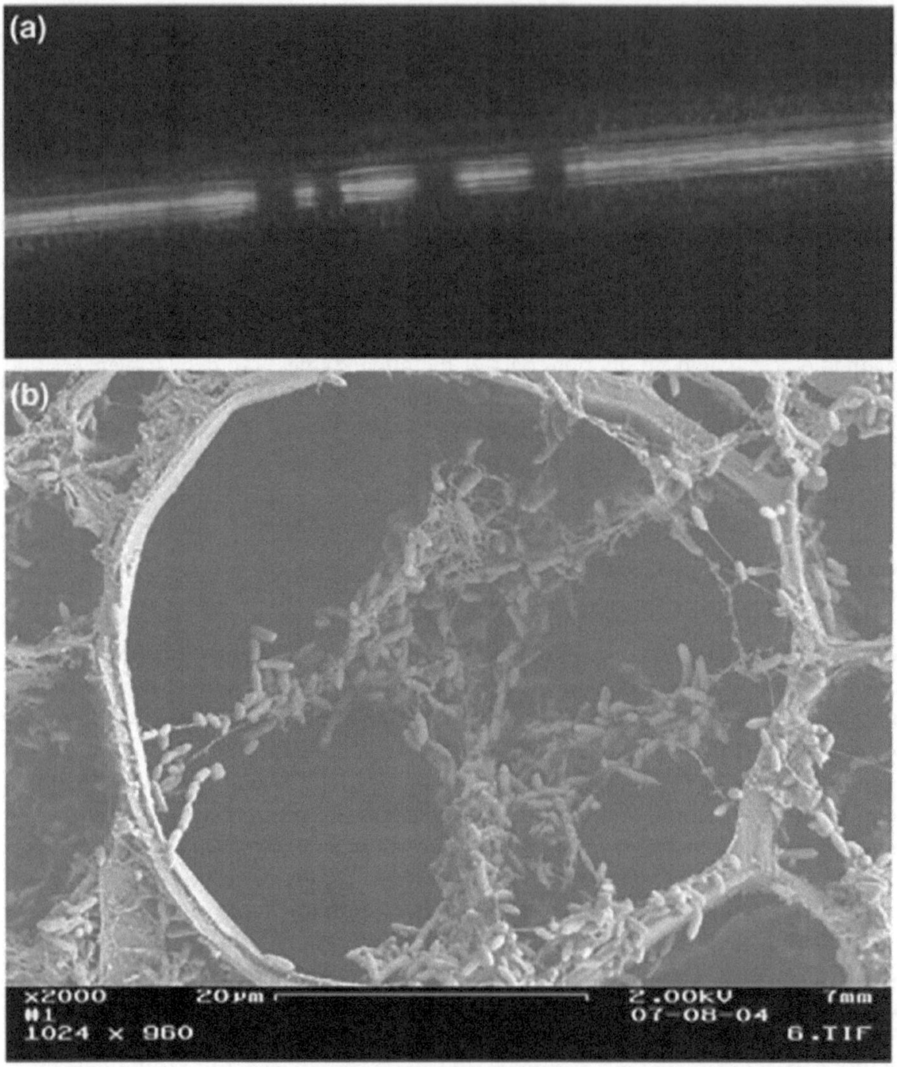

FIGURE 3.2 Colonization of the vasculature by *Pantoea stewartii* subsp. *stewartii*. Section (a) GFP-tagged wild-type strain DC283. Section (a) represents colonizing leaf xylem vessels indicating biofilm formation, while (b) shows SEM results indicating bacterial colonization. (From Ramey, B.E., et al., *Curr. Opin. Microbiol.*, 7, 602–609, 2004, Licensed Under 4721270501372.)

plants. Further such biofilms can also have effects on sulfur availability for plants (Seneviratne and Jayasinghearachchi 2005).

It has been reported that rhizobial biofilms alter the concentration of K to its optimal range in the agricultural soil, as a result of which other nutrients like nitrogen and phosphorous fixation and availability promote plant growth and development (Seneviratne and Indrasena 2006).

Optimal concentrations of necessary nutrient in the soil is significantly important. If the concentration of one nutrient increases as compared to others, it may cause antagonistic effect on the availability of other nutrients and subsequently affect plant growth. Seneviratne and Jayasinghearachchi worked on rhizobial biofilms that have the potential to enhance fixation of N_2 and P, but they also decrease Sulfate and K concentration. As a result of decrease in S and K, the enzyme nitrogenase becomes more active as it is involved in N_2 fixation (Seneviratne and Jayasinghearachchi 2005).

Moreover, a little literature has been reported for microbial biofilms, but the studies do not signify that the reported biofilms can be applied for nutrient availability and maintenance. Although some microbes have been isolated and reported for the fixation of other essential nutrients, microorganisms need proper substrate for the development of biofilms. Different parts of plants such as roots, root hairs, root meristem, stem, leaves, and shoot meristem are the interactive parts of plants and microbes (Figure 3.3). In this regard research needs to be conducted to observe the

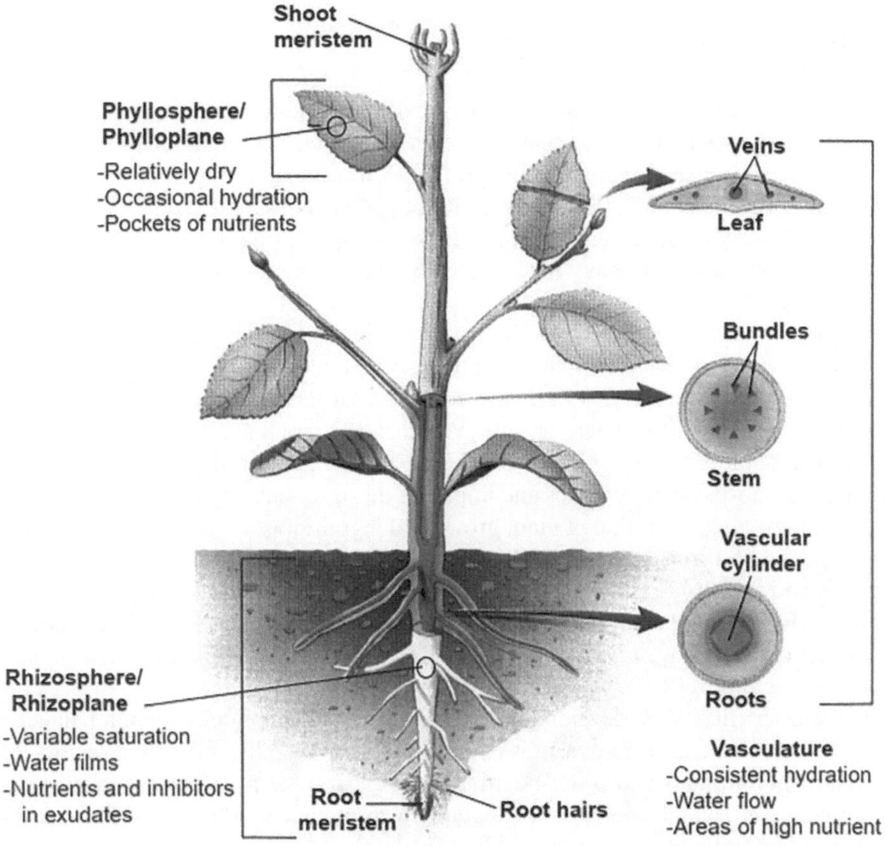

FIGURE 3.3 A terrestrial plant and sites where microbes can interact for nutrient exchange as well as form biofilms. Internal tissues have been represented and are specifically involved in nutrient transport. (From Ramey, B.E., et al., *Curr. Opin. Microbiol.*, 7, 602–609, 2004, Licensed Under 4721270501372.)

interaction and association of microbes with plants interacting parts, their biofilms formation, and their evaluation for the fixation of other nutrients not reported yet in the agriculture sector (Ramey et al. 2004).

3.4 BIOFILM IN AGRICULTURE

Biofilm plays a very important role in agriculture. For a decade researchers showed key interest in using biofilms to deal with the following aspects of agriculture:

- Soil fertility
- Soil texture improvement
- High crop yield
- Removal of heavy metals from agriculture land
- To overcome soil salinity
- To maintain soil pH

3.4.1 The Role of Biofilm in Soil Fertility

Mostly in agriculture, the fertilizers, specifically organic fertilizers and agrochemicals, have been studied for its negative effects on soil fertility maintenance and crop yield, mostly because of their worsening effects on soil containing microbes and soil fauna (Seneviratne et al. 2009). The salinity of soil also has a great effect on soil fertility and directly on seed germination, which is most important for crop production. Also, the distribution of heavy metals greatly affects the fertility rate of soil (Shen et al. 2002).

To solve these problems, microbial activity is needed to improve rhizosphere (Park et al. 2011). It has been stated that the farmers in China are practicing reduced quantity of chemical fertilizers for the production of more crop yields and to lower the adverse effects of fertilizers on soil (Vercraene-Eairmal et al. 2010).

The use of biofilm containing Trichoderma/Anabaena with various agriculturally beneficial fungi/bacteria develops and improves the agronomic behavior (soil fertility, microbiology, and promotion of plant growth) of leguminous crops (Velmourougane, Prasanna, and Saxena 2017). Trichoderma and Anabaena-based microbial biofilms enhance soil fertility (Prasanna et al. 2016). Microbial biofilms also help controlling the humidity and yet improve soil fertility rate (Mazor et al. 1996). Multispecies microbial biofilms have a vital role in sustaining the ecological equilibrium in soil (Burmølle et al. 2014).

For soil fertility, plant development and growth, and crop production, a balanced use of biofertilizers is needed, as it is environmentally friendly and a cheap source of nourishment for plants and soil. The mineral solubilization, by bacteria and fungi, is well recognized, which shows their capability as proficient biofertilizers. The microbial inoculant enhanced biomass yields directly by solubilizing soil nutrients, thus indirectly increasing the growth of plants. It has been reported that some bacterial species including potassium solubilizing bacteria (KSB) and phosphate solubilizing bacteria (PSB) enhanced soil nutrient levels significantly. The biofilms improved the uptake rate of K indirectly by promoting the growth of plants and directly by enhancing K solubilization in the soil. KSB significantly improved the soil quality and crop production (Han and Lee 2005). *B. mucilaginosus* develops biofilm efficiently on a nitrogen-free potassium-feldspar-supplemented media, and the results revealed that microbes can degrade various minerals and can fixate soil Nitrogen. K- and P-solubilizers and N-fixers based biofilms influence the growth of crop and yield. Also, these profitable biofilms also promote the quality of farmland and agriculture production (Wang 2009).

3.4.2 The Role of Biofilm in Saline Soil

High soil salinity, which is commonly increasing throughout the world, directly affects the soil fertility and yield production. Salinity is intrinsic condition at many beach areas, but in farmlands, poor irrigation systems often lead to secondary salinization and promote soil infertility. To overcome this problem, a microbial biofilm mostly consists of cyanobacteria used in the early 1950 (Roeselers, Van Loosdrecht, and Muyzer 2008).

In saline soil it is required to enhance seed germination and the growth of seedling (Lambers 2003). The best solution in such circumstance is to use salt resistant/tolerant bacterial biofilms that mainly secrete gibberellins and auxins, enhance the growth rate of seedling/plants under salinity surroundings, and develop soil fertility rate (Mayak, Tirosh, and Glick 2004). Halo tolerant cyanobacteria biofilms help to reduce the salinity of soil and enhance soil quality (Roeselers, Van Loosdrecht, and Muyzer 2008). In a report, stated by Giri and Mukerji (2004), mostly in saline soil, the availability of P in the soil or the developed absorption rate of P in plants promote their growth and lower the negative effect of salinity stress on soil. The use of PSB colonies can upgrade the P accessibility to the plants, since enhanced P nutrition improves biotic nitrogen fixation and the accessibility of other nutrients because these bacterial colonies are also well known to promote the growth of other beneficial substances and hence increase soil fertility rate (Han and Lee 2005). This improvement of land area salinity is probably affected by a temporal accumulation of Na^+ ions in the EPS sheaths of cyanobacteria, causing a delimited Na^+ influx in the crop roots. Although it is impossible to remove Na^+ from the soil, as with the death and dwindling of the cyanobacteria, Na^+ is released again to the soil. In the field of imbalanced fertilization, reduced level of potash leads to the loss of crop yield and is a major threat to farmers. Microbial biofilms can overcome this problem by improving the availability of K as they play a key role in K^- solubilization. Inappropriate use of agrochemicals for crop yield production has caused soil contamination, making the soil infertile (Atouei, Pourbabaee, and Shorafa 2019, Tan, Lal, and Wiebe 2005).

3.4.3 The Role of Biofilm in Soil pH Maintenance

Soil pH also plays an important role in soil fertility. It influences the nutrients and metals availability as well as organic matter mineralization. After extensive irrigation with wastewater, mostly up to 60 years, the pH of soil increases. In contrast, the pH of soil decreases due to the livestock fodder, lettuce, and production of orange with wastewater irrigation in agriculture land for the life interval of 15,

20, and 40 years, respectively. Hence the change in pH does affect the microbiota of soil which lead to soil infertility. Interestingly the fungal biofilm communities are not much vulnerable to the changing pH. But the difference in pH may affect the solubility of various components of soil, like metals. Irrigation with wastewater decreases the pH of soil due to the increase of free metals. Sequentially, the availability and concentration of these free metals have possible effects on the microbial biofilm communities and indirectly on soil fertility rate (Tan, Lal, and Wiebe 2005, Becerra-Castro et al. 2015).

3.4.4 The Role of Biofilm in Soil Texture Improvement

Due to continuous cultivation annually and closed environment, a dramatic depletion of nutrient and organic matter is observed followed by the deterioration in soil physical properties. Improved chemical, physical, and biological properties of such soils required the applications of organic materials. Under such conditions, it has been assumed that indole acetic acid (IAA) is a vital phytochemical for plant growth, and the activity of 1-aminocyclopropane-1-carboxylate (ACC) deaminase and siderophores support plants to obtain adequate iron with large amounts of other metals. Various studies have been conducted so far with the Biofilmed biofertilisers (BFBFs) in laboratory, and nursery and farm conditions of tea mainly in Sri Lanka have publicized promising results for soil re-establishment and enhanced crop yield. Moreover, EPS secreted by cyanobacteria and algae can help prevent soil erosion by improving the soil water-holding ability (Roeselers, Van Loosdrecht, and Muyzer 2008, Seneviratne et al. 2011).

3.4.5 The Role of Biofilm in Crop Production

With the use of cyanobacteria-based microbial biofilms, it is suggested to improve perseverance and maintenance of wheat rhizosphere. Cyanobacteria-based biofilm plays a vital role in persistent nitrogen control, and soil potency is said to be responsible for preserving sustainable crop yields in rice. Silicate bacterial biofilms have the capacity to degrade soil K, to activate K for the captivation and consumption of plants, and to enhance crop yield. The inoculation of microbial biofilm in the land comprising rice crop under waterlogged and System of Rice Intensification (SRI) displayed different effects on the parameters of plant growth and the dynamics of soil nutrient levels (Selvakumar, Panneerselvam, and Ganeshamurthy 2012, Das and Pradhan 2016).

Cyanobacterium-based microbial biofilms act as plant growth promoting mediators and sustain nitrogenous-based fertilizers in maize hybrids. Inoculation of novel biofilms helps to terminate Rhizoctonia species in cotton crop's soil. Microbial biofilms enhance the quantity of micronutrients in rice cultivars (Adak et al. 2016). In chickpea, plant growth and crop yield enhanced by the inoculation of cyanobacteria biofilms. On lettuce, maize, tea, and strawberry the effects of plant growth of biofilm-based fertilizers were checked. Cyanobacterium-based biofilms work as a matrix for agriculturally beneficial bacteria such as Mesorhizobium, Azotobacter, Pseudomonas, and Serratia. Such cyanobacterium-based biofilms promote bacterial survival rate and also plant growth promoting characters under controlled conditions. In a greenhouse environment, using fungal biofilms, Jayasinghearachchi and Seneviratne showed nitrogen accumulation and enhancement of nodulation in soybean. Biofilms such as cyanobacteria as biofertilizers are totally new concepts (Swarnalakshmi et al. 2013). Using such biofilms has an enhanced effect on crops (grain yields), particularly in Southeast Asia. Moreover, synthetic or naturally produced PGPR biofilms have many beneficial uses in agriculture and biotechnological applications. In streams, food chains are initially based upon biofilms and are the major nutrition source for various species of shrimps and fishes. These fishes and shrimps represent a significant protein source for humans. However, land sources affect the quality of such food sources (Burgos-Caraballo, Cantrell, and Ramírez 2014).

3.4.6 The Role of Biofilm in Balancing Soil Nutrient Demand

Traditional and seasonal crops affect the rate of soil fertility due to the excess use of fertilizers, especially chemical fertilizers. These chemical fertilizers initially support and enhance the growth rate of crop yields. But with the passage of time, the reservoirs of such chemicals accumulate in the bottom layer of soil. Abundance of these chemicals cause soil infertility and start affecting crop yields by their negative effect. By noticing such drawbacks, the farmers have somehow managed to enhance crop yield and soil fertility by minimizing the use of chemical fertilizers worldwide (Seneviratne et al. 2011). However, this is clear that the N containing fertilizers subdue the activity of microbes, mostly nitrogen fixer. This leads to proliferate N-poor soil bacterial communities with little biomass, due to the reduced N supply by the nitrogen fixers. Hence, fertilizer rate should be inversely proportional to the growth of microorganisms and their valuable effects on the plant and soil fertility rate. Mostly the beneficial microbial communities depend on N_2 fixers for their survival and growth (Singh, Pandey, and Singh 2011). Various studies showed the enhanced efficacy of beneficial microbial activity in N_2 exchange, P solubilization, hormonal production in plants, etc., when N_2 fixers were integrated to biofilms. For in situ fertilization of soil, Cyanobacteria may also be useful for N_2 fixation. It has been reported that N_2 fixing cyanobacteria on paddy fields of rice enhance soil fertility (Jayasinghearachchi and Seneviratne 2004).

In an active soil ecosystem, transformation and dissipation of energy arise from the plants, which has been classified as primary producers. This energy transfers to the organisms along the food chain. Thus, plants support wide-ranging heterotroph diversity including microbes. Bacteria and Fungi are famous for degradation of hazardous pollutants as well. They degrade plant's derived compounds directly or indirectly and are the primitive agents of degradation. Saprotroph-based fungal biofilms utilize a wide range of organic compounds and thus contribute to mineralization processes in an ecosystem. They have the capability to ferment carbohydrates. But mainly they inhale nonfermentable organic materials and carbohydrates and help in soil fertility (Seneviratne et al. 2011). The potential uses of biofilms to reduce the environmental pollution are another important aspect discussed in this chapter.

3.5 ENVIRONMENTAL THREATS, BIOFILMS IMPORTANCE, AND APPLICATIONS

All living organisms, plants, animals, and humans living in their environment need air, water, and food free of contaminates for efficient growth and survival (El-Shahawi et al. 2010). As air is the source of carbon dioxide, oxygen, hydrogen, and other essential elements, animals and plants use them for their life processes like photosynthesis and respiration (Whiting, Roll, and Vickerman 2003). The presence of hazardous pollutants in the environment causes severe injuries to the health of organism on accumulation in their bodies or may eventually cause death. The injuries may be directly affecting the organism or due to the reaction with their environmental components or internal body fluid cause toxicity (Malek et al. 2015).

A variety of contaminants have been reported in air, soil, and water. These pollutants affect plant growth, nutrition, and life processing in one way or the other (Malek et al. 2015). A variety of traditional methods have been developed for the removal or detoxification of such toxic components from the environments. The methods mostly include excavation of contaminated soil and its transport, thermal alkaline dichlorination, solvent extraction, landfill, or incinerations. However such treatment techniques may have some basic limitations such as high cost, laborious, and may not be feasible for a large variety of impurities (Gerhardt et al. 2009). This section includes environmental contaminants, their effects on life; special reference to agriculture and plants, and the importance of biofilms in their remediation (Campanella, Bock, and Schröder 2002).

3.5.1 Remediation of Heavy Metals

Generally, soil contain heavy metals and plants tolerate them at a low concentration, but when the concentration increases through anthropogenic and geologic activities, it then harms plants as well as animals. The harmful effects of heavy metals include enzyme inhibition found in cytoplasm, causing oxidative stress resulting in structural damage to cells. The major sources of heavy metals increase in soil are their smelting and mining, excessive use of pesticides, insecticide and fertilizers in agriculture, inappropriate burning of fossil fuels, industrial waste sewage sludge, and inadequate municipal waste disposal (Shen et al. 2002). The removal/detoxification of heavy metal is very much important, as they get accumulated in plant's parts (seed, fruits, etc.) and alternatively affect other organisms as well. The remediation of heavy metals can be accomplished either by immobilization, concentration, and separation to a specific comportment to minimize their predicted hazardous effects (Barkay and Schaefer 2001). However due to some limitations and adverse effects, the concept of "chemicals for the removal of chemicals" may be expensive and restricted to specific metals. Therefore, the alternate solution is the use of microorganisms for the cleaning of environment from such contaminants. Microbial biofilms have been reported in terms of their application for the remediation of heavy metals (Lloyd 2003). For instance, Sundar et al. (2011) reported the removal of trivalent chromium in a continuous flowing bioreactor, and using bacterial (*Bacillus subtilis* and *Bacillus cereus*) biofilm, he reported that the biofilm has the ability to remove

98% of Cr (III) from the environment (Sundar et al. 2011). Moreover biofilm of *E. coli* was used to remove Ni(II), Cr(VI), Fe(III), and Cd(II) from effluents (Quintelas et al. 2009). In the bioremediation of heavy metal contaminant substrate and supporting medium are also very much important. A group of scientists reported *E. coli* as a promising microbe for the removal of Cd (II), Cr (VI), Fe (III), Ni (II), and hexavalent chromium, supported on kaolin and granulated activated carbon (GAC), respectively. The biofilm supported on GAC acts as an excellent absorbent for the removal of Cr (VI) (Quintelas et al. 2009, Gabr et al. 2009). A study by Toner et al. (2005) reported on the sorption of zinc metal using soil and freshwater bacterium *Pseudomonas putida*. The bacterial culture was embedded in a polymeric material, and sorption experiment was performed at a pH of 6.9. Afterward, the experimental data was fitted with van Bemmelen-Freundlich model for validation (Toner et al. 2005). It has been reported that cadmium (Cd) present in agricultural water can be removed by using microbial biofilms, and their removal ability can be enhanced by using additional materials. The research conducted aims to determine the removal efficiency of biofilm, GAC, and biofilms covered with activated carbon (BAC). It was observed that Cd removal capability of biofilm covered GAC is more as compared to other materials (Dianati-Tilaki et al. 2004). Diels et al. (2003) reported a very attractive technology for the removal of heavy metals (Zn, Cu, Pb, Hg, Ni, and Co). They have used the metal resistant bacteria inoculated on the bed sand and proceeded for their biosorption and bioprecipitation capability (Diels et al. 2003). Another study conducted on the removal of copper, cadmium, and zinc from agriculture water based on multiple sorption-desorption cycles. The pattern followed was copper>zinc>cadmium, but excellent sorption capability results were recorded for copper metal (Costley and Wallis 2001a). Costly and Wallis group also studied the ability of a rotating biological contactor (RBC) for the removal of Cd, Cu, and Zn metals with percent (%) removal capability of 73, 42, and 33, respectively (Costley and Wallis 2001b). Another group of researchers reported a high rate of metal uptake by polymer entrapped *Pseudomonas* sp.National Collection of Industrial, Marine and Food Bacteria (NCIMB 11592) attached with granules of activated carbon.

3.5.2 Remediation of Plastic Wastes

The use of plastic material in industries for the transportation of materials like food, clothes, medicines as well as for their use in construction, and shelter increases due to some characteristics properties such as light weight, elasticity, strength, and durability (Mathur, Mathur, and Prasad 2011). Besides these properties, it is considered as a major, potent and nondegradable polymer, with high levels of toxicity and a key contaminant of environment. Due to the problem of degradability and xenobiotic organic nature, its annual accumulation rate is approximately 25 million tons per year (Hadad, Geresh, and Sivan 2005). Scientists are in search to develop a proper method for the degradation of plastic to clean the environment. Up to now four mechanisms have been reported for plastic degradation: thermooxidative degradation, photodegradation, hydrolytic degradation, and biodegradation using microbial cells (Andrady 2011). Biofilms development occur by the accumulation of single or multiple microorganisms on a specific surface and has been reported that polyethylene may also be used as suitable medium for microbial cell growth and biofilm formation.. Fungal species, i.e., *Aspergillus niger* (ITCC No. 6052), isolated from wastewater with residues of plastic were used for the biodegradation of plastic. After the inoculation of fungal spores, experiment was maintained up to 1 month with continuous shaking at 120 rpm at 30°C. A significant growth was observed on the plastic surface. Moreover 3.44% mass reduction and 61% reduction in the tensile strength were recorded. The results were further justified using Scanning electron microscope (SEM) analysis for the growth and hyphal penetration to the plastic material and concluded that the fungal strain can be potentially used to degrade polyethylene. Other types of biofilms based on fungal cells have been reported by Kumar et al. (2017) to study the degradation pattern and capability of microbes. Both high-density (HD) and low-density (LD) polyethylenes were treated for a specific period of time, i.e., 1 month. Initially calculated amounts of polythene sheets were incorporated in all the culturing flasks and autoclaved along with the medium for 15 min at 121°C and 15 psi. After sterilization, all the three algal strains were inoculated (i.e., 1% culture with respect to total volume), and the flasks were incubated at temperature 27°C±2°C for one month with continuous light source (12:12 h dark and night). The degradation of polyethylene was determined gravimetrically, and the results were recorded in percentage (%). The highest percent degradation of polyethylene was recorded for *Anabaena spiroides* (8.18%±0.66%) followed by Diatom *Navicula pupula* (4.44%±0.82%) and *Scenedesmus dimorphus* (3.74%±0.26%). Moreover, after the completion of experiment, algal dry weight was also determined (Kumar, Kanna, and Elumalai 2017). It was reported that a bacterial strain *Rhodococcus ruber* (C208) is capable of producing biofilm as well as degrade polyethylene at a rate of 0.86% per week (Sivan, Szanto, and Pavlov 2006). It was also reported by Orr, Hadar, and Sivan (2004) that *Rhodococcus rubber* (C208) strain has a significant capability to degrade polyethylene as compared to other isolates. Further it was recorded that bacterium can degrade 8% of polyethylene with 30 days of incubation (Orr, Hadar, and Sivan 2004).

3.5.3 Remediation of Synthetic Dyes

In various industries such as textile, tunneling and wood synthetic dyes have been used extensively. However, these dyes may have adverse effects on living organisms. Due to their carcinogenic and mutagenic nature, their removal from environment is significantly important, and thus needs an appropriate mechanism for eradication. A variety of chemical methods have been developed, but few reports based on biofilms are available for such dye eradication. *Pseudomonas luteola* was reported for their decolorizing activity of azo dye (reactive red 22) in a contaminated sample. Further the degradation pattern of dye was also investigated by observing and fluctuating a variety of parameters such as pH, temperature, and concentration. Moreover, samples were analyzed through High-performance liquid chromatography (HPLC) to validate dye degradation (Chang et al. 2001). Another study was conducted by Kapdan and Kargi (2002) for the biodegradation and adsorption of textile dye, Everzol Turquoise Blue G, in an activated sludge unit of laboratory scale. The biofilm of *Coriolus versicolor*, a white-rod fungus, was used for degradation study. From the experiment it was recorded that the highest decolorization efficiency (E) was 82% at dyestuff (D_0) 200 mg/L, adsorbent concentrations (A) 150 mg/L, and a sludge age of $\theta_c = 20$ days, respectively (Kapdan and Kargi 2002).

3.5.4 Remediation of Toxic Compounds

Environmental contamination through recalcitrant and persistent synthetic chemicals is a global issue. These synthetic chemicals have complex hydrocarbons, and phenols being highly toxic, carcinogenic, and mutagenic in nature are considered as xenobiotic. These xenobiotics include dichlorobenzene, naphthalene, chlorinated phenols, organophosphorus compounds, xylene polynucleated aromatics hydrocarbons toluene, pesticides DDT, and heavy metals. Through human activities, accidentally or intentionally, these hazardous chemicals released into environment thus cause severe water and soil pollution. Traditionally these pollutants were removed physically either by digging or damping the contaminated area through fresh soil and also chemical oxidation/stabilization approach. But both of these methods were laborious, need high capital investment, and often produce secondary pollutants in case of chemical oxidation. Further, these approaches were not ecofriendly and give a permanent solution for pollutant removal. These shortcomings have seeded more cost-effective and greener concepts of microbial remediation for the removal of these pollutants, either by stimulating the soil indigenous microbes or introducing engineered microbes into the site of pollution. Due to the ubiquitous nature of microorganisms (bacteria, fungi), they have a large number much more as compared to other living organisms in the earth. Naturally these microbes have the ability to detoxify organic compounds by producing different enzymes to obtain their nutrients, and thus act as natural scavengers. Microorganisms have a diverse genetic makeup and can live in different environmental conditions like pH and temperature, even in the absence of oxygen. Interestingly, these microbes have anthropogenic ability to form biofilm by attachment to any surface. In biofilms different microbial strains can combine with each other to induce a synergistic effect that can boost the biodegradation process. Microbial biofilm has a great potential for removal of recalcitrant and more persistent toxic compounds due to high microbial biomass and availability of pollutants. Also, the synergistic effect of microbes in biofilm, their horizontal gene transfer, and resistance to different stresses and chemotaxis make microorganisms more efficient for bioremediation (Megharaj et al. 2011).

Dasgupta et al. (2013) have reported a synthetic designed yeast biofilm on gravel which has the ability to degrade more than 97% diesel oil within 10 days. It has considered the first successful yeast-based artificial biofilm (Dasgupta, Ghosh, and Sengupta 2013). In another study, Al-Awadhi et al. have reported an artificially designed microbial biofilm on gravel particles and glass plates which have the ability to degrade hydrocarbons. This biofilm has a consortia of diverse microbes including bacteria, filamentous cyanobacteria, diatoms, and picoplankton. Among all microbes phototrophic microorganisms can form biofilms due to its ability of colonization (Al-Awadhi et al. 2003).

Immobilized bacteria and microalgae consortia on gravel particles in the intertidal zone of Arabian Gulf coast were also reported in previous literature, and this biofilm has the potential to degrade hydrocarbons. The physical analysis of this biofilm has showed that each gravel particle is coated with 100 mg of biomass. Among them, the most prominent phototrophs were *Cyanobacteria dermocarpella* species and *Acinetobacter calcoaceticus*, which degrade hydrocarbons (Radwan and Al-Hasan 2001). Chemical industry effluents have chlorinated aromatic compounds which are highly carcinogenic even in low concentrations like 2,4-dichlorophenol (DCP). For degradation of such toxic compounds a rotating porous tube biofilm reactor has been designed to contain activated sludge supplemented with *Pseudomonas putida* that degrades DCP. The degradation efficiency of this bioreactor was 70%–100%. Similarly, *Pseudomonas* sp. and *Rhodococcus* sp. contain a three-phase fluidized bed biofilm reactor for biodegradation of polychlorinated biphenyls (Kargi and Eker 2005). Commonly nitroaromatics compounds show resistance to microbial remediation or often produced toxic metabolites upon degradation. Gisi et al. reported a fixed bed column reactor having microbial consortia for degradation of 4,6-dinitro-ortho-cresol (Gisi, Stucki, and Hanselmann 1997). Kumar et al. have reported *Burkholderia vietnamiensis* containing a gas phase membrane bioreactor of biodegradation of toluene. The removal efficiency of toluene was 89% with a loading rate of 14 kg/(m^3d^1) (Kumar et al. 2009). Branched alkyl benzene sulfonates (BAS) produced severe environmental pollution. *Stenotrophomonas maltophilia* biofilm was developed and used for biodegradation of dodecylbenzene sulfonate sodium (Farzaneh et al. 2010). Thiocyanate is widely used in different insecticides and herbicides and is toxic in nature. It has been biodegraded using fluidized-carrier biofilm reactor. Maximum degradation of thiocyanate was reported up to 80% (Jeong and Chung 2006). In short, biofilm is green, cost-effective, and an easy approach for the remediation of various environmental pollutants.

3.5.5 The Role of Biofilm in Removing Pollutants from Agriculture Land

The microbial biofilm is essential to conserve soil performance, controls the catalysis and buildup of organic matter, the breakdown of organic deposits, and is used as an indicator for the management of soil, the contamination of hazardous metals, and agrochemical fertilizer practices (Triveni et al. 2015). The poor development of plant growth and lower root elongation are the key factors for plant-based metal accumulation in the metal contaminated soil. It has been reported that greater nutrient capture by seedlings biopriming along with Plant growth-promoting microorganisms PGPMs leads to the construction of plant growth regulators (PGRs) at the root border enhanced root growth, resulting in best absorption of nutrients and water from the soil. Potassium solubilizing microbes (KSMs) are very beneficial biofilms that produce minimum pollution and low energy-required approaches, broadly used in the development of accessible potassium for adaptation by plants and elimination of debris from minerals. These biofilms are soil friendly and modify soil fertility. Normally soils contain over 20,000 ppm of total potassium, of which plants can use only the transferable potassium on the crust of the soil. Mostly in soil, potassium exists in the form of water-soluble, nonexchangeable, exchangeable, and structural mineral. Among these forms exchangeable and water-soluble pools easily donate potassium for plant uptake. Sometimes nonexchangeable potassium may be uptake by plant where exchangeable potassium is in low level (George and Michael 2002). Partial negatively charged soil minerals trap available, soluble, or exchangeable potassium and carbon-based matter in soil, while unavailable or nonexchangeable potassium resides predominantly, and plants are not capable to uptake it. Further, microbial biofilms enhance its availability. Biosorption comprises of many mechanisms, mainly chelating, ion exchange, diffusion, and absorption via cell walls. Such "passive" mechanisms take place inside the cells at the microbial biofilm community level. The uptake of metal in active mode and high amounts is called bioaccumulation. This whole process is based on cellular metabolism. The various phototrophic microbiofilms have the potential to accumulate or sorb heavy metals in one way or the other. Algal biofilms have the potential to detoxify wastewater contaminated with heavy metals (Bender et al. 1994). EPS that is partially negatively charged at high pH levels may have the potential for the metal-trapping assets of phototrophic microbial associations. Mucilage sheaths isolated from the biofilm of cyanobacteria *Aphanothece halophytica* and *Microcystis aeruginosa* show solid affinity for heavy metals such as zinc, lead, and copper. Along with bioaccumulation and biosorption, the high pH of photosynthetically active microbial biofilms may promote the removal of heavy metals by precipitation. Low cost and elevated efficiency are the key benefits of using biosorption for metal removal (Diels et al. 2003).

3.5.5.1 Biofilm-Based Biofertilizers

A biofertilizer comprises of living microbes used as additive in agriculture to improve the crop yield and growth. Biofertilizer exerts a direct or indirect positive impact on plant productivity and crop yields through different techniques. It is a viable source for replacing depleted soil microbes by modifying biodiversity but is also used as a supplement (Herath et al. 2015). Rhizobium genera such as *Azospirillum, Azotobacter, Bradyrhizobium, Burkholderia, Gluconacetobacter, Herbaspirillum, Klebsiella, Pseudomonas, Azoarcus, Enterobacter,* and various other rhizobacterial genera favor the plant growth through the supply of essential nutrients (Bianciotto et al. 2001). As a whole, biofertilizers comprise of one or more of the earlier labeled bacterial species as "active ingredients." Studies show that beneficial plant growth promoting biofilm shows unique interactions with nearby host plants. A more specific example is the symbiotic interactions between Leguminosae (a plant family), *Rhizobiaceae* and bacteria. Such relationships have been studied in *Azospirillum, Herbaspirillum,* and cereal crops (Swarnalakshmi et al. 2013).

3.5.5.2 The Use of Biofilm as Biofertilizer

The side effects of agrochemicals used in agriculture opened a gateway for the scientists to search environmentally friendly additives for better crop yield and productivity. For this purpose, biofilms is of interest (to use as biofertilizers). Biofilms supply nutrients to plants and improve plant growth and yield. Besides plant health, biofilms also play a key role to sustain soil productivity and environmental health. KSM-based biofilms not only trigger the nonsoluble potassium minerals but also influence other mineral maintenance as well, such as Zn, P, and Fe. Hence, the worth of crop yield can be enhanced, the environmental effect can be lower, the physiochemical channels can be improved, and the most important is the cost of the product can be respectively decreased. Biofertilizer increases the hormonal action of plants, which in return promote plant growth (Seneviratne and Jayasinghearachchi 2005).

In recent advance researches, a developed strain of fungal rhizobial biofilm has been used in vitro for the production of biofertilizers known as BFBFs. As compared to the monocultures of microbes or mixed cultures, the BFBF biofilm exhibited improved BNF, nutrient mineral release in the soil, plant growth related hormones and organic acid, etc. when added to soil. A study reported shows the beneficial effects of BFBFs on nutrient depleted soil due to repeated tea cultivation in Sri Lanka. His results showed that reducing chemical fertilizers use by 50% increases plant growth and soil fertility in the presence of BFBF biofilm. BFBF base biofilm has been proven successful in restoration of deteriorated soil and increased crop production (Seneviratne and Jayasinghearachchi 2005). BFBF biofilms with reduced agrochemicals improve soil fertility and plant growth by helping plant roots to store more nutrients. Nutrient accumulation is necessary for plant growth and development. Thus BFBFs' use can be appreciated as a medium to improve soil fertility due to their nutrient accumulation capability. Moreover, the use of BFBFs decreases the rate of transpiration that favors plant growth and soil fertility. This is due to rhizobia which lessens the conductance of stomata. Microbial biofilms provide many nutrients to plants, due to which the demand for chemical fertilizers get to low level. As a result, soil pH and moisture remain in a suitable range, beneficial to both plant growth and soil fertility (Seneviratne et al. 2009).

3.6 CURRENT STATUS AND FUTURE PROSPECTIVE

3.6.1 Current Status

The term biofilm was first coined by Costerton and his coworkers in 1978. More specifically the term biofilms can be described as microbes (bacteria, Fungi, and mycorrhizae) associated with their synthesized metabolites, attached on biotic or abiotic surfaces, to protect themselves from any exogenous harms and stresses (Costerton, Geesey, and Cheng 1978). Thus, the major components of biofilm formation are microbial community, biotic/abiotic substratum, medium/environment, and their mutual interactions (Høiby 2014, Petrova and Sauer 2012), as well as with their EPS (Altaf, Ahmad, and Al-Thubiani 2017). Biofilm formation is the major characteristic of microbes (Stewart and Franklin 2008), and major structures involved in their formations are cellular pili, flagella, cell membrane proteins, lipopolysaccharides, etc. (Hinsa et al. 2003, Belas 2013). It has been reported that variation in environmental conditions may cause the induction of biofilm formation, as a result of which microbial cells immobilize at specific points with altered metabolic activities. Furthermore, it has been compared that in a mature biofilm, only 10% of the total is microbial cells while the remaining 90% is the surrounding matrix formed of various biopolymers (Flemming and Wingender 2010).

With the development in the field of microbiology, the applications of biofilms in various areas of biotechnology for beneficial purchases increase globally. Biofilms may be formed by a single species of microbes or it may consist of multiple communities with different organisms (bacteria-bacteria, bacteria-fungi, bacteria-algae, etc.) (Seneviratne et al. 2009) that produce extracellular substances associated with its substratum (Flemming and Wingender 2010). Certain strains of bacteria such as *Bacillus* spp. (*B. subtilis*, *B. polymyxa*, *B. brevis*, *B. amyloliquefaciens*, *B. licheniformis*, *B. thuringiensis*), *Lactobacillus* spp. (*L. acidophilus*, *L. casei*, *L. plantarum*, *L. paracasei*, *L. reuteri*), *Enterococcus* sp. (*E. faecalis*, *E. casseliflavus*, *E. faecium*), and *Pseudomonas* spp. (*P. fluorescens*, *P. chlororaphis*, *P. putida*) have been reported for biofilm formation. Furthermore, several genera of agriculturally important fungi and bacteria including cyanobacteria and anabaena have a major role in agriculture improvement due to beneficial biofilm activity (Prasanna et al. 2011, 2013, 2014, 2015).

Keeping in view the beneficial characteristics of biofilms, their applications are increasing in various fields such as fermentation, water filtration, and purification, in agriculture (i.e., biofertilizers, soil texture improver, etc.), biofouling, anticorrosive agents, antimicrobials, as microbial fuel cells as well as in environment and biomedicines. In agriculture biofilms have significant importance and can be used as nutrient mobilizers, biocontrol agents, biofertilizers, and plant growth promotors. Further biofilms may inhibit plant pathogen growth, and their pathogenicity as well has an important role in bioremediation. The potential use of biofilm in agriculture has been reviewed in fields like biofertilizers, biocontrol agent, bioremediation, plant growth promotion, etc. (Velmourougane, Prasanna, and Saxena 2017).

Agricultural importance of biofilms is due to their biotechnological aspects like fermentative capabilities and antimicrobial and biochemical characteristics. Due to these qualities, they improve plant nutrient availability and uptake by their recyclability and also involve in water contents and soil texture maintenance of agricultural land (Turhan et al. 2019). Moreover, microorganisms of biofilms form a relationship with host plants or their tissues for mutual benefits, which may be mutualism, symbiosis, and pathogenesis. Beneficial biofilms derived from fungi, mycorrhizae, rhizobacteria (PGPR), and other bacteria that promote plant growth have been reported for the improvement of various crop yields. Besides these, PGPR and PGPR-like microbes can also prevent or inhibit the proliferation of other microbes (phytopathogens) by direct or indirect interaction as well as to facilitate the uptake of nutrients such as N, O, C, H, P, Ca, S, Zn, etc. from the soil and surrounding environments. Further biofilm can overcome problems like drought, optimal growth, pH maintenance, and other growth promoting factors of the environment to enhance productivity. In this regard, to increase crop productivity, a large-scale development of biofilm and consortia of biofilm is required (Rekadwad and Khobragade 2017).

Climatic changes may cause adverse effects on soil physical (texture) and chemical (fertility) nature, as with the increase soil temperature, decomposition of soil organic matter contents increases which may be lost. Therefore, biofilm acts as biofertilizers as reported and applied (Triharyanto 2018, Karmakar et al. 2016). As compared to conventional biofertilizers, biofilm biofertilizers are more efficient because they are more resistant to environmental stresses as well as show excellent adoptability features. Moreover, the advantages of biofilm biofertilizers are nutrient availability, production of growth promoting substances (e.g., IAA), and decreasing plant diseases by inhibiting pathogen growth and enhance beneficial microbial growth (Das et al. 2017). Fungal-bacterial biofilm (FBBs) are capable to create association with plant roots (Seneviratne et al. 2009), and such biofilms have "pseudo nodules" or nodule-like structures involved in biological N_2 fixation, mostly for nonlegume plants (Jayasinghearachchi and Seneviratne 2004). Biofilm also increased production of plant growth promoting hormones like IAA for plant growth and development (Bandara, Seneviratne, and Kulasooriya 2006). FBB root associated biofilm significantly affect bean nodule production and enhance nutrient absorbance from surrounding environment (Seneviratne et al. 2009), protect plant from pathogenic attack (Das, Prasanna, and Saxena 2017), enhance cell signaling between plants and rhizobia, which ultimately leads to establish effective symbiosis (Bais et al. 2006). The application of biofilms like Trichoderma/Anabaena with other microbes that have agricultural importance can improve soil fertility, resulting in enhanced growth of legume crops (Prasanna et al. 2014, Triveni et al. 2013). Trichoderma- and Anabaena-based bacterial biofilms also have a major role as biocontrol agents and plant growth promoters in *Macrophomina phaseolina* challenged cotton crop (Triveni et al. 2015). The application of biofilms has also been reported in various other crops such as rice (Adak et al. 2016) and Chickpea (Bidyarani et al. 2016). With great concern to agricultural issues and economic crises, biofilms are the alternate, ecofriendly, easy, and cost-effective solution for crops and plant growth, development as well as recovery of agriculture productivity. Biofilms are not short-term solutions for single agriculture crop but provide long-term maintenance sources as well as show compatibility to most food producing crops. Besides these this technology gets attraction due to its beneficial potential, capability of advancement and less harmful effects, and scientist extending it to other fields of research (Velmourougane, Prasanna, and Saxena 2017, Herath et al. 2015).

3.6.2 Conclusion and Future Perspective

Research is in progress, and some beneficial results have been reported on the use of biofilms for agriculture and crop productivity such as nitrogen and other nutrients (oxygen, carbon, potassium, magnesium, manganese, etc.) fixation, soil fertility enhancement, improvement in soil, physicochemical features. However, further research in both laboratory and field is required to understand the complete mechanism of each type of biofilm formation, microbe-plant interactions, as well as their specification in terms of area of application. Moreover, as a future direction, there is the need to integrate biofilm-based technology for agricultural practices, to improve crop productivity by nutrient availability and pathogen protection. Recent investigations indicate an increased interest in the agricultural research on biofilm associated rhizobacteria as biofertilizers. This easy, economical, and ecofriendly biofertilizer technology may lead to reduce the use of chemical fertilizers in future for sustainable agriculture without hampering crop yields.

3.7 ACKNOWLEDGMENTS

The authors are thankful to the Higher Education Commission (HEC) of Pakistan for financial support under project no. 6118. Dr Asma also acknowledges the technical support from her parent institute NIBGE.

REFERENCES

Adak, Anurup, Radha Prasanna, Santosh Babu, Ngangom Bidyarani, Shikha Verma, Madan Pal, Yashbir Singh Shivay, and Lata Nain. 2016. "Micronutrient enrichment mediated by plant-microbe interactions and rice cultivation practices." *Journal of Plant Nutrition* 39 (9):1216–1232.

Agladze, Konstantin, Debra Jackson, and Tony Romeo. 2003. "Periodicity of cell attachment patterns during *Escherichia coli* biofilm development." *Journal of Bacteriology* 185 (18):5632–5638.

Al-Awadhi Husain, Al-Hasan Redha H, Sorkhoh Nasser A, Salamah Samar and Samir Radwan. 2003. "Establishing oil-degrading biofilms on gravel particles and glass plates." *International Biodeterioration & Biodegradation* 51 (3):181–185.

Altaf, Mohd Musheer, Iqbal Ahmad, and Abdullah Safar Al-Thubiani. 2017. "Rhizobacterial biofilms: diversity and role in plant health." In Kumar, V., Kumar, M., Sharma, S., and Prasad, R. (eds) *Probiotics in Agroecosystem*, pp. 145–162. Springer, Singapore.

Andrady, Anthony L. 2011. "Microplastics in the marine environment." *Marine Pollution Bulletin* 62 (8):1596–1605.

Assaf, Danielle, Doron Steinberg, and Moshe Shemesh. 2015. "Lactose triggers biofilm formation by *Streptococcus mutans*." *International Dairy Journal* 42:51–57.

Atouei, Maryam Talebi, Ahmad Ali Pourbabaee, and Mehdi Shorafa. 2019. "Alleviation of salinity stress on some growth parameters of wheat by exopolysaccharide-producing bacteria." *Iranian Journal of Science and Technology, Transactions A: Science* 1:1–9.

Bais, Harsh P., Tiffany L. Weir, Laura G. Perry, Simon Gilroy, and Jorge M. Vivanco. 2006. "The role of root exudates in rhizosphere interactions with plants and other organisms." *Annual Review of Plant Biology* 57:233–266.

Bandara, W. M. Manoj S., Gamini Seneviratne, and Sabharatna Ananda Kulasooriya. 2006. "Interactions among endophytic bacteria and fungi: effects and potentials." *Journal of Biosciences* 31 (5):645–650.

Barkay, Tamar, and Jeffra Schaefer. 2001. "Metal and radionuclide bioremediation: issues, considerations, and potentials." *Current Opinion in Microbiology* 4 (3):318–323.

Barraud, Nicolas, Daniel J. Hassett, Sung-Hei Hwang, Scott A. Rice, Staffan Kjelleberg, and Jeremy S. Webb. 2006. "Involvement of nitric oxide in biofilm dispersal of *Pseudomonas aeruginosa*." *Journal of Bacteriology* 188 (21):7344–7353.

Becerra-Castro, Cristina, Ana Rita Lopes, Ivone Vaz-Moreira, Elisabete F. Silva, Célia M. Manaia, and Olga C. Nunes. 2015. "Wastewater reuse in irrigation: a microbiological perspective on implications in soil fertility and human and environmental health." *Environment International* 75:117–135.

Belas, Robert. 2013. "When the swimming gets tough, the tough form a biofilm." *Molecular Microbiology* 90 (1):1–5.

Bender, Judith, Susana Rodriguez-Eaton, Udoudo M. Ekanemesang, and Peter Phillips. 1994. "Characterization of metal-binding bioflocculants produced by the cyanobacterial component of mixed microbial mats." *Applied and Environmental Microbiology* 60 (7):2311–2315.

Berk, Veysel, Jiunn C. N. Fong, Graham T. Dempsey, Omer N. Develioglu, Xiaowei Zhuang, Jan Liphardt, Fitnat H. Yildiz, and Steven Chu. 2012. "Molecular architecture and assembly principles of *Vibrio cholerae* biofilms." *Science* 337 (6091):236–239.

Bianciotto Valeria, Andreotti Silvia, Balestrini Raffaella Maria, Bonfante Paola, Perotto Silvia. 2001. "Extracellular polysaccharides are involved in the attachment of *Azospirillum brasilense* and *Rhizobium leguminosarum* to arbuscular mycorrhizal structures." *European Journal of Histochemistry* 45:39–50.

Bidyarani, Ngangom, Radha Prasanna, Santosh Babu, Firoz Hossain, and Anil Kumar Saxena. 2016. "Enhancement of plant growth and yields in chickpea (*Cicer arietinum*

L.) through novel cyanobacterial and biofilmed inoculants." *Microbiological Research* 188: 97–105.

Boland, Thomas, Robert A. Latour, and Fred J. Stutzenberger. 2000. "Molecular basis of bacterial adhesion." In An, Y. H., and Friedman, R. J. (eds) *Handbook of Bacterial Adhesion*, pp. 29–41. Springer, Totowa, NJ.

Burgos-Caraballo, Sofía, Sharon A. Cantrell, and Alonso Ramírez. 2014. "Diversity of benthic biofilms along a land use gradient in tropical headwater streams, Puerto Rico." *Microbial Ecology* 68 (1):47–59.

Burmølle, Mette, Dawei Ren, Thomas Bjarnsholt, and Søren J. Sørensen. 2014. "Interactions in multispecies biofilms: do they actually matter?" *Trends in Microbiology* 22 (2):84–91.

Campanella, Bruno F., Claudia Bock, and Peter Schröder. 2002. "Phytoremediation to increase the degradation of PCBs and PCDD/Fs." *Environmental Science and Pollution Research* 9 (1):73–85.

Chai, Yunrong, Pascale B. Beauregard, Hera Vlamakis, Richard Losick, and Roberto Kolter. 2012. "Galactose metabolism plays a crucial role in biofilm formation by *Bacillus subtilis*." *MBio* 3 (4):e00184–e00112.

Chambers, Jacob R., and Karin Sauer. 2013. "Small RNAs and their role in biofilm formation." *Trends in Microbiology* 21 (1):39–49.

Chang, Jo-Shu, Chien Chou, Yu-Chih Lin, Ping-Jei Lin, Jin-Yen Ho, and Tai Lee Hu. 2001. "Kinetic characteristics of bacterial azo-dye decolorization by *Pseudomonas luteola*." *Water Research* 35 (12):2841–2850.

Chen, Yun, Shugeng Cao, Yunrong Chai, Jon Clardy, Roberto Kolter, Jian-hua Guo, and Richard Losick. 2012. "A *Bacillus subtilis* sensor kinase involved in triggering biofilm formation on the roots of tomato plants." *Molecular Microbiology* 85 (3):418–430.

Costerton, J.W., Geesey Gill and Cheng K.J. 1978. "How bacteria stick." *Scientific American* 238 (1):86–95.

Costerton, J. William, Zbigniew Lewandowski, Douglas E. Caldwell, Darren R. Korber, and Hilary M. Lappin-Scott. 1995. "Microbial biofilms." *Annual Review of Microbiology* 49 (1):711–745.

Costley, Shauna and Frederick M. Wallis. 2001a. "Bioremediation of heavy metals in a synthetic wastewater using a rotating biological contactor." *Water Research* 35 (15):3715–3723.

Costley, Shauna and Frederick M. Wallis. 2001b. "Treatment of heavy metal-polluted wastewaters using the biofilms of a multistage rotating biological contactor." *World Journal of Microbiology and Biotechnology* 17 (1):71–78.

Czaczyk, Kasia, and Kamila Myszka. 2007. "Biosynthesis of extracellular polymeric substances (EPS) and its role in microbial biofilm formation." Polish Journal of Environmental Studies 16 (6):799–806.

Das, Ipsita, and Madhusmita Pradhan. 2016. "Potassium-solubilizing microorganisms and their role in enhancing soil fertility and health." In Meena, V., Maurya, B., Verma, J., and Meena, R. (eds) *Potassium Solubilizing Microorganisms for Sustainable Agriculture*, pp. 281–291. Springer, New Delhi.

Das, Krishnashis, Radha Prasanna, and Anil Kumar Saxena. 2017. "Rhizobia: a potential biocontrol agent for soilborne fungal pathogens." *Folia Microbiologica* 62 (5):425–435.

Das, Krishnashis, Mahendra Vikram Singh Rajawat, Anil Kumar Saxena, and Radha Prasanna. 2017. "Development of *Mesorhizobium ciceri*-based biofilms and analyses of their antifungal and plant growth promoting activity in chickpea challenged by *Fusarium wilt*." *Indian Journal of Microbiology* 57 (1):48–59.

Dasgupta, Debdeep, Ritabrata Ghosh, and Tapas K. Sengupta. 2013. "Biofilm-mediated enhanced crude oil degradation by newly isolated *Pseudomonas* species." *ISRN Biotechnology* 250749:1–13.

Davey, Mary Ellen, and George A. O'Toole. 2000. "Microbial biofilms: from ecology to molecular genetics." *Microbiology and Molecular Biology Reviews* 64 (4):847–867.

Dianati, tilaki Ramazan ali, Mahvi Amir, Shariat Mamak and Nasseri Simin . 2004. "Study of cadmium removal from environmental water by biofilm covered granular activated carbon." *Iranian Journal of Public Health* 33 (4):43–52.

Diels, Ludo, Spaans PH, Van Roy Sandra, Hooyberghs Liliane, Ryngaert Annemie, Wouters Hans, Walter E, Winters J, Macaskie Lynne, Finlay John, Pernfuss Barbara, Pümpel Thomas and Tsezos Marios.. 2003. "Heavy metals removal by sand filters inoculated with metal sorbing and precipitating bacteria." *Hydrometallurgy* 71 (1–2):235–241.

Donlan, Rodney M. 2002. "Biofilms: microbial life on surfaces." *Emerging Infectious Diseases* 8 (9):881.

El-Shahaw, Mohammad, Hamza Abdulhamid, Bashammakh Abdulaziz Saleh and Al-Saggaf WT. 2010. "An overview on the accumulation, distribution, transformations, toxicity, and analytical methods for the monitoring of persistent organic pollutants." *Talanta* 80 (5):1587–1597.

Ertesvåg, Helga, and Svein Valla. 1998. "Biosynthesis and applications of alginates." *Polymer Degradation and Stability* 59 (1–3):85–91.

Farzaneh, Hosseini, Malekzadeh Fereidon, Amirmozafari Noor, and Ghaemi Naser. 2010. "Biodegradation of dodecylbenzene sulfonate sodium by *Stenotrophomonas maltophilia* biofilm." *African Journal of Biotechnology* 9 (1):055–062.

Flemming, Hans-Curt, and Jost Wingender. 2001. "Relevance of microbial extracellular polymeric substances (EPSs)—Part I: structural and ecological aspects." *Water Science and Technology* 43 (6):1–8.

Flemming, Hans-Curt, and Jost Wingender. 2010. "The biofilm matrix." *Nature Reviews Microbiology* 8 (9):623.

Flemming, Hans-Curt, Jost Wingender, Ulrich Szewzyk, Peter Steinberg, Scott A. Rice, and Staffan Kjelleberg. 2016. "Biofilms: an emergent form of bacterial life." *Nature Reviews Microbiology* 14 (9):563.

Gabr, Rabei M., Sanaa M. F. Gad-Elrab, Romany N. N. Abskharon, Sedky H. A. Hassan, and Ahmed A. M. Shoreit. 2009. "Biosorption of hexavalent chromium using biofilm of *E. coli* supported on granulated activated carbon." *World Journal of Microbiology and Biotechnology* 25 (10):1695.

Gauri, Samiran S., Santi M. Mandal, and Bikas R. Pati. 2012. "Impact of *Azotobacter exopolysaccharides* on sustainable agriculture." *Applied Microbiology and Biotechnology* 95 (2):331–338.

George, Rehm, and Schmitt Michael. 2002. *Potassium for Crop Production*. Communication and Educational Technology Services, University of Minnesota Extension, Minneapolis.

Gerhardt, Karen E., Xiao-Dong Huang, Bernard R. Glick, and Bruce M. Greenberg. 2009. "Phytoremediation and rhizoremediation of organic soil contaminants: potential and challenges." *Plant Science* 176 (1):20–30.

Giri, Bhoopander, and Krishna G. Mukerji. 2004. "Mycorrhizal inoculant alleviates salt stress in *Sesbania aegyptiaca* and *Sesbania grandiflora* under field conditions: evidence for reduced sodium and improved magnesium uptake." *Mycorrhiza* 14 (5):307–312.

Gisi, D., Stucki G., and K.W. Hanselmann. 1997. "Biodegradation of the pesticide 4,6-dinitro-ortho-cresol by microorganisms in batch cultures and in fixed-bed column reactors." *Applied Microbiology and Biotechnology* 48 (4):441–448.

Glick, Rivka, Christie Gilmour, Julien Tremblay, Shirley Satanower, Ofir Avidan, Eric Déziel, E. Peter Greenberg, Keith Poole, and Ehud Banin. 2010. "Increase in rhamnolipid synthesis under iron-limiting conditions influences surface motility and biofilm formation in *Pseudomonas aeruginosa*." *Journal of Bacteriology* 192 (12):2973–2980.

Hadad, D., Geresh Shimona and Sivan Alex. 2005. "Biodegradation of polyethylene by the thermophilic bacterium *Brevibacillus borstelensis*." *Journal of Applied Microbiology* 98 (5):1093–1100.

Han, H.S. and K.D. Lee. 2005. "Phosphate and potassium solubilizing bacteria effect on mineral uptake, soil availability, and growth of eggplant." *Research Journal of Agriculture and Biological Sciences* 1 (2):176–180.

Hannan, Saad, Derren Ready, Azmiza S. Jasni, Michelle Rogers, Jonathan Pratten, and Adam P. Roberts. 2010. "Transfer of antibiotic resistance by transformation with eDNA within oral biofilms." *FEMS Immunology & Medical Microbiology* 59 (3):345–349.

Hausner, Martina, and Stefan Wuertz. 1999. "High rates of conjugation in bacterial biofilms as determined by quantitative in situ analysis." *Applied and Environmental Microbiology* 65 (8):3710–3713.

Hazell, Peter, and Stanley Wood. 2007. "Drivers of change in global agriculture." *Philosophical Transactions of the Royal Society B: Biological Sciences* 363 (1491):495–515.

Herald, Paula J., and Edmund A. Zottola. 1988. "Attachment of *Listeria monocytogenes* to stainless steel surfaces at various temperatures and pH values." *Journal of Food Science* 53 (5):1549–1562.

Herath, Lasantha Indika, Menikdiwela Kalhara, Igalavithana Avanthi Deshani, Seneviratne Gamini . 2015. "Developed fungal-bacterial biofilms having nitrogen fixers: universal biofertilizers for legumes and non-legumes." In de Bruijn, F. (ed) *Biological Nitrogen Fixation*, pp. 1041–1046. Wiley, Hoboken, NJ.

Hickman, Jason W., and Caroline S. Harwood. 2008. "Identification of FleQ from *Pseudomonas aeruginosa* as ac-di-GMP-responsive transcription factor." *Molecular Microbiology* 69 (2):376–389.

Hinsa, Shannon M., Manuel Espinosa-Urgel, Juan L. Ramos, and George A. O'Toole. 2003. "Transition from reversible to irreversible attachment during biofilm formation by *Pseudomonas fluorescens* WCS365 requires an ABC transporter and a large secreted protein." *Molecular Microbiology* 49 (4):905–918.

Høiby, Niels. 2014. "A personal history of research on microbial biofilms and biofilm infections." *Pathogens and Disease* 70 (3):205–211.

Janissen, Richard, Duber M. Murillo, Barbara Niza, Prasana K. Sahoo, Marcelo M. Nobrega, Carlos L. Cesar, Marcia L. A. Temperini, Hernandes F. Carvalho, Alessandra A. De Souza, and Monica A. Cotta. 2015. "Spatiotemporal distribution of different extracellular polymeric substances and filamentation mediate *Xylella fastidiosa* adhesion and biofilm formation." *Scientific Reports* 5: 9856.

Jayasinghe, arahchi Himali, Seneviratne Gamini . 2004. "A bradyrhizobial-penicillium spp. biofilm with nitrogenase activity improves N_2 fixing symbiosis of soybean." *Biology and Fertility of Soils* 40 (6):432–434.

Jefferson, Kimberly K. 2004. "What drives bacteria to produce a biofilm?" *FEMS Microbiology Letters* 236 (2):163–173.

Jeong, Yong-Shik, and Jong Shik Chung. 2006. "Biodegradation of thiocyanate in biofilm reactor using fluidized-carriers." *Process Biochemistry* 41 (3):701–707.

Johnston, A. Edward, Paul R. Poulton, and Kevin Coleman. 2009. "Soil organic matter: its importance in sustainable agriculture and carbon dioxide fluxes." *Advances in Agronomy* 101: 1–57.

Jonas, Rainer, and Luiz F. Farah. 1998. "Production and application of microbial cellulose." *Polymer Degradation and Stability* 59 (1–3):101–106.

Kapdan, Ilgi Karapinar, and Fikret Kargi. 2002. "Simultaneous biodegradation and adsorption of textile dyestuff in an activated sludge unit." *Process Biochemistry* 37 (9):973–981.

Kaplan, Jeffrey B, and Daniel H. Fine. 2002. "Biofilm dispersal of *Neisseria subflava* and other phylogenetically diverse oral bacteria." *Applied and Environmental Microbiology* 68 (10):4943–4950.

Karatan, Ece, and Paula Watnick. 2009. "Signals, regulatory networks, and materials that build and break bacterial biofilms." *Microbiology and Molecular Biology Reviews* 73 (2):310–347.

Karched, Maribasappa, Radhika G. Bhardwaj, and Sirkka E. Asikainen. 2015. "Coaggregation and biofilm growth of *Granulicatella* spp. with *Fusobacterium nucleatum* and *Aggregatibacter actinomycetemcomitans*." *BMC Microbiology* 15 (1):114.

Kargi, Fikret, and Serkan Eker. 2005. "Removal of 2,4-dichlorophenol and toxicity from synthetic wastewater in a rotating perforated tube biofilm reactor." *Process Biochemistry* 40 (6):2105–2111.

Karmakar, Rajib, Indranil Das, Debashis Dutta, and Amitava Rakshit. 2016. "Potential effects of climate change on soil properties: a review." *Science International* 4 (2):51–73.

Kolenbrander, Paul E., Robert J. Palmer, Jr., Saravanan Periasamy, and Nicholas S. Jakubovics. 2010. "Oral multispecies biofilm development and the key role of cell–cell distance." *Nature Reviews Microbiology* 8 (7):471.

Kolodkin-Gal, Ilana, Diego Romero, Shugeng Cao, Jon Clardy, Roberto Kolter, and Richard Losick. 2010. "D-amino acids trigger biofilm disassembly." *Science* 328 (5978):627–629.

Kumar, Amit, Jo Dewulf, Tom Van De Wiele, and Herman Van Langenhove. 2009. "Bacterial dynamics of biofilm development during toluene degradation by *Burkholderia vietnamiensis* G4 in a gas phase membrane bioreactor." *Journal of Microbiology and Biotechnology* 19 (9):1028–1033.

Kumar, Anita Suresh, and Kalpana Mody. 2009. "Microbial exopolysaccharides: variety and potential applications." In Rehm, B. H. A. (ed) *Microbial Production of Biopolymers and Polymer Precursors: Applications and Perspectives*, pp. 229–253. Caister Academic Press, Norfolk.

Kumar, C. Ganesh, and Sanjeev K. Anand. 1998. "Significance of microbial biofilms in food industry: a review." *International Journal of Food Microbiology* 42 (1–2):9–27.

Kumar Ramachandran Vimal, Kanna Gopal Rajesh and Elumalai Sanniyasi . 2017. "Biodegradation of polyethylene by green photosynthetic microalgae." *Journal of Bioremediation and Biodegradation* 8 (381):2.

Lambers, Hans. 2003. "Introduction: dryland salinity: a key environmental issue in southern Australia." *Plant and Soil* 257:v–vii.

Li, Yung-Hua, Peter C. Y. Lau, Janet H. Lee, Richard P. Ellen, and Dennis G. Cvitkovitch. 2001. "Natural genetic transformation of *Streptococcus mutans* growing in biofilms." *Journal of Bacteriology* 183 (3):897–908.

Lindow, Steven E, and Johan H. J. Leveau. 2002. "Phyllosphere microbiology." *Current Opinion in Biotechnology* 13 (3):238–243.

Lloyd, Jonathan R. 2003. "Microbial reduction of metals and radionuclides." *FEMS Microbiology Reviews* 27 (2–3):411–425.

Malek, Angela M., Aaron Barchowsky, Robert Bowser, Terry Heiman-Patterson, David Lacomis, Sandeep Rana, Ada Youk, and Evelyn O. Talbott. 2015. "Exposure to hazardous air pollutants and the risk of amyotrophic lateral sclerosis." *Environmental Pollution* 197: 181–186.

Mathur, Garima, Ashwani Mathur, and Ramasare Prasad. 2011. "Colonization and degradation of thermally oxidized high-density polyethylene by *Aspergillus niger* (ITCC No. 6052) isolated from plastic waste dumpsite." *Bioremediation Journal* 15 (2):69–76.

Matthysse, Ann G., Mazz Marry, Leonard Krall, Mitchell Kaye, Bronwyn E. Ramey, Clay Fuqua, and Alan R. White. 2005. "The effect of cellulose overproduction on binding and biofilm formation on roots by *Agrobacterium tumefaciens*." *Molecular Plant-Microbe Interactions* 18 (9):1002–1010.

Mayak, Shimon, Tsipora Tirosh, and Bernard R. Glick. 2004. "Plant growth-promoting bacteria that confer resistance to water stress in tomatoes and peppers." *Plant Science* 166 (2):525–530.

Mazor, Gideon, Giora J. Kidron, Ahuva Vonshak, and Aharon Abeliovich. 1996. "The role of cyanobacterial exopolysaccharides in structuring desert microbial crusts." *FEMS Microbiology Ecology* 21 (2):121–130.

Megharaj, Mallavarapu, Balasubramanian Ramakrishnan, Kadiyala Venkateswarlu, Nambrattil Sethunathan, and Ravi Naidu. 2011. "Bioremediation approaches for organic pollutants: a critical perspective." *Environment International* 37 (8):1362–1375.

Merighi, Massimo, Vincent T. Lee, Mamoru Hyodo, Yoshihiro Hayakawa, and Stephen Lory. 2007. "The second messenger bis-(3′-5′)-cyclic-GMP and its PilZ domain-containing receptor Alg44 are required for alginate biosynthesis in *Pseudomonas aeruginosa*." *Molecular Microbiology* 65 (4):876–895.

Monds, Russell D., Peter D. Newell, Robert H. Gross, and George A. O'Toole. 2007. "Phosphate-dependent modulation of c-di-GMP levels regulates *Pseudomonas fluorescens* Pf0-1 biofilm formation by controlling secretion of the adhesin LapA." *Molecular Microbiology* 63 (3):656–679.

Monds, Russell D., and George A. O'Toole. 2009. "The developmental model of microbial biofilms: ten years of a paradigm up for review." *Trends in Microbiology* 17 (2):73–87.

Musk, Dinty J., David A. Banko, and Paul J. Hergenrother. 2005. "Iron salts perturb biofilm formation and disrupt existing biofilms of *Pseudomonas aeruginosa*." *Chemistry & Biology* 12 (7):789–796.

Naessens, Myriam, An Cerdobbel, Wim Soetaert, and Erick J. Vandamme. 2005. "Leuconostoc dextransucrase and dextran: production, properties, and applications." *Journal of Chemical Technology & Biotechnology: International Research in Process, Environmental & Clean Technology* 80 (8):845–860.

Nisbet, B.A., Ian W. Sutherland, Ian J. Bradshaw, M. Kerr, Edwin R. Morris and W.A. Shepperson. 1984. "XM-6: a new gel-forming bacterial polysaccharide." *Carbohydrate Polymers* 4 (5):377–394.

O'Connell, Paul F. 1992. "Sustainable agriculture—A valid alternative." *Outlook on Agriculture* 21 (1):5–12.

Ohashi, A and Harada Hideki. 1994. "Adhesion strength of biofilm developed in an attached-growth reactor." *Water Science and Technology* 29 (10–11):281.

Oliveira, Rosário, Melo Luis F, Oliveira A, Salgueiro R . 1994. *Polysaccharide Production and Biofilm Formation by Pseudomonas fluorescens: Effects of pH and Surface Material.* Elsevier Science, Amsterdam.

Olson, Eric R. 1993. "Influence of pH on bacterial gene expression." *Molecular Microbiology* 8 (1):5–14.

Orr, I. Gilan, Y. Hadar, and A. Sivan. 2004. "Colonization, biofilm formation, and biodegradation of polyethylene by a strain of *Rhodococcus ruber*." *Applied Microbiology and Biotechnology* 65 (1):97–104.

Park, Jin Hee, Dane Lamb, Periyasamy Paneerselvam, Girish Choppala, Nanthi Bolan, and Jae-Woo Chung. 2011. "Role of organic amendments on enhanced bioremediation of heavy metal (loid) contaminated soils." *Journal of Hazardous Materials* 185 (2–3):549–574.

Petrova, Olga E., and Karin Sauer. 2012. "Sticky situations: key components that control bacterial surface attachment." *Journal of Bacteriology* 194 (10):2413–2425.

Prasanna, Radha, Santosh Babu, Ngangom Bidyarani, Arun Kumar, Sodimalla Triveni, Dilip Monga, Arup Kumar Mukherjee, Sandhya Kranthi, Nandini Gokte-Narkhedkar, and Anurup Adak. 2015. "Prospecting cyanobacteria-fortified composts as plant growth promoting and biocontrol agents in cotton." *Experimental Agriculture* 51 (1):42–65.

Prasanna, Radha, Amrita Kanchan, Simranjit Kaur, Balasubramanian Ramakrishnan, Kunal Ranjan, Mam Chand Singh, Murtaza Hasan, Anil Kumar Saxena, and Yashbir Singh Shivay. 2016. "Chrysanthemum growth gains from beneficial microbial interactions and fertility improvements in soil under protected cultivation." *Horticultural Plant Journal* 2 (4):229–239.

Prasanna, Radha, Arun Kumar, Santosh Babu, Gautam Chawla, Vidhi Chaudhary, Surender Singh, Vishal Gupta, Lata Nain, and Anil Kumar Saxena. 2013. "Deciphering the biochemical spectrum of novel cyanobacterium-based biofilms for use as inoculants." *Biological Agriculture & Horticulture* 29 (3):145–158.

Prasanna, Radha, S. Pattnaik, T. C. K. Sugitha, L. Nain, and Anil Kumar Saxena. 2011. "Development of cyanobacterium-based biofilms and their in vitro evaluation for agriculturally useful traits." *Folia Microbiologica* 56 (1):49–58.

Prasanna, Radha, Sodimalla Triveni, Ngangom Bidyarani, Santosh Babu, Kuldeep Yadav, Anurup Adak, Sangeeta Khetarpal, Madan Pal, Yashbir Singh Shivay, and Anil Kumar Saxena. 2014. "Evaluating the efficacy of cyanobacterial formulations and biofilmed inoculants for leguminous crops." *Archives of Agronomy and Soil Science* 60 (3):349–366.

Prigent-Combaret, Claire, Eva Brombacher, Olivier Vidal, Arnaud Ambert, Philippe Lejeune, Paolo Landini, and Corinne Dorel. 2001. "Complex regulatory network controls initial adhesion and biofilm formation in *Escherichia coli* via regulation of thecsgD gene." *Journal of Bacteriology* 183 (24):7213–7223.

Prüß, Birgit M., Christopher Besemann, Anne Denton, and Alan J. Wolfe. 2006. "A complex transcription network controls the early stages of biofilm development by *Escherichia coli*." *Journal of Bacteriology* 188 (11):3731–3739.

Quintelas, Cristina, Zélia Rocha, Bruna Silva, Bruna Fonseca, Hugo Figueiredo, and Teresa Tavares. 2009. "Biosorptive performance of an *Escherichia coli* biofilm supported on zeolite NaY for the removal of Cr (VI), Cd (II), Fe (III), and Ni (II)." *Chemical Engineering Journal* 152 (1):110–115.

Radwan, Samir and Al-Hasan Redha . "Potential application of coastal biofilm-coated gravel particles for treating oily waste." *Aquatic Microbial Ecology* 23 (2):113–117.

Ramey, Bronwyn E., Maria Koutsoudis, Susanne B. von Bodman, and Clay Fuqua. 2004. "Biofilm formation in plant–microbe associations." *Current Opinion in Microbiology* 7 (6):602–609.

Rekadwad, Bhagwan N., and Chandrahasya N. Khobragade. 2017. "Microbial biofilm: role in crop productivity." In Kalia, V. (ed) *Microbial Applications*, Vol. 2, pp. 107–118. Springer, Cham.

Rickard, Alexander H., Peter Gilbert, Nicola J. High, Paul E. Kolenbrander, and Pauline S. Handley. 2003. "Bacterial coaggregation: an integral process in the development of multi-species biofilms." *Trends in Microbiology* 11 (2):94–100.

Roberson, Emily B., and Mary K. Firestone. 1992. "Relationship between desiccation and exopolysaccharide production in a soil *Pseudomonas* sp." *Applied and Environmental Microbiology* 58 (4):1284–1291.

Roeselers, Guus, Mark C. M. Van Loosdrecht, and Gerard Muyzer. 2008. "Phototrophic biofilms and their potential applications." *Journal of Applied Phycology* 20 (3):227–235.

Rollet, Cécile, Laurent Gal, and Jean Guzzo. 2009. "Biofilm-detached cells, a transition from a sessile to a planktonic phenotype: a comparative study of adhesion and physiological characteristics in *Pseudomonas aeruginosa*." *FEMS Microbiology Letters* 290 (2):135–142.

Römling, Ute, Mark Gomelsky, and Michael Y. Galperin. 2005. "C-di-GMP: the dawning of a novel bacterial signalling system." *Molecular Microbiology* 57 (3):629–639.

Selvakumar, Govindan, Periyasamy Panneerselvam, and Arakalagud Nanjundaiah Ganeshamurthy. 2012. "Bacterial mediated alleviation of abiotic stress in crops." In Maheshwari, D. K. (ed) *Bacteria in Agrobiology: Stress Management*, pp. 205–224. Springer, Berlin.

Seneviratne, Gamini and IK Indrasena. 2006. "Nitrogen fixation in lichens is important for improved rock weathering." *Journal of Biosciences* 31 (5):639–643.
Seneviratne, Gamini, APDA Jayasekara, MSDL De Silva, UP Abeysekera . 2011. "Developed microbial biofilms can restore deteriorated conventional agricultural soils." *Soil Biology and Biochemistry* 43 (5):1059–1062.
Seneviratne, Gamini and Jayasinghe arahchi Himali . 2005. "A rhizobial biofilm with nitrogenase activity alters nutrient availability in a soil." *Soil Biology and Biochemistry* 37 (10):1975–1978.
Seneviratne, Gamini, RMMS Thilakaratne, APDA Jayasekara, KACN Seneviratne, KRE Padmathilake and MSDL De Silva. 2009. "Developing beneficial microbial biofilms on roots of non legumes: a novel biofertilizing technique." In Khan, M., Zaidi, A., and Musarrat, J. (eds) *Microbial Strategies for Crop Improvement*, pp. 51–62. Springer. Berlin.
Seneviratne, Gamini, MLMAW Weerasekara, KACN Seneviratne, JS Zavahir, ML Kecskés and IR Kennedy. 2010. "Importance of biofilm formation in plant growth promoting rhizobacterial action." In Maheshwari, D. K. (ed) *Plant Growth and Health Promoting Bacteria*, pp. 81–95. Springer, Berlin.
Shen, Zhen-Guo, Xiang-Dong Li, Chun-Chun Wang, Huai-Man Chen, and Hong Chua. 2002. "Lead phytoextraction from contaminated soil with high-biomass plant species." *Journal of Environmental Quality* 31 (6):1893–1900.
Simões, Lúcia Chaves, Manuel Simoes, and Maria Joao Vieira. 2008. "Intergeneric coaggregation among drinking water bacteria: evidence of a role for *Acinetobacter calcoaceticus* as a bridging bacterium." *Applied and Environmental Microbiology* 74 (4):1259–1263.
Singh, Jay Shankar, Vimal Chandra Pandey, and Davinder P. Singh. 2011. "Efficient soil microorganisms: a new dimension for sustainable agriculture and environmental development." *Agriculture, Ecosystems & Environment* 140 (3–4):339–353.
Sivan Alex, M. Szanto, V. Pavlov. 2006. "Biofilm development of the polyethylene-degrading bacterium *Rhodococcus ruber*." *Applied Microbiology and Biotechnology* 72 (2):346–352.
Sleytr, Uwe B. 1997. "I. Basic and applied S-layer research: an overview." *FEMS Microbiology Reviews* 20 (1–2):5–12.
Starkey, Melissa, Matthew R. Parsek, Kimberly A. Gray, and Sung Il Chang. 2004. "A sticky business: the extracellular polymeric substance matrix of bacterial biofilms." In Ghannoum, M., and O'Toole, G. A. (eds) *Microbial Biofilms*, pp. 174–191. American Society of Microbiology Press, Washington, DC.
Stepanović, Srdjan, Ćirković Ivana, Mijać Vera and Švabić-Vlahović Milena . 2003. "Influence of the incubation temperature, atmosphere and dynamic conditions on biofilm formation by *Salmonella* spp." *Food Microbiology* 20 (3):339–343.
Stewart, Philip S., and Michael J. Franklin. 2008. "Physiological heterogeneity in biofilms." *Nature Reviews Microbiology* 6 (3):199.
Stoodley, Paul, Ruth Cargo, Cory J. Rupp, Suzanne Wilson, and Isaac Klapper. 2002. "Biofilm material properties as related to shear-induced deformation and detachment phenomena." *Journal of Industrial Microbiology and Biotechnology* 29 (6):361–367.
Sundar, K, Sadiq Mohammed, Mukherjee Amitava and Chandrasekaran Natarajan . 2011. "Bioremoval of trivalent chromium using bacillus biofilms through continuous flow reactor." *Journal of Hazardous Materials* 196:44–51.
Sutherland, Ian W. 2001. "Microbial polysaccharides from gram-negative bacteria." *International Dairy Journal* 11 (9):663–674.
Swarnalakshmi, Karivaradharajan, Radha Prasanna, Arun Kumar, Sasmita Pattnaik, Kalyana Chakravarty, Yashbir Singh Shivay, Rajendra Singh, and Anil Kumar Saxena.

2013. "Evaluating the influence of novel cyanobacterial biofilmed biofertilizers on soil fertility and plant nutrition in wheat." *European Journal of Soil Biology* 55:107–116.

Tan, Zhu-Xia, Rattan Lal, and Keith D. Wiebe. 2005. "Global soil nutrient depletion and yield reduction." *Journal of Sustainable Agriculture* 26 (1):123–146.

Thormann, Kai M., Renée M. Saville, Soni Shukla, and Alfred M. Spormann. 2005. "Induction of rapid detachment in *Shewanella oneidensis* MR-1 biofilms." *Journal of Bacteriology* 187 (3):1014–1021.

Toner, Brandy, Alain Manceau, Matthew A. Marcus, Dylan B. Millet, and Garrison Sposito. 2005. "Zinc sorption by a bacterial biofilm." *Environmental Science & Technology* 39 (21):8288–8294.

Triharyanto, Eddy. 2018. "The application of biofilm biofertilizer-based organic fertilizer to increase available soil nutrients and spinach yield on dry land (a study case in Lithosol soil type)." *IOP Conference Series: Earth and Environmental Science*, Solo City, Indonesia.

Triveni, Sodimalla, Radha Prasanna, Arun Kumar, Ngangom Bidyarani, Rajendra Singh, and Anil Kumar Saxena. 2015. "Evaluating the promise of Trichoderma and Anabaena based biofilms as multifunctional agents in *Macrophomina phaseolina*-infected cotton crop." *Biocontrol Science and Technology* 25 (6):656–670.

Triveni, Sodimalla, Radha Prasanna, Livleen Shukla, and Anil Kumar Saxena. 2013. "Evaluating the biochemical traits of novel Trichoderma-based biofilms for use as plant growth-promoting inoculants." *Annals of Microbiology* 63 (3):1147–1156.

Turhan, Emel Ünal, Zerrin Erginkaya, Mihriban Korukluoğlu, and Gözde Konuray. 2019. "Beneficial biofilm applications in food and agricultural industry." In Malik, A., Erginkaya, Z., and Erten, H. (eds) *Health and Safety Aspects of Food Processing Technologies*, pp. 445–469. Springer, Cham.

Uchida, Ryusei. 2000. "Essential nutrients for plant growth: nutrient functions and deficiency symptoms." In Silva, J. A. and Uchida, R., (eds) *Plant Nutrient Management in Hawaii's Soils, Approaches for Tropical and Subtropical Agriculture*, pp. 31–55. College of Tropical Agriculture and Human Resources, University of Hawaii, Manoa.

Valle, Jaione, Sandra Da Re, Solveig Schmid, David Skurnik, Richard d'Ari, and Jean-Marc Ghigo. 2008. "The amino acid valine is secreted in continuous-flow bacterial biofilms." *Journal of Bacteriology* 190 (1):264–274.

Velmourougane, Kulandaivelu, Radha Prasanna, and Anil Kumar Saxena. 2017. "Agriculturally important microbial biofilms: present status and future prospects." *Journal of Basic Microbiology* 57 (7):548–573.

Vercraene-Eairmal, Marion, Béatrice Lauga, Stéphanie Saint Laurent, Nicolas Mazzella, Sébastien Boutry, Maryse Simon, Solange Karama, François Delmas, and Robert Duran. 2010. "Diuron biotransformation and its effects on biofilm bacterial community structure." *Chemosphere* 81 (7):837–843.

Villain-Simonnet, Agnès, Michel Milas, and Marguerite Rinaudo. 2000. "A new bacterial exopolysaccharide (YAS34). II. Influence of thermal treatments on the conformation and structure. Relation with gelation ability." *International Journal of Biological Macromolecules* 27 (1):77–87.

Vlamakis, Hera, Yunrong Chai, Pascale Beauregard, Richard Losick, and Roberto Kolter. 2013. "Sticking together: building a biofilm the *Bacillus subtilis* way." *Nature Reviews Microbiology* 11 (3):157.

Wang, Wei. 2009. "Isolation of the silicate bacteria strain and determination of the activity of releasing silicon and potassium." *Journal of Anhui Agricultural Sciences* 37 (17):7889–7891.

Weber, Harald, Christina Pesavento, Alexandra Possling, Gilbert Tischendorf, and Regine Hengge. 2006. "Cyclic-di-GMP-mediated signalling within the σS network of *Escherichia coli*." *Molecular Microbiology* 62 (4):1014–1034.

Weinhouse, Haim, Shai Sapir, Dorit Amikam, Yehudit Shilo, Gail Volman, Patricia Ohana, and Moshe Benziman. 1997. "c-di-GMP-binding protein, a new factor regulating cellulose synthesis in *Acetobacter xylinum*." *FEBS Letters* 416 (2):207–211.

Whiting, D., M. Roll and L. Vickerman. 2003. "Plant physiology. Photosynthesis, respiration and transpiration." Gardening series. Colorado master gardener; no. 7.710.

Wolfe, Alan J., and Karen L. Visick. 2008. "Get the message out: cyclic-di-GMP regulates multiple levels of flagellum-based motility." *Journal of Bacteriology* 190 (2):463–475.

Yang, Liang, Yang Liu, Hong Wu, Niels Høiby, Søren Molin, and Zhi-jun Song. 2011. "Current understanding of multi-species biofilms." *International Journal of Oral Science* 3 (2):74.

Yin, Wen, Yiting Wang, Lu Liu, and Jin He. 2019. "Biofilms: the microbial 'protective clothing' in extreme environments." *International Journal of Molecular Sciences* 20 (14):3423.

Zhang, Wenbo, Agnese Seminara, Melanie Suaris, Michael P. Brenner, David A. Weitz, and Thomas E. Angelini. 2014. "Nutrient depletion in *Bacillus subtilis* biofilms triggers matrix production." *New Journal of Physics* 16 (1):015028.

4 Biofilms in Fermentation for the Production of Value-Added Products

Ehsan Mahdinia
Albany College of Pharmacy and Health Sciences

Ali Demirci
Pennsylvania State University

CONTENTS

4.1 INTRODUCTION

Microbial cells of various types of fungi, bacteria, or even archaea growing in suspended-cell fermentations usually have the capability to migrate onto a suitable surface and start colonizing it by secreting extracellular biopolymers and creating a microbial community (Burmølle et al., 2006). This process of immobilization of cells on the support surface is not limited to specific conditions, but usually occurs due to certain physiochemical and environmental factors and requires certain and profound alterations of the planktonic cells on both genotype and phenotype levels (Kuchma and O'Toole, 2000). With these alterations the biofilm communities gain synergic attributes and with higher cell densities at hand are given extraordinary metabolic reactions and resilience (Branda et al., 2005). When these biofilm form where they are not supposed to, such as operation rooms or kitchens and food packaging rooms, they become real headaches with such survival skills (Xu et al., 2011). However, these enhanced attributes can be put to good use when microorganisms are expected to thrive and metabolize. Therefore, a biofilm reactor is simply a bioreactor where biofilm formations are allowed to form on specified suitable support surfaces in controlled environments to harness these physiochemical and metabolic superiorities for the purpose of producing value-added products or bioremediation (Qureshi et al., 2005). Specifically, a brief description of the biofilm formation, the support materials for hosting the biofilm matrices, the advantages and disadvantages of biofilm reactors, and major and most recent applications for the value-added products in fermentation technologies will be reviewed.

4.2 BIOFILM

Biofilm is the essential part of the biofilm reactor. Here biofilm formation conditions and variables in biofilm reactors along with biofilm supporting types and materials are discussed and also the pros and cons for general biofilm reactor applications are reviewed.

4.2.1 Biofilm Formation

Microorganisms in liquid state fermentation often exist in planktonic form, i.e., the cell-to-cell interactions are minimal and there are no extracellular matrices to bind them together. However, many species of microorganisms are capable of migrating into biofilm form. They colonize a suitable surface (biotic or abiotic) and form channelized extracellular matrices that we know as biofilms (Kuchma and O'Toole, 2000) (Figure 4.1).

The structure, chemistry, and physiology of biofilm formations originate from the nature of the migrating microorganisms and the properties of the surface environment hosting the biofilm (Branda et al., 2005). Planktonic cells often need incentives for the migration into biofilm. This incentive can be certain physiochemical and environmental factors, such as temperature or pH shocks, starvation or substrate inhibition, or even the presence of a hostile external factor such as an antimicrobial agent or an invasive species (Landini et al., 2010). As a reaction, planktonic cells initiate

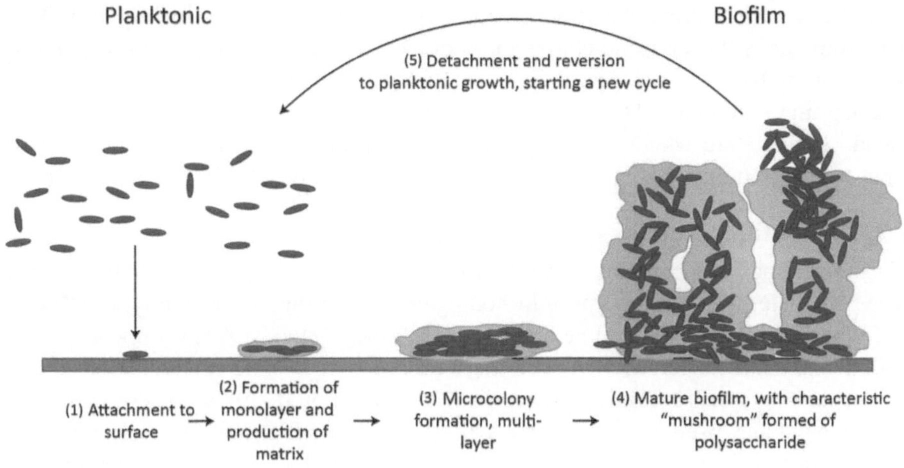

FIGURE 4.1 Schematic representation of a mushroom-type biofilm formation (Courtesy of British Society for Immunology, Hollmann, B., Perkins, M., and Walsh, D., 2020, August 30. Biofilms and their role in pathogenesis. *British Society for Immunology*. Retrieved from https://www.immunology.org/public-information/bitesized-immunology/pathogens-and-disease/biofilms-and-their-role-in).

cell-to-surface and cell-to-cell contacts resulting in the formation of microcolonies. The microcolonies signal differentiations on cellular levels to form pillar-like structures inside extracellular polysaccharides (EPS) which act as the backbone for the biofilm structures (You et al., 2015). These matrices hold fluid-filled channels inside them that enable the biofilm formation to actively interact with the bulk of the liquid around them. Planktonic cells need to undergo certain profound genetic and morphologic alterations to facilitate biofilm formation (Kuchma and O'Toole 2000). The biofilm formations are then able to generate planktonic cells and diffuse them back into the liquid bulk, rendering the migration process reversible overall.

As Figure 4.1 depicts, there are generally four steps in the migration of a planktonic cell to a mature biofilm formation. In the first step, the planktonic cell reaches the surface through diffusion, convection, or self-motility. Then, reversible attachment of the planktonic cells onto the surface and adhering to it takes place (Flemming and Wingender, 2010). As the population of the attached cells increase, the cells bridge between each other and create a monolayer with an extracellular matrix that binds them together and irreversibly immobilizing the cells. More layers of the immobilized cells stack on top of each other forming microcolonies over time. The forces that hold these layers together and to the surface include covalent, ionic, dipole-dipole, and hydrogen bonds along with hydrophobic interactions. As microcolonies increase in number and grow larger, the biofilm matures. The mature biofilm is then able to create offspring cells and migrate them back into planktonic form. Thus, although the immobilization process is irreversible, the overall migration is not (Vasudevan, 2014). Besides, planktonic cells can congregate in the liquid bulk and then join the mature biofilm formation (Demirci et al., 2007). At the same time, fluid shear stress due to movement of the liquid bulk can cause erosion that detaches the immobilized cells from the mature biofilm. These detached cells can reintegrate into the liquid bulk as planktonic cells. When rapid change or depletion of nutrients (oxygen or substrates) occurs in the biofilm, larger parts of the mature biofilm can detach, which is known as sloughing (Ercan and Demirci, 2013). Mature biofilm formations usually reach a pseudo-steady state where the biofilm growth and matrix expansion match detachment from it and thus the biofilm thickness remains steady (Demirci et al., 2007).

4.2.1.1 Biofilm Structures

The morphology of the mature biofilm formation has been categorized into three main types: the heterogeneous, the pseudo-homogenous, and the mushroom types (Keevil and Walker, 1992; Nyvad and Fejerskov, 1997; Stewart et al., 1995; Wimpenny et al., 2000). The major difference between these three formats is the cell density of them. Heterogeneous types are only exposed to the flowing liquid bulk and therefore dense and planar. The pseudo-homogeneous type contains EPS matrices separated by water channels. The mushroom type is made of a dome-like structure surrounded by water channels (Figure 4.1). All these formations are generally by volume comprised of 15% immobilized cells and 85% of the extracellular matrices (Agle, 2007).

4.2.1.2 Passive Immobilization

Passive immobilization techniques are based on cells' innate ability to attach to the surfaces. This process usually occurs in two different mechanisms: natural adsorption of planktonic cells onto a surface (such as glass or metal surfaces in the bioreactor) or migration of microbial films or flocculants around and within a solid support material and colonizing it (Fukuda, 1995). The natural adsorption of cells onto surfaces is driven by the electrostatic interactions between them while colonization is based on the porosity of the biomass support particles (BSP), where a variety of forces play roles in immobilizing the grouped cells (Atkinson et al., 1979). Colonization allows for more robust immobilization and thus porosity and specific surface of the support materials play a key role in development of biofilm in the bioreactors. Usually, the biofilm is developed through colonization in the initial batch period(s) via microbial formation EPS (Demirci et al., 2007).

4.2.1.3 Active Immobilization

Not all microbial strains tend to form biofilm formation naturally, and therefore passive immobilization is not an option for them. In those cases, bioprocess engineers employ entrapment in polymer matrices or coupling agents to immobilize cells (Demirci et al., 2007). For cell entrapment, an inert porous polymer matrix such as alginate, agar, polyacrylamide, gelatin, collagen, or chitosan is formed, as the planktonic mature cells are introduced into the polymer medium. In this form, polymeric sturdy matrices in the form of beads, cubes, or any other desired formats are formed containing the live cells. The key factor is that the matrix is porous enough to let substrates (nutrients and oxygen) in and products out and yet not too porous to remove the cells free (Shuler et al., 2017). This crucial factor then determines the polymerization material, concentration, physiochemical conditions of the polymerization reaction, and even reaction kinetics. In coupling techniques, the surface of immobilization is activated by reactive groups or coupling agents. The live cells then covalently and therefore permanently bond with the surface via these active sites. Although active immobilization can be applied to almost any microorganisms, it has two major disadvantages. The first one includes coupling and cross-link agent toxicity to cell viability, activity and/or reproduction, matrix instability, cell leakage, poor mass and heat transfer through the pores, low operational stability and control, and cost. The second and even more critical limitation is that the entrapped or cross-linked cells do not adapt to immobilization or "migrate," and therefore the profound genetic and morphological alterations that are usually beneficial to boost value-added products and occur with passive immobilization can be absent (Characklis and Marshal, 1990).

4.2.2 Biofilm Support Materials

In theory, any type of surface available to planktonic cells can serve as a biofilm support. In biofilm reactors, even smooth metallic or glass surfaces may host biofilm formations. However, porous and nutrient surfaces can host them more favorably, and the biofilm formation physiochemistry and thus productivity are certainly boosted (Demirci et al., 2007). In general, the biofilm support materials should be ideally nutrient rich, porous, and at the same time chemically nontoxic to the migration process and production stage. Also, they must be inexpensive, available on industrial scale, and mechanically durable since the formation is usually exposed to high shear stresses due to active agitation and/or aeration and particle collision, and finally compatible to the fermentation process (Melo and Oliveira, 2001). Many bio and synthetic materials have been used for creating biofilm supports from sawdust, wood chips, wool, and cotton fabrics, corn stalks, coal to steel mesh, sand, glass, polypropylene and polyesters, ceramics, and composites; especially when it comes to agriculture-based waste materials to be used, sky is the limit and literally any available cellulosic sources like banana leaves, coconut coir, groundnut shell, pea shells, etc. have been used (Kumar et al., 2015; Demirci et al., 2007). Below, we discuss the main three biological and physiochemical requirements that the support material should meet.

4.2.2.1 Nutritional Factors

In general, biofilm support can include or exclude agriculture-based materials such as flours, extracts, and food residues. For instance, synthetic polymeric materials have been widely used for hosting biofilm formation, and studies have indicated that optimizing physiochemical properties and porosity of these support can enhance cell adhesion and population in the biofilm (Voběrková et al., 2016; Martinov et al., 2010). However, the infusion of nutrients in the support has become more popular as biofilm reactor technology has gained growing attention by fermentation technologies. The primary reason is that the presence of nutrients in the support enhances cell growth that enables higher cell densities in the biofilm matrices, which is almost always a crucial factor in increasing biofilm reactor efficiencies (Ercan and Demirci, 2013). The second reason is that studies have indicated that the presence of nutrients help the migration and adhesion processes by luring the planktonic cells to selectively migrate on the support surface (Burmølle et al., 2006). This is always beneficial to constructing more efficient biofilm reactors.

4.2.2.2 Surface Properties

The most critical property that the surface of support material should have is high porosity, which translates into higher specific surface area and thus enables higher cell density of the biofilm matrix. When the biofilm matrices are formed onto smooth nonporous surfaces such as glass, metal, or plastic surfaces that are available in any bioreactor type vessels, the resulting biofilm formation has two major flaws. After several layers of subcolonies are formed onto the surface; despite the fact that microfluidic channels naturally form in the biofilm matrices, the biofilm thickness is not optimum anymore for robust mass and heat transfer to occur for optimum fermentation conditions (Kuchma and O'Toole, 2000). Moreover, the overall adhesion of the biofilm matrix onto the support surface due to limited specific surface area is generally weak, and as a result, the biofilm formation is susceptible to high liquid shear rates coming from agitation and aeration (Cheng et al., 2010a). On the other hand, pore sizes on the support surfaces must be optimum to enable colonization of the interior surfaces as well as the surface itself, without adding an extra barrier to the transport phenomena occurring between biofilm matrix and the broth liquid bulk, which is always a limiting challenge for biofilm reactors (Demirci et al., 2007). Thus, it is essential to prepare the biofilm support surfaces with optimum surface porosity and smoothness/roughness so that biofilm formation steps, i.e., adhesion, colonization, and maturation as well as fermentation process controlled by robust mass and heat transfer between biofilm matrix and liquid bulk, take place efficiently.

4.2.2.3 Support Chemistry

The support surface chemistry must not be toxic or inhibitory to either the biofilm formation steps or the main fermentation process. Many organic or synthetic materials meet this critical characteristic; however, there are numerous other features of the surface chemistry that can affect the biofilm matrix. For example, hydrophobicity or hydrophilicity of the porous surface can deeply affect the cell adhesion step, where the process is controlled by the van der Waals forces of attraction or repulsion between the support surface and the biofilm matrix (Cheng et al., 2010a). In some cases, researchers have observed that high hydrophobicity of the support surface favors better adhesion of cells and thus colonization (Pereira et al., 2000; Teixeira and Oliveira 1999; Sousa et al., 1997), while some studies found less hydrophobic surfaces with higher nutrient contents are more suitable for fermentation of certain metabolites (Ho et al., 1997a).

Besides the hydrophobic/hydrophilic interactions, many other chemical aspects of the support surface are crucial, such as surface charge (since microorganisms are usually densely charged on the cells surfaces), exposed organic groups, polymeric texture (i.e., how the monomeric groups are aligned to form the polymers used in the support matrices), and even density of the surface material can play a role in the interaction of the biofilm matrix and the support surface (Cheng et al., 2010a). When nutrients are applied in the support matrices, yet another parameter must be taken into account, that is leaching of the nutrients into the broth. This leaching can be unfavorable in two ways. If the leaching occurs slowly, as the nutrients leach from the support matrix, after several batches of fermentation, nutrients can deplete from the surface of support, and thus the benefits of the nutrients disappear. On the other hand, if the leaching happens rapidly, not only the benefits go away quickly, but the extra and foreign nutrients entering the liquid broth may interfere with the fermentation process optimality (Ho et al., 1997b; Demirci et al., 2007). Therefore, the support materials that employ nutrients must always be optimized with their formulation and texture to be able to withstand leaching of the nutrients and at the same time expose enough of the nutrients to benefit from them.

4.2.2.4 Plastic Composite Supports (PCS)

As mentioned above, it is usually beneficial for bioprocess engineers to employ nutrients in the support matrices for more robust biofilm formations. The challenge is not only to optimize the composition for the process but the fact that these biological materials that carry the nutrients such as flours, extracts, biopolymers, and proteins, usually do not have the necessary sturdiness and stability to endure biofilm reactor designs. On the other hand, the synthetic materials such as steel, ceramics, or plastic polymers have excellent durability and compatibility for biofilm support materials (Demirci et al., 1993). Therefore, an innovative approach has been devised to mix the two types of support materials and compose composite materials. Thus, these composite materials are surface activated; i.e., they have high porosity, specific surface area and roughness, carry the nutrients essential for better migration and cell adhesion, and also inherit excellent durability and stability from the nonbiological ingredients (Ercan and Demirci, 2013).

One of the most popular composites that have been formulated for this purpose are Plastic Composite Supports (PCS) created by researchers at Iowa State University (U.S. patent number: 5595893). In PCS production, polypropylene (usually 50% by weight) is mixed with agriculture-based biomaterials such as soybean hulls, soybean flours, yeast extract, bovine albumin extract, and red blood cell extracts that carry the nutrients for the biofilm support. Table 4.1 shows some of the most commonly used compositions of the PCS.

TABLE 4.1
Most Commonly Used PCS Compositions

PCS Type	Composition
SF	Polypropylene (50%), soybean hulls (45%), soybean flour (5%), salt
SFY	Polypropylene (50%), soybean hulls (40%), soybean flour (5%), salt, yeast extract (5%)
SFYB	Polypropylene (50%), soybean hulls (35%), soybean flour (5%), salt, yeast extract (5%), bovine albumin (5%)
SFYR	Polypropylene (50%), soybean hulls (35%), soybean flour (5%), salt, yeast extract (5%), bovine red blood cell (5%)

Permission obtained as: Mahdinia, E., Demirci, A., Berenjian, A. 2017. Strain and plastic composite support (PCS) selection for vitamin K (Menaquinone-7) production in biofilm reactors. *Bioprocess and Biosystems Engineering*, *40*(10): 1507–1517.

Source: Mahdnia et al., 2017a; Izmirlioglu and Demirci, 2016; Ercan and Demirci 2013; Cheng et al., 2010a.

Then, the mixture is extruded through a rotary solid-state extruder at a low rotation speed and high temperatures (barrel temperature of 200°C and a die temperature of 167°C). The results are PCS tubes with desired inner and outer diameters and lengths with good porosity and surface activity that can be fixed in biofilm reactors to host the biofilm matrices (Pometto et al. 1997). The polypropylene in the PCS serves as the solid matrix of it that grants it high durability. The biological ingredients provide the nutrient content of the support. Since their development, PCS supports have been utilized in numerous applications of biofilm reactors for a variety of value-added products. Table 4.2 summarizes these applications up to date, where a dozen of the PCS tubes have been attached onto the bioreactor shafts in a grid-like fashion.

The polypropylene content of the PCS provides excellent durability and stability for long-term utilization of the constructed biofilm reactors in batch, fed-batch, and continuous regimes (Mahdinia et al., 2019c; Cheng et al., 2010a; Ercan and Demirci,

TABLE 4.2
Biofilm Reactor Applications Utilizing PCS Technology for Producing Value-Added Products

PCS Type[a]	Value-Added Product	Microorganism	References
SFYB	Lactic acid	*Lactobacillus casei*	Ho et al. (1997a,b,c) and Velázquez et al. (2001)
SFYB	Ethanol	*Saccharomyces cerevisiae*	Demirci et al. (1997), and Germec et al. (2015, 2016, 2018, 2019)
SFYBR	Succinic acid	*Actinobacillus succinogenes*	Urbance et al. (2003)
SFYB	Nisin	*Lactococcus lactis*	Bober and Demirci (2004), and Pongtharangkul and Demirci (2006a,b,c, 2007, 2008)
SFYR	Pullulan	*Aureobasidium pullulans*	Cheng et al. (2010b, 2011b,d)
SFYR	Bacterial cellulose	*Acetobacter xylinum*	Cheng et al. (2009a,b, 2011c)
SFYB	Lysozyme	*Kluyveromyces lactis*	Ercan and Demirci (2013, 2014, 2015b,c)
SFY	Ethanol	*Saccharomyces cerevisiae*	Izmirlioglu and Demirci (2016)
SFYBR	Kojic acid	*Aspergillus oryzae*	Liu et al. (2016)
SFY	Ethanol	*Aspergillus niger and Saccharomyces cerevisiae*	Izmirlioglu and Demirci (2017)
SFYB	Menaquinone-7 (Vitamin K2)	*Bacillus subtilis natto*	Mahdinia et al. (2017a, 2018a,b,c,d, 2019a,b)

[a] SF: Soybean hulls and soybean flour.
SFY: Soybean hulls, soybean flour and yeast extract.
SFYB: Soybean hulls, soybean flour, yeast extract and bovine albumin.
SFYR: Soybean hulls, soybean flour, yeast extract and bovine red blood cells.
SFYBR: Soybean hulls, soybean flour, yeast extract, bovine albumin and bovine red blood cells.

2013). Also, the slow and controlled release of the nutrients from the PCS matrix provides ideal benefits for enhanced cell migration and population, and at the same time lowering the nitrogen requirement of the medium (Ho et al., 1997a; Cotton et al., 2001). Moreover, the PCS compositions are flexible and can be easily optimized to meet specific requirements of the fermentation process (Demirci et al., 2007).

4.2.3 Biofilm Reactor Types

As mentioned above, the basic incentive behind biofilm reactors is the fact that when microorganisms migrate into biofilm formation, one of the extraordinary characteristics that they find is enhanced metabolism that enables them to metabolize nutrients and biosynthesize metabolites more rapidly and efficiently (Characklis and Cooksey, 1983; Characklis and Marshal, 1990). Traditionally and dominantly biofilm reactors are constructed to boost fermentation conditions in liquid state fermentation regimes where the biofilm formations are beneficial to the fermentation process and are tuned to withstand the liquid state agitation and in aerobic conditions aeration stresses (Berenjian et al., 2015; Mahdinia et al., 2017b). However, in fewer cases the biofilm formations are adequate in solid state fermentation and are employed, despite the fact that solid state fermenters are more difficult to scale up (Junker, 2004; Mahdinia et al., 2019d). Nonetheless, biofilm reactors are designated to fermentation regimes where biofilm formations are allowed to form in controlled environments in order to harness the benefits of the bioprocess enhancements. Thus, different biofilm reactor types fundamentally differ in the geometry of the biofilm formation on the support surfaces versus the geometry and operational conditions of the bioreactor vessel itself. Below are the most conventional geometries used in biofilm reactors with their properties and common applications along with some innovative and unconventional novel types.

4.2.3.1 Biofilm Reactor Geometry

Of the more conventional types of biofilm reactors that have been more commonly used in fermentation technologies, continuously stirred tank biofilm reactors, trickling filter tanks, rotating disc biofilm reactors, fluidized-bed biofilm reactors, airlift biofilm reactors, membrane biofilm reactors (MBfR), and PCS biofilm reactors can be mentioned (Figure 4.2).

In batch continuously stirred biofilm reactors, the commonly used continuously stirred tank reactor (CSTR) is adapted to contain support particles with biofilm matrices on them. Obviously, they are the simplest biofilm reactor type since CSTRs have been used by chemical engineers for many decades and are very well characterized. However, to form the biofilm matrices and maintain them on the support with active agitation occurring, engineers must find optimum conditions (support size and type, agitation and aeration rates) since particle collision and physical erosion of the biofilm matrices and the support particles are critical concerns (Muffler and Ulber, 2014). After one batch is complete, the broth is harvested, and without any adjustments, fresh sterile medium replaces the broth and the next batch starts. Thus, the immobilized cells on the support particles act as an inoculum and the cells

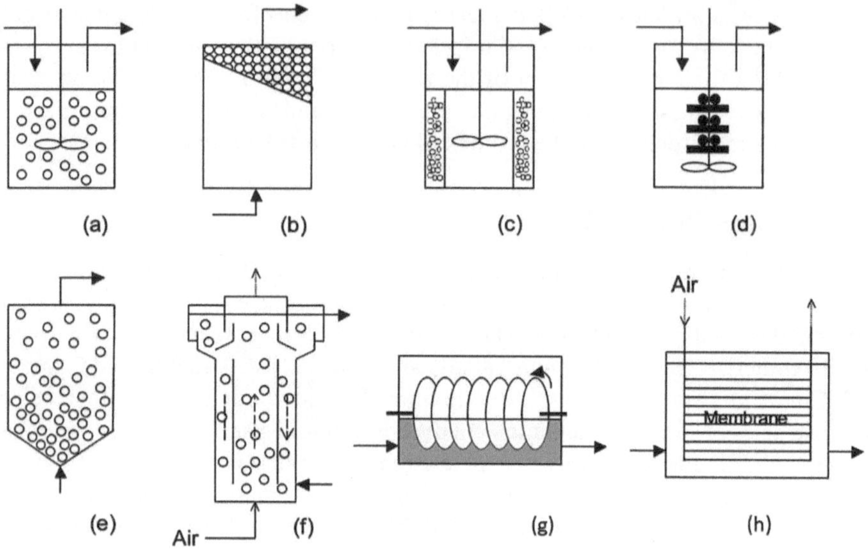

FIGURE 4.2 Various most commonly used geometries of biofilm reactors: (a) continuously stirred tank, (b) trickling filter tank, (c) PCS-rings, (d) PCS-grid, (e) fluidized bed, (f) airlift, (g) rotating disc, and (h) MBfRs.

that migrate back into planktonic form from the biofilm matrix populate the fresh medium (Demirci et al., 2007). This is quite beneficial in reducing the lag phase of the fermentation; however, if the batches extend to late stationary phases of fermentation, it can cause growth inhibition and this benefit is lost. Also, while the spent broth is being harvested and before the fresh medium is filled in, the biofilm formations are exposed in the tanks to nothing but air, and in large biofilm reactors this period of time can be significantly long. Thus, there is always a risk of starvation and death for the exposed cells that may damage the biofilm integrity (Qureshi et al., 2005). Therefore, continuous fermentation regimes for such biofilm reactors can be a good option.

Trickling filter is perhaps the most widely used biofilm reactor since they have dominated the old-fashioned aerobic tanks in wastewater treatment facilities by cutting massive aeration costs (Kornaros and Lyberatos, 2006) (Figure 4.2b). Other than that, various types have been adjusted to various needs in fermentation technologies.

The PCS biofilm reactors on the other hand are a more novel type of biofilm reactors as compared to the CST biofilm reactors or trickling filter, and unlike the other types (Figure 4.2a and b), they have been mostly used for producing value-added products rather than bioremediation and treatment purposes. The PCS that hosts the biofilm formation can be assembled in the vessel in various fashions; yet, two most common geometries are PCS ring biofilm reactors where the PCS rings are confined by a porous mesh boundary on the outer side of the vessel and PCS-grid biofilm reactors where the PCS tubes are installed onto the agitation shaft and rotate with it while hosting the biofilm (Demirci et al., 2007; Ho et al., 1997a) (Figure 4.2c and d).

With PCS ring design, the overgrowth of the biofilm matrices on the rings and fouling is a major issue as liquid shear rates are significantly lower at the outer layers to prevent biofilm overgrowth. In PCS-grid design, however, as long as agitation rates are optimum so that the biofilm matrices remain intact, the biofilm formation thickness and thus robustness can be kept optimum. As a result, the PCS-grid design has been more popular than the ring design (Ercan and Demirci, 2013; Cheng et al., 2010; Mahdinia et al., 2019c).

The moving-bed biofilm reactors (MBBR) that include designs such as fluidized-bed biofilm reactors and airlift biofilm reactors are essentially small and light support particles that host biofilm matrices which are lifted in the liquid bulk of the broth in the bioreactor vessel by the turbulence caused not by active agitation but by liquid recycling stream or air dispensed from the bottom of the bioreactor. Since the support particles are rather small and fluidized during the fermentation process, it is essentially a challenge to monitor and make sure of the microbial adhesion and community morphology of the biofilm matrices on them (Tang et al., 2016) (Figure 4.2e and f). Also, since high turbulence must be maintained in these vessels to ensure good mass and heat transfer coefficients, these designs are usually energy and therefore cost intensive (Muffler and Ulber, 2014). However, these designs are still quite popular in biotechnology for bioremediation purposes where synergic microbial communities comprised of many strains can be naturally formed on the support particles to efficiently remove tough pollutants. The trick is to gain superior compositions and maintain optimality throughout the fermentation process through novel advancing approaches such as enhancing the particle support surfaces with nanoparticle coatings or photon activations (Wang et al., 2018a; Zhang et al., 2016).

The rotating disc biofilm reactors are another design of biofilm reactors that have received comparable attention recently in biotechnology for value-added products, as they have been popular in biotechnology for a longer time. Usually, the support surfaces are in the form of large disks that are horizontally partially submerged into the fermentation liquid bulk. These designs are known as rotary biofilm contactors or rotating biological contactors (RBCs). As the disks rotate at controlled speeds, the partial contact with the broth allows for optimum contact time (Qureshi et al., 2005). This approach has rendered the RBCs popular for conditions where substrate toxicity to the biofilm matrix in highly aerobic fermentation processes does not allow full submergence. A common example is remediation of complex waste streams such as polyphenols in olive oil mill wastewater (Alemzadeh and Nazemi, 2006). In other applications where fouling or gas-liquid transfer can be an issue for value-added products, the RBC has been applied to enable semicontinuous production for enhanced productivity (Lin et al., 2014; Roukas, 2018; Converti et al., 2006) (Figure 4.2g). The discs may be installed vertically as well and completely submerged in the broth. While the rotation speeds in these cases are much lower compared to stirred tank biofilm reactors, these drip-flow rotating disk biofilm reactors can be used for fragile biofilm matrices where the main concern is the integrity of the biofilm (Schwartz et al., 2010).

The other novel design of biofilm reactors is the MBfR also known as the membrane-aerated biofilm reactor (MABR) (Figure 4.2h). These biofilm reactors are based on gas-transferring membranes which supply a gaseous electron donor or acceptor substrate, such as oxygen, hydrogen, and methane. On one side of the membrane the gaseous substrate diffuses through the membrane media into the biofilm matrices on the membrane outer surface. On the other side of the membrane from the bulk liquid, the complementary substrate (electron donor or acceptor) defuses into the biofilm, making MBfR counterdiffusional (Nerenberg, 2016; Chen and Ni, 2016). The specific characteristic of the MBfR is the specificity of the membrane that allows only for selected nutrients to pass through and reach the biofilm matrix and at the same time has the physiochemical properties necessary to host the biofilm (Muffler and Ulber, 2014). While the membrane can perfectly protect the biofilm matrix and control the biofilm and liquid bulk interactions, the biofilm overgrowth and fouling and eventually poor mass transfer efficiencies are the downside of this novel technology (Demirci et al., 2007). A more novel geometry of these biofilm reactors is the hollow-fiber membrane biofilm reactors (HFMBR) where the conventional membranes are replaced by hollow fibers that enable advanced selectivity for consumption of gas substrates with extremely low solubilities such as hydrogen (Parameswaran et al., 2019).

Another novel design of biofilm reactors that has received large amount of attention in the past decade are emerse photobioreactors (EPBR). Traditionally, in photobioreactors, there are light sources that feed the biofilm matrices of phototrophic microorganisms such as cyanobacteria and microalgae under axenic conditions for exclusive cultivation of cell suspensions rather than for cultivation of biofilms. In EPBRs, however, phototrophic organisms are used that naturally grow on surfaces exposed to air such as microalgae. These designs have created great opportunities to enhance the production of bioactive compounds, biofuel precursors, and fatty acids (Muffler and Ulber, 2014).

4.2.3.2 Biofilm Reactor: Pros & Cons

In the previous section, the most commonly used geometries of biofilm reactors have been discussed and specifically mentioned some exclusive advantages and disadvantages associated with each type. However, biofilm reactors in general due to the presence of the biofilm matrices as microbial communities are associated with certain pros and cons compared to similar designs with suspended-cell bioreactors; no matter which specific design is utilized. For instance, since biofilm formations are densely packed microbial communities, biofilm reactors provide higher cell densities that generally lead to higher substrate consumption rates and as a result higher productivity. This is perhaps the most significant advantage of biofilm reactors that was reported in numerous studies applying them (Muffler and Ulber, 2014; Qureshi et al., 2005). The other major advantage of biofilm formations present in fermentation process is the extraordinary traits that migrated cells find in the biofilm community. Alterations on genetic and phenotypic levels not only grant cells more resilience to harsh conditions and environments, but in favorable conditions provided in a biofilm reactor grant them boosted metabolism (Kuchma and O'Toole, 2000; Burmølle et al., 2006; Branda et al., 2005). Faster and more efficient metabolism leads to faster and more efficient nutrient uptake leading to faster and higher concentration of metabolites including the desired value-added product. Besides, in biofilm reactors there is no need for vessel sterilization in between batches and no need for reinoculation in repeated-batch modes, and the mature biofilms are often more potent at starting the next batch than any prepared inoculum. Meaning, the lag phase in those modes are conveniently nonexistent or at least much shorter (Demirci et al., 2007; Ercan and Demirci, 2013). This latter tribute builds on top of the higher cell densities and boosted metabolism; more of faster and stronger workers ideally and simply means more product.

Unfortunately, the amazing advantages mentioned above also come with some major limitations and disadvantages as well. The most severe limitation usually comes from the physiochemical nature of the biofilm matrices. Although the matrix microbial community is highly capable of transforming the uptaken nutrients into products, the bottleneck is usually the infusion of the nutrients into the matrix to reach the transformation sites and through a similar mechanism the product needs to diffuse out. In other words, the dense and compact nature of the biofilm matrices grants them high cell density and resilience, but at the same time creates mass and heat transfer issues. This disadvantage multiplies in severity with aerobic fermentations due to limited oxygen solubility in broth bulk (Demirci et al., 2007; Muffler and Ulber, 2014). Moreover, the biofilm overgrowth adds to its thickness and adds to mass transfer limitations, and in compact designs such as packed-bed biofilm reactor (PBBR) or MBfRs fouling becomes a major issue as well (Qureshi et al., 2005). The other barrier for biofilm reactor application is that they are generally more complex and therefore less predictable and more difficult to troubleshoot. For instance, contamination is always a risk to any fermentation process, but in biofilm reactors in case of a contamination even detection sometimes can be an issue and decontamination demands dismissing the mature biofilm and start over, which is costly and time consuming (Demirci et al., 2007; Qureshi et al., 2005). Unpredictability can also be concerning. The dynamic interactions between the biofilm matrix, the planktonic cells, and the bulk liquid add numerous variables to the effective ones. Usually, these variables are numerous for bioprocess engineers to identify them all and therefore impossible to monitor and control. For example, in an oxygen limiting fermentation with suspended-cell bioreactors, online monitoring Dissolved Oxygen (DO) levels with simply a DO probe in the liquid bulk tells a lot how the process is doing; yet, by adding the biofilm matrices in there, the DO probe does not shed a lot of light on how the dynamic oxygen transfer goes on between the bulk and the matrix. Needless to say, up to date there are no easy ways to monitor the DO levels in the biofilm matrices. To make the long story short, bioprocess engineers sometimes run into biofilm reactors behaving unpredictably; i.e., biofilm reactors built in same geometries, same support designs, and same microbial strains behave quite differently since biofilm formation conditions with its numerous variables come into play as well (Ercan and Demrici, 2013).

4.3 APPLICATIONS OF BIOFILM REACTORS FOR PRODUCTION OF VALUE-ADDED PRODUCTS

In this section, the most recent and novel applications of the biofilm reactors for the production of value-added products in recent years along with some of the historically breakthrough studies will be reviewed in detail.

4.3.1 Alcohols

Alcohols are perhaps on top of the list of production in biofilm reactors. The reason is simple. Since biofilm reactors have been originally developed for bioremediation purposes and wastewater treatment applications and the alcohol and organic acid fermentations are typical pathways in such processes; researchers have been conveniently using their familiarity of the processes and tune them for production (Del Borghi et al., 1985; Converti et al., 2006). Thus, over the past several decades numerous studies have applied different geometries of biofilm reactors for these purposes, which have been covered in previous works (Demirci et al., 2007; Ercan and Demirci, 2013; Cheng et al., 2010; Muffler and Ulber, 2014).

One of the perks of alcohols production is to produce them using waste materials. Izmirlioglu and Demirci (2016) used potato waste hydrolysate with *Saccharomyces cerevisiae* in PCS-grid biofilm reactors to produce bioethanol. By optimizing growth conditions at pH 4.2, 34°C, and 100 rpm in repeated batch biofilm reactors, the yeast produced 37.05 g/L of ethanol with a 2.31 g/L/h productivity and 92.08% theoretical yield. Scanning Electron Microscopy (SEM) analysis of the PCS exterior and interior surfaces indicated that the high porosity of the PCS matrix has enabled dense formations of the yeast biofilm matrices. Visual observations of the yeast cells showed no morphological alteration on the immobilized cells. One of the valuable findings of this study was the operational relatively low pH for optimum ethanol production, which reduces contamination risks on larger scales (Izmirlioglu and Demirci, 2016). In a subsequent study, they applied a coculture of *Aspergillus niger* and *S. cerevisiae* in the biofilm reactors to enable simultaneous saccharification and fermentation of ethanol from the potato waste directly, without needing the costly enzymatic saccharification prior to fermentation. They evaluated effects of temperature, pH, and aeration rates in biofilm reactors by response surface methodology (RSM) and found the optimal conditions of 35°C, pH 5.8, with no aeration required. Under these conditions, a maximum ethanol concentration of 37.93 g/L was achieved at the end of 72 h fermentation, with a 0.41 g ethanol/g starch yield. SEM analysis of the PCS tubes in this case also revealed dense biofilm formations of both cultures, with the interior surfaces being more favorable for *S. cerevisiae* yeast cells than *A. niger* mold since no mycelia or spores were observed; perhaps since the mold has less anaerobic nature. Yet, the mold was able to thrive and express the enzymes in biofilm form without aeration. The SEM interestingly also revealed that the hyphae of the mold provide surface area for the yeasts' attachment (Izmirlioglu and Demirci, 2017).

In another study with similar PCS tubes, Germec et al. (2015) screened through different PCS compositions for best type for producing bioethanol form carob pod extract as a waste product with *S. cerevisiae* in repeated batch biofilm reactors. Using RSM optimization, they determined an initial sugar content of 7.71°Bx, pH 5.18, and 120 rpm rendering ethanol concentration, yield, and production rate as 24.51 g/L, 48.59%, and 2.14 g/L/h, respectively, within only 12 h of fermentation. They reported higher productivity and yield values compared to suspended or immobilized cells in stirred tank bioreactors with similar media and conditions. Also, they were able to maintain repeated batches for 160 days without contamination or operational failures (Germec et al., 2015). Later, they used nonsterile enriched (NSE) and nonsterile nonenriched (NSNE) carob extracts with the yeast in the same biofilm reactor finding ethanol production and yield at 18.46 g/L and 33.76% for NSE medium and 19.57 g/L and 38.14% for the NSNE medium, respectively. These findings were valuable given the fact that nonsterile media eliminate costly and energy-intensive sterilization steps, and the yields were still standing on top of the ones from suspended or immobilized cells in a stirred tank bioreactor (Germec et al., 2016). Germec et al. (2018) were then able to apply Modified Richards Model (MRM) to fit growth, ethanol production, and sugar consumption in both nonsterile media with high accuracy ($R2 > 0.95$). They also fitted ten flexible models to describe the ethanol fermentation, finding that Weibull model well fitted the experimental data of cell growth, ethanol production, and substrate along with predicting the kinetic parameters. Combined with the high yields in nonsterile media, the authors hope that these accurate models can enable large-scale production (Germec et al., 2019).

In another recent ethanol production study, Todhanakasem et al. (2019) developed plastic and corn silk composites as a biotic/abiotic support for *Zymomonas mobilis* biofilm formation to produce ethanol from rice straw hydrolysates. The researchers applied both multistage continuous culture with two vessels in series and repeated batch processes with two different strains of *Z. mobilis.* The composite support was produced by extruding small pieces of the corn silk with the plastic content at a ratio of 1:4. Then the biofilm matrices were formed on the composite support by repeated batch fermentation cycles. SEM analysis of the biofilm matrices showed dense homogeneous structures of cell attachments distributed along the entire surface area of the supports on day 3 with a significant morphological shift on day 5 of the initial biofilm development batch fermentations. While one of the strains showed higher yield in the multistage continuous mode and the other in the batch mode, yield values were highly maintained with no significant differences among the three consecutive repeated batches (Todhanakasem et al., 2019).

In a different study, Shen et al. (2017) developed a horizontally oriented rotating packed bed (h-RPB) biofilm reactor to improve mass transfer of syngas into the biofilm matrix of a mutant strain of *C. carboxidivorans* and continuously produce bioethanol. The support materials were packed in the reactor and half submerged in the liquid and half exposed to the headspace, where mass transfer in the headspace phase compensated for the lower volumetric mass transfer coefficient (k_La) of the h-RPB reactor in the liquid phase as compared to the CSTR counterpart. Overall results showed 7.0 g/L titer and 6.7 g/L/day productivity of ethanol, respectively, which were 3.3 times higher than those observed in the CSTR under the same operational conditions. The h-RPB design features simple mechanical design, inexpensive parts for assembly, low power demand with high ethanol output, proving its utility for syngas fermentation applications (Shen et al., 2017).

In another study utilizing syngas, Wang et al. (2018b) constructed an HFMBR hosting mixed culture biofilm matrices from mesophilic digester treating starch wastewater with 86.3% majority of *Clostridium* species with a total of 20 days needed to enrich the syngas fermentation bacteria and start producing ethanol. Operating at acidic conditions of pH 4.5, sole ethanol was in batch mode at maximum concentrations of 16.9 g/L. Another interesting and practical finding was that the partial pressure of hydrogen (PH2) in the biofilm matrix and carbon monoxide (PCO) could be used for the acetate and ethanol production in HFMBR, where high PH2 and PCO favored ethanol production, while low PH2 and PCO benefited acetate production. The other observation was that while archaea dominated seed sludge after 130 days of enrichment, the high temperature of 35°C shifted it to bacteria in the HFMBR. SEM analysis of the biofilm matrices after the enrichments showed dense biofilm formation formed on the outer surface of the hollow-fiber membrane with dominantly rod-shaped bacteria. These results showed for the first time a feasible ethanol and acetate selective pathway from syngas at pH 4.5 using HFMBR technology (Wang et al., 2018b).

4.3.2 Organic Acids

Organic acids are also popular for the production in biofilm reactors, similar to bioethanol due to compatibility with mixed culture biofilm matrices that are widely used for bioremediation applications, as comprehensively discussed in previous works (Cao et al., 1997; Demirci et al., 2007; Ercan and Demirci, 2015a; Cheng et al., 2010; Muffler and Ulber, 2014).

One of the common acids produced in this way is succinic acid, which is a bicarboxylic acid with a variety of applications in food, pharmaceutical, and agricultural industries as a pH regulator, flavoring agent, and additive for the preparation of drugs and ion chelator and surfactant. In a research with PBBR, Ferone et al. (2018) used *Actinobacillus succinogenes* to produce succinic acid in anaerobic continuous mode at 37°C with carbon dioxide supply to the biofilm. The growth conditions, dilution rate, and medium composition (mixture of glucose, xylose, and arabinose) were optimized. At a dilution rate of $0.5\,h^{-1}$, 43.0 g/L of succinic acid was produced with 88% glucose conversion and volumetric productivity of 22 g/L/h. The researchers were able to maintain continuous operational conditions for over 5 months at these conditions. Also, they found out that 5-hydroxymethylfurfural (HMF) remarkably reduced succinic acid production by 22.6% when compared to furfural that inhibited production by 16% (Ferone et al., 2018).

Similarly, Bradfield and Nicol (2014) have used an external-recycle biofilm reactor with *A. succinogenes* using D-glucose and carbon dioxide as carbon substrates. The researchers achieved highest product yield with 0.91 g/g of glucose and the highest titer of 48.5 g/L. They also found out that the yields were a function of glucose consumption with the succinic acid to acetic acid ratio, which increases as glucose consumption increased. More specifically, the yield increased continuously with an increase in glucose consumption while the acetate and formate concentration flattened out and finally diminished. They also carried out a metabolic flux analysis based on the established C3 and C4 metabolic pathways in the strain, showing that an additional source of Nicotinamide Adenine Dinucleotide (NADH) was present, indicating that the increase in the succinate to acetate ratio could not be attributed to the decrease in formic acid; which could attribute to biofilm transformation (Bradfield and Nicol, 2014).

In a follow-up study, the same researchers investigated the consumption of xylose into the biofilm reactor. The results indicated that succinic acid yields on xylose were 0.55–0.68 g/g, titers reached $10.9–29.4\,L^{-1}$, and productivities were 1.5–3.4 g/L/h, which were lower than those with glucose mass balance closures on xylose and were up to 18.2% lower than those on glucose. However, product ratios (succinic acid to acetic acid ratios of 3.0–5.0 g/g) and carbohydrate consumption rates were similar. The authors concluded that xylose can also be successfully and efficiently converted to succinic acid using the biofilm reactor design same as glucose, giving the perspective that cheaper and more abundant sources can be used in the same fashion for succinic acid production (Bradfield and Nicol, 2016).

Herselman et al. (2017) used a similar biofilm reactor design and strain of *A. succinogenes* to investigate the effect of carbon dioxide availability on succinic acid production in continuous mode. The results concluded that at high carbon dioxide values between 36.8% saturation and full saturation it is not limiting productivity and flux to product is constant. But as carbon dioxide levels decreases, an upper threshold is reached where metabolic flux distributions remain constant, but productivity and substrate uptake start to decline with decreasing carbon dioxide levels. A further decrease leads to a lower threshold, where productivity continues to decrease with a concomitant shift in carbon flux away from product towards C3 fermentative pathways including ethanol. Since carbon dioxide is a cheap cosubstrate to the fermentation of succinic acid, the gas supplied to biofilm formation can be a limitation from an industrial point of view, and these results confirmed that adequate carbon dioxide supply to the fermenter can be achieved without requiring major sparging schemes and thus reducing operational expenses (Herselman et al., 2017).

Similarly, Longanesi et al. (2018) used a PBBR with cheese whey as substrate, which is another major waste byproduct of dairy industry, especially Greek yoghurt production, and investigated five commercial biofilm carriers as support materials. Statistical analysis led to selection of a sintered glass porous material to be used in the packed beds. The succinic acid productivities were similar when cheese whey or lactose was used as substrate, reaching a maximum productivity of 0.72 g/L/h, a specific production rate of 0.18 g/g/h, and a biofilm concentration of about 4 g/L volume of the packed bed. The authors have concluded that the PBBR with cheese whey as substrate for succinic acid production is feasible, but the productivity was still behind that attained with glucose-fed processes (Longanesi et al., 2017).

Mokwatlo and Nicol (2017) took a deeper look at the characteristics of *A. succinogenes* biofilm formation used in the studies mentioned above by Bradfield and Nicol (2014, 2016), which have demonstrated exceptional capabilities as biocatalysts for high productivity, titer, and yield production of succinic acid. They also used confocal scanning laser microscopy, scanning electron microscopy, and image analysis software to investigate the structure and cell viability of the biofilms. The images showed low biofilm surface area coverage near the base and thickening towards the outer layer. Water channels were present at the deeper portions of the biofilm with a greater portion of inactive cells closest to the attachment surface. Surprisingly, viability test showed that over 65% of the biofilm consisted of dead cells, with a cell viability gradient where the outer layer exhibits a greater fraction of active cells compared to the base layer. Processing of the images also revealed that the immobilized cells underwent a phenotypic alteration into a filamentous cell morphology completely distinct from rod-like morphology of planktonic cells allowing extensive cell entanglements within microcolonies. These features may not only add to the intactness of microcolonies but also explain how the biofilm matrices respond to the physiochemical and nutrient cues applied in biofilm reactors used to produce succinic acid (Mokwatlo and Nicol, 2017).

Another commonly produced organic acid in biofilm reactors is lactic acid. Narayanan and Das in 2017 assembled a semifluidized bed biofilm reactor. The substrates used were cheese whey and molasses as low-cost feedstock. The behavior of the biofilm reactor was modeled equivalent to a plug-flow dispersion reactor, and the performance equation was solved using a modified form of fourth-order Runge-Kutta method. The resulting model was able to fit data gathered in the biofilm reactor on pilot-scale experiments. The authors using the modeled data concluded that large capacities of the biofilm reactor for bioconversion of sucrose and lactose within a low reactor volume is feasible and an increase in substrate flow rate increases fractional substrate conversion. Bioreactor performance was also reported to improve with increased microbial concentration in biofilms (Narayanan and Das, 2017).

Shahab et al. (2018) composed an artificial cross-kingdom consortium and cocultivated the aerobic fungus *Trichoderma reesei* that produces cellulolytic enzymes to enable facultative anaerobic lactic acid bacteria in a biofilm matrix on an MABR to produce lactic acid from 5% (w/w) microcrystalline cellulose whole-slurry pretreated beech wood as substrate in batch and fed-batch reactors. The results demonstrated the ability of the consortium to coferment hexoses and pentoses from nondetoxified whole-slurry pretreated beech wood without carbon catabolite repression with superior product purities. The highest titer of acetic acid produced was 19.8 g/L after 200 h of fermentation. The stability of the cocultured biofilm and the ability to feed hexoses and pentoses to the biofilm reactor can have a potential for commercial production (Shahab et al., 2018).

Similar to the application for producing alcohols, syngas and methane can be economical substrates for organic acid production in biofilm reactors as well. Similar to their approach for bioethanol production from syngas, Wang et al. (2018c) used their HFMBR with syngas and mixed-culture fermentation at lower temperature of 25°C that resulted in fermentation of organic acids mostly lactic, butyrate, and caproate acids along with ethanol. The caproate titers were 3.4–5.7 g/L in batch fermentations, which were higher compared to pure culture fermentation with *Clostridium carboxidivorans*. The 16S rRNA analyses of the strains present in the biofilm matrices showed that the dominant genera were *Clostridium sensu* and *Prevotella* sp. Given that lower operating temperatures are more energy and cost efficient and the fact that caproate has higher economic value than acetate; these pathways may be preferable to the previous demonstration of the HFMBR with syngas for industrial purposes (Wang et al., 2018c).

In a follow-up study, Wang et al. (2018d) used ultrafiltration and microfiltration apparatus in addition to the HFMBR with syngas and mixed-culture fermentation at similar operation conditions to demonstrate the effects of pore sizes on the in situ utilization of synthesis gas. Results revealed that the addition of ultrafiltration led to total consumption of the syngas, whereas microfiltration addition led to a significant accumulation of syngas in the biofilm reactor. As a result, volatile fatty acids (VFAs) of acetate, butyrate, and caproate were produced at significantly higher rates with the ultrafiltration unit in both batch and continuous modes. The inhibitory effect of the microfiltration add-on eventually caused a washout on the biofilm, rendering only the ultrafiltration add-on useful for enhancing VFA production from syngas (Wang et al., 2018d).

Shen et al. (2018) also used an HFMBR with mesophilic and thermophilic mixed culture fermentation for syngas fermentation. They were able to accomplish over 95% conversation rates, and the acid profile consisted of acetate (4.22 g/L), butyrate (1.35 g/L), caproate (0.88 g/L), and caprylate (0.52 g/L) at 35°C with a significant shift towards acetate as the main metabolite at 55°C reaching acetate concentration and the production rates of 24.6 g/L and 16.4 g/(L/day), respectively, at pH 6.5 in the continuous mode. However, higher temperature would add significantly to the operating costs. Illumina high-throughput sequencing indicated a shift from *Clostridium* (41.6%) to *Thermoanaerobacterium* (92.8%) as the conditions shifted from mesophilic to thermophilic conditions (Shen et al., 2018).

Chen and Ni in 2016 used a lab-scale MBfR with syngas (60% hydrogen and 40% carbon dioxide) supplied from inside the membrane to produce fatty acids. Then they fitted the experimental data with a model integrating multiple production pathways of the fatty acids including acetate, butyrate, and caproate. The models indicated that high hydraulic retention times (HRT) are required for chain elongation to produce a higher proportion of caproate with a higher added value by allowing more of acetate produced to be further converted to butyrate and caproate, which greatly facilitates chain elongation processes. Moreover, the distribution of each fatty acid species was also significantly affected by HRT (Chen and Ni, 2016).

Later on, Chen et al. (2018) also used a mixed-culture MBfR with hollow fiber with methane as the substrate to produce fatty acids. They were able to produce 10 g of short-chain fatty acids (mostly acetate and propionate) per liter per day. The 16S rRNA gene sequencing of the biofilm matrix determined that the consortium in the biofilm was dominantly methanogens, and acid-producing bacteria were most responsible for bioconversion of methane into acids (Chen et al., 2018).

In a different approach, Liu et al. (2016) have used PCS-grid biofilm reactors of SFYBR (polypropylene (50%), soybean hulls (35%), soybean flour (5%), salt, yeast extract (5%), bovine albumin (2.5%) and bovine red blood cell (2.5%)) type selected out of five types of PCS in repeated-batch fermentations to produce kojic acid from *Aspergillus oryzae*. Kojic acid is a secondary metabolite and a lesser member of the organic acids produced in biorefineries but with strong metal chelating capacity as a potent tyrosinase inhibitor and is therefore applied widely in cosmetic and food industries. Nitrogen deficient medium was applied for higher productivity reaching 3.09 g/L/day, which was higher than the results from suspended-cell runs in batch fermentation. *Aspergillus oryzae* mycelium morphology changed under nitrogen starvation conditions where feather-like mycelia were observed with high RNA expression (*kojA* and *kojT*) and resulted in higher productivity. Under optimum conditions in the batch fermentations in the biofilm reactors, kojic acid concentrations reached 35.7 g/L (Liu et al., 2016).

4.3.3 Enzymes

Although applying biofilm reactors for producing various kinds of enzyme began in the 1980s, they definitely have not received much attention from researchers. Webb et al. (1986) as the pioneer reported that continuous cellulase production by immobilized *Trichoderma viride* on stainless steel BPS in a 10-L spouted bed fermenter can be beneficial compared to planktonic cell; the conclusion was that enzyme productivity of the immobilized cells was more than three times that of suspended cells. Furthermore, switching to continuous production mode in the same fermenter at a dilution rate of $0.15\,h^{-1}$ gained 31% higher yield and 53% greater volumetric productivity than batch operations (Webb et al., 1986). Oriel (1988) reported successful immobilization of *Escherichia coli* in simple porous silicone polymer beads in both batch and continuous modes, enhancing amylase production along with plasmid stability by immobilization (Oriel, 1988). Similarly, Nakashima et al. (1988) achieved enhanced lipase activity when they entrapped the live cells in cubic polyurethane foam particles, where the specific activity of dried cells within support was seven-fold higher than freely suspended cells (Nakashima et al., 1988). In 1995, Kang et al. used a bubble column biofilm reactor with *A. niger* immobilized on celite and polyurethane foams for enhanced production of xylanase and cellulase by using rice straw as substrate (Kang et al., 1995).

Many researchers have also focused on peroxidase production by immobilizing *Phanerochaete* fungus. Linko (1988) used nylon-web immobilized *Phanerochaete chrysosporium* in carbon limited conditions to enhance lignin peroxidase production and characterized the produced enzyme (Linko, 1988; Haapala and Linko, 1993). Then it was in 1992 when Venkatadri et al. used a biofilm membrane reactor to enhance the production of lignin peroxidase by the same species. The improved enzyme biosynthesis led to five-fold improvement in the subsequent treatment of pentachlorophenol (Venkatadri et al., 1992). In the same year, Jones and Briedis investigated lignin peroxidase production by the white-rot fungus in a rotary biological contactor (Jones and Briedis, 1992). Solomon and Petersen (2002) utilized an MBfR to immobilize the fungus and produced lignin and manganese peroxidases (Solomon and Petersen, 2002). Govender et al. in 2003 and 2010 used a membrane gradostat reactor for this purpose (Govender et al., 2003, 2010). Later on, Khiyami et al. (2006) took a different approach to the lignin peroxidase, producing white-rot fungus with the PCS-grid biofilm reactors, and were also able to produce manganese peroxidase as well (Khiyami et al., 2006).

More recently, Hui et al. (2010) immobilized the mold *Aspergillus terreus*, isolated from rotting bagasse onto woven nylon pads. They reported hydrolase, carboxymethylcellulase, and β-glucosidase activities by the immobilized cells at 1.8, 12.0, and 2.4 U/mL, respectively, while planktonic cells under the same conditions were only able to produce 2.1, 13.6, and 3.2 U/mL of the same enzymes, respectively (Hui et al., 2010). At the same time, Zhao et al. (2015) detected dense immobilization of *Irpex lacteus* onto sawdust particles when grown in a deep tray solid-state bioreactor to produce manganese peroxidase. The maximum enzyme activity was observed in the stationary phase at 84 h, reaching 950 U/L (Zhao et al., 2015).

4.3.4 Vitamins

Up until recent years, the benefits of biofilm formations for enhancing biosynthesis of vitamins were unexplored, let alone development of biofilm reactors for these purposes.

When Mitra et al. (2012) designed a polymethylmethacrylate conico-cylindrical flask with eight equidistantly spaced rectangular strips mounted radially on a circular disk and cultivated the yeast *Candida famata* biofilm on the support surfaces, and results indicated that the highest riboflavin (vitamin B_2) production with the biofilm formation could reach 290 mg/L under optimum aeration and growth conditions, whereas with planktonic cells these concentrations were never above 55 mg/L. The authors concluded that with such stunning differences, as chemical synthesis of riboflavin is replaced by fermentation methods; *C. famata* in biofilm reactors may dominate the industrial production (Mitra et al., 2011; Mitra et al., 2012).

Another breakthrough in this field came when Berenjian et al. (2013) discovered a positive and rather crucial effect from *Bacillus subtilis natto* biofilm and pellicle formation on menaquinone-7 (also known as MK-7, a form of vitamin K2) secretion (Berenjian et al., 2013). It was already known to scientists that extracellular MK-7 in the bacteria is secreted attached to a protein that solubilizes it in the aqueous broth and possibly has a direct role in the making of the poly-γ-glutamic acid (PGA) skeleton in the extracellular matrix of the biofilm (Ikeda and Doi, 1990). MK-7 is the most valuable and potent form of vitamin K for its positive effects in preventing cardiovascular diseases and osteoporosis and also fighting cancer cells (Berenjian et al., 2015; Mahdinia et al., 2017b). Also, its production in solid state fermentation and liquid state fermentation strategies without robust agitation and aeration to allow for biofilm formation imposes critical mass transfer and heat transfer issues, rendering production on large industrial scales unfeasible since *B. subtilis natto* is highly aerobic. On the other hand, attempts with planktonic cells in agitate and aerated liquid states indicated severe impairment of the vitamin biosynthesis (Mahdinia et al., 2017b).

Therefore, Mahdinia et al. (2017a) investigated the feasibility of immobilizing the bacterial cells through passive immobilization onto PCS surfaces to enhance MK-7 biosynthesis by allowing mature biofilm formations on the PCS and at the same time introduce robust agitation and aeration. They screened through combinations of a total of thirteen different *Bacillus* strains with four types of PCS for maximum MK-7 biosynthesis. At last, a strain of *B. subtilis natto* isolated form fermented natto with the SFYB PCS type was selected to construct the biofilm reactor in 2-L bioreactors (Mahdinia et al., 2017a). Once the biofilm reactors were constructed in the PCS-grid fashion onto the agitation shaft, the growth conditions were optimized using RSM for a glycerol-based and glucose-based media in repeated batch mode. Glucose is much more readily metabolized hypothesizing higher productivities by *B. subtilis* while glycerol had shown to be beneficial for MK-7 synthesis (Berenjian et al., 2011). In the glycerol-based medium the optimum conditions were at 35°C, pH 6.6, and 200 rpm, and the biofilm reactors were able to produce 12.1 mg/L of MK-7 in 144 h of fermentation, which was 58% higher than the amounts produced under similar conditions in suspended-cell bioreactors with planktonic cells (Mahdinia et al., 2018a). In the glucose-based media the trend was similar with 30°C, pH 6.5, and 234 rpm as optimum conditions, the biofilm reactors produced 18.5 mg/L of the vitamin within 144 h, which was this time 237% higher than the levels produced with planktonic cells. The other expectable observation was that glucose depletion occurred halfway through the fermentation run while in the glycerol-based medium carbon source did not occur even after 6 days (Mahdinia et al., 2018b).

Later on, the researchers optimized media components in both media with RSM statistical optimization in order to further enhance MK-7 concentrations. The optimum glycerol-based medium turned out to include 48.2 g/L of glycerol, 8.1 g/L of yeast extracts, and 13.6 g/L of soytone producing a maximum MK-7 concentration of 14.7 mg/L in biofilm reactors, which was similarly 57% higher compared to vitamin levels in the suspended-cell reactors. The notable change this time was that with even higher levels of glycerol in the optimized medium, glycerol depletion did occur within 120 h of fermentation (Mahdinia et al., 2019a). Optimization of the glucose-based medium set components of glucose at 152.6 g/L, yeast extract at 8 g/L, and casein tryptone at 17.6 g/L was slightly different from the original glucose fortified tryptic soy broth medium. With these modifications, the MK-7 maximum concentration reached 20.5 mg/L, which was 344% higher than the performance of planktonic cells. The huge gap was also observed with carbon source consumption rates, where the biofilm formations were able to deplete the initial glucose content within 72 h of fermentation when the planktonic cells could not fully utilize the glucose but left over 40 g/L of it at the end of the runs. These findings left no doubt, behind that the biofilm formations are highly more capable of metabolizing the nutrients and therefore secreting MK-7 (Mahdinia et al., 2018c).

In order to overcome the carbon source depletion in the biofilm reactors and thus sustain MK-7 secretion stability, Mahdinia et al. (2018d) investigated different combinations of fed-batch regimes at different stages of fermentation. The results indicated that starting with the readily metabolized glucose-based medium and implementing pure glucose (or similarly glycerol) additions at 72 and 144 h marks of fermentation, the MK-7 secretion can be sustained for 288 h resulting in over 28.7 mg/L of MK-7 in the broth and still 230% higher than the level from planktonic cells with the same fed-batch regimes. These high MK-7 turnouts were comparable with the highest concentrations achievable in static fermentation with heavy sporulation and pellicle formation (32 mg/L), which put the PCS biofilm reactor for industrial fermentation of MK-7 in the spotlight (Mahdinia et al., 2019e). SEM analysis of the PCS that was operational in the biofilm reactors nonstop for over 24 months indicated partial and dense biofilm formation attached to the interior and exterior surfaces of the PCS pores via PGA links. The other observation was that the morphology of *B. subtilis* cells was quite distinct in the two media and also changed throughout the runs, showing the effect of nutrient cues from the media (Mahdinia et al., 2018d).

Furthermore, the authors used modified Gompertz and Luedeking-Piret models with over 95% accuracy to fit the MK-7 biosynthesis profiles in repeated batch fermentations. Also, the modified Logistic model was able to very accurately fit the carbon consumption behaviors in both glucose- and glycerol-based media. The high accuracy of the Luedeking-Piret model indicated that the MK-7 biosynthesis in *B. subtilis* in biofilm reactors follows a mixed-metabolite behavior, and the lag phase for biosynthesis is negligible and secretion continues throughout the run. These models then can be reliable to pave the path when it comes to scale-up the biofilm reactors to larger pilot and plant scale and revolutionize MK-7 biosynthesis plants, since the biofilm reactor designs in this field are still in the preliminary steps, and also more metabolic engineering attempts are made now and then to enhance MK-7 biosynthesis by overexpression techniques (Mahdinia et al., 2019b; Yuan et al., 2019; Ren et al., 2019).

4.3.5 Antimicrobials

Another significant product produced by using biofilm reactors includes antimicrobial agents. Bober and Demirci (2004) used PCS biofilm reactors with *Lactococcus lactis* subsp. *lactis* under different reduced nitrogen content media, with glucose as carbon source in repeated-batch fermentations to produce nisin; after Sakhamuri et al. (2004) used Fourier transform mid-infrared (FTIR) spectroscopy and noninvasively detect the potential of *Lactococcus lactis* biofilm to produce nisin (Sakhamuri et al.., 2004). Nisin is a polycyclic antibacterial peptide that has a wide range of applications mostly in food industries as a food preservative. Bober and Demirci (2004) reported that nisin production with glucose fed-batch fermentations were slightly improved in the biofilm reactors compared to suspended-cell bioreactors. Later, Pongtharangkul and Demirci (2006a) used the same design of biofilm reactors and strain, but observed that replacing glucose with sucrose can increase nisin production from 579 to 1,100 IU/mL. They also found out that the biofilm system could fasten the nisin production while pH changes can have detrimental effects on nisin expression by the bacteria, and excess lactic acid produced during the fermentation can be toxic and therefore inhibitory to nisin expression as well (Pongtharangkul and Demirci, 2006b,c). Thus, when they implemented an online recovery apparatus using silicic acid coupled with a microfilter module nisin levels were significantly improved to reach 7,445 IU/mL when compared with the batch fermentation without the online recovery of 1,897 IU/mL (Pongtharangkul and Demirci, 2007).

Ercan and Demirci (2013) constructed PCS-grid biofilm reactors with *Kluyveromyces lactis* to produce human lysozyme. Lysozyme is an antimicrobial agent heavily used in cheese, wine, cosmetic, and pharmaceutical industries, and human lysozyme is superior to its conventional counterpart from chicken eggs for people allergic to egg products. For this purpose, the researchers screened through four different PCS types with test tube fermentations and selected SFYB for the biofilm reactors. Then, the growth conditions were optimized by RSM for maximum lysozyme production in repeated batch fermentations. Under these conditions of 25°C, pH 4, and no aeration, highest lysozyme concentrations of 141 U/mL was reached within 72 h of fermentation, which were higher than the levels observed in suspended-cell bioreactors without the presence of biofilm formation (Ercan and Demirci, 2013). In the next step of the study, the researchers used RSM again to optimize components of the synthetic medium used in the biofilm reactors to further improve lysozyme production. The optimum medium was composed of 16.3% lactose, 1.2% casamino acid, and 0.8% yeast nitrogen base, improving lysozyme concentrations to 173 U/mL which were 57% higher than the 110.3 U/mL of lysozyme in suspended-cell bioreactors (Ercan and Demirci, 2014). Then, the researchers investigated fed-batch and continuous fermentation modes in these biofilm reactors with optimum conditions. In fed-batch fermentation with glucose as the initial carbon source, lactose was introduced at constant 0.6 mL/min for 10 h leading to significantly higher lysozyme levels (187 U/mL). By switching to continuous mode, biofilm reactors were able to provide significant higher productivity (7.5 U/mL/h) compared to the maximum productivity in suspended-cell bioreactor (4 U/mL/h), perhaps due to higher cell density at higher dilution rates in biofilm reactors (Ercan and Demirci, 2015b). Finally, the researchers were able to couple the biofilm reactors with an online recovery system using silicic acid as absorbent of the produced lysozyme. The study determined 25°C, pH 4, and 25% silicic acid: fermentation broth volume ratio for adsorption while 25°C, pH 6.2, and 5% (wt/v) sodium dodecyl sulfate with 1M NaCl and 20% (v/v) ethanol as eluent were best desorption parameters, leading to 95.6% lysozyme adsorption and 98% desorption. Using the online recovery system, 280.4 U/mL of lysozyme was produced, which was 63% higher than the conditions without recovery (Ercan and Demirci, 2015c).

Around the same time, Sarkar (2015) employed an extended surface biofilm reactor (ESBR) with marine isolate *Streptomyces sundarbansensis* to produce antimicrobial agents. In the ESBR, eight equidistantly spaced polymethylmethacrylate rectangular strips were installed radially on a circular disk to provide enhanced surface area available for the biofilm formation to latch on. The optimum pH of 10 and temperature 30°C were determined in smaller scale fermentations, and then under optimum conditions the ESBR with 4.2 L volume was able to deliver productivities 15% higher than the small-scale values (Sarkar, 2015).

4.3.5 Other Products

So far, successful applications of biofilm reactors to ferment alcohols, acids, enzymes, vitamins, and antimicrobials have been covered. However, there are numerous other metabolites that recent studies have reported enhanced production in biofilm reactors.

Yet another more popular topic for biofilm reactors recently has been biohydrogen and biomass production, based on similar logic with alcohols and acids being the close conditions to bioremediation applications. Barca et al. (2016) developed an artificial consortium of two anaerobic bacterial strains of *Clostridium acetobutylicum* and *Desulfovibrio vulgaris* in an upflow anaerobic packed-bed reactor (APBR). The authors investigated the scale-up conditions from batch to continuous mode, where the APBR was continuously fed with a glucose medium as a synthetic wastewater source for the production of biohydrogen. The pH and alkalinity conditions were optimized along with void HRT, and after 3–4 days of stabilization, the APBR was able to reach a stable hydrogen output of 2.3 NL (Normal Liter) hydrogen/day/L (Barca et al., 2016).

In a most recent study, Kongjan et al. (2019) also employed biofilm upflow anaerobic reactors with plastic biofilm support and mixed cultures of dominant *Thermoanaerobacterium* species with acetate and butyrate as main fermentation products and xylose as the substrate. Under optimum mesophilic conditions, a maximum production rate of 13.3 L hydrogen/L/day with a yield of 221 mL hydrogen/g xylose was achieved in the biofilm with the organic loading rate of 60 g xylose/L/day and HRT of 4 h. The results were compared with productions in an upflow anaerobic sludge blanket bioreactor under similar conditions (Kongjan et al., 2019).

Prakash et al. (2018) utilized a biodiesel industrial effluent rich in crude glycerol in a biofilm reactor with *Bacillus amyloliquefaciens* on cartridges made of dried coconut coir packed in PVC tubes as support. The biofilm reactors were operated in continuous mode yielding 165 L of hydrogen/L of the liquid feed, which was 1.18-fold higher than that observed with the similar but nonbiofilm forming *B. thuringiensis* strain under similar conditions. Furthermore, the researchers were able to improve the efficiency of the process further by partially recycling the outgoing effluent from the biofilm reactor (Prakash et al., 2018).

Carrillo-Reyes et al. (2016) constructed an anaerobic sequencing batch biofilm reactor (ASBBR) inoculated with a microbial consortium of mostly hydrogen-producing bacteria composed of *Clostridium pasteurianum* and *Clostridium beijerinckii* strains. The ASBBR was then operated over 6 months with unsterile lactose-rich effluent and various HRTs ranging from 1 to 34 h. Under these conditions the highest hydrogen productivity was 519 mmole hydrogen/L/day, and the highest yield was 7.11 mole hydrogen/mole lactose without any significant methane production, concluding that specific preparation of the inoculum may help enhance the long-term performance of these mixed culture biofilm reactors (Carrillo-Reyes et al., 2016).

Tomczak et al. (2018) also concluded that HRT plays a key role in determining the output of biohydrogen production when they used an APBR with recirculation flow of the liquid phase and inoculated mesophilic sludge from an urban wastewater treatment plant pretreated at 100°C for 1 h to inhibit the methane-producing bacteria activity and harvest anaerobic spore-forming bacteria. They used a synthetic wastewater with glucose and sucrose and achieved the highest average hydrogen yield of 2.35 mol hydrogen/mol substrate and hydrogen productivity of 0.085 L/h/L at an HRT of 2 h (Tomczak et al., 2018).

Unlike the dark fermentations applied in the above studies, Wen et al. (2017) used a small tube biofilm reactor on a shaker with photofermentation of *Rhodopseudomonas* sp. to produce biohydrogen from a synthetic acetate and glutamate medium. SEM analysis of the biofilm surface showed the formation of a dense biofilm after 72 h of inoculation which boosted cumulative hydrogen volume, hydrogen concentration, and substrate utilization efficiencies by 75%, 10%, and 18%, respectively (Wen et al., 2017).

Another field of studies that have employed biofilm reactors on large scales with mixed culture fermentations are for biomass production such as algae and microalgae productions (Gross et al., 2015). Wood et al. (2015) utilized undiluted produced water from oil and natural gas extraction in a rotating algal biofilm reactor to produce cyanobacterial biofilm biomass and phycocyanin. They were able to achieve dry weight biomass productivity of 4.8 g/m^2-day along with phycocyanin productivity of 84.6 mg/m^2-day. The produced phycocyanin blue pigments have applications in cosmetics, foods, medicine, and biotechnology (Wood et al., 2015).

Other miscellaneous products produced by biofilm reactors include the work of Baeza et al. (2016) that used an MBBR with sludge inoculum using paper mill wastewater rich with nutrients to produce polyhydroxyalkanoates (PHA) which is used as bioplastics in industry. The maximum percentage of PHA-accumulating cells were 85.1% with the paper mill wastewater, whereas with the effluent from thermomechanical pulping the accumulation percentage was 89.4%. These results were yet another example of simultaneous wastewater treatment and value-added production by biofilm reactors (Baeza et al., 2016).

Jiang et al. (2016) also used an MBBR but with pure culture of *B. subtilis* on polypropylene support particles to produce PGA using a glucose and glutamate synthetic medium in batch mode where highest product concentrations of 42.7 g/L and productivity of 0.59 g/L/h were achieved. By implementing dissolved oxygen-stat feeding and repeated fed-batch cultures, production stability was enhanced, and the product concentration and productivity values were boosted to 74.2 g/L and 1.24 g/L/h. PGA is a biodegradable and water-soluble biopolymer which has external applications in food additives, drug carriers, biological adhesives, adsorption of heavy metals, and as fertilizer synergists (Jiang et al., 2016).

Zune et al. (2015) assembled a biofilm reactor based on metal structured packing inside with biofilm matrices of a mutant *Aspergillus oryzae* to produce Gla green fluorescent protein as a recombinant protein. The production in biofilm reactors showed improvement in stability, possibly due to the high shear stress in suspended-cell bioreactors, causing biomass autolysis and leakage of intracellular fusion protein into the extracellular medium. Also, protein integrity was higher in biofilm reactors (Zune et al., 2015).

Other biopolymers produced by PCS biofilm reactors are pullulan and bacterial cellulose. Pullulan is biosynthesized by the polymorphic fungus *Aureobasidium pullulans* and as a linear glucosic polysaccharide has various applications from food additives to bioremediation agents (Cheng et al., 2009a, 2011c). Cheng et al. (2009a) investigated different PCS types and pH profiles on pullulan production in PCS biofilm reactor with *A. pullulans* and were able to produce 32.9 g/L of pullulan after 7 days of fermentation, which was 1.8-fold higher than the amounts in suspended-cell bioreactors (Cheng et al., 2009a). Using RSM optimization with Box-Behnken design, they were able to increase pullulan concentrations to 60.7 g/L in 7 days under optimum sucrose and nitrogen source concentrations (Cheng et al., 2010b; Mahdinia et al., 2020). They also studied the effect of various initial ammonium ion concentrations from ammonium sulfate and fed-batch addition of 10 g/L of sucrose with the conclusion that gradual sucrose addition can suppress the expression of pullulan-degrading enzyme (Cheng et al., 2011a). In continuous mode, ammonium sulfate and sucrose concentrations and dilution rate were optimized leading to 1.33 g/L/h of pullulan productivity and 93% purity (Cheng et al., 2011b). Finally, Cheng et al. (2010c) utilized mathematical models to fit biomass, pullulan, and sucrose profiles during the fermentation in the suspended-cell bioreactors. The results showed that a modified Gompertz model can be considered as a universal equation for all three responses as it can fit to them with high accuracy (Cheng et al., 2010c).

Bacterial cellulose, unlike plant-based cellulose, is a pure form without the unwanted impurities and contaminants such as lignin, pectin, and hemicellulose (Lin et al, 2013). Cheng et al. (2009b) used the PCS-grid biofilm reactors with *Acetobacter xylinum* and were able to produce bacterial cellulose with higher crystallinity of 93% (Cheng et al., 2009b). When 1.5% carboxylmethylcellulose (CMC) was added to the starting medium, results indicated that production was enhanced to 13 g/L which was 170% higher than the result obtained without CMC (Cheng et al., 2009c, 2011d). Lin et al. (2014) constructed a horizontal rotating disk bioreactor with PCS as solid support to form *Gluconacetobacter xylinum* biofilm on them and produce bacterial cellulose from a modified corn steep liquor medium. Under semicontinuous fermentations, 0.24 g/L/day of productivity was achieved in 5 days and could be sustained for at least five consecutive runs. The bacterial cellulose obtained from the RBC had lower crystallinity (66.9%) and mechanical property (Young's modulus of 372.5 MPa) compared to those from static culture fermentations (Lin et al., 2014).

Roukas in 2018 also used a modified horizontal rotary biofilm reactor with polypropylene discs hosting the fungus *B. trispora* biofilm formation to enhance the production of carotene with glucose and corn steep liquor medium. Under optimum conditions in a culture grown in 1 L of the synthetic medium with an initial pH 11, an aeration rate of 4 vvm and disc rotation speed of 13 rpm, maximum carotene productivity reached 57.5 mg/L/day which was staggeringly six times higher than the amounts produced in conventional stirred tank bioreactors (Roukas, 2018).

4.4 CONCLUSIONS AND FUTURE TRENDS

In this chapter, biofilm characteristics that are utilized in biofilm reactors and how the biofilm matrices form as well as how biofilm reactors are designed, constructed, and what types of support materials have been used for hosting the biofilms have been summarized. Then, the most up to date and major applications of biofilm reactors in producing value-added products including alcohols, organic acids, biofuels, enzymes, vitamins, antimicrobials, biopolymers, and biomass were reviewed. Furthermore, the main advantages of applying biofilm reactors over conventional fermentation strategies with planktonic cells and the disadvantages that biofilm reactors face were discussed. One of these limitations are general sophistications that applying biofilm formations create for the fermentation processes. Biological systems are complicated on their own, and many bioreactor designs and optimizations even today rely on empirical and trial-and-error approaches. Biofilm formations bring even more variables to the table with their dynamic and complex interactions with the liquid bulk, rendering biofilm reactors inconveniently unpredictable, inflexible, and tedious to operate. Naturally, bioprocess engineers would rather stick to their suspended-cell reactor roots and forfeit the magnificent metabolic superiorities that biofilm reactors can offer. With more powerful modeling tools such as machine learning techniques and the emerging artificial intelligence (AI) technologies, these uncertainties about the complex biofilm reactors may be gradually addressed in the upcoming years. As a result, one can expect to see more accurate models developed exclusively for biofilm reactors and applied to facilitate more occasions where they will be applied for finer and higher value products.

REFERENCES

Agle, M.E., 2007. Biofilms in the food industry. In: Wang, H.H., Agle, M., Meredith, E., eds. *Biofilms in the Food Environment*. Ames, IA: Blackwell Publishing and The Institute of Food Technologists, 3–19. doi: 10.1002/9780470277782.ch1.

Alemzadeh, I. and Nazemi, A.R., 2006. Physicochemical and biological treatment of olive mill wastewater by rotating biological contactor (RBC) reactors. *Iranian Journal of Chemistry and Chemical Engineering*, *25*(4):47–53.

Atkinson, B., Black, G.M., Lewis, P.J.S. and Pinches, A., 1979. Biological particles of given size, shape, and density for use in biological reactors. *Biotechnology and Bioengineering*, *21*(2):193–200. doi: 10.1002/bit.260210206.

Baeza, R., Jarpa, M. and Vidal, G., 2016. Polyhydroxyalkanoate biosynthesis from paper mill wastewater treated by a moving bed biofilm reactor. *Water, Air, & Soil Pollution*, *227*(9):299. doi: 10.1007/s11270-016-2969-x.

Barca, C., Ranava, D., Bauzan, M., Ferrasse, J.H., Giudici-Orticoni, M.T. and Soric, A., 2016. Fermentative hydrogen production in an upflow anaerobic biofilm reactor inoculated with a coculture of *Clostridium acetobutylicum* and *Desulfovibrio vulgaris*. *Bioresource Technology*, *221*:526–533. doi: 10.1016/j.biortech.2016.09.072.

Berenjian, A., Mahanama, R., Talbot, A., Biffin, R., Regtop, H., Valtchev, P., Kavanagh, J. and Dehghani, F., 2011. Efficient media for high menaquinone-7 production: response surface methodology approach. *New Biotechnology*, *28*(6):665–672. doi: 10.1016/j.nbt.2011.07.007.

Berenjian, A., Chan, N.L.C., Mahanama, R., Talbot, A., Regtop, H., Kavanagh, J. and Dehghani, F., 2013. Effect of biofilm formation by *Bacillus subtilis natto* on menaquinone-7 biosynthesis. *Molecular Biotechnology*, *54*(2):371–378. doi: 10.1007/s12033-012-9576-x.

Berenjian, A., Mahanama, R., Kavanagh, J. and Dehghani, F., 2015. Vitamin K series: current status and future prospects. *Critical Reviews in Biotechnology*, *35*(2):199–208. doi: 10.3109/07388551.2013.832142.

Bober, J.A. and Demirci, A., 2004. Nisin fermentation by *Lactoccocus lactis* subsp. lactis using plastic composite supports in biofilm reactors. *Agricultural Engineering International: CIGR Journal*, *6*:1–15.

Bradfield, M.F.A. and Nicol, W., 2014. Continuous succinic acid production by *Actinobacillus succinogenes* in a biofilm reactor: steady-state metabolic flux variation. *Biochemical Engineering Journal*, *85*:1–7. doi: 10.1016/j.bej.2014.01.009.

Bradfield, M.F. and Nicol, W., 2016. Continuous succinic acid production from xylose by *Actinobacillus succinogenes*. *Bioprocess and Biosystems Engineering*, *39*(2):233–244. doi: 10.1007/s00449-015-1507-3.

Branda, S.S., Vik, Å., Friedman, L. and Kolter, R., 2005. Biofilms: the matrix revisited. *Trends in Microbiology*, *13*(1):20–26. doi: 10.1016/j.tim.2004.11.006.

Burmølle, M., Webb, J.S., Rao, D., Hansen, L.H., Sørensen, S.J. and Kjelleberg, S., 2006. Enhanced biofilm formation and increased resistance to antimicrobial agents and bacterial invasion are caused by synergistic interactions in multispecies biofilms. *Applied Environmental Microbiology*, *72*(6):3916–3923. doi: 10.1128/AEM.03022-05.

Cao, N., Du, J., Chen, C., Gong, C.S. and Tsao, G.T., 1997. Production of fumaric acid by immobilized *Rhizopus* using rotary biofilm contactor. *Applied Biochemistry and Biotechnology*, *63*(1):387–394. doi: 10.1007/BF02920440.

Carrillo-Reyes, J., Trably, E., Bernet, N., Latrille, E. and Razo-Flores, E., 2016. High robustness of a simplified microbial consortium producing hydrogen in long-term operation of a biofilm fermentative reactor. *International Journal of Hydrogen Energy*, *41*(4):2367–2376. doi: 10.1016/j.ijhydene.2015.11.131.

Characklis, W.G. and Cooksey, K.E., 1983. Biofilms and microbial fouling. In: Laskin, A.I., ed. *Advances in Applied Microbiology*. Academic Press, 29, 93–138. doi: 10.1016/S0065-2164(08)70355-1.

Characklis, W.G. and Marshall, K.C., 1990. Biofilms: a basis for an interdisciplinary approach. In: Characklis, W.G., Marshall, K.C., eds. *Biofilms*. New York, NY: John Wiley & Sons, Inc., 3–17.

Chen, X. and Ni, B.J., 2016. Anaerobic conversion of hydrogen and carbon dioxide to fatty acids production in a membrane biofilm reactor: a modeling approach. *Chemical Engineering Journal*, *306*:1092–1098. doi: 10.1016/j.cej.2016.08.049.

Chen, H., Zhao, L., Hu, S., Yuan, Z. and Guo, J., 2018. High-rate production of short-chain fatty acids from methane in a mixed-culture membrane biofilm reactor. *Environmental Science & Technology Letters*, *5*(11):662–667. doi: 10.1021/acs.estlett.8b00460.

Cheng, K.C., Demirci, A. and Catchmark, J.M., 2009a. Effects of plastic composite support and pH profiles on pullulan production in a biofilm reactor. *Applied Microbiology and Biotechnology*, *86*(3):853–861. doi: 10.1007/s00253-009-2332-x.

Cheng, K.C., Catchmark, J.M. and Demirci, A., 2009b. Enhanced production of bacterial cellulose by using a biofilm reactor and its material property analysis. *Journal of Biological Engineering*, *3*(1):12. doi: 10.1186/1754-1611-3-12.

Cheng, K.C., Catchmark, J.M. and Demirci, A., 2009c. Effect of different additives on bacterial cellulose production by *Acetobacter xylinum* and analysis of material property. *Cellulose*, *16*(6):1033–1045. doi: 10.1007/s10570-009-9346-5.

Cheng, K.C., Demirci, A. and Catchmark, J.M., 2010a. Advances in biofilm reactors for production of value-added products. *Applied Microbiology and Biotechnology*, *87*(2):445–456. doi: 10.1007/s00253-010-2622-3.

Cheng, K.C., Demirci, A. and Catchmark, J.M., 2010b. Enhanced pullulan production in a biofilm reactor by using response surface methodology. *Journal of Industrial Microbiology & Biotechnology*, *37*(6):587–594. doi: 10.1007/s10295-010-0705-x.

Cheng, K.C., Demirci, A., Catchmark, J.M. and Puri, V.M., 2010c. Modeling of pullulan fermentation by using a color variant strain of *Aureobasidium pullulans*. *Journal of Food Engineering*, *98*(3):353–359.

Cheng, K.C., Demirci, A., Catchmark, J.M. and Puri, V.M., 2011a. Effects of initial ammonium ion concentration on pullulan production by *Aureobasidium pullulans* and its modeling. *Journal of Food Engineering*, *103*(2):115–122. doi: 10.1016/j.jfoodeng.2010.10.004.

Cheng, K.C., Demirci, A. and Catchmark, J.M., 2011b. Continuous pullulan fermentation in a biofilm reactor. *Applied Microbiology and Biotechnology*, *90*(3):921–927. doi: 10.1007/s00253-011-3151-4.

Cheng, K.C., Demirci, A. and Catchmark, J.M., 2011c. Pullulan: biosynthesis, production, and applications. *Applied Microbiology and Biotechnology*, *92*(1):29–44. doi: 10.1007/s00253-011-3477-y.

Cheng, K.C., Catchmark, J.M. and Demirci, A., 2011d. Effects of CMC addition on bacterial cellulose production in a biofilm reactor and its paper sheets analysis. *Biomacromolecules*, *12*(3):730–736. doi: 10.1021/bm101363t.

Converti, A., De Faveri, D., Perego, P., Dominiguez, J.M., Carvalho, J.C.M., Palma, M.S.A. and Del Borghi, M., 2006. Investigation on the transient conditions of a rotating biological contactor for bioethanol production. *Chemical and Biochemical Engineering Quarterly*, *20*(4):401–406.

Cotton, J., Pometto III., A.L. and Gvozdenovic-Jeremic, J., 2001. Continuous lactic acid fermentation using a plastic composite support biofilm reactor. *Applied Microbiology and Biotechnology*, *57*(5–6):626–630. doi: 10.1007/s002530100820.

Del Borghi, M., Converti, A., Parisi, F. and Ferraiolo, G., 1985. Continuous alcohol fermentation in an immobilized cell rotating disk reactor. *Biotechnology and Bioengineering*, *27*(6):761–768. doi: 10.1002/bit.260270602.

Demirci, A., Pometto, A.L. and Johnson, K.E., 1993. Evaluation of biofilm reactor solid support for mixed-culture lactic acid production. *Applied Microbiology and Biotechnology*, *38*(6):728–733. doi: 10.1007/BF00167135.

Demirci, A., Pometto III., A.L. and Ho, K.G., 1997. Ethanol production by *Saccharomyces cerevisiae* in biofilm reactors. *Journal of Industrial Microbiology and Biotechnology*, *19*(4):299–304. doi: 10.1038/sj.jim.2900464.

Demirci, A., Pongtharangkul, T. and Pometto III., A.L., 2007. Applications of biofilm reactors for production of value-added products by microbial fermentation. In: Blaschek, H.P., Wang, H.H., Agle, M.E., eds. *Biofilms in the Food Environment*. Ames, IA: Blackwell Publishing and The Institute of Food Technologists, 167–189. doi: 10.1002/9780470277782.ch8.

Ercan, D. and Demirci, A., 2013. Production of human lysozyme in biofilm reactor and optimization of growth parameters of *Kluyveromyces lactis* K7. *Applied Microbiology and Biotechnology*, *97*(14):6211–6221. doi: 10.1007/s00253-013-4944-4.

Ercan, D. and Demirci, A., 2014. Enhanced human lysozyme production in biofilm reactor by *Kluyveromyces lactis* K7. *Biochemical Engineering Journal*, *92*:2–8. doi: 10.1016/j.bej.2014.04.013.

Ercan, D. and Demirci, A., 2015a. Current and future trends for biofilm reactors for fermentation processes. *Critical Reviews in Biotechnology*, *35*(1):1–14. doi: 10.3109/07388551.2013.793170.

Ercan, D. and Demirci, A., 2015b. Effects of fed-batch and continuous fermentations on human lysozyme production by *Kluyveromyces lactis* K7 in biofilm reactors. *Bioprocess and Biosystems Engineering*, *38*(12):2461–2468. doi: 10.1007/s00449-015-1483-7.

Ercan, D. and Demirci, A., 2015c. Enhanced human lysozyme production by *Kluyveromyces lactis* K7 in biofilm reactor coupled with online recovery system. *Biochemical Engineering Journal*, *98*:68–74. doi: 10.1016/j.bej.2015.02.032.

Ferone, M., Raganati, F., Ercole, A., Olivieri, G., Salatino, P. and Marzocchella, A., 2018. Continuous succinic acid fermentation by *Actinobacillus succinogenes* in a packed-bed biofilm reactor. *Biotechnology for Biofuels*, *11*(1):138. doi: 10.1186/s13068-018-1143-7.

Flemming, H.C. and Wingender, J., 2010. The biofilm matrix. *Nature Reviews Microbiology*, *8*(9):623–633. doi: 10.1038/nrmicro2415.

Fukuda H., 1995. Immobilized microorganism bioreactors. In: Asenjo, J.A., Merchuk, J.C., eds. *Bioreactor System Design*. New York, NY: Marcel Dekker, 339–375.

Germec, M., Turhan, I., Karhan, M. and Demirci, A., 2015. Ethanol production via repeated-batch fermentation from carob pod extract by using *Saccharomyces cerevisiae* in biofilm reactor. *Fuel*, *161*:304–311. doi: 10.1016/j.fuel.2015.08.060.

Germec, M., Turhan, I., Demirci, A. and Karhan, M., 2016. Effect of media sterilization and enrichment on ethanol production from carob extract in a biofilm reactor. *Energy Sources, Part A: Recovery, Utilization, and Environmental Effects*, *38*(21):3268–3272. doi: 10.1080/15567036.2015.1138004.

Germec, M., Karhan, M., Demirci, A. and Turhan, I., 2018. Ethanol production in a biofilm reactor with non-sterile carob extract media and its modeling. *Energy Sources, Part A: Recovery, Utilization, and Environmental Effects*, *40*(22):2726–2734. doi: 10.1080/15567036.2018.1511643.

Germec, M., Cheng, K.C., Karhan, M., Demirci, A. and Turhan, I., 2019. Application of mathematical models to ethanol fermentation in biofilm reactor with carob extract. *Biomass Conversion and Biorefinery*, 1–16. doi: 10.1007/s13399-019-00425-1.

Govender, S., Jacobs, E.P., Leukes, W.D. and Pillay, V.L., 2003. A scalable membrane gradostat reactor for enzyme production using *Phanerochaete chrysosporium*. *Biotechnology Letters*, *25*(2):127–131. doi: 10.1023/A:1021963201340.

Govender, S., Pillay, V.L. and Odhav, B., 2010. Nutrient manipulation as a basis for enzyme production in a gradostat bioreactor. *Enzyme and Microbial Technology*, *46*(7):603–609. doi: 10.1016/j.enzmictec.2010.03.007.

Gross, M., Jarboe, D. and Wen, Z., 2015. Biofilm-based algal cultivation systems. *Applied Microbiology and Biotechnology*, *99*(14):5781–5789. doi: 10.1007/s00253-015-6736-5.

Haapala, R. and Linko, S., 1993. Production of *Phanerochaete chrysosporium* lignin peroxidase under various culture conditions. *Applied Microbiology and Biotechnology*, *40*(4):494–498. doi: 10.1007/bf00175737.

Herselman, J., Bradfield, M.F., Vijayan, U. and Nicol, W., 2017. The effect of carbon dioxide availability on succinic acid production with biofilms of *Actinobacillus succinogenes*. *Biochemical Engineering Journal*, *117*:218–225. doi: 10.1016/j.bej.2016.10.018.

Ho, K.G., Pometto, A.I., Hinz, P.N. and Demirci, A., 1997a. Nutrient leaching and end product accumulation in plastic composite supports for L-(+)-lactic Acid biofilm fermentation. *Applied Environmental Microbiology*, *63*(7):2524–2532.

Ho, K.L., Pometto, A.L. and Hinz, P.N., 1997b. Optimization of L-(+)-lactic acid production by ring and disc plastic composite supports through repeated-batch biofilm fermentation. *Applied Environmental Microbiology*, *63*(7):2533–2542.

Hui, Y.S., Amirul, A.A., Yahya, A.R. and Azizan, M.N.M., 2010. Cellulase production by free and immobilized *Aspergillus terreus. World Journal of Microbiology and Biotechnology*, *26*(1):79. doi: 10.1007/s11274-009-0145-9.

Ikeda, H. and Doi, Y., 1990. A vitamin-K2-binding factor secreted from *Bacillus subtilis. European Journal of Biochemistry*, *192*(1):219–224. doi: 10.1111/j.1432-1033.1990.tb19218.x.

Izmirlioglu, G. and Demirci, A., 2016. Ethanol production in biofilm reactors from potato waste hydrolysate and optimization of growth parameters for *Saccharomyces cerevisiae. Fuel*, *181*:643–651. doi: 10.1016/j.fuel.2016.05.047.

Izmirlioglu, G. and Demirci, A., 2017. Simultaneous saccharification and fermentation of ethanol from potato waste by cocultures of *Aspergillus niger* and *Saccharomyces cerevisiae* in biofilm reactors. *Fuel*, *202*:260–270. doi: 10.1016/j.fuel.2017.04.047.

Jiang, Y., Tang, B., Xu, Z., Liu, K., Xu, Z., Feng, X. and Xu, H., 2016. Improvement of poly-γ-glutamic acid biosynthesis in a moving bed biofilm reactor by *Bacillus subtilis* NX-2. *Bioresource Technology*, *218*:360–366. doi: 10.1016/j.biortech.2016.06.103.

Jones, S.C. and Briedis, D.M., 1992. Adhesion and lignin peroxidase production by the white-rot fungus *Phanerochaete chrysosporium* in a rotating biological contactor. *Journal of Biotechnology*, *24*(3):277–290. doi: 10.1016/0168–1656(92)90037-A.

Junker, B.H., 2004. Scale-up methodologies for *Escherichia coli* and yeast fermentation processes. *Journal of Bioscience and Bioengineering*, *97*(6):347–364. doi: 10.1016/S1389-1723(04)70218-2.

Kang, S.W., Kim, S.W. and Lee, J.S., 1995. Production of cellulase and xylanase in a bubble column using immobilized *Aspergillus Niger* KKS. *Applied Biochemistry and Biotechnology*, *53*(2):101–106. doi: 10.1007/BF02788601.

Keevil, C.W. and Walker, J.T., 1992. Nomarsky DIC microscopy and image analysis of bio-film. *Binary Computational Microbiology*, *4*:93–95.

Khiyami, M.A., Pometto, A.L. and Kennedy, W.J., 2006. Ligninolytic enzyme production by *Phanerochaete chrysosporium* in plastic composite support biofilm stirred tank bioreactors. *Journal of Agricultural and Food Chemistry*, *54*(5):1693–1698. doi: 10.1021/jf051424l.

Kongjan, P., Inchan, S., Chanthong, S., Jariyaboon, R., Reungsang, A. and Sompong, O., 2019. Hydrogen production from xylose by moderate thermophilic mixed cultures using granules and biofilm upflow anaerobic reactors. *International Journal of Hydrogen Energy*, *44*(6):3317–3324. doi: 10.1016/j.ijhydene.2018.09.066.

Kornaros, M. and Lyberatos, G., 2006. Biological treatment of wastewaters from a dye manufacturing company using a trickling filter. *Journal of Hazardous Materials*, *136*(1):95–102. doi: 10.1016/j.jhazmat.2005.11.018.

Kuchma, S.L. and O'Toole, G.A., 2000. Surface-induced and biofilm-induced changes in gene expression. *Current Opinion in Biotechnology*, *11*(5):429–433. doi: 10.1016/S0958-1669(00)00123-3.

Kumar, P., Sharma, R., Ray, S., Mehariya, S., Patel, S.K., Lee, J.K. and Kalia, V.C., 2015. Dark fermentative bioconversion of glycerol to hydrogen by *Bacillus thuringiensis. Bioresource Technology*, 182:383–388. doi: 10.1016/j.biortech.2015.01.138.

Landini, P., Antoniani, D., Burgess, J.G. and Nijland, R., 2010. Molecular mechanisms of compounds affecting bacterial biofilm formation and dispersal. *Applied Microbiology and Biotechnology*, *86*(3):813–823. doi: 10.1007/s00253-010-2468-8.

Lin, S.P., Calvar, I.L., Catchmark, J.M., Liu, J.R., Demirci, A. and Cheng, K.C., 2013. Biosynthesis, production, and applications of bacterial cellulose. *Cellulose*, *20*(5):2191–2219. doi: 10.1007/s10570-013-9994-3.

Lin, S.P., Hsieh, S.C., Chen, K.I., Demirci, A. and Cheng, K.C., 2014. Semicontinuous bacterial cellulose production in a rotating disk bioreactor and its materials properties analysis. *Cellulose*, *21*(1):835–844. doi: 10.1007/s10570-013-0136-8.

Linko, S., 1988. Production and characterization of extracellular lignin peroxidase from immobilized *Phanerochaete chrysosporium* in a 10-l bioreactor. *Enzyme and Microbial Technology*, *10*(7):410–417. doi: 10.1016/0141-0229(88)90035-X.

Liu, J.M., Yu, T.C., Lin, S.P., Hsu, R.J., Hsu, K.D. and Cheng, K.C., 2016. Evaluation of kojic acid production in a repeated-batch PCS biofilm reactor. *Journal of Biotechnology*, *218*:41–48. doi: 10.1016/j.jbiotec.2015.11.023.

Longanesi, L., Frascari, D., Spagni, C., DeWever, H. and Pinelli, D., 2018. Succinic acid production from cheese whey by biofilms of *Actinobacillus succinogenes*: packed bed bioreactor tests. *Journal of Chemical Technology & Biotechnology*, *93*(1):246–256. doi: 10.1002/jctb.5347.

Mahdinia, E., Demirci, A. and Berenjian, A., 2017a. Strain and plastic composite support (PCS) selection for vitamin K (menaquinone-7) production in biofilm reactors. *Bioprocess and Biosystems Engineering*, *40*(10):1507–1517. doi: 10.1007/s00449-017-1807-x.

Mahdinia, E., Demirci, A. and Berenjian, A., 2017b. Production and application of menaquinone-7 (vitamin K2): a new perspective. *World Journal of Microbiology and Biotechnology*, *33*(1):2. doi: 10.1007/s11274-016-2169-2.

Mahdinia, E., Demirci, A. and Berenjian, A., 2018a. Optimization of *Bacillus subtilis* natto growth parameters in glycerol-based medium for vitamin K (menaquinone-7) production in biofilm reactors. *Bioprocess and Biosystems Engineering*, *41*(2):195–204. doi: 10.1007/s00449-017-1857-0.

Mahdinia, E., Demirci, A. and Berenjian, A., 2018b. Utilization of glucose-based medium and optimization of *Bacillus subtilis* natto growth parameters for vitamin K (menaquinone-7) production in biofilm reactors. *Biocatalysis and Agricultural Biotechnology*, *13*:219–224. doi: 10.1016/j.bcab.2017.12.009.

Mahdinia, E., Demirci, A. and Berenjian, A., 2018c. Enhanced vitamin K (menaquinone-7) production by *Bacillus subtilis* natto in biofilm reactors by optimization of glucose-based medium. *Current Pharmaceutical Biotechnology*, *19*(11):917–924. doi: 10.2174/13892 01020666181126120401.

Mahdinia, E., Demirci, A. and Berenjian, A., 2018d. Implementation of fed-batch strategies for vitamin K (menaquinone-7) production by *Bacillus subtilis* natto in biofilm reactors. *Applied Microbiology and Biotechnology*, *102*(21):9147–9157. doi: 10.1007/s00253-018-9340-7.

Mahdinia, E., Demirci, A. and Berenjian, A., 2019a. Effects of medium components in a glycerol-based medium on vitamin K (menaquinone-7) production by *Bacillus subtilis* natto in biofilm reactors. *Bioprocess and Biosystems Engineering*, *42*(2):223–232. doi: 10.1007/s00449-018-2027-8.

Mahdinia, E., Mamouri, S.J., Puri, V.M., Demirci, A. and Berenjian, A., 2019b. Modeling of vitamin K (menaquinoe-7) fermentation by *Bacillus subtilis* natto in biofilm reactors. *Biocatalysis and Agricultural Biotechnology*, 17:196–202. doi: 10.1016/j.bcab.2018.11.022.

Mahdinia, E., Demirci, A. and Berenjian, A., 2019c. Biofilm reactors as a promising method for vitamin K (menaquinone-7) production. *Applied Microbiology and Biotechnology*, *103*(14):5583–5592. doi: 10.1007/s00253-019-09913-w.

Mahdinia, E., Cekmecelioglu, D. and Demirci, A., 2019d. Bioreactor scale-up. In: Berenjian, A., ed. *Essentials in Fermentation Technology. Learning Materials in Biosciences*. Springer, Cham, 213–236. doi: 10.1007/978-3-030-16230-6_7.

Mahdinia, E., Demirci, A. and Berenjian, A., 2019e. Evaluation of vitamin K (menaquinone-7) stability and secretion in glucose and glycerol-based media by *Bacillus subtilis* natto. *Acta Alimentaria*, *48*(4):405–414. doi: 10.1556/066.2019.48.4.1.
Mahdinia, E., Liu, S., Demirci, A. and Puri, V.M., 2020. Microbial Growth Models. In: Demirci, A., Feng, H. and Krishnamurthy, K., eds. *Food Safety Engineering*. Cham, Switzerland: Springer. 357-398. doi: 10.1007/978-3-030-42660-6_14.
Martinov, M., Hadjiev, D. and Vlaev, S., 2010. Gas–liquid dispersion in a fibrous fixed bed biofilm reactor at growth and nongrowth conditions. *Process Biochemistry*, *45*(7):1023–1029. doi: 10.1016/j.procbio.2010.03.008.
Melo, L.F. and Oliveira, R., 2001. Biofilm reactors. In: Cabral, J.M.S., Mota, M. and Tramper, J., eds. *Multiphase Bioreactor Design*. New York, NY: Taylor & Francis, 271–309. doi: 10.1201/b12644.
Mitra, S., Banerjee, P., Gachhui, R. and Mukherjee, J., 2011. Cellulase and xylanase activity in relation to biofilm formation by two intertidal filamentous fungi in a novel polymethylmethacrylate conico-cylindrical flask. *Bioprocess and Biosystems Engineering*, *34*(9):1087–1101. doi: 10.1007/s00449-011-0559-2.
Mitra, S., Thawrani, D., Banerjee, P., Gachhui, R. and Mukherjee, J., 2012. Induced biofilm cultivation enhances riboflavin production by an intertidally derived *Candida famata*. *Applied Biochemistry and Biotechnology*, *166*(8):1991–2006. doi: 10.1007/s12010-012-9626-7.
Mokwatlo, S.C. and Nicol, W., 2017. Structure and cell viability analysis of *Actinobacillus succinogenes* biofilms as biocatalysts for succinic acid production. *Biochemical Engineering Journal*, *128*:134–140. doi: 10.1016/j.bej.2017.09.013.
Muffler, K. and Ulber, R. eds., 2014. *Productive Biofilms. Advances in Biochemical Engineering/Biotechnology*. Berlin: Springer, 146. doi: 10.1007/978-3-319-09695-7.
Nakashima, T., Fukuda, H., Kyotani, S. and Morikawa, H., 1988. Culture conditions for intracellular lipase production by *Rhizopus chinensis* and its immobilization within biomass support particles. *Journal of Fermentation Technology*, *66*(4):441–448. doi: 10.1016/0385-6380(88)90012-X.
Narayanan, C.M. and Das, S., 2017. Studies on synthesis of lactic acid from molasses and cheese whey in semi-fluidized bed biofilm reactors. *International Journal of Environment and Waste Management*, *19*(1):1–20. doi: 10.1504/ijewm.2017.083555.
Nerenberg, R., 2016. The membrane-biofilm reactor (MBfR) as a counterdiffusional biofilm process. *Current Opinion in Biotechnology*, *38*:131–136. doi: 10.1016/j.copbio.2016.01.015.
Nyvad, B. and Fejerskov, O., 1997. Assessing the stage of caries lesion activity on the basis of clinical and microbiological examination. *Community Dentistry and Oral Epidemiology*, *25*(1):69–75. doi: 10.1111/j.1600-0528.1997.tb00901.x.
Oriel, P., 1988. Immobilization of recombinant *Escherichia coli* in silicone polymer beads. *Enzyme and Microbial Technology*, *10*(9):518–523. doi: 10.1016/0141-0229(88)90043-9.
Parameswaran, P., Krajmalnik-Brown, R., Popat, S., Rittmann, B., Torres, C., and Arizona State University, 2019. *Membrane Biofilm Reactors, Systems, and Methods for Producing Organic Products*. U.S. Patent 10435659.
Pereira, M.A., Alves, M.M., Azeredo, J., Mota, M. and Oliveira, R., 2000. Influence of physicochemical properties of porous microcarriers on the adhesion of an anaerobic consortium. *Journal of Industrial Microbiology and Biotechnology*, *24*(3):181–186. doi: 10.1038/sj.jim.2900799.
Pongtharangkul, T. and Demirci, A., 2006a. Evaluation of culture medium for nisin production in a repeated-batch biofilm reactor. *Biotechnology Progress*, *22*(1):217–224. doi: 10.1021/bp050295q.
Pongtharangkul, T. and Demirci, A., 2006b. Effects of pH profiles on nisin production in biofilm reactor. *Applied Microbiology and Biotechnology*, *71*(6):804–811. doi: 10.1007/s00253-005-0220-6.

Pongtharangkul, T. and Demirci, A., 2006c. Effects of fed-batch fermentation and pH profiles on nisin production in suspended-cell and biofilm reactors. *Applied Microbiology and Biotechnology*, *73*(1):73–79. doi: 10.1007/s00253-006-0459-6.

Pongtharangku, T. and Demirci, A., 2007. Online recovery of nisin during fermentation and its effect on nisin production in biofilm reactor. *Applied Microbiology and Biotechnology*, *74*(3):555–562. doi: 10.1007/s00253-006-0697-7.

Pongtharangkul, T., Demirci, A. and Puri, V.M., 2008. Modeling of growth and nisin production by *Lactococcus lactis* during batch fermentation. *Biological Engineering Transactions*, *1*(3):265–275. doi: 10.13031/2013.25335.

Pometto III, A.L., Demirci, A., Johnson, K.E., and Iowa State University Research Foundation (ISURF), 1997. *Immobilization of Microorganisms on a Support Made of Synthetic Polymer and Plant Material*, U.S. Patent 5595893.

Prakash, J., Gupta, R.K., Priyanka, X.X. and Kalia, V.C., 2018. Bioprocessing of biodiesel industry effluent by immobilized bacteria to produce value-added products. *Applied Biochemistry and Biotechnology*, *185*(1):179–190. doi: 10.1007/s12010-017-2637-7.

Qureshi, N., Annous, B.A., Ezeji, T.C., Karcher, P. and Maddox, I.S., 2005. Biofilm reactors for industrial bioconversion processes: employing potential of enhanced reaction rates. *Microbial Cell Factories*, *4*(24):1–21. doi: 10.1186/1475-2859-4-24.

Ren, L., Peng, C., Hu, X., Han, Y. and Huang, H., 2019. Microbial production of vitamin K2: current status and future prospects. *Biotechnology Advances*, *39*:107453. doi: 10.1016/j.biotechadv.2019.107453.

Roukas, T., 2018. Modified rotary biofilm reactor: A new tool for enhanced carotene productivity by *Blakeslea trispora*. *Journal of Cleaner Production*, *174*:1114–1121. doi: 10.1016/j.jclepro.2017.11.048.

Sakhamuri, S., Bober, J., Irudayaraj, J. and Demirci, A., 2004. Simultaneous determination of multiple components in nisin fermentation using FTIR spectroscopy. *Agricultural Engineering International: CIGR Journal*, *6*:1–16.

Sarkar, S., 2015. Enhanced antimicrobials production by *Streptomyces sundarbansensis* sp. nov. in a novel extended surface biofilm reactor. *International Journal of Advanced Biotechnology and Research*, *6*(1):12–20.

Schwartz, K., Stephenson, R., Hernandez, M., Jambang, N. and Boles, B.R., 2010. The use of drip flow and rotating disk reactors for *Staphylococcus aureus* biofilm analysis. *Journal of Visualized Experiments*, *46*:e2470. doi: 10.3791/2470.

Shahab, R.L., Luterbacher, J.S., Brethauer, S. and Studer, M.H., 2018. Consolidated bioprocessing of lignocellulosic biomass to lactic acid by a synthetic fungal-bacterial consortium. *Biotechnology and Bioengineering*, *115*(5):1207–1215. doi: 10.1002/bit.26541.

Shen, Y., Brown, R.C. and Wen, Z., 2017. Syngas fermentation by Clostridium carboxidivorans P7 in a horizontal rotating packed bed biofilm reactor with enhanced ethanol production. *Applied Energy*, *187*:585–594. doi: 10.1016/j.apenergy.2016.11.084.

Shen, N., Dai, K., Xia, X.Y., Zeng, R.J. and Zhang, F., 2018. Conversion of syngas (CO and H_2) to biochemicals by mixed culture fermentation in mesophilic and thermophilic hollow-fiber membrane biofilm reactors. *Journal of Cleaner Production*, *202*:536–542. doi: 10.1016/j.jclepro.2018.08.162.

Shuler, M.L., Kargi, F. and DeLisa, M., 2017. *Bioprocess Engineering: Basic Concepts*. Englewood Cliffs, NJ: Prentice Hall; 576.

Solomon, M.S. and Petersen, F.W., 2002. Membrane bioreactor production of lignin and manganese peroxidase. *Membrane Technology*, 2002(4):6–8. doi: 10.1016/S0958-2118(02)80131-8.

Sousa, M., Azeredo, J., Feijo, J. and Oliveira, R., 1997. Polymeric supports for the adhesion of a consortium of autotrophic nitrifying bacteria. *Biotechnology Techniques*, *11*(10):751–754. doi: 10.1023/A:1018400619440.

Stewart, P.S., Murga, R., Srinivasan, R. and de Beer, D., 1995. Biofilm structural heterogeneity visualized by three microscopic methods. *Water Research*, *29*(8):2006–2009. doi: 10.1016/0043-1354(94)00339-9.

Tang, B., Yu, C., Bin, L., Zhao, Y., Feng, X., Huang, S., Fu, F., Ding, J., Chen, C., Li, P. and Chen, Q., 2016. Essential factors of an integrated moving bed biofilm reactor–membrane bioreactor: adhesion characteristics and microbial community of the biofilm. *Bioresource Technology*, 211:574–583. doi: 10.1016/j.biortech.2016.03.136.

Teixeira, P. and Oliveira, R., 1999. Influence of surface characteristics on the adhesion of *Alcaligenes denitrificans* to polymeric substrates. *Journal of Adhesion Science and Technology*, *13*(11):1287–1294. doi: 10.1163/156856199X00190.

Todhanakasem, T., Salangsing, O.L., Koomphongse, P., Kanokratana, P. and Champreda, V., 2019. *Zymomonas mobilis* biofilm reactor for ethanol production using rice straw hydrolysate under continuous and repeated batch processes. *Frontiers in Microbiology*, *10*:1777. doi: 10.3389/fmicb.2019.01777.

Tomczak, W., Ferrasse, J.H., Giudici-Orticoni, M.T. and Soric, A., 2018. Effect of hydraulic retention time on a continuous biohydrogen production in a packed bed biofilm reactor with recirculation flow of the liquid phase. *International Journal of Hydrogen Energy*, *43*(41):18883–18895. doi: 10.1016/j.ijhydene.2018.08.094.

Urbance, S.E., Pometto III, A.L., DiSpirito, A.A. and Demirci, A., 2003. Medium evaluation and plastic composite support ingredient selection for biofilm formation and succinic acid production by *Actinobacillus succinogenes*. *Food Biotechnology*, *17*(1):53–65. doi: 10.1081/FBT-120019984

Vasudevan, R., 2014. Biofilms: microbial cities of scientific significance. *Journal of Microbiology & Experimentation*, *1*(3):84–98. doi: 10.15406/jmen.2014.01.00014.

Velázquez, A., Pometto Iii, A.L., Ho, K.L.G. and Demirci, A., 2001. Evaluation of plastic-composite supports in repeated fed-batch biofilm lactic acid fermentation by *Lactobacillus casei*. *Applied Microbiology and Biotechnology*, *55*(4):434–441. doi: 10.1007/s002530000530.

Venkatadri, R., Tsai, S.P., Vukanic, N. and Hein, L.B., 1992. Use of a biofilm membrane reactor for the production of lignin peroxidase and treatment of pentachlorophenol by *Phanerochaete chrysosporium*. *Hazardous Waste and Hazardous Materials*, *9*(3):231–243. doi: 10.1089/hwm.1992.9.231.

Voběrková, S., Hermanová, S., Hrubanová, K. and Krzyžánek, V., 2016. Biofilm formation and extracellular polymeric substances (EPS) production by *Bacillus subtilis* depending on nutritional conditions in the presence of polyester film. *Folia Microbiologica*, *61*(2):91–100. doi: 10.1007/s12223-015-0406-y.

Wang, X., Bi, X., Hem, L.J. and Ratnaweera, H., 2018a. Microbial community composition of a multistage moving bed biofilm reactor and its interaction with kinetic model parameters estimation. *Journal of Environmental Management*, *218*:340–347. doi: 10.1016/j.jenvman.2018.04.015.

Wang, H.J., Dai, K., Xia, X.Y., Wang, Y.Q., Zeng, R.J. and Zhang, F., 2018b. Tunable production of ethanol and acetate from synthesis gas by mesophilic mixed culture fermentation in a hollow fiber membrane biofilm reactor. *Journal of Cleaner Production*, *187*:165–170. doi: 10.1016/j.jclepro.2018.03.193.

Wang, Y.Q., Zhang, F., Zhang, W., Dai, K., Wang, H.J., Li, X. and Zeng, R.J., 2018c. Hydrogen and carbon dioxide mixed culture fermentation in a hollow-fiber membrane biofilm reactor at 25°C. *Bioresource Technology*, *249*:659–665. doi: 10.1016/j.biortech.2017.10.054.

Wang, H.J., Dai, K., Wang, Y.Q., Wang, H.F., Zhang, F. and Zeng, R.J., 2018d. Mixed culture fermentation of synthesis gas in the microfiltration and ultrafiltration hollow-fiber membrane biofilm reactors. *Bioresource Technology*, *267*:650–656. doi: 10.1016/j.biortech.2018.07.098.

Webb, C., Fukuda, H. and Atkinson, B., 1986. The production of cellulase in a spouted bed fermentor using cells immobilized in biomass support particles. *Biotechnology and Bioengineering*, *28*(1):41–50. doi: 10.1002/bit.260280107.

Wen, H.Q., Du, J., Xing, D.F., Ding, J., Ren, N.Q. and Liu, B.F., 2017. Enhanced photofermentative hydrogen production of *Rhodopseudomonas* sp. nov. strain A7 by biofilm reactor. *International Journal of Hydrogen Energy*, *42*(29):18288–18294. doi: 10.1016/j.ijhydene.2017.04.150.

Wimpenny, J., Manz, W. and Szewzyk, U., 2000. Heterogeneity in biofilms. *FEMS Microbiology Reviews*, *24*(5):661–671. doi: 10.1111/j.1574-6976.2000.tb00565.x.

Wood, J.L., Miller, C.D., Sims, R.C. and Takemoto, J.Y., 2015. Biomass and phycocyanin production from cyanobacteria dominated biofilm reactors cultured using oilfield and natural gas extraction produced water. *Algal Research*, *11*:165–168. doi: 10.1016/j.algal.2015.06.015.

Xu, H., Lee, H.Y. and Ahn, J., 2011. Characteristics of biofilm formation by selected foodborne pathogens. *Journal of Food Safety*, *31*(1), 91–97. doi: 10.1111/j.1745-4565.2010.00271.x.

You, G., Hou, J., Xu, Y., Wang, C., Wang, P., Miao, L., Ao, Y., Li, Y. and Lv, B., 2015. Effects of CeO_2 nanoparticles on production and physicochemical characteristics of extracellular polymeric substances in biofilms in sequencing batch biofilm reactor. *Bioresource Technology*, *194*:91–98. doi: 10.1016/j.biortech.2015.07.006.

Yuan, P., Cui, S., Liu, Y., Li, J., Du, G. and Liu, L., 2019. Metabolic engineering for the production of fat-soluble vitamins: advances and perspectives. *Applied Microbiology and Biotechnology*, *104*:935–951. doi: 10.1007/s00253-019-10157-x.

Zhang, L., Xing, Z., Zhang, H., Li, Z., Wu, X., Zhang, X., Zhang, Y. and Zhou, W., 2016. High thermostable ordered mesoporous SiO_2–TiO_2 coated circulating-bed biofilm reactor for unpredictable photocatalytic and biocatalytic performance. *Applied Catalysis B: Environmental*, *180*:521–529. doi: 10.1016/j.apcatb.2015.07.002.

Zhao, X.S., Huang, X.J., Yao, J.T., Zhou, Y. and Jia, R., 2015. Fungal growth and manganese peroxidase production in a deep tray solid-state bioreactor, and in vitro decolorization of poly R-478 by MnP. *Journal of Microbiology and Biotechnology*, *25*:803–813. doi: 10.4014/jmb.1410.10054.

Zune, Q., Delepierre, A., Gofflot, S., Bauwens, J., Twizere, J.C., Punt, P.J., Francis, F., Toye, D., Bawin, T. and Delvigne, F., 2015. A fungal biofilm reactor based on metal structured packing improves the quality of a Gla::GFP fusion protein produced by *Aspergillus oryzae*. *Applied Microbiology and Biotechnology*, *99*(15):6241–6254. doi: 10.1007/s00253-015-6608-z.

5 Biofilms in Aquaculture

Mahroze Fatima
University of Veterinary and Animal Sciences

Syed Zakir Hussain Shah
University of Gujrat

Muhammad Bilal
Huaiyin Institute of Technology

CONTENTS

5.1 INTRODUCTION

Aquaculture is the fastest growing sector after agriculture and poultry farming and this trend is expected to continue (FAO, 2017). The global Fisheries and Aquaculture production has reached approximately 2 billion tons of fish, of which 47% came from aquaculture (FAO, 2018). Aquaculture has been playing a very important role in underdeveloping countries by providing livelihood and good quality protein (El-gayar and Leung, 2000). Therefore, aquaculturists are focusing on enhanced yield and sustainable aquaproduction, for which, the expansion of already existing ponds and intensive culture (fully controlled culture) in the ponds, throughout the world, has a great potential. Nevertheless, frequent water replacement, the high stocking density of fish, and remaining feed and excessive fertilizer, which are used to intensify the fish production rate, create a large number of aquatic wastes (Beveridge et al., 1997).

Moreover, the cost of intensive aquaproduction, in terms of technical expertise and huge investments, is unaffordable for a poor farmer, which implies that along with the environmental conditions, the financial status of a farmer is the major constraint in the growth of aquaculture. In addition, intensive aquaculture also leads to high aquatic pollution due to heavy loads of organic matter and nutrients coming from dead fish and remaining feed, which are likely to cause water toxicity and environmental risks in the long run (Beveridge et al., 1997; Piedrahita, 2003). The only way to reduce the aforesaid dissolved nutrients was to frequently exchange water with clean freshwater; however, this will also increase the production cost of culture.

On the other hand, a new alternative to maintain the water quality in aquaculture is the application of biofloc and biofilm techniques (Avnimelech, 2006). This new alternative way involves the biological treatment of water, where excessive nutrients are absorbed from water by microorganisms precolonized on filters having a high surface to volume ratio (Wheaton, 1977). In biofilms, microorganisms are associated with the matrix of extracellular polymeric substances (EPS) linked with submerged surfaces that are responsible for the aquatic nitrogen cycle and other biochemical cycles in water (Meyer-Reil, 1994). The nitrogen cycling is done by heterotrophic bacteria, which intensively grow on the overloading unwanted nutrients in the aquaculture. The microorganisms convert overloading waste nutrients into organic matter which are consumed by aquatic animals for growth. The occurrence of biofilms in aquaculture reduced the cost of shrimp culture by decreasing the water exchange rate (Thompson et al., 2002). Hopkins et al. (1995) and McIntosh et al. (2000) observed that the reduced or suppressed water exchange rate, due to the presence of biofilms, did not cause any stress and harm to cultured animals). It was also demonstrated that nitrogen uptake by biofilms reduces the pathogenic bacteria in the water body, as enormously high values of nitrogenous compounds intensify the production of pathogenic bacteria (Austin and Austin, 1999; Brock and Main, 1994; Thompson et al., 2002). Moreover, the protozoans present in the biofilms also play a vital part in the elimination of pathogenic bacteria by grazing them (Thompson et al., 1999). Furthermore, the microalgae present in biofilms produce some antibodies which inhibit the growth of harmful pathogenic bacteria (Alabi al., 1999). Consequently, it is assumed that the removal of biofilms will enhance the chances of production and multiplication of pathogenic bacteria (Thompson et al., 2002).

Apart from intensive fish culture, the substratum attached biofilms significantly increases the production of finfish and shellfish in extensive freshwater culture by manipulating C:N ratio (van Dam et al., 2002; Hargreaves, 2006). For the development of biofilms, the hard substrata and cheap carbohydrates resources are necessary (Asaduzzaman et al., 2008), which can be very easily installed in farmer's typical agriculture system. Various studies reported a significantly increased growth performance and survival rate of freshwater prawns in a substratum-provided culture system, compared to the substratum-less culture system (Tidwell and Bratvold, 2005; Uddin et al., 2006). The biofilms improve the water quality by trapping suspended solids and organic matter, speed up the process of nitrification with the help of microorganisms, provide natural food to culture animals, increase the production of heterotrophic unicellular proteins, and offer shelter to aquatic organisms by minimizing territorial effects (McIntosh et al., 2000; Hari et al., 2004; Crab et al., 2007;

Avnimelech, 2007). Hence, the major objective of this chapter is to highlight the importance of biofilms in the sustainable production of aquaculture.

5.2 FORMATION OF BIOFILMS IN WATER

Biofilm formation starts when certain microorganisms adhere to any submerged surface by producing a slimy glue-like secretion and reproduce there. After a few hours, the colonies of that microorganism (bacteria, yeast, protozoa, arthropods) establish on the surface and form a macromolecular film (Whal, 1989; Johnson, 2008). The major advantage of biofilm formation is the protection of microorganisms from the adverse aquatic environment. A biofilm community can be formed by a single species or by multiple species; however, a community based on multiple species provides an optimum chemical and physical environment for an organism's survival and growth (King et al., 2008). Biofilms can be formed on a wide range of surfaces including medical devices, water supply pipes, living tissues, natural materials like rocks, aquatic plants, metals, and plastics (Donlan, 2002; Kordmahaleh and Shalke, 2013). The formation of biofilms is a dynamic response to environmental stimuli, where cells organize themselves into microcolonies first, then cell division and recruitment take place on the extracellular matrix. While attaching to the extracellular matrix, cells develop complex, well-differentiated associations to facilitate uptake of nutrients (Toutain et al., 2004). According to Hall-Stoodley and Stoodley (2002), the motility of surface and redistribution of attached cells are the most significant parts of biofilm formation (Figures 5.1 and 5.2).

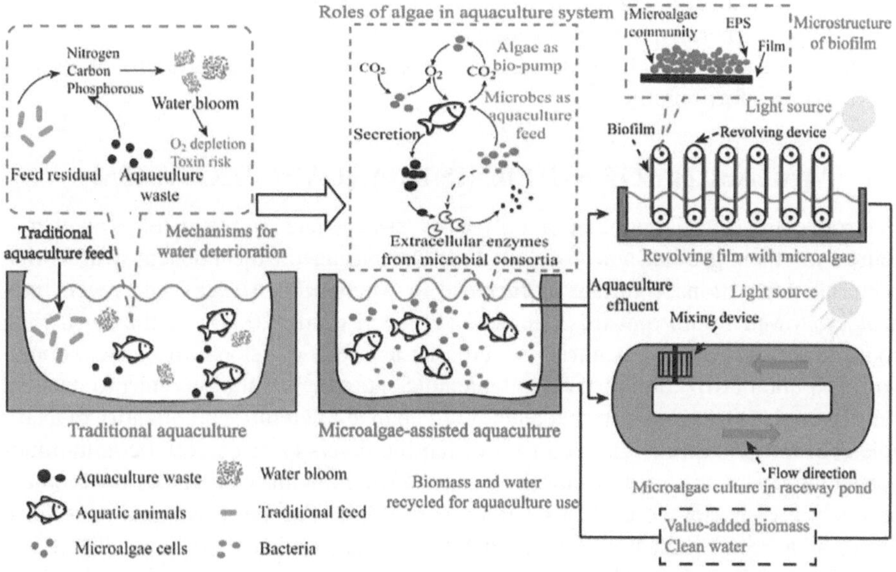

FIGURE 5.1 A comparison of traditional and biofilm-based aquaculture. (Reprinted with permission from Han, P., et al., *Appl. Sci.*, 9, 2377–2397, 2019.)

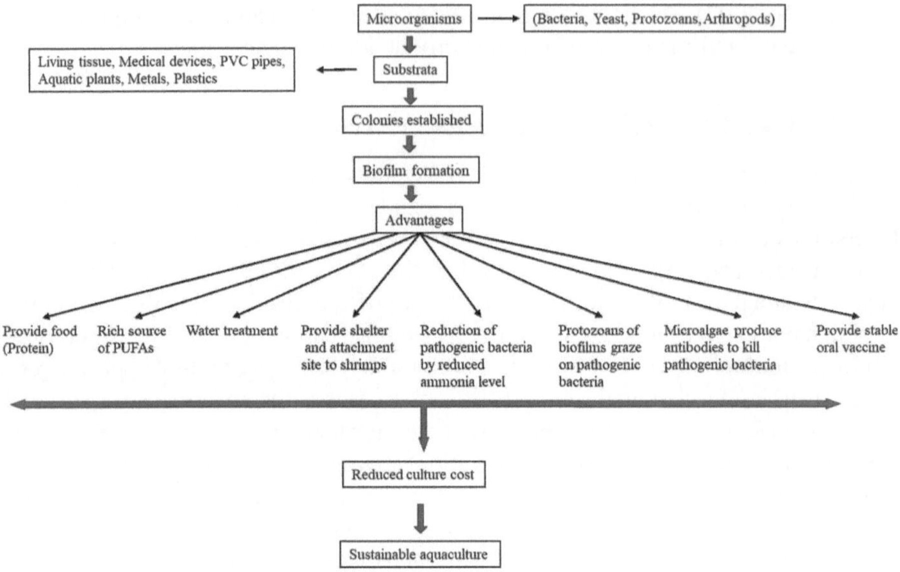

FIGURE 5.2 Biofilm formation and benefits to aquaculture.

The biofilm development depends on (1) the properties of the submerged surface to be colonized, (2) constituents of the organisms, and (3) physicochemical parameters of aquatic medium. The detachment of biofilms includes both external stimuli as well as internal factors, although many species took advantage of dispersal and consider it a chance for colonizing in new habitats (Sauer et al., 2002; Hall-Stoodley et al., 2004). Generally, biofilm formation proceeds as

Microorganisms attach to the surface → microorganism form microcolonies → microcolonies form biofilms

5.3 AVAILABILITY OF NUTRIENTS IN AQUATIC ECOSYSTEM

Heterotrophic bacterial population utilizes excess nutrients in the water as an energy source for their growth and development. Depending on the nutrient availability, bacterial populations continue to increase in water until 4 weeks and reach their peak, followed by an onwards decrease (Pradeep et al., 2004). At the start, high concentrations of available nutrients increase the rate of division of bacterial population, and upon utilization of the available nutrients, the populations start to decline. Gatune et al. (2012) reported that a temporal change in quality and quantity of available organic substrata is the factor on which the diversity of bacterial communities depends on. The presence of inorganic nutrients significantly enhances the biomass production of periphytons (Ghosh and Gaur, 1994). The periphyton biomass also serves as a strong indicator of eutrophication in aquatic water bodies (Mattila and Raeisaenen, 1998). In freshwater aquatic environment phosphorus (Vymazal et al., 1994, carbon (Sherman and Fairchild, 1989), and nitrogen (Barnese and Schelske, 1994) are the most limiting nutrients, of which phosphorus is of prime consideration

TABLE 5.1
Mineral Ratios and Organic Matter of a Water Body

Mineral Ratios	Biomass	References
High Si:P/N:P	Diatoms	Baffico and Pedroso (1996)
Low Si:P/N:P	Cyanophytes	Baffico and Pedroso (1996)
High Si:N	Diatoms	Sommer (1996)
High N:P	Chlorophytes	Sommer (1996)

Si = silicon, P = phosphorus, N = nitrogen.

(Ghosh and Gaur, 1994). The limiting role of C and N depends on the hardness, acidity, and presence of algal species in the aquatic environment. The presence of different types of organic matter depends on the ratios of different minerals, as shown in Table 5.1.

Phosphorus and nitrogen are supplemented in the form of fertilizer in fishponds, and the excessive concentration of these fertilizers also provides nutrients for biofilms. The structure of a biofilm depends on the concentrations of dissolved organic matter. Wetzel et al. (1997) reported that the mucilage concentration was lower in the biofilm community, which was treated with dissolved organic carbon, compared to an untreated community. Moreover, a boost in the growth rate of biofilm cells grown on EPS was observed in the presence of excessive nutrients (Molobela et al., 2010).

5.4 GRAZING EFFECT

Grazing enforced by ciliates and flagellates is an important determinant of biofilm communities. The effect of grazing is stronger in the control of a biofilm population as compared to the effect of nutrients as periphyton recycle and take nutrients from substratum as well. Huchette et al. (2000) highlighted the importance of grazing with a case example of tilapia culture in plastic cages in Bangladesh, where Myxophyceae and Chlorophyceae concentrations were higher before the stocking of fish, and after fish stocking the diatoms dominated the periphyton community, which implies that fish conveniently grazed on Myxophyceae and Chlorophyceae, leaving the diatom ratio higher. Hence, it is evident from the case example that biofilms enhance the chances of natural food availability to cultured species, leading to their increased production.

5.5 AQUATIC SPECIES SUITABLE FOR BIOFILM-BASED CULTURE

It has been evidenced from previous reports that fish species showed more growth potential on biofilm compared to other crustacean species due to their high grazing efficacy Asaduzzaman et al., 2010). Table 5.2 shows a list of aquaculture species that showed an efficient culture potential on biofilms.

TABLE 5.2
Aquaculture Species Suitable for Biofilm-Based Culture

Species	References
Nile tilapia	Shrestha and Knud-Hansen (1994)
Carp	Ramesh et al. (1999)
Rohu (*Labeo rohita*)	Azim et al. (2001)
Orange fin labeo (*Morulius calbasu*)	Azim et al. (2001)
Kuria labeo (*L. gonius*)	Azim et al. (2001)
Penaeid shrimp, *Fenneropenaeus paulensis*	Ballester et al. (2007) and Thompson et al. (2002)
Penaeusesculentus	Burford et al. (2004)
L. vannamei	Audelo-Naranjo et al. (2011) and Moss and Moss (2004)
P. monodon	Anand et al. (2012), Arnold et al. (2009), and Khatoon et al. (2007)
Acrobrachium rosenbergii	Tidwell et al. (1998)
Farfantepenaeus merguiensis	Erler et al. (2004)
Tropical rock lobster (*Panulirusornatus*)	Bourne et al. (2006)
Cirrhinus mrigala	Bharti et al. (2013) and Mridula et al. (2006)
Fusitriton brasiliensis	Viau et al. (2013)

5.6 SUBSTRATES SUITABLE FOR BIOFILM-BASED CULTURE

The substrates, which are easily degradable and have high fiber contents, are more suitable for biofilm-based aquaculture, ultimately resulting in a yield of cultured species. A review of different substrates with their production potential is given in Table 5.3.

5.7 WATER QUALITY IMPROVEMENT

The attached and suspended growth are two terms in which biological nitrification can be accomplished. In suspended growth, the microorganisms are in direct contact with water and are mobile in liquid media. In contrast, in the process of attached growth or fixed film, the bacteria do not move freely and are attached to a medium of the solid support with the help of a viscoelastic layer of biofilm. Therefore, when the process of attached growth is compared with the process of suspended growth, it shows many advantages like (1) easy handling, (2) wash off safety of bacterial populations, and (3) enhanced process stability due to increased shock resistance (Fitch et al., 1998; Nogueira et al., 1998). Biofilm and probiotic applications efficiently enhanced the quality of water during the culture of *Catla catla* fingerlings (Pradeep et al. 2003). In the substrata-based treatments, the concentration of total ammonia was decreased in comparison to control and feed-based treatments, which were perhaps due to increased nitrification level in the substrata-based treatments (Azim et al., 2002a). From research studies, it was observed that in cultured water the level of ammonium was reduced by the occurrence of nitrifying bacteria in the biofilm (; Ramesh et al., 1999; Bharti et al., 2013). More clearly, the concentrations

TABLE 5.3
Potential of Different Substrates for Biofilm-Based Culture

Culture Species	Substrate	Increase in Production (%)	References
Labeo rohita	Sugarcane bagasse	47.5	Ramesh et al. (1999)
	Paddy straw	29.1	
	Dried *Eichhornea* for	17.6	
Cyprinus carpio	Sugarcane bagasse	47.4	Ramesh et al. (1999)
	Paddy straw	32.9	
	Dried *Eichhornea* for	20.7	
Cyprinus carpio, Labeo rohita, and *Oreochromis mossambicus*	Sugarcane bagasse	<50	Umesh et al. (1999)
L. rohita	Sugarcane bagasse	93.69	Keshavanath et al. (2012)
	Palm leaf	103	
	Bamboo mate	44	
Common carp	Sugarcane bagasse	98	Keshavanath et al. (2012)
	Palm leaf	74	
	Coconut leaf	100	
	Bamboo mate	20	
Litopenaeus vannamei	Molasses	31.4	Schveitzer et al. (2013)
Freshwater prawn	Bamboo shoots	23	Asaduzzaman et al. (2008)
Polyculture of *L. rohita, Catla catla,* and *L. calbasu*	Bamboo	71	Azim et al. (2002)
	Jutestick	67	
	Kanchi	66	

of nitrifying bacteria leading to a rise in concentration of nitrates and nitrites play a vigorous role in the management of water quality parameters (Kaiser and Wheaton, 1983). On the contrary, ammonium is also absorbed by microalgae, which utilize it in the formation of new biomass (Thompson et al., 2002).

Nitrogen uptake by biofilms also reduces the levels of pathogenic bacteria, because in normal conditions these pathogens occur only when the values of nitrogenous compounds become extremely high (Austin and Austin, 1999; Brock and Main, 1994). The major factor behind the eutrophication of coastal water and rivers is the excretion of wastewater directly from hatcheries in huge concentrations, which can be minimized by various periphytic microalgae used to significantly lower the levels of nitrite and ammonia from the system (Ziemann et al., 1992). By using biofilm technology, the levels of NO_2-N, total ammonia nitrogen (TAN) and soluble reactive phosphorus (SRP) were decreased to more than 80% in the Oscillatoria tanks, while their levels were reduced to 60% in Cymbella, Navicula, and Amphora tanks, without water exchange during a culture period of 60 days. In hatchery tanks, the use of biofilms has an additional benefit beyond the reduction of excessive nutrients, which is the minimized risk of pathogen invasion resulting from incoming water during water exchange (Khatoon et al., 2007). Hence, it was observed that, without the exchange of water, in biofilm-based experimental tanks the water was clear compared to control tanks in which the water was turbid. Biofilms showed the potential to decrease the level of phosphorus and other spare nutrients present in water (Bratvold and Browdy, 2001; Hansson, 1989). Furthermore, the communities of biofilm trap the organic matter and decrease the level of water turbidity (van Dam et al., 2002).

The water quality parameters were in the permissible range during shrimp (*P. monodon*) culture when bamboo was used as substrata (Anand et al., 2013). Cohen et al. (2005) reported a low nitrite concentration in the raceway system due to the complete oxidation of ammonia to nitrates. In aquaculture, a reduced concentration of nitrites and ammonia can be obtained without water exchange through reduction of nitrification process by microbial (Ebeling et al., 2006). In freshwater fishponds, the concentration of total ammonia was reduced by substrate addition, which enhanced the quality of water through the formation of biofilm on that substratum (Ramesh et al., 1999; Dharmaraj et al., 2002). The average concentration of total ammonia was high (0.95 mg/L) in the ponds in which substratum was absent, while the concentration was less (0.56 mg/L) in the ponds in which substratum was present (Azim et al. 2002b); moreover, the nitrite-nitrogen concentration in the column of water was also decreased by the biofilm on substratum (Asaduzzaman et al., 2008). Keshavanath et al. (2012) reported that in aquaculture, both the biodegradable and the natural substrata provide optimum water quality conditions for carp culture. Additionally, Viau et al. (2013) studied a significantly low level of nitrites in biofilm-based treatments, when compared to feed-based treatments.

5.8 BIOFILM AS FOOD SOURCE

The beneficial role of the use of submerged substratum in fishponds for enhanced production rate has been proven from many research experiments (Keshavanath et al., 2002; Jana et al., 2004; Azim et al., 2005; Asaduzzaman et al., 2008). Based on the nutritive value, the products of probiotic bacteria and microalgae are well known and mostly used as a stimulant in the diet of shrimp juveniles (Wang, 2007; Ju et al., 2009). Burford et al. (2004) observed that the 39%–53% of nitrogen and carbon requirement of shrimps was fulfilled through epiphytes present in water. Periphytons were also used as an additional food source in substratum added freshwater ponds (Azim and Wahab 2005), and due to the presence of substratum in the freshwater ponds, the specific growth rate of postlarval shrimp was enhanced up to 28% (Khatoon et al. 2007). Ballester et al. (2003) highlighted the significance of biofilm as a food source by examining the postlarvae of *F. paulensis*, where survival and growth rates were not increased when the artificial substrata with periodically removed biofilms were present. Although the synergism of physical and biological factors is associated with the use of artificial substrata, the most significant aspect of the use of artificial substrate is the provision of food in the cage culture of postlarvae of *F. paulensis*.

The biofilm, which is present on the surface of the substrate, is made up of organisms that belong to the natural diet of penaeid shrimp and serve an additional source of high-quality proteins (Ballester et al., 2007). Keshavanath et al. (2002) reported an increased production of *L. fimbriatus* and *Tor khudree* by 30%–59% during biofilm-based culture compared to fish culture without a substratum. Similarly, Radhakrishnan and Sugumaran (2010) reported an increased growth performance of *Heteropneustes fossilis* when sugarcane bagasse was used as substratum, compared to substratum-free culture. In the same way, during *L. vannamei* culture, the high production rate and improved feed conversion ratio (FCR) was observed using artificial substrata, in spite of high stocking density of fish, and a decrease in NH_3 nitrogen concentration was also observed due to increase of nitrification (Bratvold and Browdy 2001). Moss and Moss (2004) determined that the production rate of postlarvae of *L. vannamei* has increased flow-through of the system in the presence of artificial substrata having different stocking densities. They also recommended that the growth of shrimp in substratum-based culture was enhanced because of the presence of attachment sites for particulate organic matter, and in nursery ponds, the negative effects of high stocking density were reduced by using the artificial substrata.

5.9 NUTRITIONAL QUALITY OF BIOFILM

Various studies have broadly reported that the nutritional profile of biofilm is considered appropriate according to the requirement of fish (Dempster et al., 1993; Makarevich et al., 1993; Azim et al., 2002a). The proximate composition of biofilm cells ranges between 2% and 9% lipids, 23%–30% protein, 16%–42% ash, and 25%–28% nitrogen-free extract (NFE) (van Dam et al., 2002; Thompson et al., 2002; Azim et al., 2005). A diet containing 18%–50% proteins, 10%–25% lipids, <8.5% ash, <20% carbohydrates, <1.5% phosphorus, <10% water, and trace amounts of vitamins and minerals is considered nutritionally balanced for fish farmers (Craig and Helfrich, 2002). Hence, exploiting the nutritional quality of biofilms, they can be efficiently used as a feed supplement in the diet of cultured shrimps and finfish, which require lipid content up to 10% (Akiyama et al., 1992) and protein content up to 40% (Shiau, 1998). Renukaradhya and Varghese (1986) reported a 30% protein requirement for Indian major carps, which can be optimally provided by the microbial communities of biofilm (Wasielesky et al., 2006).

Biofilm is considered an effective source of protein; hence, these attributes improve the growth of shrimp and fish species (Anand et al., 2013; Oser, 1959). In addition to the ability of heterotrophic bacteria and microalgae as macronutrient source, they efficiently work for enhancement of immune system, growth promotion, dietary stimulation, and bioactivation of compounds (Supamattaya et al., 2005; Ju et al., 2008; Kuhn et al., 2010; Xu et al., 2012). Thus, in juveniles of tiger shrimp, growth performance was enhanced due to the beneficial role of microbes and algae in biofilm. The shrimp larvae and fish are very sensitive to some fatty acids like polyunsaturated fatty acids (PUFA) (Watanabe et al., 1983; Sorgeloos and Lavens, 2000), which are derived from natural food sources like macroinvertebrates, zooplankton, and phytoplankton (Parrish, 2009). Bacteria are abundantly available in nature and also act as potential food sources for the culture of various aquatic species (Burford et al., 2004; Azim and Wahab, 2005; Keshavanath and Gangadhar, 2005).

It is evident from previous studies, that in extensive and semi-intensive culture system, the bacteria performed a crucial role as a source of nutrition when appearing as biofilm, which enhanced the ability of grazing, survival, and growth of penaeid shrimps (Bratvold and Browdy, 2001; Azim and Wahab, 2005; Keshavanath and Gangadhar, 2005). Ballester et al. (2007) reported that the growth of shrimps was probably enhanced due to the nematodes and protozoans present in the biofilm. Furthermore, these organisms are rich in protein to energy ratio because of their capability of synthesis, elongation, and saturation of long-chain fatty acids, and hence they improve the quality of microbial aggregates in the form of biofilm (Zhukova and Kharlamenko, 1999). The substrata of jute stick periphyton contained fewer lipid contents, while the periphytons obtained from Kanchi and bamboo had higher lipid levels. However, the jute stick periphytons contained high ash contents, and the energy content was the same in both periphytons obtained from Kanchi and bamboo, which was higher than the jute stick. Nevertheless, the sediments present at the bottom consisted of more than 90% ash and extremely low amounts of lipids and proteins (Azim et al., 2002b).

5.10 SUBSTRATA AND SURVIVAL RATE

In the rearing system with high stocking density, the mortality of shrimps is increased because of cannibalism, and it was also reported that in newly molted specimens, the rate of cannibalism is great (Abdussamad and Thampy 1994). The effective method used to improve the survival rate and lower the cannibalism is the addition of substrata. The substrata provide protection and shelter from predators besides the provision of biofilm. Many studies demonstrated an inverse relationship between the growth of shrimp and stocking density (Martin et al., 1998; Wasielesky et al., 2001; Preto et al., 2005). Khatoon et al. (2007) reported the presence of postlarvae in the pipes (PVC coated), which might work as a refugium for molting postlarvae, necessary for its survival. In the nursery rearing tanks of *L. vannamei,* the survival rate was increased up to 24% when the window screen of fiberglass was given to them as substratum (Sandifer et al. 1987). Keshavanath et al. (2002) reported that with the increase in the food supply, the substratum present in the tank also acts as a shelter for fish and reduces its stress level. In *L. vannamei* the growth and survival rates were significantly improved with microalgae supplementation in the diet (Ju et al., 2009). The postlarvae of *F. paulensis* (shrimp) reared in cages showed enhanced survival and biomass by the addition of polyethylene substratum (Ballester et al., 2007).

It was examined in an experimental trial that the substratum was constantly inhabited by the shrimps for the use of biofilm as a food source. The area for distribution of shrimps was also increased in the culture system due to the presence of artificial substratum, besides the provision of food by biofilms (Ballester et al., 2007). In an intensive culture system, the *L. vannamei* showed increased performance due to the presence of artificial substratum in the rearing system because it hindered overcrowding effects (Bratvold and Browdy, 2001). In addition, the artificial substrata were used in the culture of *M. rosenbergii* (freshwater prawn) for increased growth, and the presence of substrata also reduce the negative territorial competition (Tidwell et al. (1998). Asaduzzaman et al. (2008) reported that the survival rate of prawns was enhanced to 63%–72% by the provision of periphyton substrata. Viau et al. (2013) found that the survival rate of *F. brasiliensis* was increased significantly in biofilm-based treatment as compared to feed-based treatment.

5.11 BIOFILM-BASED VACCINATION

The bacterial biofilm has a natural resistant property; therefore, they can be used in the development of useful, cheap oral vaccines. Among different vaccination methods, the oral vaccination method is the most ideal, easy, and cheap technique for mass administration to all sizes of fish (Azad et al., 1999). Quentel and Vegneulle (1997) reported that in aquaculture the oral vaccination is more beneficial due to its easy application to any size, age, and number of fish without causing any stress. Nevertheless, the reduced response was observed when the antigens were directly administered in oral vaccination due to the digestive degradation of antigens in the foregut before their access to hindgut (immune system) and other lymphoid organs (Johnson and Amend, 1983; Rombout et al., 1985). Therefore, different strategies have been developed to overcome this gastric destruction to improve the oral vaccination technique, especially the use of microsphere-encapsulated antigens (Piganelli et al., 1994; Dalmo et al., 1995). A biofilm of *Aeromonas hydrophila* was developed and studied by Azad et al. (1997) for oral vaccination of carp species in which the significantly increased protection, and the antibody titer was observed in contrast with the free cell-based vaccine. A protective layer of the glycocalyx on bacterial vaccination formed on the substratum makes it antibiotic resistant (Anwar and Costerton, 1990) and protects from serum and whole blood killing effect and phagocytosis (Anwar et al., 1992). The bacterial glycocalyx is a compound originated from hexose polymer, and hence it protects such vaccines from digestion in the intestinal tract (Costerton and Irvin, 1981). The biofilm vaccines facilitate the property of prolonged retention of antigens in lymphoid tissues and gut, which triggers the early and increased response of primary antibodies (Azad et al., 2000).

Nayak et al. (2004) performed an experiment and found that the biofilm vaccine was more significant for oral vaccination of carnivorous fish in which a well-developed stomach and digestive system is present. Pradeep et al. (2004) demonstrated that the prominent bacteria present in the biofilm were bacillus species, and Cohen et al. (2003) reported that in fish *B. subtilis* influences the immunomodulatory and antitumor activities. Research has demonstrated that the *B. subtilis* and their spore function as probiotics because they enhance the growth and survival of beneficial lactic acid bacteria in the intestinal tract of humans as well as animals (Hoa et al., 2000). Alya et al. (2008) examined that in *O. niloticus* the *Aeromonas hydrophila* growth was reduced due to *Lactobacillus acidophilus* and *B. subtilis*. In *O. niloticus*, the production of *P. fluorescens* was also reduced due to *B. subtilis*. The two strains of Bacillus (*B. licheniformis* 2336 and *B. subtilis* 2335) are well characterized by their proven probiotic role by several clinical studies (Bilev, 2002).

5.12 CONCLUDING REMARKS

Biofilms improve the growth and survival of aquaculture species by providing them a source of nutrition and clean good quality water. The substrata-based culture provides not only food but also shelter and protection to finfish and shellfish species. The cultured heterotrophic bacteria provide natural protein according to the requirement of different aquatic species. Biofilms enhance the immune responses of aquaculture species by the provision of oral vaccines by the microorganisms attached to substrata. A vast variety of substrata can be used in biofilm techniques, and locally available biomass can be efficiently converted into a valuable nutrient source for culture species, which will ultimately enhance the sustainability of culture, leading to the improved financial status of a poor farmer.

ACKNOWLEDGMENT

The authors are grateful to their representative universities/institutes for providing literature facilities.

CONFLICT OF INTERESTS

The authors declare no conflict of interest in any capacity including financial and competing.

REFERENCES

Abdussamad EM, Thampy DM (1994). Cannibalism in the tiger shrimp *Penaeusmonodon* fabricius in nursery rearing phase. *J. Aquacult. Trop.* 9:67–75.

Akiyama DM, Dominy WG, Lawrence AL (1992). Penaeid shrimp nutrition. In: Fast AW and Lester AL (eds) *Marine Shrimp Culture: Principles and Practices.* Developments in aquaculture and fisheries science. Elseiver, Amsterdam, vol. 3, pp. 535–568.

Alabi AO, Cob ZC, Jones DA, Latchford JW (1999). Influence of algal exudates and bacteria on growth and survival of white shrimp larvae fed entirely on microencapsulated diets. *Aquacult. Int.* 7 (3):137–158.

Alya SM, Ahmed YAA, Ghareebb AA, Mohameda MF (2008). Studies on *Bacillus subtilis* and *Lactobacillus acidophilus*, as potential probiotics, on the immune response and resistance of *Tilapia nilotica* (*Oreochromis niloticus*) to challenge infections. *Fish Shellfish Immunol.* 25:128–136.

Anand PSS, Kohli MPS, Dam Roy S, Sundaray JK, Kumar S, Sinha A, Pailan GH, Sukham MK (2013). Effect of dietary supplementation of periphyton on growth performance and digestive enzyme activities in *Penaeus monodon. Aquaculture* 392–395:59–68.

Anand PSS, Kumar S, Panigrahi A, Ghoshal TK, Dayal JS, Biswas G, Sundaray JK, De D, Ananda, Raja R, Deo AD, Pillai SM, Ravichandran P (2012). Effects of C:N ratio and substrate integration on periphyton biomass, microbial dynamics, and growth of *Penaeusmonodon juveniles. Aquacult. Int.* Retrieved from: http://link.springer.com/article/10.1007%2Fs10499-012-9585-6.

Anwar H, Costerton JW (1990). Enhanced activity of combination of tobramycin and piperacillin for eradication of sessile biofilm cells of *Pseudomonas aeruginosa. Antimicrob. Agents Chemother.* 34:1666–1671.

Anwar H, Strap JL, Costerton JW (1992). Susceptibility of biofilm cells of *Pseudomonas aeruginosa* to bactericidal activities of whole blood and serum. *FEMS Microb. Lett.* 92:235–242.

Arnold SJ, Coman FE, Jackson CJ, Groves SA (2009). High-intensity, zero water exchange production of juvenile tiger shrimp, *Penaeus monodon*: an evaluation of artificial substrates and stocking density. *Aquaculture* 293:42–48.

Asaduzzaman M, Rahman MM, Azim ME, Islam MA, Wahab MA, Verdegem MCJ, Verreth JAJ (2010). Effects of C/N ratio and substrate addition on natural food communities in freshwater prawn monoculture ponds. *Aquaculture* 306:127–136.

Asaduzzaman M, Wahab MA, Verdegem MCJ, Benerjee S, Akter T, Hasan MM, Azim ME (2008). Effects of addition of tilapia *Oreochromis niloticus* and substrates for periphyton developments on pond ecology and production in C/N-controlled freshwater prawn *Macrobrachium rosenbergii* farming systems. *Aquaculture* 287(34):371–380.

Audelo-Naranjo JM, Martínez-Córdova LR, Voltolina D, Gómez-Jiménez S (2011). Water quality, production parameters, and nutritional condition of *Litopenaeus vannamei* (Boone, 1931) grown intensively in zero water exchange mesocosms with artificial substrates. *Aquacult. Res.* 42:1371–1377.

Austin B, Austin D (1999). *Bacterial Fish Pathogens: Disease of Farmed and Wild Fish*, 3rd edn., p. 457. Springer, Chichester.

Avnimelech Y (2006). Biofilters: the need for a new comprehensive approach. *Aquacult. Eng.* 34(3):172–178.

Avnimelech Y (2007). Feeding with microbial flocs by tilapia in minimal discharge bioflocs technology ponds. *Aquaculture* 264:140–147.

Azad IS, Shankar KM, Mohan CV (1997). Evaluation of biofilm of *Aeromonas hydrophila* for oral vaccination of carps. In: Flegel TW and Macrae IH (eds) *Diseases in Asian Aquaculture III Fish Health Section*. AFS, Manila, pp. 181–186.

Azad IS, Shankar KM, Mohan CV, Kalita B, (1999). Biofilm vaccine of *Aeromonashydrophila*–standardization of dose and duration for oral vaccination of carps. *Fish Shellfish Immunol.* 9:519–528.

Azad IS, Shankar KM, Mohan CV, Kalita B (2000). Uptake and processing of biofilm and free-cell vaccines of *Aeromonas hydrophila* in Indian major carps and common carp following oral vaccination antigen localization by a monoclonal antibody. *Dis. Aquat. Organ.* 43:103–108.

Azim ME, Verdegem MCJ, Khatoon HM, Wahab A, van Dam AA, Beveridge MCM (2002a). A comparison of fertilization, feeding, and three periphyton substrates for increasing fish production in freshwater pond aquaculture in Bangladesh. *Aquaculture* 212:227243.

Azim ME, Wahab MA, Verdegem MCJ, van Dam AA, van Rooij JM, Beveridge MCM (2002b). The effects of artificial substrates on freshwater pond productivity and water quality and the implications for periphyton-based aquaculture. *Aquat. Living Resour.* 15:231–241.

Azim ME, Wahab MA (2005). Periphyton-based pond polyculture. In: Azim ME, Verdegem MCJ, van Dam AA, Beveridge MCM (eds) *Periphyton: Ecology, Exploitation and Management*. CABI Publishing, Wallingford, pp. 207–222.

Azim ME, Wahab MA, van Dam AA, Beveridge MCM, Verdegem MCJ (2001). The potential of periphyton-based culture of two Indian major carps, rohu *Labeorohita* (Hamilton) and gonia *Labeogonius* (Linnaeus). *Aquacult. Res.* 32:209–216.

Baffico GD, Pedrozo FL (1996). Growth factors controlling periphyton production in a temperate reservoir in Patagonia used for fish farming. *Lakes Reserv. Res. Manage.* 2:243–249.

Ballester ELC, Wasielesky W, Cavalli RO, Abreu PC (2007). Nursery of the pink shrimp *Farfantepenaeus paulensis* in cages with artificial substrates: biofilm composition and shrimp performance. *Aquaculture* 269(1–4):355–362.

Ballester ELC, Wasielesky WJ, Cavalli RO, Santos MHS, Abreu PC (2003). Influência do biofilme no crescimento do camarão-rosa Farfantepenaeus paulensis em sistemas de berçário. *Atlântica* 25:117–122.

Barnese LE, Schelske CL (1994). Effects of nitrogen, phosphorous, and carbon enrichment on planktonic and periphytic algae in a softwater, oligotrohic lake in Florida, USA. *Hydrobiologia* 277:159–170.

Beveridge MCM, Phillips MJ, Macintosh DJ (1997). Aquaculture and the environment: the supply of and demand for environmental goods and services by Asian aquaculture and the implications for sustainability. *Aquacult. Res.* 28:797–807.

Bharti V, Pandey PK, Purushothaman CS, Rajkumar M (2013). Bacterial communities in biofilm and their effect on production of *Cirrhinus mrigala* (Hamilton, 1822). *Aquaculture* (Communicated).

Bilev AE (2002). Comparative evaluation of probiotic activity in respect to in vitro pneumotropic bacteria and pharmacodynamics of biosporin-strain producers in patients with chronic obstructive pulmonary diseases. *Voen. Med. Zh.* 323:54–57.

Bourne DG, HØj L, Webster NS, Swan J, Hall MR (2006). Biofilm development within a larval rearing tank of the tropical rock lobster, *Panulirus ornatus. Aquaculture* 260:27–38.

Bratvold D, Browdy CL (2001). Effect of sand sediment and vertical surfaces (AquaMats™) on production, water quality, and microbial ecology in an intensive *Litopenaeus vannamei* culture system. *Aquaculture* 195:81–94.

Brock JA, Main KL (1994). *A Guide to the Common Problems and Diseases of Cultured* (Penaeus vannamei), p. 242. The World Aquaculture Society, Baton Rouge.

Burford MA, Sellars MJ, Arnold SJ, Keys SJ, Crocos PJ, Preston NP (2004). Contribution of the natural biota associated with substrates to the nutritional requirements of the post-larval shrimp, *Penaeus esculentus* (Haswell) in high density rearing systems. *Aquacult. Res.* 35:508–515.

Cohen JM, Samocha TM, Fox JM, Gandy RL, Lawrence J (2005). Characterization of water quality factors during intensive raceway production of juvenile *Littopenaeus vannamei* using limited discharge and biosecure management tools. *Aquacult. Eng.* 32:425–442.

Cohen LA, Zhao Z, Pittman B., Scimeca J (2003). Effect of soya protein isolate and conjugated linoleic acid on the growth of dunning R-3327AT-1 rat prostate tumors. *Prostate* 54:169–180.

Costerton JW, Irvin RT (1981). The bacterial glycocalyx in nature and disease. *Ann. Rev. Microbiol.* 35:299–324.

Crab R, Avnimelech Y, Defoirdl T, Bossier P, Verstraete W (2007). Nitrogen removal techniques in aquaculture for a sustainable production. *Aquaculture* 270:1–14.

Craig S, Helfrich LA (2002). *Understanding Fish Nutrition, Feeds, and Feeding (Publication 420–256)*, p. 4. Virginia Cooperative Extension, Yorktown, VA.

Dalmo RA, Leifson RM, Bogwald J (1995). Microspheres as antigen carriers: Studies on intestinal absorption and tissue localization of polystyrene microspheres in Atlantic salmon, *Salmo salar* L. *J. Fish Dis.* 18:87–91.

Dempster PW, Beveridge MCM, Baird DJ (1993). Herbivory in the tilapia *Oreochromis niloticus*: a comparison of feeding rates on phytoplankton and periphyton. *J. Fish Biol.* 43:385–392.

Dharmaraj M, Manissery JK, Keshavanath P (2002). Effects of a biodegradable substrate, sugarcane bagasse and supplemental feed on growth and production of fringe-lipped peninsula carp, *Labeo fimbriatus* (Bloch). *Acta. Ichthyol. Piscat.* 32:137–144.

Donlan RM (2002). Biofilms: Microbial life on surfaces. *Emerg. Infect. Dis.* 8(9):881–890.

Ebeling JM, Timmos MB, Bisogni JJ (2006). Engineering analysis of the stoichiometry of photoautotrophic, autrophic, and heterotrophic removal of ammonia-nitrogen in aquaculture systems. *Aquaculture* 257:346–358.

El-Gayar OF, Leung P (2000). ADDSS: a tool for regional aquaculture development. *Aquacult. Eng.* 23(1–3):181–202.

Erler D, Pollard PC, Knibb W (2004). Effects of secondary crops on bacterial growth and nitrogen removal in shrimp farm effluent treatment systems. *Aquacult. Eng.* 30:103–114.

FAO, 2017. *FAOSTAT.* Food and Agriculture Organization of the United Nations, Rome.

FAO. 2018. *The State of Food and Agriculture 2018. Migration, Agriculture, and Rural Development.* Food and Agriculture Organization of the United Nations, Rome.

Fitch MW, Pearson N, Richards G, Burken JG (1998). Biological fixed film systems. *Water Environ. Res.* 70:495–518.
Gatune C, Vanreusel A, Cnudde C, Ruwa R, Bossier P, De Troch M (2012). Decomposing mangrove litter supports a microbial biofilm with potential nutritive value to penaeid shrimp postlarvae. *J. Exp. Marine Biol. Ecol.* 426–427:28–38.
Ghosh M, Gaur JP (1994). Algal periphyton of an unshaded stream in relation to in situ nutrient enrichment and current velocity. *Aquat. Bot.* 47:185–189.
Hall-Stoodley L, Costerton JW, Stoodley P (2004). Bacterial biofilms: from the natural environment to infectious diseases. *Nat. Rev. Microbiol.* 2:95–107.
Hall-Stoodley L, Stoodley P (2002). Developmental regulation of microbial biofilms. *Curr. Opin. Biotechnol.* 13:228–233.
Han P, Lu Q, Fan L, Zhou W (2019). A review on the use of microalgae for sustainable aquaculture. *Appl. Sci.* 9: 2377–2397.
Hansson LA (1989). The influence of a periphytic biolayer on phosphorous exchange between substrate and water. *Arch. Hydrobiol.* 115:21–26.
Hargreaves JA (2006). Photosynthetic suspended-growth system in aquaculture. *Aquacult. Eng.* 34:344–363.
Hari B, Kurup MB, Varghese JT, Sharma JW, Verdegem MCJ (2004). Effects of carbohydrate addition on production in extensive shrimp culture systems. *Aquaculture* 241:179–194.
Hoa NT, Baccigalupi L, Huxham A (2000). Characterization of Bacillus species used for oral bacteriotherapy and bacterioprophylaxis of gastrointestinal disorders. *Appl. Environ. Microbiol.* 66:5241–5247.
Hopkins JS, Sandifer PA, Browdy CL (1995). A review of water management regimes which abate the environmental impacts of shrimp farming. In: Browdy CL, Hopkins, JS (eds) *Swimming through Troubled Water. Proceedings of the Special Session Shrimp Farming.* The World Aquaculture Society, Baton Rouge, pp. 157–166.
Huchette SMH, Beveridge MCM, Baird DJ, Ireland M (2000). The impacts of grazing by tilapias (*Oreochromis niloticus* L.) on periphyton communities growing on artificial substrate in cages. *Aquaculture* 186:45–60.
Jana SN, Garg SK, Patra BC (2004). Effect of periphyton on growth performance of grey mullet, *Mugilcephalus* (Linn.), in inland saline groundwater ponds. *J. Appl. Ichthyol.* 20:110–117.
Johnson KA, Amend DF (1983). Escacy of *Vibrio anguillarum*and *Yersinia ruckeri* bacterin applied by oral and anal intubation of salmonids. *J. Fish Dis.* 6:473–476.
Johnson LR (2008). Microcolony and biofilm formation as a survival strategy for bacteria. *J. Theor. Biol.* 251:24–34.
Ju ZY, Forster IP, Conquest L, Dominy W (2008). Enhanced growth effects on shrimp (*Litopenaeus vannamei*) from inclusion of whole shrimp floc or floc fractions to a formulated diet. *Aquacult. Nutr.* 14:533–543.
Ju ZY, Forster IP, Dominy WG (2009). Effects of supplementing two species of marine algae or their fractions to a formulated diet on growth, survival, and composition of shrimp (*Litopenaeus vannamei*). *Aquaculture* 29:237–243.
Kaiser GE, Wheaton FW (1983). Nitrification filters for aquatic culture systems: state of the art. *J. World Maric. Soc.* 14:302–324.
Keshavanath P, Gangadhar B (2005). Research on periphyton-based aquaculture in India. In: Azim ME, Verdegem MCJ, Van Dam AA, Beveridge MCM (eds) *Periphyton and Ecology, Exploitation and Management.* CABI Publishing, Wallingford, pp. 223–236.
Keshavanath P, Gangadhar B, Ramesh TJ, van Damb AA, Beveridge MCM, Verdegem, MCJ (2002). The effect of periphyton and supplemental feeding on the production of the indigenous carps *Tor khudree* and *Labeo fimbriatus. Aquaculture* 213:207–218.
Keshavanath P, Manissery JK, Bhat AG, Gangadhara B (2012). Evaluation of four biodegradable substrates for periphyton and fish production. *J. Appl. Aquac.* 24:2012.

Khatoon H (2006). Use of selected periphyton species to improve the water quality and shrimp postlarval production. PhD Thesis. University Putra Malaysia, Malaysia, p. 180.

Khatoon H, Yusoff FM., Banerjee S, Shariff M, Mohamed S (2007). Use of periphytic cyanobacterium and mixed diatoms coated substrate for improving water quality, survival, and growth of *Penaeus monodon* Fabricius postlarvae. *Aquaculture* 271:196–205.

King RK, Flick Jr. GJ, Smith SA, Pierson MD, Boardman GD, Coale Jr. CW (2008). Comparison of bacterial presence in biofilms on different materials commonly found in recirculating aquaculture systems. *J. Appl. Aquacult.* 18(1):79–88.

Kordmahaleh FA, Shalke SE (2013). Bacterial biofilms: microbial life on surfaces. *J. of Biol.*, 2(5), 242–248.

Kuhn DD, Lawrence AL, Boardman GD, Patnaik S, Marsh L, Flick GJ (2010). Evaluation of two types of bioflocs derived from biological treatment of fish effluent as feed ingredients for Pacific white shrimp, *Litopenaeus vannamei*. *Aquaculture* 303: 28–33.

Lohman K, Jones JR, Perkins BD (1992). Effects of nutrient enrichment and flood frequency on periphyton biomass in northern Ozark streams. *Can. J. Fish. Aquat. Sci.* 49:1198–1205.

Makarevich TA, Zhukova TV, Ostapenya AP (1993). Chemical composition and energy value of periphyton in a mesotrophic lake. *Hydrobiol. J.* 29:34–38.

Martin JLM, Veran Y, Guelorget O, Pham D (1998). Shrimp rearing: stocking density, growth, impact on sediment, waste output, and their relationships studied through the nitrogen budget in rearing ponds. *Aquaculture* 164:135–149.

Mattila J, Raesiaenen R (1998). Periphyton growth as an indicator of eutrophication; an experimental approach. *Hydrobiologia* 377:15–23.

McIntosh D, Samocha TM, Jones ER, Lawrence AL, McKee DA, Horowitz S, Horowitz A (2000). The effect of a commercial bacterial supplement on the high-density culturing of *Litopenaeus vannamei* with a low-protein diet in an outdoor tank system and no water exchange. *Aquacult. Eng.* 21(3):215–227.

McIntosh PR (2000). Changing paradigms in shrimp farming: IV. Low protein feeds and feeding strategies. *Global Aquacult. Advoc.* 3:44–50.

Meyer-Reil M (1994). Microbial life in sedimentary biofilms—the challenge to microbial ecologists. *Mar. Ecol. Prog. Ser.* 112:303–311.

Molobela P, Cloete TE, Beukes M (2010). Protease and amylase enzymes for biofilm removal and degradation of extracellular polymeric substances (EPS) produced by *Pseudomonas fluorescens* bacteria. *Afr. J. Microbiol. Res.* 4(14):515–1524.

Moss KRK, Moss SM (2004). Effects of artificial substrate and stocking density on the nursery production of Pacific white shrimp *Litopenaeus vannamei*. *J. World Aquac. Soc.* 35:536–542.

Mridula RM, Manissery JK, Rajesh KM, Keshavanath P, Shankar KM, Nandeesha MC (2006). Effect of microbial biofilm in the nursery phase of mrigal, *Cirrhinus mrigala*. *J. Indian Fish. Assoc.* 33:103–112.

Nayak DK, Asha A, Shankar KM, Mohan CV (2004). Evaluation of biofilm of *Aeromonas hydrophila* for oral vaccination of *Clarias batrachus*a carnivore model. *Fish Shellfish Immunol.* 16:613–619.

Nogueira R, Lazarova V, Manem J, Melo LF (1998). Influence of dissolved oxygen on the nitrification kinetics in a circulating bed biofilm reactor. *Bioprocess. Eng.* 19:441.

Oser BL (1959). An integrated essential amino acid index for predicting the biological value of proteins. In: Albanese AA (eds) *Protein and Amino Acid Nutrition*. Academic Press, New York, NY, pp. 281–295.

Parrish CC (2009). Essential fatty acids in aquatic food webs. In: Arts MT, Brett MT, Kainz MJ (eds) *Lipids in Aquatic Ecosystems*. Springer, Dordrecht, pp. 309–326.

Piedrahita RH (2003). Reducing the potential environmental impact of tank aquaculture effluents through intensification and recirculation. *Aquaculture* 226(1–4):35–44.

Piganelli JD, Zhang JA, Christensen JM, Kaattari SL (1994). Enteric coated microspheres as an oral method for antigen delivery to salmonids. *Fish Shellfish Immunol.* 4:179–188.

Pradeep B, Pandey PK, Ayyappan S (2003). Effect of probiotic and antibiotics on water quality and bacterial flora. *J. Inland Fish. Soc. India* 35(2):68–72.

Pradeep B, Pandey PK, Ayyappan S (2004). Variations in bacterial biofilm with reference to nutrient levels. *Asian J. Microb. Biotech. Environ. Sci.* 6(4):649–652.

Preto AL, Cavalli R, Pissetti T, Abreu PC, Wasielesky WJ (2005). Efeito da densidade de estocagemsobre o biofilme e odesempenho de póslarvas do camarão-rosa *Farfantepenaeus paulensis* cultivados emgaiolas. *Cienc. Rural* 35:1417–1423.

Quentel C, Vegneulle M (1997). Antigen uptake and immune responses after oral vaccination. *Developments in Biological Standardization*, 90: 69–78.

Radhakrishnan MV, Sugumaran E (2010). Effect of sugarcane bagasse on growth performance of *Hereropneustes fossilis* (Bloch.) fingerlings. *J. Exp. Sci.* 1(1):1–2.

Ramesh MR, Shankar KM, Mohan CV, Varghese TJ (1999). Comparison of three plant substrates for enhancing carp growth through bacterial biofilm. *Aquacult. Eng.* 19:119–131.

Renukaradhya KM, Varghese TJ (1986). Protein requirement of the carps. *Catlacatla* (Ham.). *Proc. Indian Acad. Sci.* 95:103–107.

Rombout JWHM, Lamers CHJ, Helfrich MH, Dekker A, Taverne-Thiele AJ (1985).Uptake and transport of intact macromolecules in the intestinal epithelium of carp (*Cyprinuscarpio* L.) and the possible immunological implications. *Cell Tissue Res.* 239:519–530.

Sandifer PA, Hopkins JS, Stokes AD (1987). Intensive culture potential of *Penaeus vannamei. J. World Aquacult. Soc.* 18:94–100.

Sauer K, Camper AK, Ehrlich GD, Costerton JE, Davies DG (2002). *Pseudomonas aeruginosa* displays multiple phenotypes during development as a biofilm. *J. Bacteriol.* 184:1140–1154.

Schveitzer R, Arantes R, Baloi MF, Costodio PS, Arana LV, Seiffert WQ Andreatta ER (2013). Use of artificial substrates in the culture of *Litopenaeus vannamei* (biofloc system) at different stocking densities: effects on microbial activity, water quality, and production rates. *Aquacult. Eng.* 54:93–103.

Sherman JW, Fairchild GW (1989). Algal periphyton community response to nutrient manipulation in softwater lakes. *J. Phycol.* 25(2):13.

Shiau SY (1998). Nutrient requirements of penaeid shrimps. *Aquaculture* 164:77–93.

Shrestha MK, Knud-Hansen CF (1994). Increasing attached microorganism biomass as a management strategy for Nile Tilapia (*Oreochromis niloticus*) production. *Aquac. Eng.* 13:101–108.

Sommer U (1996). Nutrient competition experiments with periphyton from the Baltic Sea. *Mar. Ecol. Progr. Ser.* 140:161–167.

Sorgeloos P, Lavens P (2000). Experiences on importance of diet for shrimp postlarvae quality. *Aquaculture* 191:169–176.

Supamattaya K, Kiriratnikom S, Boonyaratpalin M, Borowitzka L (2005). Effect of a Dunaliella extract on growth performance, health condition, immune response, and disease resistance in black tiger shrimp (*Penaeus monodon*). *Aquaculture* 248:207–216.

Thompson FL, Abreu PC, Cavalli RO (1999). The use of microorganisms as food source for *Penaeus paulensis* larvae. *Aquaculture* 174:139–153.

Thompson FL, Abreu PC, Wasielesky W (2002). Importance of biofilm for water quality and nourishment in intensive shrimp culture. *Aquaculture* 203:263–278.

Tidwell JH, Bratvold D (2005). Utility of added substrates in shrimp culture. In: Azim ME, Verdegem MCJ, van Dam AA, Beveridge MCM (eds) *Periphyton: Ecology, Exploitation and Management.* CABI Publishing, Wallingford, pp. 247–268.

Tidwell JH, Coyle SD, Schulmeister G (1998). Effects of added substrate on the production and population characteristics of freshwater prawns (*Macrobrachium rosenbergii*) to increasing amounts of artificial substrate in ponds. *J. World Aquacult. Soc.* 31: 174–179.

Toutain CM, Caiazza NC, O'Toole GA (2004). Molecular basis of biofilm development by *Pseudomonads*. In: Ghannoum M, O'Toole GA (eds) *Bacterial Biofilms*. ASM Press, Washington, DC, pp. 43–63.

Uddin MS, Azim ME, Wahab MA, Verdegem MCJ (2006). The potential of mixed culture of genetically improved farmed tilapia (GIFT, *Oreochromis niloticus*) and freshwater prawn (*Macrobrachium rosenbergii*) in periphyton-based systems. *Aquacult. Res.* 37:241–247.

Umesh NR, Shankar KM, Mohan CV (1999). Enhancing growth of common carp, rohu and Mozambique tilapia through plant substrate: the role of bacterial biofilm. *Aquacul. Int.*, 7(4):251–260.

Van Dam AA, Beveridge MCM, Azim ME, Verdegem MCJ (2002). The potential of fish production based on periphyton. *Rev. Fish Biol. Fish.* 12:1–31.

Viau VE, de Souza DM, Rodriguez EM, Jr. WW, Abreu PC, Ballester ELC (2013). Biofilm feeding by postlarvae of the pink shrimp *Farfentepenaeu sbtasiliensis* (Decapoda, Penaidae). *Aquacult. Res.* 44:783–794.

Vymazal J, Craft CB Richardson CJ (1994). Periphyton response to nitrogen and phosphorous additions in Florida Everglades. *Arch. Hydrobiol. Suppl.* 103:75–97.

Wang YB (2007). Effect of probiotics on growth performance and digestive enzyme activity of the shrimp *Penaeus vannamei*. *Aquaculture* 269:259–264.

Wasielesky W, Atwood H, Stokes A, Browdy CL (2006). Effect of natural production in a zero exchange suspended microbial floc based super intensive culture system for white shrimp *Littopenaeus vannamei*. *Aquaculture* 258:396–403.

Wasielesky WJ, Poersch LH, Bianchini A (2001). Effect of stocking density on pen reared pink shrimp *Farfantepenaeus paulensis* (Pérez-Farfante, 1967) (Decapoda, Penaeidae). *Nauplius* 9:163–167.

Watanabe T, Kitajima C, Fujita S (1983). Nutritional values of live organisms used in Japan for mass propagation of fish. *Rev. Aquacult.* 34:115–143.

Wetzel RG, Ward AK, Stock M (1997). Effects of natural dissolved organic matter on mucilaginous matrices of biofilm communities. *Arch. Hydrobiol.* 139:289–299.

Whal M (1989). Marine epibiosis I. Fouling and antifouling: Some basic aspects. *Mar. Ecol.: Prog. Ser.* 58:175–189.

Wheaton FW (1977). *Aquacultural Engineering*. Wiley-Interscience, New York, NY, p.708.

Xu WJ, Pan LQ, Sun XH, Huang J (2012). Effects of bioflocs on water quality, growth, and digestive enzyme activities of *Litopenaeus vannamei* (Boone) in zero-water exchange tanks. *Aquacult. Res.* doi: 10.1111/j.1365-2109.2012.03115.x.

Zhukova NV, Kharlamenko VI (1999). Sources of essential fatty acids in the marine microbial loop. *Aquat. Microb. Ecol.* 17:153–157.

Ziemann DA, Walsh WA, Saphore EG, Fulton-Bennet K (1992). A survey of water quality characteristics of effluent from Hawaiian aquaculture facilities. *J. World Aquacult. Soc.* 23:180–191.

Part III

Microbial Biofilms: Chemical Sciences, Natural Products, and Biotechnological Approaches

6 Natural Products Affecting Biofilm Formation

Mechanisms, Effects, and Applications

Jacqueline Cosmo Andrade Pinheiro
Federal University of Cariri - UFCA

Maria Audilene Freitas
Federal University of Pernambuco - UFPE

Bárbara de Azevedo Ramos
Federal University of Pernambuco - UFPE

Luciene Ferreira de Lima
Regional University of Cariri - URCA

Henrique Douglas Melo Coutinho
Regional University of Cariri - URCA

CONTENTS

6.1 INTRODUCTION

The use of natural products for the treatment and cure of diseases dates back to ancient times where species, especially flora, were used by various civilizations due to their therapeutic and medicinal properties. The complexity of the phytochemical composition of these products, which consist of a variety of active principle categories and different functional groups such as secondary metabolites, characterizes their bioactivities (Yunes and Filho Cechine 2014).

Natural products are characterized by their structural heterogeneity and biological or pharmacological actions, which contribute to a high interaction capacity with the metabolism, proteins, and/or biomolecules of living beings. This biosynthetic character of natural products attracted the attention and interest of pharmaceutical companies aiming to obtain active molecules that may constitute drugs with high activity and low toxicity (Simões et al. 2016).

The antimicrobial potential of natural products is greatly represented among the arsenal of biological actions found in these due to the search for new therapeutic agents with greater effectiveness, especially due to the growing number of microorganisms with reduced susceptibility to antibacterial, antifungal, and antiparasitic drugs (; Thakare et al. 2019).

The microbial resistance phenomenon represents a serious health problem, being responsible for infectious outbreaks and common treatment failure, which are often followed by death. The dissemination of multiresistant microorganisms occurs through various resistance mechanisms, including the formation of biofilms which represents one of the main forms of microbial resistance, since it confers evolutionary advantages to microorganisms (Queiroz and Ferreira 2010).

When comparing planktonic microorganisms with adhered and biofilm-associated sessile microorganisms, those found in biofilms are more likely to resist antimicrobial treatments; thus, microbial infections caused by biofilms such as chronic, nosocomial, and medical device-associated infections are difficult to treat with conventional antibiotic doses and increasing mortality rate, compared to isolates from the same species in their planktonic form (Gressler et al. 2015).

Faced with this problem, there is a need for research and development of agents that have several mechanisms of antimicrobial action, able to minimize the effects of microbial resistance, as well as to eradicate the formation of biofilms, thus, natural products are characterized as a source of high antimicrobial activity and low toxicity.

6.2 MICROBIAL BIOFILM STRUCTURE AND RESISTANCE MECHANISMS

Bacterial biofilms that have been described since the 1990s were first defined by Costerton et al. (1978) as "Bacteria that adhere tenaciously and specifically to various surfaces. This adhesion is achieved by a mass of tangled polysaccharide fibers, or branched sugar molecules, that extend from the bacterial surface and form a felt-like 'glycocalyx' that surrounds an individual cell or a cell colony."

A few years later, researchers were already observing *Pseudomonas aeruginosa* on the surfaces of medical devices, as well as in an exopolysaccharide matrix in the adjacent tissue that was linked to previous infections (Costerton et al. 1999). In his studies, Costerton stated the glycocalyx is essential for biological success in the varied natural environments in which biofilms are observed. Since then, it has been shown that the formation of sessile communities and antimicrobial resistance are at the root of chronic and persistent infections (Costerton et al. 1999).

Currently, according to the International Union of Pure and Applied Chemistry (IUPAC) standards, biofilms are defined as "Microorganism clusters in which cells are embedded in an auto produced matrix formed by extracellular polymeric substances (EPS) that are adherent to each other/or to a surface," a definition which does not differ greatly from what was already known in the 1990s (Vert et al. 2012).

Biofilm composition plays an important role in its physiology and differs according to the species involved. While biofilms are composed of high cell densities, ranging from 10^8 to 10^{11} cells/wet weight, their exopolysaccharide matrix, also known as EPS, represents roughly 90% of their composition and confers several special characteristics, such as spatial organization, increased biodiversity, synergistic interactions and gene transfers, in addition to being beneficial for hydration, digestion, and protection against external agents such as antimicrobials (Morgan-Sastume et al. 2008; Balzer et al. 2010; Flemming et al. 2016). The EPS is mainly composed of polysaccharides, proteins, extracellular DNA (eDNA), lipids, and water as shown in Figure 6.1.

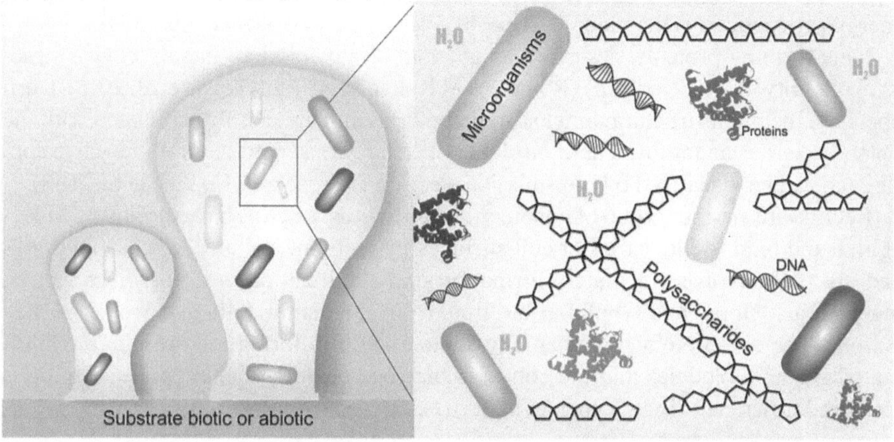

FIGURE 6.1 Biofilm structure with its matrix components. On the left is the biofilm structure adhered to a surface that may be biotic or abiotic. On the right are the main components that structure the matrix that includes microorganisms, polysaccharides, proteins, extracellular DNA, lipids, and water.

6.2.1 Polysaccharides

Polysaccharides compose most of the biofilm matrix, with scanning electron microscopy (SEM), for example, showing the polysaccharide structures forming a tangled network housing the microorganisms (Wingender et al. 2001; Trentin et al. 2011). Most exopolysaccharides are heteropolysaccharides; however, some homopolysaccharides such as sucrose and cellulose also exist (Flemming and Wingender 2010).

The most well-known exopolysaccharides are polysaccharide intercellular adhesin (PIA), synthesized by *Staphylococcus* sp., and alginate, Pel, and Psl, synthesized by *Pseudomonas aeruginosa*. The alginate in *Pseudomonas* sp. structure (1→4) may present b-D-mannuronic (ManA) and a-L-guluronic acid (GulA) residue substitutions (Leone et al. 2006; Rohde et al. 2010; Franklin et al. 2011).

Members from the *Burkholderia* genus (species associated with patients with a respiratory deficiency in cystic fibrosis), for example, are able to synthesize various types of exopolysaccharides, including poly-b-1,6-N-acetylglucosamine (PNAG) and the recently described rhamno-mannan, derived from 3-O-methyl-Rha, and termed exopolysaccharide (EPOL) (Dolfi et al. 2015).

Polysaccharides are the main matrix components in most microorganisms since they are responsible for bacterial cell aggregation through biofilm cohesion, energy retention, and for providing a protective barrier, as well as the absorption and retention of water and nutrients (Flemming and Wingender 2010).

6.2.2 Proteins

Proteins with enzymatic or structural functions are present in biofilms and play a crucial role in their metabolism. Enzymes are generally associated with biopolymer degradation, such as by degrading matrix polysaccharides or insoluble substances, to produce low molecular weight compounds and use these as a carbon source for energy production (Wingender et al. 1999; Flemming and Wingender 2010).

Fiber-forming proteins, described as amyloid or amyloid-like fibers, provide structural integrity to many biofilms (Romero and Kolter 2014; Taglialegna et al. 2016). Curli fibers are the main structural amyloid-type proteins and are closely associated with the Enterobacteriaceae family. These insoluble fibers are conserved in Gram-negative species and can be visualized by transmission electron microscopy (Erskine et al. 2018).

Lectins are another group of nonenzymatic proteins. Lectins have the ability to recognize and bind to molecules or cell surfaces by recognizing specific carbohydrates and are thus involved in matrix formation and stabilization, such as glucan binding proteins (GbpA and GbpD) in biofilms from the dental pathogen *Streptococcus mutans*. The two proteins together contribute to biofilm formation through the binding of glucan molecules and the cohesion between bacteria and exopolysaccharides (Neu and Lawrence 1999; Lynch et al. 2007).

6.2.3 Extracellular DNA (eDNA)

The presence of eDNA in biofilms plays an essential role in adherence, structural integrity, maintenance, and the transfer of virulence and resistance genes in biofilms

(Das et al. 2013; Okshevsky and Meyer 2013; Flemming et al. 2016). According to Das et al. (2011, 2013), DNA plays a role in the interaction between the microorganism and the surface through a loop structure that extends through the bacterium and can thus culminate in attractive forces, in addition to having an important role in bacterial surface hydrophobicity.

In addition to the initial adhesion, eDNA is also structurally responsible for interacting with other molecules present in the biofilm matrix, such as proteins and polysaccharides (Das et al. 2013). In *Pseudomonas aeruginosa* biofilms, for example, eDNA is the main matrix component and facilitates the connection between cells (Yang et al. 2007). Kawarai et al. (2016) demonstrate in their study the *Streptococcus mutans* biofilm induced in the presence of eDNA.

However, the presence of eDNA is also closely associated with horizontal gene transfer between the bacteria present in the biofilms. It is well known that cells present in the biofilm are at different physiological states, in addition to different populations; thus, gene acquisition is of paramount importance for biofilm physiology and maintenance (Stewart and Franklin 2008; Flemming et al. 2016).

6.2.4 Lipids and Water

While lipids can be found in the biofilm matrix, the generally hydrophobic character of the matrix is attributable to substituents such as methyl and acetyl groups attached to polysaccharides, where substances such as surfactin, viscosin, and emulsan may also be present in the matrix, replacing hydrophobic substances (Flemming and Wingender 2010).

However, water is the main component among all those present in the biofilm, and thus houses the structural and functional macromolecules of the matrix. Through a "rudimentary circulatory system" composed of pores and channels in the biofilm matrix, water facilitates the transport of nutrients, enzymes, eDNA, and other components needed to maintain biofilm physiology (Costerton et al. 1999; Flemming et al. 2016).

Each of these described substances is structured in an organized and coordinated manner to obtain an appropriate environment with physical and social interactions, where biofilms from single or multiple species are found in the environment. In this environment, there may be competition or beneficial interaction for the species. In this environment, competition or beneficial interactions between the species may exist, where for example, an exopolysaccharide-producing microorganism may integrate in its biofilm another microorganism that does not have the ability to synthesize exopolysaccharides, but which may be capable of synthesizing proteins or enzymes essential for biofilm physiology (Stewart and Franklin 2008; Flemming and Wingender 2010; Flemming et al. 2016).

However, biofilms are a very complex network, and such organization is indispensable for the survival and perpetuation of these microorganisms. Biofilm formation occurs in very important stages which can be didactically divided into three: cell adhesion, maturation, and cell release and dispersion (Trentin et al. 2013; Koo et al. 2017).

Initially, a surface attack, either biotic or abiotic, occurs randomly through Brownian motions and microorganismal motility. Physicochemical interactions are of paramount importance to abiotic surfaces; however, the initial adhesion is accompanied by specific interactions mediated by receptor-ligand interactions in biotic surfaces. Additionally, electrostatic interactions, van der Waals forces, hydrophobic interactions, hydrogen bonds, and adhesin-mediated bonds may also be associated with these forces (Dunne 2002; Trentin et al. 2013).

During this early stage, microorganisms adhere to the surface, and to each other, forming aggregates on the substrate. This adhesion is the first step that confers virulence and resistance to a microorganism since these produce adhesins and other virulence factors through transcriptional regulation to, for example, overcome tissue barriers or aggregate on a surface, and thus initiate the production of a matrix that will provide protection to various external agents (Jett et al. 1994; Flemming and Wingender 2010).

Adhesion strengthening occurs mainly through the expression of surface adhesins such as PIA in *S. epidermidis* and Pili in *Pseudomonas* sp. (Rohde et al. 2010; Lee et al. 2018). Another very important factor for the initial biofilm formation is the production of substances signal to increase cell density. This chemical communication is known as quorum sensing (QS) and occurs in several species; however, these are different in Gram-positive and Gram-negative species.

Gram-positive species produce a precursor molecule that is cleaved into a functional signaling molecule transported outside the cell. Upon reaching a concentration threshold, this binds to a receptor protein on the cell surface and thus activates several other factors associated with QS. On the other hand, Gram-negative species mainly produce acyl-homoserine lactone (AHL) molecules, which bind to regulatory proteins within the cell and act as transcription factors for various enzymes (Dickschat 2010; Banerjee and Ray 2017). QS is a crucial step in biofilm initiation, since microorganisms can "recruit" other microorganisms using this chemical communication.

Biofilm maturation is defined by the moment the cells begin to synthesize matrix components following the initial adhesion process, where this maturation step presents high metabolic activity and cellular replication. Moreover, several factors such as pH, available oxygen levels, carbon source, and inorganic substances influence maturation (Dunne 2002).

After the biofilm structure is completely formed, these resemble mushrooms, enveloped by EPS and permeated by water channels, where, as previously mentioned, these channels function as a circulatory system, responsible for transporting molecules such as nutrients or toxins (Costerton et al. 1999).

The presence of an EPS is the main factor conferring resistance to a biofilm since antimicrobials cannot penetrate the structure effectively. Another important factor is that given bacterial cells are embedded in this network, antigens are unable to recognize microorganisms, and thus the exopolysaccharides are able to "deceive" the host's immune system, allowing these to colonize and persist in the host's system for a prolonged time. Additionally, the presence of antimicrobial resistance genes is also responsible for decreasing antimicrobial perfusion by inactivating or degrading these (Costerton et al. 1999; Stewart and Franklin 2008; Flemming and Wingender 2010).

Mature biofilm physiology is very heterogeneous and varies according to the species involved. For example, aerobic cells are found on the surface where oxygen concentration is higher, while the fermenters or anaerobic cells are found in the innermost layers. Nutrient disposal follows the same pattern, despite the presence of channels that allow it to be perfused throughout the biofilm structure, with surface cells having a direct contact with ions, minerals, and other nutrients available in the environment (Flemming et al. 2016).

It has been previously reported that cells found more inwardly in this structure have very low metabolic activity and are known as "persisters"; this being another important factor for the resistance of microorganisms in a biofilm. Because these cells have a lower metabolism, they are less susceptible to antimicrobials, thus even though the antimicrobial agent may penetrate the matrix, these microorganisms most often persist and produce the biofilm again (Costerton 1999; Dunne 2002).

The detachment of surface cells and generation of planktonic organisms, or groups of microorganisms joined by EPS, is the last important step in biofilm infections becoming more chronic. This detachment, that occurs under certain situations or events programmed through QS, allows the microorganism to colonize another environment or surface. Thus, antimicrobial use reduces the symptoms caused by planktonic cells that are released in a manner almost always programmed by the biofilm; however, the biofilm itself cannot be killed given the aforementioned mechanisms, and thus, infections often present recurring and persistent symptoms (Costerton 1999; Trentin et al. 2013; Flemming et al. 2016).

6.3 NATURAL PRODUCTS WITH ANTIMICROBIAL ACTION

Natural products are chemical substances with biological activity found in plants, animals, and fungi, where these compounds are used by these living beings for their defense, metabolism, and development (Cragg and Newman 2013). Natural products have been used for thousands of years by populations from different countries to treat various pathologies, either as an alternative or as complementary to synthetic drugs (Calixto 2005; Veiga-Junior and Mello 2008).

Plants are considered as a source of various substances that are used in the treatment of various diseases (Montovani et al. 2009). Most compounds found in plants with biological activity are useful for medicinal purposes and are termed secondary metabolites (Simões 2007). These secondary metabolites synthesize various chemical compounds with complex structures such as alkaloids, flavonoids, tannins, and phenylpropanoids, including others, where the antimicrobial activity of plants has been attributed to these compounds (Sokovic et al. 2013).

These plants may be used in the form of plant extracts, which are concentrated preparations with various possible compositions and consistencies obtained from plant materials dried by steam distillation, cold pressing, or solvent extraction. The most common processes used for extraction may also involve maceration, infusion, decoction, digestion, percolation, distillation, and drying (FIB 2010). Several therapeutic potentials presented by certain plant extracts may exist, with an antimicrobial activity against a series of pathogens standing out. Many plants have shown remarkable *in vitro* and *in vivo* antimicrobial activities, and thus an intense search in traditional medicine exists directed at the antimicrobial characterization of plants (Diaz et al. 2010).

The antimicrobial activity of extracts has been investigated in studies such as Machado et al. (2003) who analyzed 14 plant extracts used to treat infectious diseases and found the *Punica granatum* extract inhibited methicillin-sensitive and -resistant *Staphylococcus aureus* strains, concluding that these substances are potential agents for the treatment of infections caused for this microorganism. Rodrigues et al. (2019) evaluated the anti-*Candida* and modulatory effects of the *Tarenaya spinosa* (Jacq.) ethanolic and aqueous extracts, where both presented an antifungal action and modulatory effect with most of the isolates.

The study by Andrade et al. (2019) evaluated the microbial biofilm biomass inhibitory potential of the *Ziziphus juazeiro* aqueous leaf and bark extracts, where the results presented bacterial and fungal biofilm eradication, with the *Z. juazeiro* aqueous extract presenting a biofilm inhibitory effect against both microorganisms. Morais-Braga et al. (2016) demonstrated in their study that extracts obtained from *Psidium guajava* and *Psidium brownianum* possessed an inhibitory activity against *Staphylococcus aureus* ATCC25923, with both extracts potentiating the effect of the antibiotic against resistant bacterial strains, thus presenting a significant synergistic action.

In addition to plant extracts, essential oils also stand out due to their recognized antibacterial and antioxidant activities (Figueiredo et al. 2014). Essential oils are volatile products present in various plant organs (shoots, bark, trunks, roots, fruits, flowers, seeds, and resins) and are related to the secondary metabolism of plants, performing several important functions for plant survival, such as defense against microorganisms (Lima et al. 2006).

The chemical composition of essential oils depends on several factors such as weather, predatory action, and plant age (Gobbo Neto and Lopes 2007). The chemical components present in most essential oils are terpenoid and phenylpropanoid derivatives (Simões et al. 2007). In addition to the traditional steam distillation extraction methodology, new extraction methods such as ultrasound, microwave, and supercritical fluid exist (Dima and Dima 2015). Essential oils have been widely employed for their previously observed properties in nature, i.e., for their antibacterial action and antifungal activities (Bakkali et al. 2008).

Packer and Luz (2007) observed a fungistatic activity for the *Rosmarinus officinalis* L. essential oil through the solid medium diffusion technique, identifying inhibition halos greater than 60 mm. The *Cinnamomum cassia* essential oil, together with 74 samples from other natural products, was evaluated against *Aspergillus niger* using the agar diffusion technique, and a 40 mm inhibition halo value was obtained with this essential oil against the fungus (Pawar and Thaker 2006). Nuryastuti et al. (2009) observed that the cinnamon essential oil possesses antimicrobial activity against clinical *S. epidermidis* planktonic and biofilm cultures, including those with gentamicin resistance.

The antifungal activity of the *Cymbopogon winterianus* essential oil against filamentous fungi from the *Aspergillus*, *Rhizopus*, *Penicillium,* and *Fusarium* genera was investigated by Souza et al. (2005), while Prabuseenivasan et al. (2006) investigated its activity against Gram-negative and Gram-positive bacteria. For both studies, the *C. winterianus* essential oil presented an antimicrobial activity.

Sokovic et al. (2010) demonstrated the oregano essential oil (OEO), thyme essential oil, and its main compounds were active against the bacteria *Bacillus subtilis*, *Staphylococcus epidermidis*, *S. aureus*, *Salmonella Enteritidis*, *S. Typhimurium*,

Escherichia coli, *Proteus mirabilis*, *Pseudomonas aeruginosa*, and *Listeria monocytogenes*. The study by Silva et al. (2013) also showed that, from the evaluated essential oils, the highest efficacy was obtained with thyme and oregano, which presented activity against all tested bacterial strains.

Fixed oils are among the natural products group with antimicrobial activity, which according to the Brazilian Society of Pharmacognosy (SBF 2017), fixed oils and fats can be obtained from plants or animals. Plant oils and fats can exist in various plant parts such as seeds, where these accumulate in greater quantities (Pereira 2009). As aforementioned, fixed oils can also be obtained from animals, with the use of animals specifically for medicinal purposes being known as zootherapy (Alves and Rosa 2005), which can be understood as the use of medicines made from animal body parts, products of their metabolism (such as body secretions and excretions), or the materials built by them, such as nests and cocoons (Costa Neto and Alves 2010).

The antimicrobial action of these compounds has been described in studies such as Oliveira et al. (2014), where the authors found the combination of *Spilotes pullatus* fixed oil with antibiotics demonstrated a synergistic effect when combined with gentamicin against all tested strains in their study.

A similar work was done by Sales et al. (2014), using the *Rhinella jimi* fixed oil against bacteria and *Candida* yeasts, where the authors observed that fixed oil alone did not possess a significant activity; however, when this was associated with aminoglycosides, a synergistic effect reducing the Minimal inhibitory concentration (MIC) of these drugs was observed. De Queiroz et al. (2019) evaluated the modulatory action of the *Scrofa domesticus* fixed oil in association with amikacin and amoxicillin antibiotics, observing synergism with all tested bacterial strains.

Natural products have increasingly shown to be important alternatives in the search for bioactive substances with antimicrobial properties; however, bioprospecting of these products has been scarcely explored, requiring further studies to obtain more natural compounds to treat infections caused by pathogenic microorganisms.

6.4 NATURAL PRODUCT MECHANISMS OF ACTION OVER MICROBIAL BIOFILMS

In general, diseases caused by microorganisms play a vital role in morbidity and mortality worldwide, with bacteria and fungi being microorganisms that have acquired resistance over the decades, promoting the prolonged treatment of infected patients by colonizing medical devices, putting patients at risk of both local and systemic infectious complications, such as catheter-associated bloodstream infections and endocarditis (Maki et al. 2002).

Antibiotic resistance and the recurrence of infections reflect the failure of antibiotics used to treat persistent biofilm-associated infections, and thus finding a strategy that can control biofilm formation continues to challenge microbiologists. The use of natural products in association with commercial medicines demonstrates beneficial properties for the control of biofilms from health-threatening pathogens, where this association proves to be a promising strategy that may be used in personal care and pharmaceutical applications (Algburi et al. 2017). Several studies describing the potent antimicrobial activities of medicinal plants and their essential oils, extracts, and their individual constituents against microbial virulence exist.

Biofilms can be beneficially used in several areas, in addition to causing health damages and its evolution revealing the growth of contaminations and persistent infections, especially in healthcare sectors (Padoveze 2013). Biofilm formation is a resistance strategy in which prokaryotes and eukaryotes adhere to any surface, forming a complex community that impacts humans naturally, medicinally, and industrially. In addition, its formation is dynamic involving several elements in a single multicellular process associated with adhesion mechanisms, the synthesis of exopolysaccharides that make up the extracellular polymeric matrix (EPS), microbial motility, and cell-to-cell or cell-surface interactions (*QS*) (López et al. 2010; Abinaya and Gayathri 2019; Borges et al. 2014).

EPS is responsible for the morphology, structure, cohesion, and functional integrity of the biofilm, and its composition determines most of the physicochemical properties, promoting the persistence of infections and protecting microorganisms inside it against the action of antibiotics, as well as providing resistance to UV radiation, predators, and dehydration. A biofilm consists of 80%–95% water, making it porous and adsorbent, with 10% of the biofilm mass being composed of microorganisms, while the polymeric extracellular matrix (ECM) represents roughly 70%–95% of the biofilm dry mass organic matter. This distribution demonstrates that EPS is a determinant in the organization, virulence, and exchange of genetic material (by horizontal transfer of resistance genes, transcription of specific proteins) and enzymes that induce antimicrobial inefficiency (Langer et al. 2018; Giordani et al. 2013).

Plants and fungi are notable resources for drug research and QS inhibition (QSI) studies, given their natural products (Gökalsın and Sesal 2016). QS is the interaction between cell-to-cell and cell-to-surface attachments through the dispersion of signaling molecules and thus coordinates their behavior in a manner dependent on cell density. These molecules regulate the expression of genetic factors of virulence, bioluminescence, sporulation, motility, mating, and biofilm formation (Castillo-Juárez et al. 2015; Brackman and Coenye 2015). These molecules regulate the expression of virulence genetic factors, bioluminescence, sporulation, motility, mating, and biofilm formation (Castillo-Juárez et al. 2015; Brackman and Coenye 2015). When the regulating molecules reach a threshold, a genetic program initiates biofilm formation followed by virulence factor expression (Estrela et al. 2016). In order for QS inactivation to occur, a natural product must act as follows: (i) inhibit signal molecule biosynthesis or AHL signal reception, where this signal is produced by roughly 70 Gram-negative bacterial species, and/or (ii) enzymatically inactivate and biodegrade QS molecules (Alvarez et al. 2014). QS complexity is further enhanced by the multiplicity of signals: (1) peptides, (2) AHLs, and (3) diketopiperazines. QS interference affects biofilm development and increases the susceptibility of the microorganism to the antimicrobial (Koul et al. 2016).

P. aeruginosa is a Gram-negative, biofilm-forming QS example which uses two lactone-based (AHL) pathways: the rhlI/rhlR pathway using butyryl-AHL (C4-HSL), and the lasI/lasR pathway using 3-oxo-dodecanoyl lactone (3-oxo C12-HSL). A third signaling molecule, 2-heptyl-3-hydroxy-4 (1H) -quinolone, is also identified, which plays a role in *P. aeruginosa* virulence and possibly interspecies communication (Cady et al. 2012).

The study by Mary and Banu (2015) evaluated the antiquorum activity of the *Vitex trifolia* leaf methanol extract against AHL-dependent virulence factor in *Pseudomonas aeruginosa*, where the effect of the extract on phenotypic characteristics such as pigment production, EPS production, motility, and biofilm formation was tested. The *V. trifolia* extract inhibited 69% of the protease and 49% of the elastase production, as well as 85% of the pyocyanin production, indicating the presence of an anti-QS activity by the extract. This extract reduced 96% of EPS production and flagellate motility compared to the control in the *P. aeruginosa* model.

Evernic acid obtained from lichens presented QSI for las and rhl systems at concentrations between 7.25 and 116 μM. The results were expressed as green fluorescent protein (GFP)/bacterial cell growth to attest the reduced fluorescence due to rhlA-gfp and lasB-gfp QSI in *P. aeruginosa* (POA1). In this study, evernic acid inhibited approximately 54% of lasB-GFP and 50% of 116 μM rhlA-GFP GFP expression compared to the untreated control group (Gökalsın and Sesal 2016).

The compound 3,5,7-trihydroxyfavone (TF) isolated from *A. scholaris* leaves can inhibit QS-regulated virulence factor and biofilm formation in *P. aeruginosa* by inhibiting planktonic cell growth and reducing 74.5% of the exopolysaccharide (EPS) production at a concentration of 100 μg/mL. EPS provides cell stability, mediates surface adhesion, and serves as a support for cells, enzymes, and antibiotics (Algburi et al. 2017). This compound was also effective at reducing pycocyanin, a secondary metabolite produced by *P. aeruginosa* in the form of a blue-green phenazine pigment present in the QS system. TF presented significant results in *P. aeruginosa* swimming motility reduction, consequently impeding QS at 25, 50, 75, and 100 μg/mL concentrations. In silico molecular interaction studies showed the TF-LasR complex expresses greater binding attraction than the standard drug gentamicin. Therefore, TF may inhibit the efficacy of QS signaling molecules (3-oxo-C12-HSL and C4-HSL), which were regulated to form a virulence factor during biofilm formation. Therefore, the isolated flavonol successfully suppressed *P. aeruginosa* biofilm formation, inhibiting the QS mechanism (Abinaya and Gayathri 2019).

The *Staphylococcus aureus* biofilm, a human Gram-positive commensal bacterium, along with *S. epidermitis*, is one of the most common causes of colonization in the adult population and is a risk for chronic infections. This biofilm's mechanism of formation is cyclic and formed by dispersive agents such as eDNA, dispersin B, phenol soluble modulins (PSMs), proteases, and nucleases, which act by releasing previously formed biofilm cells to expand the infection. The release of planktonic cells occurs through cell disaggregation followed by surface adhesion and is associated with proteins, the production of ECM, and microcolony formation. This growth model is dependent on matrix composition which has the polysaccharide intracellular adhesin (PIA). PIA is produced by enzymes encoded in the *icaADBC* locus which is responsible for PIA synthesis, export, and modification (Lister and Horswill 2014).

Methicillin-sensitive *S. aureus* (MSSA) strains generally produce an intercellular adhesin-dependent biofilm encoded by the *icaADBC* (PIA) polysaccharide operon. Autolysin is implicated in the biofilm phenotype of methicillin-resistant *S. aureus* (MRSA) isolates that form PIA-independent biofilms and require surface proteins such as fibronectin binding proteins, Atl-mediated sis, and eDNA for surface colonization and biofilm accumulation. Multiresistant biofilm-associated *S. aureus* requires the bacteria being inhibited from adhering to living or nonliving surfaces at an early stage to reduce biofilm formation. Disruption of the biofilm architecture during the maturation process, QS signal interference, or QSI affect the expression and dissemination of virulence factors (McCarthy et al. 2015).

The study by Qin et al. (2014) verified the use of resveratrol and ursolic acid against MRSA biofilm formation using the RNA-seq transcriptome analysis methodology. The use of resveratrol inhibits biofilm formation by altering QS, surface protein synthesis, and capsular polysaccharides, whereas ursolic acid-induced biofilm formation inhibition may reduce the metabolism of some amino acids and adhesin expression. RNA-seq-based transcriptome analysis demonstrated the studied compounds affect protein synthesis in cells associated with biofilms, modifying the expression of several genes that encode the major virulence factors involved in *S. aureus* pathogenesis.

In addition to *S. aureus*, cis-2-decenoic acid (C2DA) produced by *Pseudomonas aeruginosa* has been reported to act as a fatty acid messenger inducing the dispersion of *Escherichia coli*, *Klebsiella pneumoniae*, *Proteus mirabilis*, *Streptococcus pyogenes*, *Bacillus subtilis*, and the yeast *Candida albicans*; thus, these species may encode stress regulators involved in biofilm dispersion (Flemming and Wingender 2010). C2DA potentially controls the onset of biofilm formation in addition to the dispersion of existing biofilms with additive or synergistic effects on biofilm formation (Jennings et al. 2012).

Proanthocyanidin A2-phosphatidylcholine from *Aesculus hippocastanum*, dihydroxybenzofuran from *Krameria lappacea*, and chelerythrine and sanguinarine from *Macleya cordata* were isolated based on microbiological screening tests from their respective crude plant extracts, which showed potentially interesting antimicrobial activity. The four compounds affected biofilm formation with comparable efficacy, with their inhibition ranging from 1.3 to 5.5 times the strongest inhibitory effect against *S. aureus* and *S. epidermitis* (Artini et al. 2012).

The pectin OEO at a concentration of 25 mg/mL demonstrated an antibacterial activity along with an anti-QS activity with a significance of $p<0.05$ using the *Chromobacterium violaceum* bacterial model. The OEO action is associated with one of its components, carvacrol, which in *C. violaceum* inhibits the production of violacein which is linked to the expression of the gene responsible for the synthesis of AHLs, a key signaling molecule for QS formation (Alvarez et al. 2014).

Eucaliptus globulus and *Eucaliptus radiata* essential oils were tested for anti-QS activity using the *C. violaceum* ATCC 12472 biomonitor strain, where oleodines were found to inhibit *C. violaceum* violacein production, in which the *E. radiata* oil presented the most anti-QS activity, since the violacein inhibition measurement was twice as high as that of the *E. globulus* oil. Therefore, the loss of the purple pigment by the model bacteria is indicative of QSI by eucalyptus essential oils (Luís et al. 2016).

In Ethiopia, 18 extracts were tested, where two presented activity against QS. The *Albiza schimperiana* methanolic root extract (ASRM) and *Justica schimperiana* seed oil ether extract (JSSP) exhibited observable quorum quenching (QQ) activities, suggesting the presence of AHL interfering molecules in the extracts against the

E. coli AI1-QQ.1 strain (Bacha et al. 2016). QS signaling molecules can be enzymatically degraded, known as QQ, preventing the accumulation and subsequent activation of the QS system. In addition to initiating AHL degradation through AHL lactones or via AHL acylases, which hydrolyze the HSL ring and the amide bonds of the AHL molecule, respectively (Brackman and Coenye 2015).

The *Candida* genus which has worldwide medical importance possesses several species that express the formation of biofilms, these being *C. albicans*, *C. dubliniensis*, *C. glabrata*, *C. krusei*, *C. tropicalis*, and *C. parapsilosis* in tissues, prostheses, catheters, and other surfaces. This characteristic makes the *Candida* genus resistant to the most common antifungal classes such as azoles, pyrimidines, echinocandins, and polyenes (Sardi et al. 2013). *Candida* spp. in immunosuppressed patients, cancer patients undergoing treatment, and those with HIV trigger major virulence mechanisms and their invasive capacity through (i) the ability to form biofilms reinforced by a quorum detection mechanism and (ii) phenotypic switching, i.e., when yeasts change to the filamentous form that determines tissue invasion (Nobile and Johnson 2015; Bennett 2015).

Candida species are capable of forming structured communities with other *Candida* species and/or bacteria, where infections caused by this association are more difficult to treat and eradicate, resulting in increased mortality and morbidity. It has been speculated that these infections are linked to competitive factors such as initial adhesion sites, nutrients, physical differences in microorganismal properties, and the production of QS molecules (Cerca and Azevedo 2012). Fungal QS, including in the *Candida* genus, has very different structures including tyrosol, farnesol, and 3R-hydroxy-tetradecanoic acid compounds that are not present in bacteria (Estrela and Abraham 2016).

These features reinforce the ability of *Candida* spp. to form biofilms that are often involved in infections and increase in yeast resistance to antimicrobials and immune defenses. In this context, 38 lichen acetone extracts were prepared and evaluated for their activity against *C. albicans* planktonic and sessile cells. The tetrazolium salt assay (2,3-bis-(2-methoxy-4-nitro-5-sulfophenyl)-2H-tetrazolium-5-carboxanilide (XTT)) was used to evaluate the antibiofilm activity of the extracts in terms of antimaturation or antibiofilm action. Terpene, sterol, depsidone, depside, dibenzofuran, and xanthone secondary metabolites were qualified in the extracts. From the 38 extracts, 7 presented antimaturation and antibiofilm activities with the best results being obtained with the *Evernia prunastri* and *Ramalina fastigiata* extracts. This antibiofilm activity may be associated with the presence of the secondary metabolite depsides, found in high quantities in the extracts through thin layer chromatography (TLC) and high-performance liquid chromatography (HPLC) analyses (Millot et al. 2017).

Oral diseases with a microbial etiology have a high incidence worldwide. The development of dental biofilms is usually due to *Streptococcus mutans* and its initial adhesin-mediated binding followed by EPS synthesis-mediated aggregation. This aggregation promotes carbohydrate metabolism, for example sucrose and starch, resulting in acid production that demineralizes and favors tooth cavitation. *Andrographis paniculata* (Acanthaceae), *Cassia alata* (Leguminosae), *Camellia sinensis*, *Psidium guava*, and *Harrisonia perforata* (Simaroubaceae) are medicinal plants traditionally used against *S. mutans* in an aqueous crude extract or organic solvent form with activity against the formation of dental biofilms which adhere to enamel and saliva-coated hydroxyapatite (SHA) (Palombo 2011).

S. mutans associated with *C. albicans* form biofilms associated with severe carious lesions. Treatment with tt-farnesol was unable to reduce *C. albicans* populations, while combination therapy with myricetin, farnesol, and fluoride decreased the reduction of *S. mutans*-produced exopolysaccharides in this colony; however, combined therapy negatively influenced virulence production by cariogenic biofilms, with increasing concentrations being required to verify product efficacy (Rocha et al. 2018). Monolaurin, a natural coconut oil product, unlike tt-farnesol, has been shown to be effective at treating mice with oral candidiasis. Following five days of oral candidiasis, Balb/c mice were treated with monolaurin, and a significant ($p<0.05$) reduction in total photon flux compared to the control group was observed (Seleem et al. 2018).

The *Solidago virgaurea* crude extract is a potent *C. albicans* biofilm formation inhibitor that inhibits adhesion and hyphae associated genes. The analysis Polymerase chain reaction quantitative real time (qRT-PCR) showed the expression of HWP1, ALS3, ECE1, and SAP6 was strongly inhibited by the *S. virgaurea* extract (4 h), with this effect remaining significant after 24 h. In summary, this genetic analysis indicated that planktonic yeast adhesion and biofilm formation were inhibited, and the mechanism may be transcriptional, preventing the expression of key factors and regulated yeast adhesins (Chevalier et al. 2019).

Results from the study by Choi et al. (2017) in the Republic of Korea show that *Camelia japonica* and *Thuja orientalis* methanolic extracts not only have distinct antibacterial activity against *S. mutans* and *C. albicans* but also present an antibiofilm activity through their glucosyltransferase (GTase) inhibitory activity, inhibiting biofilm formation by more than 92.4% and 98.0%, respectively, using the same concentration of both extracts to prevent dental infections.

Ziziphus joazeiro aqueous leaf and stem bark extracts were tested against *C. albicans* (INCQS 40006 and URM 4387) and *C. tropicalis* (INCQS 40042 and 4262) for the treatment of biofilms using the crystal violet (CV) treatment. The treatment of biofilms formed by *C. albicans* isolates presented a significant inhibition of 58.8% against the INCQS 40006 strain following 24 h at low concentrations of the *Z. joazeiro* aqueous stem extract. At 48 h, a 94.4% biofilm inhibition was obtained within 48 h at high concentrations of the leaf extract. For the URM 4387 isolate, inhibition only occurred at the 128 μg/mL concentration of the *Z. joazeiro* aqueous leaf extract, with an inhibition of 61.1%. A reduction in *C. tropicalis* INCQS 40042 biofilm formation occurred at low and high concentrations of the juazeiro stem extract by 59.1 and 74.4%, respectively, over 24 h. The lower concentrations of the juazeiro leaf extract over 48 h induced the reduction of biofilm formation by 75.5%. This study demonstrated the effectiveness of a natural product in reducing *Candida* biofilm formation (Andrade et al. 2019).

Melaleuca alternifolia essential oils (composed mainly of terpinen-4-ol, a type of monoterpene), lactoferrin (a peptide isolated from milk) and chitosan (a chitin copolymer) have demonstrated to be viable options for treating fungal infections, including against *Candida* spp. Recombinant human lactoferrin (talactoferrin - TLF) in association with amphotericin B or fluconazole prevents biofilm formation through synergism (Felipe et al. 2018).

The use of natural products and biomolecules through bioprospecting, both in the production of new drugs for the treatment of different diseases and for use in association with conventional antimicrobial therapy, is a very promising strategy for overcoming microbial multidrug resistance, which given the risks and benefits, natural products, especially in the form of phytotherapics, have promising evidence in the treatment of various microbial diseases as proven by basic research.

6.5 NATURAL PRODUCT PHYTOCHEMICAL COMPOSITION AND ANTIBIOFILM ACTION

Secondary metabolites from medicinal plants have gained importance in the scientific community, and knowledge surrounding their pharmacological properties is developing, generating solutions in various healthcare areas, including microbiology, with respect to microbial resistance. Secondary metabolites are chemically divided into three groups: terpenes, divided into monoterpenes (C10), sesquiterpenes (C15), diterpenes (C20), triterpenes (C30), and tetraterpenes (C40). Phenolic compounds with an aromatic ring are obtained through two metabolic pathways, the shikimic acid pathway and the mevalonic acid pathway that can give rise to the following groups: flavonoids, flavones, flavanones, flavonols, flavononols, isoflavones, catechins, anthocyanidins, stilbenes, lignans, and tannins; and alkaloids, with at least one N atom in their ring (Simões et al. 2016).

The elucidation of various plant-derived compounds in recent years has been of great importance for reducing bacterial and fungal biofilm formation, where substances such as emodin, garlic, and isolimonic acid interfere with the QS system. The flavonoid phloretin inhibits fimbriae by affecting planktonic cells while ginger extract reduces EPS formation. Hyperforin, 7-epiclusianone, carvacrol, the alkaloid bgugaine and diterpene casbane inhibit biofilm formation.

Plant-derived polyphenol compounds represent a broad class with biological activity that includes flavonoids, tannins, anthocyanins, phenolic acids, stilbenes, coumarins, lignans, and lignins. In the anthocyanidin group, malvidin, petunidin, and cyanidin inhibit biofilm formation and EPS production. Flavonoids, specifically, xanthohumol, naringenin, hesperidin, neohesperidin, neoeriocitrin, 8-prenylnaringeni, apigenin, fisetin, chrysin, luteolin, quercitrin, quercetin, kaempferol, rutin, daidzein, and genistein act by inhibiting the formation of biofilms from different bacteria, including *E. coli*, *S. aureus*, *V. harvey*, *S. mutans*, and *L. monocytogenes*. The compounds that inhibit biofilm formation in the tannin group are catechin, gallic acid, methyl gallate, (-)-epigallocatechin gallate, ellagic acid, tannic acid, rosmarinic acid, and 1,2,3,4,6-penta-*O*-galloyl-b-D-glucopyranose (Slobodnikova et al. 2016).

The phenolic compound carvacrol has proven antibiotic activity against *S. epidermidis* and together with eugenol inhibited and inactivated *L. monocytogenes* and *E. coli* O157:H7 in their biofilm colonies. *Trans*-cinnamaldehyde, an aromatic aldehyde obtained from cinnamon bark, has proven its antibiofilm activity against *Cronobacter sakazakii* by modulating gene transcription for biofilm formation, motility, adhesion, and QS. Terpenes such as carvacrol, thymol, and geraniol from *Cymbopogon citratus* and *Syzygium aromaticum* have shown marked antibiofilm activity against fungi and bacteria (Upadhyay et al. 2014).

According to Reen et al. (2018), members of the coumarin family are secondary metabolites found in plants and demonstrate antibiofilm and anti-QS activity. Coumarins known as ellagic acid, warfarin, scopoletin, nodakenetin, and fraxin act with an antibiofilm activity, while bergamottin, 6′,7′-dihydroxybergamottin, esculetin, coumarin-3-carboxylic acid, esculin, dephnetin, coladonin, and umbelliferone act with both antibiofilm and anti-QS activity.

Vanillin, a secondary metabolite extracted from *Aeromonas hydrophila*, is a promising QS inhibitory compound as it acts against different acyl chains present in mixed wild-type biofilms in the environment (Ponnusamy et al. 2009). Similar to vanillin, phytol, an acyclic monounsaturated diterpene alcohol, is also a compound used in the food industry, and demonstrably indicates a potential chemical with an anti-QS activity against *P. aeruginosa* infections (Pejin et al. 2015). Equisetin, a secondary metabolite isolated from marine fungi, demonstrated that it could attenuate the virulence phenotypes that regulate QS in the *P. aeruginosa* POA1 model that expresses a quorum system, with the possibility of optimizing its molecular structure for its anti-QS activity in the future (Zhang et al. 2018).

Fruit extracts presenting phenolic compounds possess a greater ability to more effectively inhibit *S. epidermidis* biofilm. Anthocyanidins, total flavonoids, and phenolic compounds present in different *Vaccinium virgatum* cultures, chlorogenic acid, and caffeic acid strongly prevented the formation of the *S. epidermidis* pathogen, possibly with antiadhesion activity. This activity is reported in several studies in the literature confirming the action of the *Vaccinium* genus in *E. coli* antiadhesion. Quercetin, a compound from the flavonoid class, did not present significant biofilm inhibition (Zimmer et al. 2014).

Melilotus albus and *Dorycnium herbaceum* acetone, ethyl acetate, and ethanol extracts rich in phenolic compounds were tested in a CV assay against *Bacillus subtilis, Staphylococcus aureus* ATCC 25923, *Pseudomonas aeruginosa,* and *Proteus mirabilis* bacterial strains. The plant extracts inhibited *P. aeruginosa* biofilm formation at concentrations from 5 to 20 mg/mL, demonstrating their antibiofilm potential (Stefanović et al. 2015). *Actinidia deliciosa* (sanguinarine alkaloid, hydroxyflavone flavonoid, phenolics, phytosterol, quinone, steroids, glycosides, and tannins), and *Syzygium aromaticum* (flavonoids, alkaloids, phenolics, saponin, phytosterol, chalcone, quinone, coumarin, glycosides, tannin) extracts exhibited antibiofilm activity against carbapenem-resistant *Acinetobacter baumannii*, presenting extract susceptibility reducing EPS, proteins involved in biofilm maturation, and the amyloidogenic ECM protein, as well as eDNA, demonstrating a therapeutic alternative for the control of biofilm formation. The authors attribute this result to both sanquinarine and hydroxyflavone compounds that may be associated with an antimicrobial activity (Tiwari et al. 2017).

The methanolic extracts from *Camellia japonica*, *Chelidonium majus* var. Asiaticum, *Chrysosplenium flagelliferum*, *Geranium sibiricum*, *Lindera glauca*, and *Thuja orientalis* showed the presence of phenolic and flavonoid compounds in the study by Choi et al. (2017) with Korean medicinal plants used by the population, where among the species, *C. japonica* and *T. orientalis* presented a high antibiofilm performance at a concentration of 0.5 mg/mL, inhibiting GTase inhibitory activity in *S. mutans* by roughly 92.4% and 98%, respectively.

In Serbia, a study using *Vinca minor* L. leaves collected in the Balkan Mountains analyzed the composition of secondary metabolites in the ethyl acetate, aqueous acetate, and acetone extracts. All extracts presented total phenols and flavonoids, with the aqueous extract obtaining the highest total phenol quantity, roughly 68.44 GA/g, followed by the acetone extract with 46.05 GA/g and ethyl acetate with 28.98 GA/g. On the other hand, the extract presenting the most flavonoids was the *V. minor* acetone extract with 91.45 RU/g, followed by the ethyl acetate extract with 87.07 RU/g. The inhibitory biofilm concentration (IBC50) was measured in the CV assay by *Proteus mirabilis*, where only the ethyl acetate extract was effective against biofilm formation with a *P. mirabilis* IBC50 of 22.8 mg/mL (Grujić et al. 2015).

Borges et al. (2014) tested the following secondary metabolites in the *Chromobacterium violaceum* CV12472 model: three isothiocyanates (allylisothiocyanate (AITC), benzylisothiocyanate (BITC), and 2-phenylethylisothiocyanate (PEITC) and six phenolic products: gallic acid (GA), ferulic acid (FA), caffeic acid (CA), phloridzin (PHL), (-) epicatechin (EPI), and oleuropein glucoside (OG). AITC, BITC, and PEITC were found with QSI activity. These ITCs affected the *C. violaceum* QS system (CviI/CviR system—LuxI/LuxR homolog) not only interfering with AHL activity but also modulating AHL synthesis. The tested phenolic compounds did not show QSI potential. Antimicrobial tests suggest the potential use of phloridzin, (-) epicatechin, and ferulic acid as therapeutic antimicrobials, where the latter compound stands out for its antimicrobial activity against strains with pathogenic potential in planktonic and biofilm states and its ability to potentiate the action of antibiotics.

Alphitolic acid, a lupane triterpenoid, is the major secondary metabolite present in *Ziziphus jujuba* leaves. The aqueous leaf extract presented several polyphenols, which may have acted synergistically with the alphitolic acid antibiofilm activity, against *S. mutans* colonization, with this study being the first to report this activity by this metabolite (Damiano et al. 2017).

A group of monoterpenes, sesquiterpenes, and other compounds extracted from essential oils and described by Ahmad et al. (2015) were quantitatively evaluated for their anti-QS activity in *C. violaceum* and *P. aeruginosa*, where these compounds exhibited ≥50% AHL-mediated QSI at lower concentrations (0.125– 0.5 mg/mL). Differences in the inhibition of violacein and pyocyanin by essential oil constituents attributable to different QS system effects were observed in the two tested organisms. From the tested compounds, thyone and citral monoterpenes were more effective at inhibiting pyocyanin in the *Pseudomonas* pqs model, while alpha-terpineol and cis-3-nonen-1-ol were more effective at inhibiting violacein in the *C. violaceum* model. Some monoterpenes have shown an opposing effect, increasing the production of violacein, such as (+) -carvone, (+) -borned, (+) -limonene, beta-pinene, and (+) and (−) -alpha-pinene.

The diterpene ent-kaur-16-in-18-oic acid isolated from *Croton antisyphiliticus* root chloroform extract reduced the *Staphylococcus aureus* biomass formed by 56% at a 250 µg/mL concentration using the CV technique, while the comparison drug, gentamicin, inhibited only 13.9% of the biomass at the 30 mg/mL concentration (Nader et al. 2014). Isosteviol is a diterpenoid isolated from *Pittosporum tetraspermum* in the study by Abdullah-Al-Dhabi et al. (2015), which demonstrated a concentration-dependent in vitro antibiofilm activity against *E. coli*, *S. typhi*, and *P. aeruginosa*, where the results indicated that maximum reductions in cell binding were observed in *P. aeruginosa* at a concentration of 100 μg/mL, while the strains exhibited a comparatively lower biofilm activity at the 20 μg/mL concentration.

Oxyprenylated natural products (isopentenyloxy, geranyloxy, and the less widespread arnesyloxy compounds and their biosynthetic derivatives) are considered of utmost phytochemical and biological importance. Several isolated natural products have an oxyprenylated side chain, including alkaloids, coumarins, flavonoids, cinnamic acids, benzoic acids, phenols, alcohols, aldehydes, anthraquinones, chalcones, lignans, and aceto and benzophenones (Epifano et al. 2007).

Furthermore, this group of eight numbered compounds were studied for their antibiofilm activity by Di Giulio et al. (2015): p-Isopentenyloxybenzaldehyde (1), geranyloxyvanillin (2), 5-dimethoxy-4-isopentenyloxybenzyl alcohol (3), 3-(4-geranyloxyphenyl)-1-ethanol (4), 3-(4-isopentenyloxyphenyl)-1-propanol (5), (2E)-3-(4-((E)3,7-Dimethylocta-2,6-dienyloxy)-3-methoxyphenyl)acrylaldehyde (6), 4-isopentenyloxyeugenol (7), and 4-isopentenyloxyisoeugenol (8). Compounds (4) and (5) were the most active against mature biofilms and their formation in the *Staphylococcus aureus* ATCC 29213, *S. epidermidis* ATCC 35984, *Escherichia coli* ATCC 8739, *Pseudomonas aeruginosa* ATCC 9027, and *Candida albicans* ATCC 10231 microorganisms from the aforementioned compounds. With the exception of *S. epidermidis*, both compounds significantly reduced ($p<0.5$) microbial biofilm formation, compared to the control, at 1/2 MIC and 1/4 MIC for compounds 4 and 5, where each concentration inhibited *E. coli* biofilm formation by 44%.

Secondary metabolites acquired from marine fungi and soil from the *Neosartorya* genus, such as *Neosartorya paulistensis*, *N. laciniosa*, *N. siamensis*, *N. tsunodae*, and *N. fischeri*, demonstrated antibiofilm activity. Of the eight extracted compounds, aszonapyrone A and sartorypirone A isolated from *N. fischeri* prevented biofilm growth in *Enterococcus* spp., *Bacillus subtilis,* and *S. aureus* MRSA isolates in the presence of their MIC×2 and MIC. However, at the aszonapyrone A and sartorypirone A ½ MIC subinhibitory concentrations, an increase in biofilm production was observed in *Enterococcus* spp. and *S. aureus* MRSA (Gomes et al. 2014). The metabolites patulin and emodin from the *Plectosphaerella cucumerina* fungus inhibited biofilm formation and impeded preformed biofilms in the *P. aeruginosa* PAO1 bacterial strain, in addition to attenuating QS-dependent virulence factors at subinhibitory concentrations (Zhou et al. 2017).

(+)-usnic acid, a metabolite present in lichens, has proven activity in the initial adhesion of marine bacteria involving their biofilm, prevents *P. aeruginosa* QS and presents antibiofilm activity against *S. aureus* (Salta et al. 2013; Francolini et al. 2004; Thenmozhi et al. 2011). This compound was also tested against the *Streptococcus* Group A serotypes M, SP5, SP7, SP9, SP31, ES199, SF370, SP11, SP24, and SP30, which form substantial biofilms, with maximum biofilm inhibition being observed at concentrations ranging from 55 to 70 μg/mL for this compound. (+)-usnic acid obtained variable results for the prevention of bacterial aggregation, preventing the aggregation of all strains after 60 min. In contrast, usnic acid completely inhibited aggregation in SP9 strains, as only minimal aggregation or no aggregation was observed, while in the case of SP31, a maximum time of 105 min was observed. According to Fourier transfer infrared (FTIR) analysis of *S. pyogenes* biofilms, usnic acid acts by reducing fatty acids and causing cellular dehydration (Nithyanand et al. 2015).

Several studies have shown positive results in biofilm reduction using established models, however, compounds which do not reduce biofilm formation exist, such as the following substances studied by Helaly et al. (2017). The akanthol, akanthozine, and three amide derivatives including one hydroxamic acid derivative were taken from a spider-associated fungus, *Akanthomyces novoguineensis*, and tested for various biological activities, including antibiofilm activity. However, these compounds did not present any preventive effects against biofilm formation in *Staphylococcus aureus* DSM1104 (ATCC25923) and *Pseudomonas aeruginosa* PA14, nor did they show significant results in antimicrobial, cytotoxic, or nematicide activities.

The medicinal potential of plants, fungi, animals, and algae among other beings is an essential resource for society, since the exploitation of these resources shows their potential in pharmaceutical, alternative medicine, and natural therapies. Likewise, the investigation of isolated compounds is a promising pathway for the development of new alternatives, as seen with the antibiofilm activity elucidated by scientists who demonstrated the efficacy of isolated secondary metabolites.

6.6 STUDY METHODOLOGIES FOR BIOFILM ANTIFORMATION AND ERADICATION BY NATURAL PRODUCT

Several methods and technologies used to study biofilms ranging from simple techniques that quantify biofilms in microplates using spectrophotometers and dyes to high resolution microscopy techniques that are able to visualize biofilm structures or the interaction of microorganisms with molecules of interest such as SEM, FTIR spectroscopy, and Atomic Force Microscopy (AFM) exist, where the appropriate choice depends on the study objective (Bosch et al. 2006; Arnal et al. 2015; Azeredo et al. 2017).

It is noteworthy that each of the biofilm formation steps (surface adhesion, biofilm maturation, and EPS development and biofilm dispersion) as well as the macromolecules present in the matrix (exopolysaccharides, proteins, eDNA, and lipids) represent a target in the investigation of substances with antibiofilm action. Thus, the target choice coupled with techniques that will be able to capture the activity of such substances are of paramount importance (Koo et al. 2017).

The simplest and most cited methodology in biofilm studies in the literature is the quantification method using the CV dye. In this methodology biofilm forming microorganisms adhere to polystyrene microtiter plates, and after the removal of planktonic cells and fixation in ethanol (99%), methanol, and DMSO, or temperatures between 55°C and 60°C, the adhered microorganisms are stained with 0.4% CV, and are then measured using a spectrophotometer with wavelengths between 550 and 570 nm (Stepanović et al. 2007;).

Using this microplate technique, Borges et al. (2012) proved the effectiveness of ferulic acid against *Staphylococcus aureus, Escherichia coli, Listeria monocytogenes,* and pathogenic *Pseudomonas aeruginosa*, while on the other hand, gallic acid did not present an effect against these microorganisms. Tentin et al. (2011) compared 24 plants collected from the semiarid region of Pernambuco, Brazil, including *Allamanda blanchetii*, *Commiphora leptophloeos*, *Anadenanthera colubrina* var *cebil*, *Myracrodruoun urundeuva*, *Parkinsonia aculeata*, and *Senna macranthera*, which showed a significant decrease in *S. epidermidis* biofilm. However, CV staining only quantifies the total biofilm and is not able to differentiate whether the microorganisms present are viable or not.

Accordingly, dyes such as 2-(4,5-dimethyl-2-thiazolyl)-3,5-diphenyl-2H-tetrazolium bromide (MTT) and XTT can be used to quantify the presence of biofilms and to identify the viability of microorganisms. These tetrazole salts can be oxidized by enzymes and reduced to formazane, and thus the dosage is based on color changes and intensities read by a spectrophotometer (Altman 1974; Wilson et al. 2017).

The microplate technique can also be used to analyze substances that inhibit biofilm formation or destroy a formed biofilm. In order to analyze whether a substance has an activity that inhibits the initial biofilm formation, first the inhibitory concentration of the compounds needs to be known to use concentrations below these known as sub-MIC, since the objective is to observe if the compound inhibits the biofilm and has no bactericidal or bacteriostatic activity, and thus viable microorganisms for biofilm formation are necessary (Liu et al. 2019).

In this test the compound, culture medium, and microorganism inoculum are normally incubated together at a temperature and time period specific for each studied microorganism. Thereafter, this can be stained with 0.4% CV, for example, and according to a microorganism growth control the percentage of biofilm inhibition is calculated. However, sometimes the objective is to investigate if a substance has an action over the mature biofilm, in which case the microorganism inoculum and culture medium are first incubated to form the biofilm, and only following this step can the planktonic cells be removed and the substance added with more culture medium, followed by another incubation at the specific temperature and time period. Quantification can then be performed with 0.4% CV as previously mentioned and compared to ascertain if the compound was able to disrupt the biofilm structure (Trentin et al. 2011).

Moreover, the microorganism QS inhibitory activity of compounds can also be analyzed to ascertain if the compound of interest is interfering with cellular communication and thus preventing them from being able to "sense" cell density and recruit other cells. The simplest and best-known method for detecting QSI is through the inhibition of violacein produced by *Chromobacterium violaceum* (Kothari et al. 2017). D'Almeida et al. (2017) showed the efficacy of certain coumarins at inhibiting *C. violaceum* QS, as well as elastase production in *P. aeruginosa*, while Asfour (2018) summarizes several naturally occurring compounds with anti-QS activity against Gram-positive and Gram-negative bacteria.

All of these simple methodologies require validation, such as the use of microscopy or gene expression analysis of genes associated with the study. Several microscopes can be used to confirm the antibiofilm action of a compound, such as SEM, where the degradation of the biofilm exopolysaccharide matrix can be observed, compared to the control biofilm which does not have the compound interference (Tentin et al. 2011). Macé et al. (2017) confirmed by SEM the effect of 1,2-naphthaquinone, 5-hydroxy-1,4-naphthoquinone, and 4-hexylresorcinol inhibitory concentrations against clinical biofilms from *S. pyogenes*, while Sajeevan et al. (2018) proved the effect of anacardic acids isolated from cashew nuts (*Anacardium occidentale*) against *S. aureus* biofilms.

Another example is Epifluorescence Microscopy, in which specific dyes can be used to stain matrix (Calcofluor White), viable and nonviable microorganisms (dyes SYBR Green/Syto9 and propidium iodide), eDNA (DAPI—4′,6-diamidine-2-phenylindole dilactate), among others, to analyze in what structure the compound is acting on, as well as to check viable or unviable cell populations in a biofilm (Chimileski et al. 2014; Feng et al. 2014; Adamus-Białek et al. 2015; Soler-Arango et al. 2019). Confocal Laser Scanning Microscopy (CLSM) can also be used alongside the aforementioned fluorophores to visualize the different components, allowing the visualization of the three-dimensional structure and, through the use of imaging software, quantify the biofilm mass (Palmer Jr. et al. 2006; Yang et al. 2015).

Some techniques such as flow cytometry, FTIR spectroscopy, and AFM are less commonly used but may also be applied for biofilm analysis. Flow cytometry mainly evaluates viable and nonviable cell populations; however, it can be used to analyze cellular metabolism through specific markers, while FTIR spectroscopy has previously been used to analyze EPS variations from *Bordetella pertussis* biofilms (Bosch et al. 2006; Cerca and Azevedo 2012). AFM through cantilever molecular interactions with the surface of interest has been used to analyze the biofilm surface or identify specific points on a bacterial membrane, such as in Arnal et al. (2015), which identified surface adhesins associated with biofilm formation in *B. pertussis*.

When it comes to proteomics and transcriptomics, most authors research the expression or suppression of different genes or proteins by comparing planktonic microorganisms with biofilm microorganisms; however, some authors use biofilm-related genes or proteins to analyze the influence of substances over these (Qayyum et al. 2016; Azeredo et al. 2017). Liu et al. (2019) confirmed the antibiofilm effect of phenylactic acid through the downregulation of Ebp and Epa genes, which are responsible for transcribing pili and polysaccharides, respectively.

From the many methodologies which can be applied, ultimately the choice must follow according to what the study aims to investigate and prove, taking into account the advances, as well as advantages, disadvantages, and reproducibility of each methodology.

6.7 LIMITATIONS AND NEGATIVE ASPECTS OF NATURAL PRODUCTS IN BIOFILM ERADICATION

As previously mentioned, microbial biofilm is typically composed of water, microorganisms, EPS, dissolved and adsorbed substances, and trapped particles (Pereira 2001). These represent a reduction in the susceptibility of microorganisms to the action of most antimicrobial agents, contributing to the permanence of the infection (Chandra and Mukherjee 2015).

Natural products have shown promise in reducing or inhibiting microbial biofilms; however, many limitations regarding their use still exist, with these limitations being observed in studies such as Andrade et al. (2019) who carried out eradication assays using the *Ziziphus joazeiro* aqueous extract against bacterial and fungal biofilms. An increase in biofilm biomass was observed for some of the yeast isolates treated with this extract, where this increase may be attributed to the phytochemical composition of the extract, since the compounds present in the extract may contribute nutrients to the development of the ECM, ensuring greater resistance to the biofilm. The matrix present in *Candida* biofilms consists mainly of carbohydrates, proteins, phosphate, and hexosamines (Dantas et al. 2014). In this study, the bacterial biofilm eradication results showed greater inhibition at a lower concentration compared to a higher concentration, probably due to a greater quantity of nutrients present in the higher extract concentrations, such as proteins and polysaccharides which, instead of reducing the biofilm, contribute to the formation and development of the biofilm (Cerca and Azevedo 2012).

A similar effect was reported in the study by Furletti (2009), who applied *Coriandrum sativum* essential oil to mature *Candida* biofilms (after 24 h), where this essential oil contributed to an increase in biofilm formation tendency in both *Candida tropicalis* CBS 94 and *Candida tropicalis* 53M7. It is noteworthy that mature biofilms are much more resistant to antifungal therapy and host immune defense mechanisms compared to planktonic yeast cells (Fanning and Mitchell 2012).

Another limitation was shown in the study by Cordeiro (2014), where *Mimosa tenuiflora* (Jurema-preta) and *Pityrocarpa moniliformis* (Angico-de-bezerro) extracts presented no action against the mature biofilm. This ineffectiveness may be justified by the strong surface adhesion generated by bacteria, hindering the removal of previously formed biofilms (Medonline 2008). Furthermore, the composition of biofilm communities gives more protection to sessile microorganisms, making them more resistant to agents employed in sanitation procedures, being up to 1,000 times more resistant than planktonic cells (Drenkard 2003), where the EPS network is considered to be one of the main characteristic responsible for providing this protection, acting as a physical barrier, preventing antimicrobial drugs from reaching their action sites, such as the outer membrane of Gram-negative bacteria, for example (Herrera et al. al. 2007).

Freitas et al. (2019) analyzed the action of the *Persea americana* leaf ethanolic extract against *Candida* yeast biofilms; however, this did not show a significant biofilm reduction activity, with an increase in biofilm growth being observed instead. The biofilm structure may have contributed to the ineffectiveness of the extract, as the tolerance of these microorganisms to antimicrobial agents and the ECM acts as a barrier, preventing the penetration of substances. When a biofilm is established, microbial communities can tolerate antimicrobial agents at concentrations 10–1,000 times higher than those used to eliminate planktonic cells (Bueno 2014; Akbari and Kjellerup 2015).

The limiting factors regarded as a negative in biofilm eradication using natural products such as plant extracts and essential oils range from the difficulty of penetrating the barrier formed by the extracellular polymeric matrix to the presence of nutrients that favor the growth of this biofilm. Despite their limitations, natural products are still the main alternative sources of bioactive compounds with antimicrobial activity. Collectively, the moderate and rational use of natural products, either as a support for the development of new drugs or as a therapy in itself, is simple, inexpensive, and very effective in promoting healthcare.

6.8 FINAL CONSIDERATIONS

With the incidence of microbial resistance, the need to investigate and develop new mechanisms for microbial control, both in their planktonic form and in association, in the sessile form, has expanded.

The challenges and limitations are numerous, especially when it comes to natural products, which today are considered a fundamental object in scientific research. Moreover, preliminary assays aimed at screening products with antimicrobial or antibiofilm activity are widely observed in studies in this area.

In this context, more clinical investigations using active biomolecules isolated from natural products, aiming to improve healthcare through the exploitation of their biopharmaceutical properties to broaden biological action, efficacy, safety, and therapeutic effectiveness have emerged (Simões et al. 2016).

The potential of assays performed with natural product is observable, in terms of biofilm treatment and eradication. The driving forces for using these products as antimicrobial agents is rather diverse with the possibility of performing complexations, combinations with other products, with antibiotics and disinfectants, as well as biocides, in order to verify synergistic or additive effects (Cerca and Azevedo 2012).

Importantly, many natural products are already greatly represented in the pharmaceutical industry, being used in compositions marketed as antimicrobial agents, to integrate toothpaste, as well as in elixirs and various cosmetic products, which further drives natural product research in the antimicrobial field.

Collectively, even with the antimicrobial and antibiofilm potential, the limitations of natural product applications and the small number of studies, especially in the area of biofilms are verified. There are still many gaps to be filled and clarified in this scientific research process.

REFERENCES

Abdullah Al-Dhabi, N., Valan Arasu, M., Rejiniemon, T. S. 2015. In vitro antibacterial, antifungal, antibiofilm, antioxidant, and anticancer properties of isosteviol isolated from endangered medicinal plant *Pittosporum tetraspermum*. *Evidence-Based Complementary and Alternative Medicine* 2015:1–11.

Abinaya, M., Gayathri, M. 2109. Inhibition of biofilm formation, quorum sensing activity, and molecular docking study of isolated 3, 5, 7-trihydroxyflavone from *Alstonia scholaris* leaf against *P. aeruginosa*. *Bioorganic Chemistry* 87:291–301.

Adamus-Białek, W., Kubiak, A., Czerwonka, G. 2015. Analysis of uropathogenic *Escherichia coli* biofilm formation under different growth conditions. *Acta Biochimica Polonica* 62(4):765–771.

Ahmad, A., Viljoen, A. M., Chenia, H. Y. 2015. The impact of plant volatiles on bacterial quorum sensing. *Letters in Applied Microbiology* 60(1):8–19.

Akbari, F., Kjellerup, B. V., 2015. Elimination of bloodstream infections associated with *Candida albicans* biofilm in intravascular catheters. *Pathogens* 4:457–469.
Algburi, A, Comito, N., Kashtanov, D., Dicks, L. M., Chikindas, M. L. 2017. Control of biofilm formation: Antibiotics and beyond. *Applied and Environmental Microbiology* 83(3):e02508–e02516.
Altman, F. P. 1974. Studies on the reduction of tetrazolium salts. III. The products of chemical and enzymic reduction. *Histochemie* 38:155–171.
Alvarez, M. V., Ortega-Ramirez, L. A., Gutierrez-Pacheco, M. M., Bernal-Mercado, A. T., Rodriguez-Garcia, I., Gonzalez-Aguilar, G. A., Ponce, A., Moreira, M. D. R., Roura, S. I., Ayala-Zavala, J. F. 2014. Oregano essential oil-pectin edible films as anti-quorum sensing and food antimicrobial agents. *Frontiers in Microbiology* 5:699.
Alves, R. R. N., Rosa, I. L. 2005. Why study the use of animal products in traditional medicines? *Journal of Ethnobiology and Ethnomedicine* 1:1–5.
Andrade, J. C., Silva, A. R. P., Freitas, M. A., Azevedo, B. R., Freitas, T. S., Santos, F.A. G., Coutinho, H. D. M. 2019. Control of bacterial and fungal biofilms by natural products of *Ziziphus joazeiro* Mart. (Rhamnaceae). *Comparative Immunology, Microbiology and Infectious Diseases* 65:226–233.
Arnal, L., Longo, G., Stupar, P., Castez, M. F., Cattelan, N., Salvarezza, R. C., Yantorno, O. M., Kasas, S., Vela, M. E. 2015. Localization of adhesins on the surface of a pathogenic bacterial envelope through atomic force microscopy. *Nanoscale* 1–13.
Artini, M., Papa, R., Barbato, G., Scoarughi, G.L., Cellini, A., Morazzoni, P., Bombardelli, E., Selan, L. 2012. Bacterial biofilm formation inhibitory activity revealed for plant derived natural compounds. *Bioorganic and Medicinal Chemistry* 20:920–926.
Asfour, H. Z. 2018. Anti-quorum sensing natural compounds. *The Journal of Microscopy e Ultrastructure* 6:1–10.
Azeredo, J., Azevedo, N. F., Briandet, R., Cerca, N., Coenye, T., Costa, A. R., Desvaux, M., Di Bonaventura, G., Hébraud, M., Jaglic, Z., Kačániová, M. 2017. Critical review on biofilm methods. Critical review on biofilm methods. *Critical Reviews in Microbiology* 43(3) 313–351.
Bacha, K., Tariku, Y., Gebreyesus, F., Zerihun, S., Mohammed, A., Weiland-Bräuer, N., Schmitz, R. A., Mulat, M. 2016. Antimicrobial and anti-quorum sensing activities of selected medicinal plants of Ethiopia: Implication for development of potent antimicrobial agents. *BMC Microbiology* 16(1):139.
Bakkali F, Averbeck S, Averbeck D, Idaomar M. 2008. Biological effects of essential oils—A review. *Food and Chemical Toxicology* 46:446–475.
Balzer, M., Witt, N., Flemming, H. C., Wingender, J. 2010. Accumulation of fecal indicator bacteria in river biofilms. *Water Science and Technology* 61:1105–1111.
Banerjee, G., Ray, A. K. 2017. Quorum-sensing network-associated gene regulation in gram-positive bacteria. *Acta Microbiologica et Immunologica Hungarica* 64(4):439–453.
Bennett, R. J. 2015. O estilo de vida parassexual de *Candida albicans*. *Current Opinion Microbiology* 28:10–17.
Borges, A. Saavedra, M. J., Simões, M. 2012. The activity of ferulic and gallic acids in biofilm prevention and control of pathogenic bacteria, Biofouling. *The Journal of Bioadhesion and Biofilm Research* 28(7):755–767.
Borges, A., Serra, S., Cristina Abreu, A., Saavedra, M. J., Salgado, A., Simões, M. 2014. Evaluation of the effects of selected phytochemicals on quorum sensing inhibition and in vitro cytotoxicity. *Biofouling* 30(2):183–195.
Bosch, A., Serra, D., Prieto, C., Schmitt, J., Naumann, D., Yantorno, O. 2006. Characterization of *Bordetella pertussis* growing as biofilm by chemical analysis and FT-IR spectroscopy. *Applied Microbiology and Biotechnology* 71:736–747.

Brackman, G., Coenye, T. 2015. Quorum sensing inhibitors as antibiofilm agents. *Current Pharmaceutical Design* 21(1):5–11.

Bueno, J. 2014. Anti-biofilm Drug Susceptibility Testing Methods: Looking for new strategies against resistance mechanism. *Journal of Microbial and Biochemical Technology S* 3:2.

Cady, N. C., McKean, K. A., Behnke, J., Kubec, R., Mosier, A. P., Kasper, S. H., Burz, D. S., Musah, R. A. 2012. Inhibition of biofilm formation, quorum sensing, and infection in *Pseudomonas aeruginosa* by natural products-inspired organosulfur compounds. *PLoS One* 7(6):38492.

Calixto, J. B. 2005. Twenty-five years of research on medicinal plants in Latin America: A personal review. *Journal of Ethnofarmacology* 100:131–134.

Castillo-Juárez, I., Maeda, T., Mandujano-Tinoco, E. A., Tomás, M., Pérez-Eretza, B., García-Contreras, S. J., Wood, T. K., García-Contreras, R. 2015. Role of quorum sensing in bacterial infections. *World Journal of Clinical Cases: WJCC* 3(7):575.

Cerca, N., Azevedo, N. F. 2012. *Biofilmes na saúde, no meio ambiente, na indústria*, Publindústria, Braga, Porto.

Chandra, J, Mukherjee, P. K. 2015. Candida biofilms: Development, architecture, and resistance. *Microbiology Spectrum* 3(4):1–14.

Chevalier, M., Doglio, A., Rajendran, R., Ramage, G., Prêcheur, I., Ranque, S. 2019. Inhibition of adhesion-specific genes by *Solidago virgaurea* extract causes loss of *Candida albicans* biofilm integrity. *Journal of Applied Microbiology* 127(1):68–77.

Chimileski, S., Franklin, M. J., Papke, R. T. 2014. Biofilms formed by the archaeon *Haloferax volcanii* exhibit cellular differentiation and social motility, and facilitate horizontal gene transfer. *BMC Biology* 12: 1–15.

Choi, H. A., Cheong, D. E., Lim, H. D., Kim, W. H., Ham, M. H., Oh, M. H., Kim, G. J. 2017. Antimicrobial and antibiofilm activities of the methanol extracts of medicinal plants against dental pathogens *Streptococcus mutans* and *Candida albicans. Journal of Microbiology Biotechnology* 27(7):1242–1248.

Cordeiro, J. C. P. 2014. Antimicrobial activity of plant extracts and biofilm formation by isolates of Salmonella spp. from goats and sheep from the San Francisco Valley. In: *Dissertação (Mestrado em Ciência Animal)*, Universidade Federal do Vale do São Francisco, Univasf, Petrolina – PE.

Costa Neto, E. M., Alves, R. R. N. 2010. Estado da arte da zooterapia popular no Brasil. In: Costa Neto, E. M., Alves, R. R. N. (eds.) *Zooterapia: os animais na medicina popular brasileira*, NUPEEA, Recife.

Costerton, J. W., Geesey, G. G., Cheng, K-J. 1978. How bacteria stick. *Scientific American* 238(1):87–95.

Costerton, J. W., Stewart, P. S., Greenberg, E. P. 1999. Bacterial biofilms: A common cause of persistent infections. *Science* 284:1318–1322.

Cragg, G. M., Newman, D. J. 2013, Natural products: A continuing source of novel drug leads, *Biochimica et Biophysica Acta* 1830:3670–3695.

D'Almeida, R. E., Molina, R. D. I., Viola, C. M., Luciardi, M. C., Penalver, C. N., Bardon, A., Arena, M. E. 2017. Comparison of seven structurally related coumarins on the inhibition of *Quorum sensing* of *Pseudomonas aeruginosa* and *Chromobacterium violaceum. Bioorganic Chemistry* 73:37–42.

Damiano, S., Forino, M., De, A., Vitali, L. A., Lupidi, G., Taglialatela-Scafati, O. 2017. Antioxidant and antibiofilm activities of secondary metabolites from *Ziziphus jujuba* leaves used for infusion preparation. *Food Chemistry* 230:24–29.

Dantas, F. C. P., Tavares, M. L. R., Targino, M. S., Costa, A. P., Dantas, F. O. 2014. *Ziziphus joazeiro* Mart—Rhamnaceae: biogeochemical characteristics and importance in the *Caatinga biome. RES Principal* 25:51–57.

Das, T., Krom, B. P., van der Mei, H. C., Busscher, H. J., Sharma, P. K. 2011. DNA-mediated bacterial aggregation is dictated by acid-base interactions. *Soft Matter* 7 2927–2935.

Das, T., Sehar, S., Manefield, M. 2013. The roles of extracellular DNA in the structural integrity of extracellular polymeric substance and bacterial biofilm development. *Environmental Microbiology Reports* 5(6):778–786.

De Queiroz Dias, D., Lima Sales, D., Cosmo Andrade, J., da Silva, A. R. P., Relison Tintino, S., de Morais Oliveira-Tintino, C. D., de Araújo Delmondes, G., Rocha, M. F. G., da Costa, J. G. M., da Nóbrega Alves, R. R., Ferreira, F. S. 2019. GC–MS analysis of the fixed oil from *Sus scrofa domesticus* Linneaus (1758) and antimicrobial activity against bacteria with veterinary interest. *Chemistry and Physics of Lipids* 219:23–27.

Diaz, M. A. N., Rossi, C. C., Mendonça, V. R., Silva, D. M., Ribon, A. O. B., Aguilar, A. P., Muñoz, G. D. 2010. Screening of medicinal plants for antibacterial activities on *Staphylococcus aureus* strains isolated from bovine mastitis. *Revista Brasileira de Farmacognosia* 20(5):724–728.

Di Giulio, M., Genovese, S., Fiorito, S., Epifano, F., Nostro, A., Cellini, L. 2015. Antimicrobial evaluation of selected naturally occurring oxyprenylated secondary metabolites. *Natural Product Research* 30(16):1870–1874.

Dickschat, J. S. 2010. Quorum sensing and bacterial biofilms. *Natural Product Reports* 27:343–369.

Dima, C., Dima, S. 2015. Essential oils in foods: Extraction, stabilization, and toxicity. *Current Opinion in Food Science* 5:29–35.

Dolfi, S., Sveronis, A., Silipo, A., Rizzo, R., Cescutti, P. 2015. A novel rhamno-mannan exopolysaccharide isolated from biofilms of *Burkholderia multivorans* C1576. *Carbohydrate Research* 411:42–48.

Drenkard, E. 2003. Antimicrobial resistance of *Pseudomonas aeruginosa* biofilms. *Microbes and Infection* 5(3):1213–1219.

Dunne, W. M. J. 2002. Bacterial adhesion: Seen any good biofilms lately? *Clinical Microbiology Reviews* 15(2):155–166.

Epifano, F., Genovese, S., Menghini, L., Curini, M. 2007. Chemistry and pharmacology of oxyprenylated secondary plant metabolites. *Phytochemistry* 68(7):939–953.

Erskine, E., MacPhee, C. E., Stanley-Wall, N. R. 2018. Functional amyloid and other protein fibers in the biofilm matrix. *Journal of Molecular Biology* 430:3642–3656.

Estrela, A. B., Abraham, W. R. 2016. Fungal metabolites for the control of biofilm infections. *Agriculture* 6(3):37.

Fanning, S., Mitchell, A. P. 2012. Fungal biofilms. *PLoS Pathogens* 8(4):e1002585.

Felipe, L. D. O., Silva Júnior, W. F. D., Araújo, K. C. D., Fabrino, D. L. 2018. Lactoferrin, chitosan, and *Melaleuca alternifolia*—natural products that show promise in candidiasis treatment. *Brazilian Journal of Microbiology* 49(2):212–219.

Feng, J., Wang, T., Zhang, S., Shi, W., Zhang, Y. 2014. An optimized SYBR Green I/PI assay for rapid viability assessment and antibiotic susceptibility testing for *Borrelia burgdorferi*. *PLoS One* 9(11):e111809.

FIB. 2010. Extratos Vegetais. Revista Food Ingredients Brasil, *São Paulo* 11:16–20.

Figueiredo, A. C., Pedro, L. G., Barroso, J. G. 2014. Aromatic and medicinal plants: Essential and volatile oils. *Magazine of the Portuguese Horticulture Association* 114(3):29–33.

Flemming, H. C., Wingender, J. 2010. The biofilm matrix. *Nature Reviews Microbiology* 8(9):623.

Flemming, H. C., Wingender, J., Szewzyk, U., Steinberg, P., Rice, S. A., Kjelleberg, S. 2016. Biofilms: An emergent form of bacterial life. *Nature Reviews – Microbiology* 14:563–575.

Francolini, I, Norris, P, Piozzi, A, Donelli, G, Stoodley, P. 2004. Usnic acid, a natural antimicrobial agent able to inhibit bacterial biofilm formation on polymer surfaces. *Antimicrob Agents Chemother* 48:4360–4365.

Franklin, M. J., Nivens, D. E., Weadge, J. T., Howell, P. L. 2011. Biosynthesis of the *Pseudomonas aeruginosa* extracellular polysaccharides, alginate, Pel, and Psl. *Frontiers in Microbiology* 2:1–16.

Freitas, M. A., Andrade, J. C., Alves, A. I. S., Santos, F. A. G. Alves, A. I. S., Santos, F.A. G., Leite-Andrade, M. C., Sales, D. L., Nunes, M., Ribeiro, P. R. V., Coutinho, H. D. M., Morais-Braga, M. F., Neves, R. P. 2019. Use of the natural products from the leaves of the fruitfull tree *Persea americana* against *Candida* sp. Biofilms using Acrylic Resin discs. *Science of the Total Environment* 703:134–779.

Furletti, V. F. 2009. Action of extracts, essential oils, and isolated fractions of medicinal plants on the formation of biofilm in *Candida* spp. In: *Tese (Doutorado)*, Universidade Estadual de Campinas – Unicamp, São Paulo – SP.

Giordani, R. B., Macedo, A. J., Trentin, D. S. 2013. Biofilmes bacterianos patogênicos: aspectos gerais, importância clínica e estratégias de combate. *Revista Liberato. dez* 14(22):113–238.

Gobbo Neto, L., Lopes, N. P. 2007. Medicinal plants: Factors influencing the content of secondary metabolites. *Química Nova* 30(2):374–381.

Gökalsın, B., Sesal, N. C. 2016. Lichen secondary metabolite evernic acid as potential quorum sensing inhibitor against *Pseudomonas aeruginosa. World Journal of Microbiology and Biotechnology* 32(9):150.

Gomes, N., Bessa, L., Buttachon, S., Costa, P., Buaruang, J., Dethoup, T., Kijjoa, A. 2014. Antibacterial and antibiofilm activities of tryptoquivalines and meroditerpenes isolated from the marine-derived fungi *Neosartorya paulistensis, N. laciniosa, N. tsunodae*, and the soil fungi *N. fischeri* and *N. siamensis. Marine Drugs* 12(2):822–839.

Gressler, L.T., Vargas, A.C., Costa, M.M., Sutili, F.J., Schwab, M., Pereira, D.I.B., Sangioni, L.A., Botton, S.A. 2015. Biofilm formation by *Rhodococcus equi* and putative association with macrolide resistance. *Pesquisa Veterinária Brasileira* 35:835–841.

Grujić, S. M., Radojević, I. D., Vasić, S. M., Čomić, L. R., Topuzović, M. D. 2015. Antimicrobial and antibiofilm activities of secondary metabolites from *Vinca minor* L. *Applied Biochemistry and Microbiology* 51(5):572–578.

Helaly, S. E., Kuephadungphan, W., Phongpaichit, S., Luangsa-Ard, J. J., Rukachaisirikul, V., Stadler, M. 2017. Five unprecedented secondary metabolites from the spider parasitic fungus *Akanthomyces novoguineensis. Molecules* 22(6):991.

Herrera, J. J. R., Cabo, M. L., González, A., Pazos, I., Pastoriza, L. 2007. Adhesion and detachment kinetics of several strains of *Staphylococcus aureus* subsp. *aureus* under three different experimental conditions. *Food Microbiology* 24(6):585–591.

Jennings, E., Courtney, H. S., Haggard, W. O. 2012. O ácido cis-2-decenóico inibe o crescimento de S. aureus e o biofilme in vitro: um estudo piloto. *Clinical Orthopaedics and Related Research* 470:2663–2670.

Jett, B.D., Huycke, M.M., Gilmore, M.S. 1994. Virulence of Enterococci. *Clinical Microbiology Reviews* 7(4):462–478.

Kawarai, T., Narisawa, N., Suzuki, Y., Nagasawa, R., Senpuku, H. 2016. *Streptococcus mutans* biofilm formation is dependent on extracellular DNA in primary low pH conditions. *Journal of Oral Biosciences* 58(2):55–61.

Koo, H., Allan, R. N., Howlin, R. P., Stoodley, P., Hall-Stoodley, L. 2017. Targeting microbial biofilms: current and prospective therapeutic strategies. *Nature Reviews Microbiology* 15(12):740–755.

Kothari, V., Sharma, S., Padia, D. 2017. Recent research advances on *Chromobacterium violaceum. Asian Pacific Journal of Tropical Medicine* 10(8):744–752.

Koul, S., Prakash, J., Mishra, A., Kalia, V. C. 2016. Potential emergence of multi-quorum sensing inhibitor resistant (MQSIR) bacteria. *Indian Journal of Microbiology* 56(1):1–18.

Langer, L. T. A., do Carmo, R. L., Staudt, K. J., Alves, I. A. (2018). Biofilmes em infecção por Candida: uma revisão da literatura. *Revista interdisciplinar em ciências da saúde e biológicas–RICSB* 2(2):1–15.

Lee, C. K., de Anda, J., Baker, A. E., Bennett, R. R., Luo, Y., Lee, E. Y., Keefe, J. A., Helali, J. S., Ma, J., Zhao, K., Golestanian, R. 2018. Multigenerational memory and adaptive adhesion in early bacterial biofilm communities. *Proceedings of the National Academy of Sciences* 115(17):1–6.

Leone, S., Molinaro, A., Alfieri, F., Cafaro, V., Lanzetta, R., Di Donato, A., Parrilli, M. 2006. The biofilm matrix of *Pseudomonas* sp. OX1 grown on phenol is mainly constituted by alginate oligosaccharides. *Carbohydrate Research* 341:2456–2461.

Lima, I.O., Oliveira, R. A. G., Lima, E. O., Farias, N. M. P., Souza, E. L. 2006. Antifungal activity and essential oils on species of *Candida. Revista Brasileira Farmacognosia* 16(2):197–201.

Lister, J. L., Horswill, A. R. 2014. *Staphylococcus aureus* biofilms: Recent developments in biofilm dispersal. *Frontiers in Cellular and Infection Microbiology* 4:178.

Liu, F., Sun, Z., Wang, F., Liu, Y., Zhu, Y., Du, L., Xu, W. 2019. Inhibition of biofilm formation and exopolysaccharide synthesis of *Enterococcus faecalis* by phenyl lactic acid. *Food Microbiology* 86:103344.

Lopez, D., Vlamakis, H., Kolter, R. 2010. Biofilms. *Cold Spring Harbor Perspectives in Biology* 2(7):a000398–a000398.

Luís, Â., Duarte, A., Gominho, J., Domingues, F., Duarte, A. P. 2016. Chemical composition, antioxidant, antibacterial and anti-quorum sensing activities of *Eucalyptus globulus* and *Eucalyptus radiata* essential oils. *Industrial Crops and Products* 79:274–282.

Lynch, D. J., Fountain, T. L., Mazurkiewicz, J. E., Banas, J. A. 2007. Glucan-binding proteins are essential for shaping *Streptococcus mutans* biofilm architecture. *FEMS Microbiology Letters* 268:158–165.

Macé, S., Hansen, L. T., Rupasinghe, H. P. V. 2017. Anti-bacterial activity of phenolic compounds against *Streptococcus pyogenes. Medicines* 4(25):1–9.

Machado, T. B., Pinto, A. V., Pinto, M. C. F. R., Leal, I. C. R., Silva, M. G., Amaral, A. C. F., Kuster, R. M., Netto-dos Santos, K. R. 2003. In vitro activity of Brazilian medicinal plants, naturally occurring naphthoquinones and their analogues, against methicillin-resistant *Staphilococcus aureus. International Journal of Antimicrobial Agents* 21(3):279–284.

Maki, D. G., Masur, H., McCormick, R. D., Mermel, L. A., Pearson, M. L., Raad, I. I., Randolph, A., Weinstein, R. A. 2002. Guidelines for the prevention of intravascular catheter-related infections. Centers for disease control and prevention. *Morbidity and Mortality Weekly Report* 9:1–29.

Mary, R. N. I., Banu, N. 2015. Screening of antibiofilm and anti-quorum sensing potential of *Vitex trifolia* in *Pseudomonas aeruginosa. International Journal of Pharmacy and Pharmaceutical Sciences* 7(8):242–245.

McCarthy, H., Rudkin, J. K., Black, N. S., Gallagher, L., O'Neill, E., O'Gara, J. P. 2015. Methicillin resistance and the biofilm phenotype in *Staphylococcus aureus. Frontiers in Cellular and Infection Microbiology* 5:1.

Medonline. Medicina on-line. Biofilme: um velho problema, uma nova batalha. 2008. Revista Virtual de Medicina.

Millot, M., Girardot, M., Dutreix, L., Mambu, L., Imbert, C. 2017. Antifungal and anti-biofilm activities of acetone lichen extracts against *Candida albicans. Molecules* 22(4):651.

Montovani, P. A. B., Gonçalves Júnior, A. C., Moraes, A., Mondardo, D., Meinerz, C., Shikida, S. 2009. Antimicrobial activity of the extract of capororoca (*Rapanea* sp.). *Revista Brasileiro de Agroecologia* 4(2):3764–3767.

Morais-Braga, M. F. B., Sales, D. L., dos Santos Silva, F., Chaves, T. P., de Carvalho Nilo Bitu, V., Avilez, W. M. T., Coutinho, H. D. M. 2016. *Psidium guajava* L. and *Psidium brownianum* Mart ex DC. potentiate the effect of antibiotics against Gram-positive and Gram-negative bacteria. *European Journal of Integrative Medicine* 8(5):683–687.

Morgan-Sastume, F., Larsen, P., Nielsen, J. L., Nielsen, P. H. 2008. Characterization of the loosely attached fraction of activated sludge bacteria. *Water Research* 42:843–854.

Nader, T. T., Coppede, J. S., Amaral, L. A., Pereira, A. M. S. 2014. Atividade antibiofilme de diterpeno isolado de *Croton antisyphiliticus* frente *Staphylococcus aureus. Ars Veterinaria* 30(1):32–37.

Neu, T. R., Lawrence, J. R. 1999. Lectin-binding analysis in biofilm systems. *Methods in Enzymology* 310:145–150.

Nithyanand, P., Shafreen, R. M. B., Muthamil, S., Pandian, S. K. 2015. Usnic acid, a lichen secondary metabolite inhibits Group A Streptococcus biofilm. *Antonie Van Leeuwenhoek* 107(1):263–272.

Nobile, C. J., Johnson, A. D. 2015. *Candida albicans* biofilmes e doenças humanas. *Annual Review of Microbiology* 69:71–92.

Nuryastuti, T., van der Mei, H., Busscher, H. J., Iravati, S, Aman, A. T., Krom, B. P. 2009. Effect of cinnamon oil on icaA expression and biofilm formation by *Staphylococcus epidermidis. Applied Environmental Microbiology* 75:6850–6855.

Okshevsky, M., Meyer, R. L. 2013. The role of extracellular DNA in the establishment, maintenance, and perpetuation of bacterial biofilms. *Critical Reviews in Microbiology*:1–11.

Oliveira, O. P., Sales, D. L., Dias, D. Q., Cabral, M. E. S., Araújo Filho, J. A., Teles, D. A. 2014. Antimicrobial activity and chemical composition of fixed oil extracted from the body fat of the snake *Spilotes pullatus. Pharmaceutical Biology* 52(6):740–744.

Packer, J. F., Luz, M. M. S. 2007. Method for evaluation and research of antimicrobial activity of products of natural origin. *Revista Brasileira de Farmacognosia* 17(1):102–107.

Padoveze, M.C. 2013. Biofilme: o inimigo invisível, Parte I.

Palmer, Jr., R. J. Haagensen, J. A. J., Neu, T. R., Sternberg, C. 2006. Confocal microscopy of biofilms—Spatiotemporal approaches. In: Pawley, J. B. (ed.) *Handbook of Biological Confocal Microscopy*, 2nd Edition, Springer Science and Business Media, New York.

Palombo, E. A. 2011. Traditional medicinal plant extracts and natural products with activity against oral bacteria: potential application in the prevention and treatment of oral diseases. *Evidence-Based Complementary and Alternative Medicine* 2011:1–15.

Pawar, V. C., Thaker, V. S. 2006. In vitro efficacy of 75 essential oils against *Aspergillus niger. Mycoses* 49(4):316–323.

Pejin, B., Ciric, A., Glamoclija, J., Nikolic, M., Sokovic, M. 2015. In vitro anti-quorum sensing activity of phytol. *Natural Product Research* 29(4):374–377.

Pereira, C. S. S. 2009. Evaluation of different technologies in the extraction of Pinhão-manso Oil (*Jatropha curcas* L). PhD diss. Universidade Federal Rural do Rio de Janeiro (UFRRJ).

Pereira, O. B. O. 2001. Comparison of the efficacy of two biocides (carbamate and glutaraldehyde) in biofilm systems. PhD thesis. Universidade do Minho: Braga.

Ponnusamy, K., Paul, D., Kweon, J. H. 2009. Inhibition of quorum sensing mechanism and *Aeromonas hydrophila* biofilm formation by vanillin. *Environmental Engineering Science* 26(8):1359–1363.

Prabuseenivasan, S, Jayakumar, M, Ignacimuthu, S. 2006. In vitro antibacterial activity of some plant essential oils. *BMC Complementary and Alternative Medicine* 39(11):486–494.

Qayyum, S., Sharma, D., Bisht, D., Khan, A. U. 2016. Protein translation machinery holds a key for transition of planktonic cells to biofilm state in *Enterococcus faecalis*: A proteomic approach. *Biochemical and Biophysical Research Communications*. 474:652–659.

Qin, N., Tan, X., Jiao, Y., Liu, L., Zhao, W., Yang, S., Jia, A. 2014. RNA-Seq-based transcriptome analysis of methicillin-resistant *Staphylococcus aureus* biofilm inhibition by ursolic acid and resveratrol. *Scientific Reports* 4:1–9.

Queiroz, A.M.P., Ferreira, C.E.F. 2010. Sensibilidade bacteriana a antimicrobianos de primeira escolha prescritos no tratamento de pneumonias em clínica e UTI pediátrica do Município de campos dos goytacazes, RJ. *Infarma* 22(5/6):32–37.

abin, N., Zheng, Y., Opoku-Temeng, C., Du, Y., Bonsu, E., Sintim, H. O. 2015. Agents that inhibit bacterial biofilm formation. *Future medicinal chemistry* 7(5):647–671.

Reen, F. J., Gutiérrez-Barranquero, J. A., Parages, M. L. 2018. Coumarin: a novel player in microbial quorum sensing and biofilm formation inhibition. *Applied Microbiology and Biotechnology* 102(5):2063–2073.

Rishton, G. M. 2008. Natural products as a Robust Source of new drugs and drug leads: Past successes and present day issues. *The American Journal of Cardiology* 101(10):43–49.

Rocha, G. R., Salamanca, E. J. F., de Barros, A. L., Lobo, C. I. V., Klein, M. I. 2018. Effect of tt-farnesol and myricetin on in vitro biofilm formed by *Streptococcus mutans* and *Candida albicans. BMC Complementary and Alternative Medicine* 18(1):61.

Rodrigues, F. C., dos Santos, A. T. L., Machado, A. J. T., Bezerra, C. F., de Freitas, T. S., Coutinho, H. D. M., Morais-Braga, M. F. B., Bezerra, J. W. A., Duarte, A. E., Kamdem, J. P., Boligon, A. A. 2019. Chemical composition and anti-Candida potential of the extracts of *Tarenaya spinosa* (Jacq.) Raf. (Cleomaceae). *Comparative Immunology, Microbiology and Infectious Diseases* 64:14–19.

Rohde, H., Frankenberger, S., Zähringer, U., Mack, D. 2010. Structure, function, and contribution of polysaccharide intercellular adhesin (PIA) to *Staphylococcus epidermidis* biofilm formation and pathogenesis of biomaterial associated infections. *European Journal of Cell Biology* 89:103–111.

Romero, D., Kolter, R. 2014. Functional amyloids in bacteria. *International Microbiology* 17:65–73.

Sajeevan, S. E., Chatterjee, M., Paul, V., Baranwal, G., Kumar, V. A., Bose, C., Banerji, A., Nair, B. G., Prasanth, B. P., Biswas, R. 2018. Impregnation of catheters with anacardic acid from cashew nut shell prevents *Staphylococcus aureus* biofilm development. *Journal of Applied Microbiology* 125:1286–1295.

Sales, D. L., Oliveira, O. P., Cabral, M. E. S., Dias, D. Q., Kerntopf, M. R., Coutinho, H. D. M. 2014. Chemical identification and evaluation of the antimicrobial activity of fixed oil extracted from *Rhinella jimi. Pharmaceutical Biology* 53:98–103.

Salta, M., Wharton, J. A., Dennington, S. P., Stoodley, P, Stokes, K. R. 2013. Antibiofilm performance of three natural products against initial bacterial attachment. *International Journal of Molecular Sciences* 14:21757–21780

Sardi, J. C. O., Mendes Giannini, M. J. S., Bernardi, T., Scorzoni, L., Fusco-Almeida, A. M. 2013. *Candida species*: current epidemiology, pathogenicity, biofilm formation, natural antifungal products and new therapeutic options. *Journal of Medical Microbiology* 62(1), 10–24.

SBF (Sociedade Brasileira de Farmacognosia). 2017. Óleos fixos e ceras. Apostila de Aula Prática de Farmacognosia UFBA. SBFgnosia.

Seleem, D., Freitas-Blanco, V. S., Noguti, J., Zancope, B. R., Pardi, V., Murata, R. M. 2018. In vivo antifungal activity of monolaurin against *Candida albicans* biofilms. *Biological and Pharmaceutical Bulletin* 41(8):1299–1302.

Silva, N., Alves, S., Gonçalves, A., Amaral, J.S., Poeta, P. 2013. Antimicrobial activity of essential oils from Mediterranean aromatic plants against several foodborne and spoilage bacteria. *Food Science and Technology International* 19:503–510.

Simões, C. M. O., Schenkel, E. P., de Mello, J. C. P., Mentz, L. A., Petrovick, P. R. 2016. *Farmacognosia: do produto natural ao medicamento.* Artmed Editora.

Simões, C. M. O., Schenkel, E. P., Gosmann, G., Mell, J. C. P., Mentz, L. A., Petrovick, P. R. 2007. *Farmacognosia: da planta ao medicamento*, 6a Ed, UFRGS, Porto Alegre.

Slobodníková, L., Fialová, S., Rendeková, K., Kováč, J., Mučaji, P. 2016. Antibiofilm activity of plant polyphenols. *Molecules* 21(12), 1717.

Sokovic, M., Glamoclija, J., Ciric, A. 2013. Natural products from plants and fungi as fungicides. In: Nita, M (ed.) *Fungicides-Showcases of Integrated Plant Disease Management from Around the World*, InTech, New York, pp. 185–232.

Sokovic, M., Glamoclija, J., Marin, P. D., Brkie, D., Van Griensven, L. J. L. D. 2010. Antibacterial effects of the essential oils of commonly consumed medicinal herbs using an in vitro model. *Molecules* 15:7532–7546.

Soler-Arango, J., Figoli, C., Muraca, G., Bosch, A., Brelles-Marino, G. 2019. The *Pseudomonas aeruginosa* biofilm matrix and cells are drastically impacted by gas discharge plasma treatment: A comprehensive model explaining plasma-mediated biofilm eradication. *PLoS One* 14(6):1–27.

Souza, E. L., Lima, E. O., Freire, K. R. L., Sousa, C. P. 2005. Inhibitory action of some essential oils and phytochemicals on the growth of various molds isolated from foods. *Brazilian Archives Biology Technology* 48(2):245–250.

Stefanović, O. D., Tešić, J. D., Čomić, L. R. 2015. *Melilotus albus* and *Dorycnium herbaceum* extracts as source of phenolic compounds and their antimicrobial, antibiofilm, and antioxidant potentials. *Journal of Food and Drug Analysis* 23(3):417–424.

Stepanović, S., Vuković, D., Hola, V., Bonaventura, G. D., Djukić, S., Ćirković, I., Ruzicka, F. 2007. Quantification of biofilm in microtiter plates: overview of testing conditions and practical recommendations for assessment of biofilm production by staphylococci. *APMIS* 115:891–899.

Stewart, P. S., Franklin, M. J. 2008. Physiological heterogeneity in biofilms. *Nature Reviews – Microbiology* 6:199–210.

Taglialegna, A., Lasa, I., Valle, J. 2016. Amyloid structures as biofilm matrix scaffolds. *Journal of Bacteriology* 198:2579–2588.

Thakare, R., Shukla, M., Kaul, G., Dasgupta, A., Chopra, S. 2019. Repurposing disulfiram for treatment of Staphylococcus aureus infections. *International Journal of Antimicrobial Agents* 53(6):709–715.

Thenmozhi, R., Balaji, K., Kumar R., Rao, T. S., Pandian, S. K. 2011. Characterization of biofilms in different clinical M serotypes of *Streptococcus pyogenes*. *Journal of Basic Microbiology* 51:196–204

Tiwari, V., Meena, K., Tiwari, M. 2017. Differential anti-microbial secondary metabolites in different ESKAPE pathogens explain their adaptation in the hospital setup. *Infection, Genetics and Evolution* 66:57–65.

Trentin, D. S., Giordani, R. B., Macedo, A. J. 2013. Biofilmes bacterianos patogênicos: Aspectos gerais, importância clínica e estratégias de combate. *Revista Liberato* 14(22):113–238.

Trentin, D. S., Giordani, R. B., Zimmer, K. R., Da Silva, A. G., Da Silva, M. V., dos Santos Correia, M. T., Baumvol, I. J. R., Macedo, A. J. 2011. Potential of medicinal plants from the Brazilian semiarid region (Caatinga) against *Staphylococcus epidermidis* planktonic and biofilm lifestyles. *Journal of Ethnopharmacology* 137:327–335.

Upadhyay, A., Upadhyaya, I., Kollanoor-Johny, A., Venkitanarayanan, K. 2014. Combating pathogenic microorganisms using plant-derived antimicrobials: a minireview of the mechanistic basis. *BioMed Research International* 2014.

Veiga-Junior, V. F., Mello, J. C. P. 2008. Monographs on medicinal plants. *Revista Brasileira de Farmacognosia* 18:464–471.

Vert, M., Doi, Y., Hellwich, K. H., Hess, M., Hodge, P., Kubisa, P., Rinaudo, M., Schué, F. 2012. Terminology for biorelated polymers and applications (IUPAC Recommendations 2012). *Pure and Applied Chemistry* 84(2):377–410.

Wilson, C., Lukowicz, R., Merchant, S., Valquier-Flynn, H., Caballero, J., Sandoval, J., Okuom, M., Huber, C., Brooks, T. D., Wilson, E., Clement, B. 2017. Quantitative and qualitative assessment methods for biofilm growth: A mini-review. *Research & Reviews: Journal of Engineering and Technology* 6(4):1–25.

Wingender, J., Jaeger, K-E., Flemming, H-C. 1999. Microbial extracellular polymeric substances In: Wingender, J., Neu, T., Flemming, H-C. (eds.) *Microbial Extracellular Polymeric Substances*, Springer, Berlin, Heidelberg, pp. 231–251.

Wingender, J., Strathmann, M., Rode, A., Leis, A., Flemming, H. C. 2001. Isolation and biochemical characterization of extracellular polymeric substances from *Pseudomonas aeruginosa. Methods in Enzymology* 336:302–314.

Wu, X., Santos, R.R., Fink-Gremmels, J. 2014. *Staphylococcus epidermidis* biofilm quantification: Effect of different solvents and dyes. *Journal of Microbiological Methods* 101:63–66.

Yang, L., Barken, K. B., Skindersoe, M. E., Christensen, A. B., Givskov, M., Tolker-Nielsen, T. 2007. Effects of iron on DNA release and biofilm development by *Pseudomonas aeruginosa. Microbiology* 153:1318–1328.

Yang, Y., Xiang, Y., Xu, M. 2015. From red to green: The propidium iodide-permeable membrane of *Shewanella decolorationis* S12 is repairable. *Scientific Reports* 5:18583.

Yunes, R. A., Filho Cechinel, V. 2014. Novas perspectivas dos produtos naturais na química medicinal moderna. In: Yunes, R. A.; Filho Cechinel, V. (eds.) Química de produtos naturais novos fármacos e a moderna farmacognosia. *UNIVALI.* 4:11–34.

Zhang, M., Wang, M., Zhu, X., Yu, W. Gong, Q. 2018. Equisetin as potential quorum sensing inhibitor of *Pseudomonas aeruginosa. Biotechnology Letters* 40(5):865–870.

Zhou, J., Bi, S., Chen, H., Chen, T., Yang, R., Li, M., Jia, A. Q. 2017. Anti-biofilm and antivirulence activities of metabolites from *Plectosphaerella cucumerina against Pseudomonas aeruginosa. Frontiers in Microbiology* 8(769): 1–17.

Zimmer, K. R., Blum-Silva, C. H., Souza, A. L. K., Wulff Schuch, M., Reginatto, F. H., Pereira, C. M. P., Lencina, C. L. 2014. The antibiofilm effect of blueberry fruit cultivars against *Staphylococcus epidermidis* and *Pseudomonas aeruginosa. Journal of Medicinal Food* 17(3):324–331.

7 Plant Growth Promoting Rhizobacteria

Biofilm and Role in Sustainable Agriculture

Mohd. Musheer Altaf
Institute of Information Management and Technology

Mohd Sajjad Ahmad Khan
Imam Abdulrahman Bin Faisal University

CONTENTS

7.1 INTRODUCTION

Soil is a main component of earth which is dynamic, full of life, and highly dissimilar. It shelters thousands of living organisms both prokaryotic and eukaryotic. The populations of microorganisms are affected by the neighboring soil atmosphere. Inside this niche the plant roots support the diversified global ecological unit and make available a large portion of carbon to the below ground soil ecological system. Moreover, much like terrestrial ecosystem, the soil microorganisms also depend on several methods for their survival such as respiration, competition for nutrients, efficient communication, and quick response to the ever-changing environments. As an innate source, soil acts as a key character in biodiversity and is implicated in several additional environmental repairs. On the other hand, soil is believed to be productive with the action of soil microbes at a sufficient pace for quick plant growth and development. Even if the soil nutrients along with other essential elements are present in high quantity, these components are rarely accessible to the soil microbes, due to the bonding of soil particles with water and nutrients. This problem of aggregation is resolved by the plant roots for their use of resources (Lambers et al., 1998; Nayak et al., 2019). Plant roots connected bacteria both within and outside participate in a key function of soil production and development. The interrelation of environmental accumulation and spatial detachment is the main determinant for the large quantity of soil bacteria, particularly rhizobacteria (Ramette and Tiedje, 2007). It is anticipated that each gram of soil holds approximately 10^7–10^{12} bacteria, together with protozoa and fungal hyphae (Young and Crawford, 2004). In soil, one of the main intriguing hotspotsof microbiological behavior and variety is the rhizosphere. Within the rhizosphere, microbiological consortiums are prosperous and plentiful in species diversity, together with several uncultured forms. The rhizosphere comprises the adjoining (about 50 μm wide) surface area straight forwardly linking the plant roots, trailing by the subsequent 1–5mm around the root (Angus and Hirsch, 2013).

The rhizosphere corresponds to the highly vivacious niche, in addition to the most significant region for the amount and quality of food for humans. It has close association in the form of physical, chemical, and biological relations. In agreement with Hiltner (1904), the rhizosphere is the space covering the live plant roots, affected by the root behavior. Relying upon the action and events such as discharge of sensitive mixtures, absorption of different nutrients, water, respiration speed, and quantitative expansion of the rhizosphere might increase from the sub-micrometer to supra-centimeter levels. The ecology of rhizosphere is largely decided through a mixture of soil medium, together with the spatial and temporal sharing of rhizodeposits, protons, gases, and the function of roots intended for soil nutrients. Rhizosphere deposition is the most important metabolic procedure changing nourishment large quantity within the rhizospheric soil. The carbon movement makes possible carbon-rich aggregates which additionally hinder the development of non-heterotrophic microorganisms (Jones et al., 2009). The rhizosphere therefore comes up as improved soil hot spot together with plant waste and dead material intended for microbiological actions. Also, the large part of soil is carbon which hinders the microbial growth and functions, whereas the rhizospheric soil is devoid of nitrogen. The main outcome might take place in the rhizosphere which leads to improved soil organic material decay and enhancement of plant growth, development, and output. On the other hand, the existence of leguminous crop plants promotes the soil nitrogen accessibility and has a considerable role in the makeup of plant associated bacterial population. Not only has the competition for nutrients, collaboration, and complementarities but also influenced the efficiency of an ecological unit (Callaway and Walker, 1997).

Root exudate within the rhizosphere establishes a large quantity of rhizospheric bacteria as the highly nutrient competent set of microorganisms which reside particularly within the rhizosphere habitat (Robin et al., 2008). Rhizobacteria participates in a key function of sustainable agriculture all the way through plant development, accessing and absorbing nourishments and biological control of plant pathogens on its own or by collection of different methods. Species belonging to *Burkholderia, Bacillus, Phyllobacterium,* and *Ralstonia* genera are the main agents of rhizospheric bacteria, whereas at species stage, *Burkholderia cepacia*, *Burkholderia gladioli*, and *Phyllobacterium rubiacearum* are in abundance (Gooden et al., 2004; Nayak et al., 2019). Soil pH, obtainable oxygen, moisture concentration, and water accessibility are the only reasons influencing rhizospheric bacteria endurance and actions.

A biofilm is a collection, aggregation, or population of microorganisms surrounded by means of a polymeric medium consisting of polysaccharides which are connected to a lifeless or living objects (Costerton et al., 1995; Vlamakis et al., 2013). The word "biofilm" was created and explained by Costerton et al. (1978). The bacterial cells, the biotic or abiotic substratum, and the adjoining matrix and their connections are the main apparatus implicated in biofilm construction. Biofilm development is a general feature, displayed by bacteria/fungi, while emerging connected to living and nonliving area. The flagella, fimbriae, pili, lipopolysaccharides, and membrane proteins, together with adhesins, of microorganisms are implicated in biofilm development (Hinsa et al., 2003; Stewart and Franklin, 2008). Biofilm construction presents a productive health benefit in conditions of sluggish development and physiological heterogeneity, in contrast to planktonic cells. Microbes build biofilms as an approach to conquer strain, for example, nutrient exhaustion, alteration in pH, or the existence of oxygen radicals, bactericidal and bacteriostatic agents in their surroundings. Biofilm development is in addition a means to persist in a good position via efficient establishment. The speed of movement of genetic matter between the microbes is fast within biofilms in contrast to the planktonic counterparts owing to closer cell association. The transformational frequencies are 10–600 times higher in biofilms as compared to planktonic cells (Yung-Hua et al., 2001; Velmourougane et al., 2017). Alteration in gene interpretation and control builds the biofilm cells phenotypically and metabolically different from the planktonic cells. Biofilm growth is a procedure, helped through characteristics together with cell-to-cell messaging, cell segregation, and prototype creation. Due to these procedures, biofilm organization is a growth method. An established five-step replica involving reversible sticking of planktonic cells, irreversible bonding, extracellular polymeric substance (EPS) development, biofilm growth, and scattering of biofilm development was projected via Sauer (Sauer, 2003; Lopez and Kolter, 2010; Asally et al., 2012; Dietrich et al., 2013).

Regarding mixed species rhizospheric bacterial biofilm, substantial data has established that rhizobacteria exist in consortium, working together and assisting continued existence on rhizospheric habitat. The conservation of the rhizobacterial population is a benefit of community dependent biofilms. As a result multispecies biofilm participates in key functions related to sustainable agriculture (Angus and Hirsch, 2013). Therefore, rhizospheric bacteria together with their biofilms have been achieving significance owing to their task in soil development and plant health. This chapter focuses on rhizobacteria and their biofilms as the basis of sustainable agriculture in the present era.

7.2 PLANT GROWTH PROMOTING RHIZOBACTERIAL DIVERSITY AND THEIR BIOFILM

A diversity of microbes is found in the rhizosphere, where a bacterium forms the leading group. Generally, the plants choose these components that are deeply implicated in plant health and growth by secreting root mucilage in the form of organic components and build a helpful situation for plant development (Sinha et al., 2001). Rhizospheric bacteria forms the most leading population of microbes surviving within the close premises of root, typically via forming a complex network of inter- and intra species communications, building biofilms and playing highly important functions related to soil fitness, plant growth, and development in order to be accepted as environmentally and metabolically proficient rhizospheric bacteria within soil oblige to possess exceptional metabolic and useful features. Colonization capability, a continued struggle through discharging biologically dynamic components, is a substantial quality along with other traits. Furthermore they also affect the constancy and livelihood of foreign population. These exact habitats are also repaired for ferocious microbiological influences and have been traditionally evaluated since the start of nineteenth century. On the other hand, the accessibility of current technologies strengthens their research, which leads to the comprehensive detection of their multiplicity and function.

The word "rhizosphere" indicates the continuation of a comparatively huge microbiological inhabitant in the region of the roots of leguminous plants that is additionally continued to other plants as the region of influence at the root section (Lynch, 1990). Owing to the existence of diverse bacterial types within the soil environment of rhizosphere, varieties learning at taxonomic, practical, and genetic levels are more intricate. The region of domination harbors 10–100 times high bacterial members creating tough contest for area and nourishments. A rhizospheric bacterium has numerous features of metabolic elasticity, functional variety, and genetic unpredictability (Sinha et al., 2001). Practically, the distinctiveness of rhizospheric bacteria largely depends on the discharge of several components such as nitrogenase, phosphatase, soil dehydrogenase enzymes, antimicrobials, phosphate solubilizers, iron chelators, and hydrogen cyanide and encourage disease resistance via induce systemic resistance (ISR) along with the biosynthesis of indole acetic acid, ethylene, etc.

The rhizosphere is additionally separated into two parts: the endorhizosphere and the ectorhizosphere. The interior root region, together with the cortical area is described as the endorhizosphere, and it comprises a considerably elevated load of bacterial inhabitants with different tasks and metabolism (Lynch, 1990). A small number of investigators also distinguish the ectorhizosphere into two regions: rhizosphere, i.e. root bordering soil section, and the rhizoplane, i.e., soil on the exterior of the root—other workers believed this as a range. In additional terms, the root exterior region at which the prokaryotes frequently connects themselves via cell surface extracellular polysaccharides or flagella or fimbriae like tools are considered as the rhizosphere (Johri et al., 2003). Root hairs are also compatible biological area for establishment, chemotaxis, and biofilm development (Pothier et al., 2007).

In the current time the center of attention is mostly on learning the bacterial variety and their biofilms. Polyphasic move has been employed to analyze the rhizospheric bacterial population, and it starts by means of the structural path. This method comprises an effort to analyze the complete rhizospheric soil bacterial population, pursued by examining the exchange of inhabitants and procedures implicated in its control. Lastly it terminates by means of the functional method. In examining the effects of the rhizobacterial type richness on an ecological unit, two issues have to be measured: the differences in inhabitants via region and point in time and their significance to the rhizosphere (Ekschmitt and Griffiths, 1998). The working of soil microbial population is dependent on suitable species range for the proficient utilization of resources and preservation of ever shifting environment (Baliyarsingh et al., 2017).Numerous research laboratories in India and other places of the world have been implicated in learning possibly efficient rhizobacterial systemics and their biofilms of native source from the rhizosphere for plant development improvement and biological control analysis (Tilak et al., 1999). Amongst the variety of bacterial genera *Acetobacter*, *Alcaligenes*, *Azospirillium*, *Azotobacter*, *Bacillus*, *Burkholderia*, *Flavobacter*, *Pseudomonas*, *Rhizobium*, *Serratia*, *Xanthomonas*, *Stenotrophomonas*, and *Streptomyces* are establish in large quantities on the rhizosphere as a population and develop biofilms (Barriuso et al., 2008).

Rhizobacterial biofilms are mainly concerned with plant development, consequently enhancing the crop production (Figure 7.1). The comparative fraction of different types of microbes in rhizosphere changes constantly. Nonsymbiotic nitrogen fixing bacteria are recognized as the primary rhizospheric bacteria intended for promotion of plant development and have agricultural significance. As successful nitrogen fixer, *Azospirillium* spp. has been analyzed for the last 50 years (Bashan et al., 2004). Additional plant growth enhancing nitrogen fixers are *Azoarcus* additionally separated into *Azovibrio* sp., *Azospira* sp., *Azonexus* sp., *Burkholderia* spp., *Gluconacetobacter diazotrophicus*, *Herbaspirillum* spp., *Azotobacter* spp., and *Paenibacillus* previously *Bacillus polymyxa*, all of which have related activities (Vessey, 2003). *Azoarcus* spp. has also in recent times achieved interest owing to its genetic diversity and metabolic flexibility. Additionally, these are also classified as diazotrophicus plant growth promoting rhizobacteria (PGPR). Three types of nitrogen fixing bacteria are plentiful: legumes *Rhizobium* symbiosis (symbiotic), *Azotobacter* (asymbiotic or free living), and *Azospirillium* (associative symbiosis) in rhizospheric soil. Someway this consequence occurs in rhizospheric community following cropping with nonlegumes and contributes to help the succeeding produce (Jha et al., 2009). *Bacillus*, *Enterobacter*, and *Pseudomonas* together with some

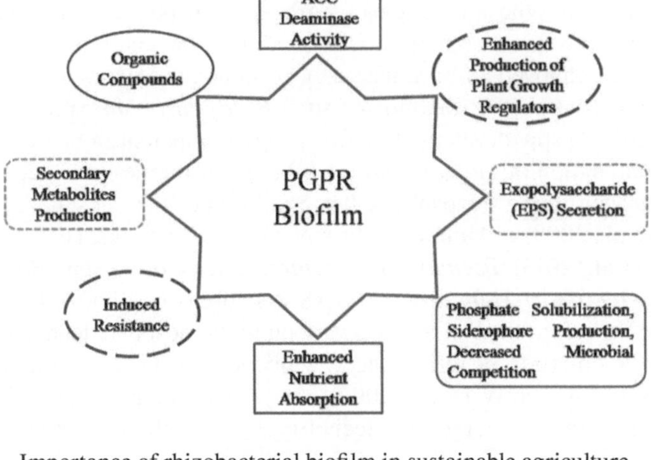

FIGURE 7.1 Importance of rhizobacterial biofilm in sustainable agriculture.

others from the rhizosphere have been analyzed systemically both as planktonic and biofilms (Lynch, 1990). *Bacillus* biofilms are an outstanding illustration to reveal at this point (Altaf and Ahmad, 2016). Over 95% of rod-shaped Gram-positive soil bacteria related to the *Bacillus* spp. has the capability to endure in unfavorable situations via forming endospores and also reveal different development abilities (Nayak et al., 2017). Likewise *Pseudomonas* biofilms have been extensively evaluated as they are the Gram-negative soil bacteria found in large quantities in the rhizosphere and having recognized plant growth promoting characters for decades (Patten and Glick,2002; Altaf and Ahmad, 2019).

7.3 RHIZOSPHERIC SOIL HEALTH AND NUTRIENT MANAGEMENT

Rhizospheric bacterial populations could extensively affect the nutrient organization, primarily for generating soil compactness together with constancy and later inhibiting phytopathogens, each of them is acknowledged toward improving soil fitness and enhancing crop production (Whipps, 2001). The discharge of helpful positive ions through soil natural resources is necessary for plant development and population nourishment. On the other hand, the populations are components of a several layer food network that employs the huge volume of rhizosphere reserves which are the main strengths of microbiological multiplicity and trait control. The rhizospheric bacteria fight among themselves for essential supplies and frequently swap to microorganism-plant relations as a consequence of enhanced competition. This additionally guides to the thought of microevolution in nutrient-deprived sites and situations (Schloter et al., 2000). The rhizosphere reserves also draw an excess of microbes to larger degree. Not only the microbiological actions but also the community inhabitants are straight forwardly related to the composition and prototype of root secretions.

Absorption of organic components provides power for standard rhizospheric bacterial biofilms, and somewhere else acknowledged as organotrophs. The accessibility of carbon is the highly wide spread inhibiting reason for the population development which additionally influences the plant fitness and yield (Alden et al., 2001). During rhizosphere functions, the utilization of nitrogen, phosphorus, and potassium (N, P, K) enhances more than one nutrient/enzyme, or the biocatalyst within the rhizosphere has been employed as an innate sign of biological function through different rhizospheric bacteria. *Acromobacter* spp., *Bradyrhizobium* spp., *Burkholderia* spp., and *Rhizobium* spp. developing biofilms were competent in biological nitrogen fixation and nodulation in *Vigna radiate* (L) Walp (Guimarães et al., 2012). A small number of nitrogen-fixing rhizospheric bacteria initiates the phosphorus absorption (Richardson et al., 2009) and improves the soil enzyme functions, i.e., dehydrogenase activity (Lee et al., 2013). *Pseudomonas striata*, *Enterobacter* spp., *Bacillus megateruium*, and *Ochrobacterium anthropi* TRS 2 solubilize $Ca(PO_4)$, $FePO_4$, $AlPO_2$, and discharge them into the soil via decay along with phosphate loaded components and through the emission of formic, acetic, gluconic, glycolic, and 2-ketogluconic organic acids (Chakraborty et al., 2009). An enhancement in phosphate solubilization guides to the discharge of catechol type siderophore in *Pseudomona* ssp.

EM85. Siderophores were produced to control/chelate/sequestrate intracellular iron levels in low iron situations. *Ochrobacterium anthropi* TRS-2 from tea rhizosphere secretes siderophore in vitro. Fluorescent siderophores are also generated through *Pseudomonas* spp. possessing biocontrol and growth augmenting capacity. The secretion of IAA by rhizospheric bacteria is a key feature for the development of plant growth and encourages stress toleration during drought, water logging, heavy metals, and pathogenicity in plants.

7.4 ROLE OF BIOFILM IN CROP PLANT DISEASE MANAGEMENT

Disease regulation and enhancement is a good method of defense for phytopathogens. It is a well-recognized activity for the rhizospheric bacteria from Proteobacteria and Firmicutes phyla. Owing to the secretion of metabolites, the development and functions of contending disease causing microbes is repressed. Metabolites having antibiotic nature impart various consequences at subinhibitory levels, and their probable role consists of protection against predatory protozoa, motility, biofilm, and uptake of nutrients (Raaijmakers and Mazzola, 2012). Through one or more methods, the rhizospheric bacteria manage the pathogen multiplication and enhance the production.

The discharge of extracellular polymeric material to construct biofilms involves microbial collaboration (Davies et al., 1998). These emissions, which are the consequence of joint procedures from clonal or multiple species microbial associations, offers elective benefits for microorganisms, for example safety from contenders and antimicrobial compounds, starting from enzymatic procedures that involve high cellular concentration and acquirement of novel genes by means of horizontal gene movement. In recent times, it has been revealed that biofilm intervened microcolonies created on root hairs of finger millet through a root-colonizing bacterial endophyte (*Enterobacter* sp.) grant a physical and chemical obstacle that puts a stop to root inhabitation by the phytopathogen *Fusarium graminearum* (Mousa et al., 2016; Hassani et al., 2018). Notably, bacterial characters associated toward motility, attachment, and biofilm formation are required for the anti-*Fusarium* functions in planta. These findings advocated that an intricate interaction occurs among the bacterium and root-hair cells, guiding to the development of this specific association micro niche. While the construction of biofilm has been largely illustrated for plant-connected bacteria (Bogino et al., 2013; Altaf and Ahmad, 2017), assorted bacterial-fungal biofilms or bacterial biofilm created on the exterior of fungal hyphae looks wide spread on plant tissues (van Overbeek and Saikkonen, 2016). In recent times, it has been revealed that bacterial capability to create a biofilm on fungal hyphae is extensively common between soil bacteria; however, it seldom takes place on the hyphae of ascomycete fungi. Particularly, the capability of *Pseudomonas fluorescens* BBc6 to develop a biofilm on the hyphae of the ectomycorrhizal fungus *Laccaría bicolor* is improved at the surrounding area of the ectomycorrhizal root tip, signifying that the organization of the ectomycorrhizal symbiosis encourages bacterial biofilm development on fungal host surfaces. Taken jointly, these findings point out that biofilm development on plant tissues correspond to a hotspot for microbial connections that locally nurture microbial collection (Hassani et al., 2018).

7.5 BIOFILM AND ENVIRONMENTAL STRESS

Undesirable climatic situations producing abiotic pressure are one of the main deterrent causes for decrease in agricultural yield (Padgham, 2009; Grayson, 2013). As per the Food and Agriculture Organization of the United Nations information (2007), barely 3.5% of the world land area has been left untouched by any ecological constraint. Key nonliving stress includes drought, low/elevated temperature, high salt level and acidic situations, light concentration, flood, anaerobic conditions, and nourishment deprivation (Wang et al., 2003; Chaves and Oliveira, 2004; Nakashima and Yamaguchi-Shinozaki, 2006; Hirel et al., 2007; Bailey-Serres and Voesenek, 2008). Water scarcity (drought) has influenced 64% of the world land area, flood (anoxia) 13% of the land area, high salt 6%, mineral scarcity 9%, acidic soils 15%, and low temperature 57% (Cramer et al., 2011). Further,out of the world's 5.2 billion ha of dryland agriculture, 3.6 billion ha is influenced by the tribulations of wearing away, soil dilapidation, and high salt (Riadh et al., 2010). Ruan et al. (2010) projected those high salt influenced soils up to 50% of entire irrigated land in the globe costing US$12 billion in the form of loss (Flowers et al., 2010). In the same way, world yearly price of land dilapidation through high salt in irrigated lands might be US$ 27.3 billion owing to failure in crop yield (Qadir et al., 2014). The harmful impact of high salt on plant development is well recognized. The region under rising salinization has approximately arrived at 62 million irrigated hectares (Hamilton, 2018). Even though some precise assessment of agricultural loss in the form of decrease in crop yield and soil fitness in conditions of agricultural environment turbulence owing to nonliving strains might not be made, it is obvious that these stresses impact large land regions and considerably influenced excellence and amount of loss in crop yield (Cramer et al., 2011).

Plants normally handle the quick variations and hardship of ecological situations due to their inherent metabolic potential (Simontacchi et al., 2015). Differences in the external surroundings might place the plant metabolism beyond homeostasis (Foyer and Noctor, 2005), and make it necessary for the plant to shelter several highly developed genetic and metabolic methods inside its cellular organization (Gill and Tuteja, 2010). Plants have a collection of defensive systems obtained at some point in the course of development to fight unfavorable ecological conditions (Yolcu et al., 2016). These methods are the basis of metabolic reprogramming in the cells (Shao et al., 2008; Bolton, 2009; Massad et al., 2012) to make possible custom biological, physical, and chemical procedures irrespective of the outer circumstances (Mickelbart et al., 2015). On several occasions plants receive aid in plummeting the load of ecological stresses with the help of the microbiome they colonized (Turner et al., 2013; Ngumbi and Kloepper, 2014).

Biofilm development by root-inhabiting microorganisms might participate in vital functions in the defense of the plant in addition to the bacteria. Improved tolerance to drought stress is observed among those plants colonized through definite bacteria (Kim et al., 2012; Naylor and Coleman-Derr, 2018). It is recognized that EPS creates the biofilm and retains moisture coating on the root region and mitigates drought stress impact within plants (Kaushal and Wani, 2016). Famine could stimulate surviving microorganisms to create a mixture of components that influenced population strength, for example, drought supported soils possesses additional antibiotics, which are assumed to be formed by drought-tolerant bacteria as a physiological reply to outnumber other bacteria for restricted nutrients, or probably as indications to encourage drought reaction methods like biofilm development (Bouskill et al., 2016). Timmusk et al. (2015) reported that a mutant of *Paenibacillus polymyxa* deficient in Sfptype 40-phosphopantetheinyl transferase had increased biofilm creation, which upon treatment to drought-stressed wheat plants was revealed to improve plant endurance and biomass creation two to three times, respectively. The methods through which biofilm development assists in plant defense are at the exploratory phase. The construction of the hydrated medium of the biofilm might develop water withholding capacity to preserve equally microbial and plant cell performance under drought (Bouskill et al., 2016). Additionally, the hydrated biofilm medium, during the limitation of dispersion, could aggregate microbial metabolism products, for instance osmolytes or biocontrol-active organizations, for a larger influence on plants as inducers of systemic stress tolerance and as limiting factor for other rhizobacteria (Wright et al., 2016). Kasim et al. (2016) established that the treatment by means of the biofilm forming PGPR (*Bacillus amyloliquefaciens*) had improved impact on the development of barley plants cultivated in high salt conditions and accomplished that the bacteria might be useful in supporting barley plants to bear salt stress next to its function in growth encouragement. Zubair et al (2019) accounted that the biofilm creating capability of psychrophilic PGPR *Bacillus* is a distinctive feature facilitating them to stay alive and create vital metabolic products under low-temperature stress. This capability is associated with the prospective of inoculated bacteria to inhabit plant roots and assist the plants in lessening low-temperature stress.

7.6 FUTURE POSSIBILITIES

Microbes develop biofilms as an approach to conquer unfavorable environmental conditions and to continue in a complimentary habitat by means of inhabitation, since they contain a collection of microbiological cells (single, double, or multiple species) surrounded through a polysaccharide medium that is related with living and nonliving surfaces. Microbial biofilms are recognized to possess wide-range of applications in several areas together with human wellbeing, farming, food, biopolymers, crude oil processing plant, metal mining, mineral discovery, biofuel cells, bioremediation, etc. As far as agriculture is concerned, emerging biofilms in situ or in vivo are future tools, because they contain huge capability to offer several profits during their application as a lone inoculant. This could make possible development of plant, soil, and atmosphere through decreasing the reliance on synthetic chemical application in agriculture. A number of features together with ecological, predominantly soil and climatic reasons, nourishment, host resulting indicators, and so on affect biofilm development. Biofilm development through plant related microbes have a key function within the organization of concerned inoculant in soil or inhabitation of root. In view of the fact that bacterial biofilms chiefly consist of polysaccharides, there is a chance of progress in soil physical characters, for example better soil collection, decrease in bulkiness, improved withholding capacity of applied moisture, and nourishments in crop rhizosphere during the use of microbial biofilm apparatus. The use of biofilm-based microbial polysaccharides whether crude or purified could also be employed in the form of spray to save crop plants from living and nonliving stresses. Gene expression analysis along with the function of electrical communication between associates within multispecies biofilms are not as much explored as compared to single species biofilms. Cracking signaling systems are achieving attention as they are the main characters managing the metabolic behavior. In future, bacterial biofilm originated compounds should be included for crop nourishment, health, and quality enhancement for sustainable agriculture.

REFERENCES

Alden, L., Demoling, F., Baath, E. 2001. Rapid method of determining factors limiting bacterial growth in soil. *Applied and Environmental Microbiology* 67: 1830–38.

Altaf, M.M., Ahmad, I. 2016. Plant growth promoting activities, biofilm formation, and root colonization by *Bacillus* spp. isolated from rhizospheric soils. *Journal of Pure and Applied Microbiology* 10: 109–20.

Altaf, M.M., Ahmad, I. 2017. *In vitro* and *in vivo* biofilm formation by *Azotobacter* isolates and its relevance to rhizosphere colonization. *Rhizosphere* 3:138–42.

Altaf, M.M., Ahmad, I. 2019. *In vitro* biofilm development and enhanced rhizosphere colonization of *Triticum aestivum* by fluorescent *Pseudomonas* sp. *Journal of Pure and Applied Microbiology* 13(3):1441–49.

Angus, A.A., Hirsch, A.M. 2013. Biofilm formation in the rhizosphere: multispecies interactions and implications for plant growth. In: *Molecular Microbial Ecology of the Rhizosphere*, ed. F.J. de Bruijn, pp. 701–712. Hoboken, NJ: John Wiley and Sons Inc.

Asally, M., Kittisopikul, M., Rue, P., Du, Y. 2012. Localized cell death focuses mechanical forces during 3D patterning in a biofilm. *Proceedings of the National Academy of Sciences of the United States of America* 109:18891–96.

Bailey-Serres, J., Voesenek, L.A. 2008. Flooding stress: acclimations and genetic diversity. *Annual Review of Plant Biology* 59: 313–39.

Baliyarsingh, B., Nayak, S.K., Mishra, B.B. 2017. Soil microbial diversity: an ecophysiological study and role in plant productivity. In: *Advances in Soil Microbiology: Recent Trends and Future Prospects. Microorganisms for Sustainability*, eds. T. Adhya, B. Mishra, K. Annapurna, D. Verma, U. Kumar. Singapore: Springer.

Barriuso, J., Solano, B.R., Lucas, J.A., Lobo, A.P., Villaraco, A.G. Manero, F.J.G. 2008. Ecology, genetic diversity, and screening strategies of plant growth promoting rhizobacteria (PGPR). In: *Plant-Bacteria Interactions: Strategies and Techniques to Promote Plant Growth*, eds. I. Ahmad, J. Pichtel, S. Hayat, pp. 1–17. Weinheim: Wiley-Vch.

Bashan, Y., Holguin, G., de-Bashan, L. E. 2004. *Azospirillum*-plant relationships: Physiological, molecular, agricultural, and environmental advances (1997–2003). *Canadian Journal of Microbiology* 50:521–77.

Bogino, P.C., de las Mercedes Oliva, M., Sorroche, F.G., Giordano, W. 2013. The role of bacterial biofilms and surface components in plant-bacterial associations. *International Journal of Molecular Sciences* 14: 15838–59.

Bolton, M.V. 2009. Primary metabolism and plant defense—Fuel for the fire. *Molecular Plant-Microbe Interactions* 22: 487–97.

Bouskill, N.J., Wood, T.E., Baran, R., Hao, Z., Ye, Z., Bowen, B.P., et al. 2016. Belowground response to drought in a tropical forest soil. II. Change in microbial function impacts carbon composition. *Frontiers in Microbiology* 7:323. doi:10.3389/fmicb.2016.00323.

Callaway, R.M., Walker, L.R. 1997. Competition and facilitation: a synthetic approach to interactions in plant communities. *Ecology* 78: 1958–65.

Chakraborty, U., Chakraborty, B.N., Basnet, M., Chakraborty, A.P. 2009. Evaluation of *Ochrobactrum anthropi* TRS-2 and its talc based formulation for enhancement of growth of tea plants and management of brown root rot disease. *Journal of Applied Microbiology* 107(2):625–34.

Chaves, M.M., Oliveira, M.M. 2004. Mechanisms underlying plant resilience to water deficits: prospects for water-saving agriculture. *Journal of Experimental Botany* 55: 2365–84. doi:10.1093/jxb/erh269.

Costerton, J.W., Geesey, G.G., Cheng, K.J. 1978. How bacteria stick. *Scientific American* 238:86–95.

Costerton, J.W., Lewandowski, Z., Caldwell, D.E., Korber, D.R., Lappin-Scott, H.M. 1995. Microbial biofilms. *Annual Review of Microbiology* 49:711–45.

Cramer, G. R., Urano, K., Delrot, S., Pezzotti, M., Shinozaki, K. 2011. Effects of abiotic stress on plants: A systems biology perspective. *BMC Plant Biology* 11:163. doi:10.1186/1471-2229-11-163.

Davies, D.G., Parsek, M.R., Pearson, J.P., Iglewski, B.H., Costerton, J.W., Greenberg, E.P. 1998. The involvement of cell-to-cell signals in the development of a bacterial biofilm. *Science* 280:295–98.

Dietrich, L.E., Okegbe, C., Price-Whelan, A., Sakhtah, H. 2013. Bacterial community morphogenesis is intimately linked to the intracellular redox state. *Journal of Bacteriology* 195:1371–80.

Ekschmitt, K., Griffiths, B.S., 1998. Soil biodiversity and its implications for ecosystem functioning in a heterogeneous and variable environment. *Applied Soil Ecology* 10: 201–15.

Flowers, T.J., Galal, H.K., Bromham, L. 2010. Evolution of halophytes: Multiple origins of salt tolerance in land plants. *Functional Plant Biology* 37: 604–12. doi:10.1007/s10142-011-0218-0.

Foyer, C.H., Noctor, G. 2005. Oxidant and antioxidant signalling in plants: A re-evaluation of the concept of oxidative stress in a physiological context. *Plant Cell Environment* 28: 1056–71. doi:10.1111/j.1365-3040.2005.01327.x.

Gill, S. S., Tuteja, N. 2010. Reactive oxygen species and antioxidant machinery in abiotic stress tolerance in crop plants. *Plant Physiology and Biochemistry* 48:909–30. doi:10.1016/j.plaphy.2010.08.016.

Gooden, D.T., Skipper, H.D., Kim, J.H., Xiong, K. 2004. Diversity of root bacteria from peanut cropping systems. *Peanut Science* 31:86–91.

Grayson, M. 2013. Agriculture and drought. *Nature* 501:S1. doi:10.1038/501S1a.

Guimarães, A.A., Jaramillo, P.M.D., Obrega, R.S.A.N., Florentino, L.A., Silva, K.B., de Souza Moreira, F.M., 2012. Genetic and symbiotic diversity of nitrogen-fixing bacteria isolated from agricultural soils in the western Amazon by using cowpea as the trap plant. *Applied and Environmental Microbiology* 78: 6726–33.

Hamilton. 2018. World Losing 2,000 Hectares of Farm Soil Daily to Salt-Induced Degradation. https://unu.edu/media-relations/releases/world-losing-2000-hectares-of-farm-soil-daily-to-salt-induced-degradation.html (accessed December 6, 2019).

Hassani, M.A., Durán, P., Hacquard, S., 2018. Microbial interactions within the plant holobiont. *Microbiome* 6:58. doi:10.1186/s40168-018-0445-0.

Hiltner, L.T. 1904. On recent experiences and problems in the field of soil bacteriology with particular regard to the founding and fallow. Work of the German Agricultural Society. 98: 59–78.

Hinsa, S.M., Espinosa-Urgel, M., Ramos, J.L., O'Toole, G.A. 2003. Transition from reversible to irreversible attachment during biofilm formation by *Pseudomonas fluorescens* WCS365 requires an ABC transporter and a large secreted protein. *Molecular Microbiology* 49:905–18.

Hirel, B., Le Gouis, J., Ney, B., Gallais, A. 2007. The challenge of improving nitrogen use efficiency in crop plants: towards a more central role for genetic variability and quantitative genetics within integrated approaches. *Journal of Experimental Botany* 58: 2369–87. doi:10.1093/jxb/erm097.

Jha, B.K., Pragash, M.G., Cletus, J., Raman, G., Sakthivel, N. 2009. Simultaneous phosphate solubilization potential and antifungal activity of new fluorescent pseudomonad strains, *Pseudomonas aeruginosa, P. plecoglossicida* and *P. mosselii. World Journal of Microbiology & Biotechnology* 25:573–81.

Johri, B.N., Sharma, A., Virdi, J.S. 2003. Rhizobacterial diversity in India and its influence on soil and plant health. *Advances in Biochemical Engineering/Biotechnology* 84:49–89.

Jones, D., Nguyen, C., Finlay, R. 2009. Carbon flow in the rhizosphere: carbon trading at the soil–root interface. *Plant Soil* 321: 5–33. doi: 10.1007/s11104-009-9925-0.

Kasim, W.A., Gaafar, R.M., Abou-Ali, R.M., Omar, M.N., Hewait, H.M. 2016. Effect of biofilm forming plant growth promoting rhizobacteria on salinity tolerance in barley. *Annals of Agricultural Sciences* 61(2):217–27.

Kaushal, M., Wani, S.P. 2016. Plant-growth-promoting rhizobacteria: Drought stress alleviators to ameliorate crop production in drylands. *Annals of Microbiology* 66: 35–42. doi:10.1007/s13213-015-1112-3.

Kim, Y.-C., Glick, B.R., Bashan, Y., Ryu, C.-M. 2012. Enhancement of plant drought tolerance by microbes. In: *Plant Responses to Drought Stress*, ed. R. Aroca, pp. 383–413. Berlin: Springer.

Lambers, H., Chapin, F. SIII, Pons, T.L. 1998. *Plant Physiological Ecology.* New York: Springer.

Lee, K.W., Periasamy, S., Mukherjee, M., Xie, C., Kjelleberg, S., Rice, S.A., 2013. Biofilm development and enhanced stress resistance of a model, mixed species community biofilm. *ISME Journal* 8(4):894–907.

Lopez, D., Kolter, R. 2010. Extracellular signals that define distinct and coexisting cell fates in *Bacillus subtilis. FEMS Microbiology Reiews* 34:134–49.

Lynch, J.M. 1990. *The Rhizosphere.* New York: John Wiley.

Massad, T.J., Dyer, L.A., Vega, C.G. 2012. Cost of defense and a test of the carbon-nutrient balance and growth-differentiation balance hypotheses for two co-occurring classes of plant defense. *PLoSOne* 7:7554. doi:10.1371/journal.pone.0047554.

Mickelbart, M.V., Paul, M., Hasegawa, P.M., Bailey-Serres, J. 2015. Genetic mechanisms of abiotic stress tolerance that translate to crop yield stability. *Nature Reviews Genetics* 16: 237–51. doi:10.1038/nrg3901.

Mousa, W.K., Shearer, C., Limay-Rios, V., Ettinger, C.L., Eisen, J.A., Raizada, M.N. 2016. Root-hair endophyte stacking in finger millet creates a physicochemical barrier to trap the fungal pathogen *Fusarium graminearum. Nature Microbiology* 1:16167.

Nakashima, K., Yamaguchi-Shinozaki, K. 2006. Regulons involved in osmotic stress-responsive and cold stress-responsive gene expression in plants. *Physiologia Plantarum* 126: 62–71. doi:10.1111/j.1399-3054.2005.00592.x.

Nayak, S.K., Nayak, S., Mishra, B.B., 2017. Antimycotic role of soil *Bacillus* sp. against Rice pathogens: a biocontrol prospective. In: *Microbial Biotechnology*, eds. J.K. Patra, C. Vishnuprasad, G. Das, pp. 29–60. Singapore: Springer.

Nayak, S.K., Nayak, S., Patra, J.K. 2019. Rhizobacteria and its biofilm for sustainable agriculture: A concise review. In: *New and Future Developments in Microbial Biotechnology and Bioengineering*, eds. M.K. Yadav, B.P. Singh, pp. 165–175. Amsterdam: Elsevier.

Naylor, D., Coleman-Derr, D. 2018. Drought stress and root-associated bacterial communities. *Frontiers in Plant Science* 8:2223. doi:10.3389/fpls.2017.02223.

Ngumbi, E., Kloepper, J. 2014. Bacterial-mediated drought tolerance: Current and future prospects. *Applied Soil Ecology* 105: 109–25. doi:10.1016/j.apsoil.2016.04.009.

Padgham, J. 2009. *Agricultural Development under a Changing Climate: Opportunities and Challenges for Adaptation*. Washington, DC: Agriculture and rural development & environmental departments, The World Bank.

Patten, C.L., Glick, B.R. 2002. Role of *Pseudomonas putida* indole acetic acid in development of the host plant root system. *Applied and Environmental Microbiology* 68: 3795–01.

Pothier, J.F., Wisniewski-Dyé, F., Weiss-Gayet, M., Moënne-Loccoz, Y., Prigent-Combaret, C. 2007. Promoter-trap identification of wheat seed extract-induced genes in the plant-growth-promoting rhizobacterium *Azospirillum brasilense* Sp245. *Microbiology* 153:3608–22.

Qadir, M., Quillérou, E., Nangia, V., Murtaza, G., Singh, M., Thomas, R. J., et al. 2014. Economics of salt-induced land degradation and restoration. *Natural Resources Forum* 38: 282–95. doi:10.1111/1477-8947.12054.

Raaijmakers, J., Mazzola, M., 2012. Diversity and natural functions of antibiotics produced by beneficial and pathogenic soil bacteria. *Annual Review of Phytopathology* 50: 403–24.

Ramette, A., Tiedje, J.M. 2007. Multiscale responses of microbial life to spatial distance and environmental heterogeneity in a patchy ecosystem. *Proceedings of the National Academy of Sciences of the United States of America* 104: 2761–66.

Riadh, K., Wided, M., Hans-Werner, K., Chedly, A. 2010. Responses of halophytes to environmental stresses with special emphasis to salinity. *Advances in Botanical Research* 53: 117–45. doi:10.1016/S0065-2296(10)53004-0.

Richardson, A.E., Barea, J.M., McNeill, A.M., Prigent-Combaret, C., 2009. Acquisition of phosphorus and nitrogen in the rhizosphere and plant growth promotion by microorganisms. *Plant Soil* 321: 305–39.

Robin, A., Vansuyt, G., Hinsinger, P., Meyer, J. M., Briat, J. F., Lemanceau, P. 2008. Iron dynamics in the rhizosphere: Consequences for plant health and nutrition. *Advances in Agronomy* 99: 183–25.

Ruan, C. J., da Silva, J. A. T., Mopper, S., Qin, P., Lutts, S. 2010. Halophyte improvement for a salinized world. *Critical Reviews in Plant Sciences* 29: 329–59. doi:10. 1080/07352689. 2010.524517.

Sauer, K. 2003. The genomics and proteomics of biofilm formation. *Genome Biology* 4(6):219.

Schloter, M., Lebuhn, M., Heulin, T., Hartmann, A., 2000. Ecology and evolution of bacterial microdiversity. *FEMS Microbiology Reviews* 24: 647–60.

Shao, H.B., Chu, L.Y., Jaleel, C.A., Zhao, C.X. 2008. Water-deficit stress—Induced anatomical changes in higher plants. *C. R. Biologies* 331: 215–25. doi:10.1016/j.crvi.2008.01.002.

Simontacchi, M., Galatro, A., Ramos-Artuso, F., Santa-Maria, G.E. 2015. Plant survival in a changing environment: The role of nitric oxide in plant responses to abiotic stress. *Frontiers in Plant Science* 6:977. doi:10.3389/fpls.2015.00977.

Sinha, R., Goel, R., Johri, B.N. 2001. Molecular markers in rhizosphere microbiology. In: *Innovative Approaches in Microbiology*, eds. D.K. Maheshwari, R.C. Dubey, p. 255. Dehradun: BSMPS.

Stewart, P.S., Franklin, M.J. 2008. Physiological heterogeneity in biofilms. *Nature Reviews Microbiology* 6:199–210.

Tilak, K.V.B.R., Singh, G., Mukkherji, K.G. 1999. Biocontrol-plant growth promoting bacteria: mechanisms of action. In: *Biotechnological Approaches in Biocontrol of Plant Pathogens*, eds. K.G. Mukkherji., B.P. Chamola., R.K. Upadhyay, pp. 115–132. New York: Kluwer/Plenum.

Timmusk, S., Kim, S.-B., Nevo, E., Abd El Daim, I., Ek, B., Bergquist, J., et al. 2015. Sfp-type PPTase inactivation promotes bacterial biofilm formation and ability to enhance wheat drought tolerance. *Frontiers in Microbiology* 6:387. doi:10.3389/fmicb.2015.00387.

Turner, T.R., James, E.K., Poole, P.S. 2013. The plant microbiome. *Genome Biology* 14:209. doi:10.1186/gb-2013-14-6-209.

van Overbeek, L.S., Saikkonen, K. 2016. Impact of bacterial–fungal interactions on the colonization of the endosphere. *Trends Plant Science* 21:230–42.

Velmourougane, K., Prasanna, R., Saxena, A.K. 2017. Agriculturally important microbial biofilms: Present status and future prospects. *Journal of Basic Microbiology* 57:548–73. doi: 10.1002/jobm.201700046.

Vessey, J.K. 2003. Plant growth promoting rhizobacteria as biofertilizers. *Plant Soil* 255: 571–86.

Vlamakis, H., Chai, Y., Beauregard, P., Losick, R. 2013. Sticking together: Building a biofilm the *Bacillus subtilis* way. *Nature Reviews Microbiology* 11:157–68.

Wang, W., Vinocur, B., Altman, A. 2003. Plant responses to drought, salinity, and extreme temperatures: Towards genetic engineering for stress tolerance. *Planta* 218: 1–14. doi:10.1007/s00425-003-1105-5

Whipps, J.M. 2001. Ecological and biotechnological considerations in enhancing disease biocontrol. In: *Enhancing Biocontrol Agents and Handling Risks*, eds. M., Vurro, J., Gressel, T.Butt, et al., pp. 43–51. Amsterdam: IOP Press.

Wright, M., Adams, J., Yang, K., McManus, P., Jacobson, A., Gade, A., et al. 2016. A root colonizing Pseudomonad lessens stress responses in wheat imposed by CuO nanoparticles. *PLoS One* 11: e0164635.

Yolcu, S., Ozdemir, F., Güler, A., Bor, M. 2016. Histone acetylation influences the transcriptional activation of POX in *Beta vulgaris* L. and *Beta maritima* L. under salt stress. *Plant Physiology and Biochemistry* 100: 37–46. doi: 10.1016/j.plaphy.2015.12.019.

Young, I.M., Crawford, J.W. 2004. Interactions and self-organization in the soil–microbe complex. *Science*. 304:1634–37.

Yung-Hua, L., Lau, P.C.Y., Lee, J.H., Ellen, R.P. 2001. Natural genetic transformation of *Streptococcus mutans* growing in biofilms. *Journal of Bacteriology* 183:897–08.

Zubair, M., Hanif, A., Farzand, A., Sheikh, T., Khan, A. R., Suleman, M., et al. 2019. Genetic screening and expression analysis of Psychrophilic*Bacillus* spp. reveal their potential to alleviate cold stress and modulate phytohormones in wheat. *Microorganisms* 7(9):337. doi:10.3390/microorganisms7090337.

8 New Trends to Eliminate *Listeria monocytogenes* in Dairy Industry

Daniel Kuhn, Leandra Andressa Pacheco, Ytan Andreine Schweizer, Sabrina Grando Cordeiro, Peterson Haas, Camila Roberta de Castro, Aluisie Picolotto, Rafaela Ziem, Talita Scheibel, Bárbara Buhl, Bruna Costa, Carla Kauffmann, Elisete Maria de Freitas, Eduardo Miranda Ethur, and Lucélia Hoehne
Vale do Taquari University- Univates

CONTENTS

8.1 INTRODUCTION

Listeria spp. are Gram-positive, psychotropic, facultative anaerobic, and nonsporulating bacteria, which are widely distributed in the natural environment and are also capable to live as an intracellular pathogen in humans and animals (Navratilova et al. 2004; Williams et al. 2011).

The genus Listeria is composed of 17 species, of which two are recognized to be pathogenic. Therefore, *L. monocytogenes* is considered as the only human pathogen causing listeriosis, a highly lethal, opportunistic infection induced by ingestion of contaminated food. The second pathogenic species, *L. ivanovii*, is mostly responsible for abortions, stillbirths, and neonatal septicemia in sheep and cattle (Vázquez-boland et al. 2001).

Generally, *Listeria monocytogenes* and *Listeria innocua* are food-borne pathogens (Moshtaghi and Mohamadpour 2007; Scallan et al. 2011; Simmons et al. 2014; Vongkamjan et al. 2016) often found in food and food processing environments (Chambel et al. 2007; Rørvik et al. 1995); however, this species were also recognized to be associated with urban environments (Sauders et al. 2012).

In most cases, listeriosis is expressed as a mild, febrile illness, but can also be present as a systemic (invasive) disease with more severe symptoms resulting in high hospitalization and case fatality rates. The incidence of systemic listeriosis is much higher in pregnant women, older people, and individuals with a weakened immune system (Buchanan et al. 2017).

Additionally, these pathogens show an extended ability to grow over a broad temperature range, due to the aptitude of the bacteria to modify its membrane composition, and by doing so maintaining the membrane fluidity (Jones et al. 1997).

Listeria monocytogenes adheres to many materials present in food processing environments, for example, metals, rubbers, and polymers. However, adhesion depends strongly on the length of time the bacteria are in contact with the material (Beresford et al. 2001).

To date, some reviews were published describing conventional methodologies utilized to eliminate this pathogen in dairy industry (Martin et al. 2018; Rankin et al. 2017). Nevertheless, some studies evidenced that *L. monocytogenes* exhibited tolerance to quaternary ammonium containing disinfectants, resistance to benzalkonium chloride and to toxic compounds—such as heavy metals and triphenylmethane dyes—through acquisition of new genes, as well as mutations in existing genes (Carpentier and Cerf 2011; Dutta et al. 2014, 2017; Elhanafi et al. 2010; Müller et al. 2013). Such genetically determined resistance is the reason why *Listeria monocytogenes* strains have been found to persist for years or decades in food processing plants (Ferreira et al. 2014).

Another possible survival mechanism is the occurrence of persistent cells in the form of a biofilm. Such a highly organized matrix consists of multiple cells and extracellular polymeric materials that protect individual cells from environmental stress and foster interactions between cells—in relation to nutrients, toxic metabolites, and genetic material—that may lead to enhanced survival and growth (Da Silva and De Martinis 2013; Ferreira et al. 2014).

Table 8.1 lists recent studies showing contamination of dairy products or dairy industry samples with *Listeria monocytogenes*. Another serious problem in food industry, especially the manufacture of dairy food, is contamination with virulent phages. They reduce product quality by infecting bacteria, and thus represent a significant risk for milk fermentation failures during the production of cheese or a variety of other fermented dairy products (Coffey and Ross 2002; Émond and Moineau 2007).

Up to date, most research on strain persistence examined the role of biofilm formation and physiological tolerance to sanitation or processing hurdles (Da Silva and De Martinis 2013). To actualize new obtained knowledge about tackling that stubborn challenge this review thematizes new research trends in the elimination of *Listeria monocytogenes* contaminations in the form of single cells as well as biofilm in dairy plants.

TABLE 8.1
***Listeria monocytogenes* Presence in Dairy Products**

	Dairy Product and Number of Collected Samples	Number of Positive Samples/Presences	Research Period	References
Iran	Raw milk: 140 Traditional cheese: 120 Traditional curd: 100 Traditional butter: 100 Traditional ice cream: 85	Raw milk: 11 (7.8%) Traditional cheese: 9 (7.5%) Traditional curd: 1 (1%) Traditional butter: 1 (1%) Traditional ice cream: –	2016	Akrami-Mohajeri et al. (2018)
Egypt	Raw milk: 100 Milking equipment: 100 Hand swabs: 100	Raw milk: 25 (25%) Milking equipment: 28 (28%) Hand swabs: 16 (16%)	2015–2016	Tahoun et al. (2017)
Brazil	Raw material and product: 180	Raw material and product: 3 (0.7%)	2013–2014	Oxaran et al. (2017)
Finland	Bottled raw milk: 105 Bulk tank milk: 115 In-line milk filtersocks: 23 Environmental: 50	Bottled raw milk: 5 (4.8%) Bulk tank milk: 2 (1.7%) In-line milk filtersocks: 9 (39%) Environmental: absence	2013–2015	Castro et al. (2017)
Morocco	Dairy products: 404	Dairy products: 3 (0.7%)	2009–2015	Amajoud et al. (2018)
Italy	Dairy products: 8716	Dairy products: 145 (1.66%)	2010–2013	Dalzini et al. (2016)
Ireland	Dairy products: 408	Dairy products: 9 (2.2%)	2013–2015	Leong et al. (2017)
Uruguay	Cheese: 195	Cheese: 19 (10%)	2011–2013	Braga et al. (2017)

8.2 NEW DISINFECTING AGENTS

Ibarra-Sánchez et al. (2018) studied the evaluation of effectiveness, behavior, and potential synergism of nisin with PlyP100 against *L. monocytogenes*, when incorporated into Queso Fresco (QF). The product showed similar bacterial reduction regardless of varying size of *L. monocytogenes* inoculum in QF, and thus when the inoculation size was 1 Log colony forming unit per gram (CFU/g), no pathogen recovery after cheese enrichment could be observed. PlyP100 was stable in QF in cold storage for up to 28 days exhibiting similar antilisteria activity regardless if contamination with *L. monocytogenes* occurred. The product alone exhibited a strong listeriostatic effect in QF, but on the contrary, nisin alone was not effective to control the pathogen in QF during cold storage. Interestingly, when a combination of nisin and PlyP100 was made, results showed a strong synergy in QF with nonenumerable levels of *L. monocytogenes* after 1 month at refrigerated storage. Additionally, *L. monocytogenes* isolates from cheese, treated with nisin, PlyP100, and their combination, did not develop resistance to nisin or PlyP100. This way of combined treatment can be an alternative form applicable in QF production.

Lourenço et al. (2017) tested a mix of Generally Recognized As Safe (GRAS) antimicrobial ingredients to control *L. monocytogenes* in Hispanic nonfermented cheese. The mix was composed of caprylic acid (CA), Nisaplin® (N, 2.5% nisin), a proportion of sodium lactate and sodium diacetate (SL/SD), *Lactococcus lactis* sbp. *lactis* DPC 3147, monolaurin (ML), and lactic acid (LA). Samples of QF curds were inoculated with 10^4 CFU/g and stored at 4°C for 21 days. During storage the count of *L. monocytogenes* reached 7–8 Log CFU/g in control samples. Using individual antimicrobial treatments, the results showed that such a strategy is not effective against the pathogen, but the application of a mixture of ingredients was impactful in the inhibition of *L. monocytogenes* growth. Thus, treatments with N and CA consistently delivered 6 Log CFU/g less counts than the controls. Supplementation of LA to treatments with SL/SD caused differences of more than 4 Log CFU/g in final Listeria populations. Samples treated with binary mixtures of N and CA were evaluated in a sensory test by consumers, and all samples on average were graded between "slightly liking" and "moderately liking." These results indicated that the combined use of antimicrobial ingredients may be an effective way to control the population of *Listeria monocytogenes* in QF plants. Other promising treatments included combinations of SL/SD supplemented with LA. The addition of N and CA had a noticeable influence to consumers, but flavor was not markedly affected by making the product unacceptable.

Kozak et al. (2018) evaluated the use of fluid milk as a model system to identify listeristatic or listericidal treatments. CA, ε-polylysine, hydrogen peroxide, lauric arginate, and sodium caprylate were tested and evaluated individually or in combination in milk, inoculated with *L. monocytogenes* at ~4 Log10 CFU/mL. The results showed a reduction of the pathogen using each product separately. However, application of a combination of ε-polylysine and sodium caprylate was observed to possess synergistic characteristics: allowing a reduction of the whole product concentration, while achieving similar pathogen inhibition. This type of application can be a promising antimicrobial treatment strategy to control *L. monocytogenes* in dairy industry. Further studies about changes of sensory properties will be realized, in the near future, to guarantee consumer acceptance of treated dairy products.

Lotfi et al. (2018) evaluated the use of the sol-gel method with ML containing antimicrobial cellulose-chitosan (CC) films against *Listeria monocytogenes* in vitro and in ultrafiltered cheese during storage at 4°C for 2 weeks. Microscopic characteristics, structure, swelling, water solubility, and antimicrobial properties were evaluated. The results of antimicrobial activity tests revealed that the addition of ML significantly increased the diameter of the inhibition zone. However, the CC film did not show any inhibitory activity against *L. monocytogenes*. The results of scanning electron microscope images and Fourier-transform infrared spectroscopy analysis confirmed a successful introduction and binding of the ML in the double layer film. The addition of ML in CC film decreased the swelling index in a significant manner, whereas the solubility of the film increased. A significant increase in solubility with ML addition was also found for both CC incorporated ML films. The addition of 0.5 and 1% ML into CC films resulted in a 2.4–2.3 log reduction of *L. monocytogenes* population in cheese after 14 days. The results revealed a suitable antibacterial activity of ML incorporated CC film, which can be for example, an applicable system for cheese packaging, to control *L. monocytogenes* in the final process of the dairy industry.

Kondrotiene et al. (2018) selected nisin producing *L. lactis* strains possessing exceptional technological characteristics and antibacterial activity, which were isolated from regional collected goat and cow milk, as well as fermented wheat and buckwheat samples—in order to apply them for more effective control of *Listeria monocytogenes* in fresh cheese manufacturing. Within whole isolates 20 out of 181 *Lactococcus* spp. strains could be identified as *L. lactis* strains harboring nisin A, Z, or the novel nisin variant GLc03 genes. This paper showed for the first time technological characteristics of *L. lactis* strains harboring the nisin variant GLc03. Altogether, 12 isolated strains showed antagonistic activity against tested food spoilage and pathogenic bacteria. Moreover, strains encoding nisin Z presented favorable enzymatic activities of acid phosphatase, esterase lipase, and phosphohydrolase. The best three strains were selected according to safety and technological criteria and were examined in fresh cheese production for the control of *Listeria monocytogenes* growth. The results showed that the number of Listeria could be significantly reduced ($P<0.001$) in model cheese, thus suggesting the application of these strains for fresh cheese safety improvement.

8.3 APPLICATION OF MICROORGANISMS

Costa et al. (2018) analyzed five isolates of *L. innocua* and *L. monocytogenes*, obtained from Gorgonzola cheese processing plants and milk supplying farms, located in Piemont and Lombardy, Italy. Biofilm formation ability and biofilm susceptibility to two hydrogen peroxide-based disinfectants, both used at the plants, were evaluated. Two disinfectants, one containing acetic acid and peracetic acid and the other containing citric acid, were tested. As a result, *L. innocua* and *L. monocytogenes* showed similar susceptibility to the tested disinfectants regarding biofilm formation. Moreover, the susceptibility to disinfectants of biofilms, which were grown at conditions mimicking clean or soiled environments, showed no significant differences in terms of log reduction between persistent and nonpersistent isolates. According to these results, *L. innocua* could be used as a surrogate for *L. monocytogenes*, not only regarding the biofilm production but also the biofilm susceptibility to disinfectants. The convenience and safety in using a nonpathogenic surrogate will certainly contribute to the clarification of factors that cause persistent colonization of *L. monocytogenes*, not only in Gorgonzola producing plants but also in other food industry environments.

Another strategy was evaluated by Haubert et al. (2018), using a food derived isolate of *Listeria monocytogenes* harboring tetM gene for plasmid-mediated exchange to *Enterococcus faecalis* on the surface of processed cheese. Therefore, the genetic basis of tetracycline resistance in a food isolate Listeria monocytogenes (Lm16) was evaluated. The results evidenced that the sequence of tetM showed 100% similarity to the *Enterococcus faecalis* sequences; thus, it was suggested that Lm16 received this gene from *E. faecalis*. In addition, transferability of the tetM gene was achieved in vitro by agar mattings between Lm16 and *E. faecalis* JH2-2, and by doing so, examining the potential for the spreading of tetM by horizontal gene transfer. Afterwards, the conjugation experiments were performed on the surface of processed cheese, confirming the transferability of the gene in a food matrix. So that is the first report of a food isolate *L.*

monocytogenes carrying the tetM gene in plasmid DNA, thus underlining the potential risk of spreading antimicrobial resistance genes between different bacteria. The allover results indicated that this method can be a promising strategy against *Listeria monocytogenes* and therefore has to be tested in cheese production in the future.

Yang et al. (2018) evaluated the inhibitory activities of Se-enriched LA bacteria (LAB), *Lactobacillus delbrueckii* ssp. *bulgaricus* and *Streptococcus thermophilus*, against pathogens *Listeria monocytogenes*, *Salmonella typhimurium*, *Escherichia coli*, and *Staphylococcus aureus* in vitro. By doing so, sodium selenite was added to de Man, Rogosa and Sharpe (MRS) roth in concentrations of 0, 20, 40, 60, 80, 100, 120, 140, and 160 μg/mL. At a concentration of 80 μg/mL the amount of accumulated Se by *Lactobacillus delbrueckii ssp. bulgaricus* and *Streptococcus thermophilus* reached 12.05±0.43 and 11.56±0.25 μg/mL, respectively, accompanied by the relative maximum living cells. It could be evidenced that bacterial culture solution and cell-free culture supernatant (CFCS) from Se-enriched LAB exerted stronger antibacterial activity than those from the non-Se strains. Scanning electron microscopy equipped with energy dispersion X-ray spectrometry showed that elemental Se nanoparticles were deposited on the cell surfaces of *Lactobacillus delbrueckii* ssp. *bulgaricus*. Additionally, CFCS of Se-enriched LAB induced more serious cell structure damage in pathogenic bacteria than non-Se LAB. Moreover, these bacteria have the potential to bring commercial advantages when included into fermented products; thus, the product not only provides probiotic bacteria and absorbable elemental or organic Se but also endows higher antibacterial activity to pathogens.

During processing, food antimicrobials have been traditionally used to prevent food spoilage, as well as for cleaning and sanitation of equipment surfaces and surrounding environments. Moreover, in modern food industries, minimal processing methods are encouraged in order to obtain safe products having a natural or "green image" (Burt 2004).

LAB isolated from herbs, fruits, and vegetables were also screened for their antilisteria activity using the MRS agar overlay assay. Ho et al. (2018) evaluated a large collection of LAB isolated from herbs, fruits, and vegetables for antilisteria activity and examined their technological properties with respect to future application in cheese production. This was achieved through characterizing bacteriocin genes, metabolic properties, and whole-genome sequencing analysis of LAB strains. The results showed that 14 strains with strong activity could be identified by 16S rRNA gene sequencing as *Lactococcus lactis* and all these strains harbored the nisin gene cluster. Within these 14 isolates, four strains also showed detectable antilisteria activity using a milk medium. Thus these four isolates have the potential to serve as biopreservatives and/or adjunct cultures in cheese manufacture: to prevent growth of the food-borne pathogen *L. monocytogenes*, enhance flavor development, and accelerate cheese ripening. However, further studies are required to carefully evaluate the presence of virulence factors and formation of biogenic amines before these selected wild strains can be used as biopreservatives.

As a further possibility to eliminate food-borne pathogens, some studies investigated the use of bacteriophages. Goodridge et al. (2018) showed new trends to combat bacterial enemies including *Salmonella enterica*, *Shigella* spp., *Listeria monocytogenes*, and Shiga-toxin producing *E. coli*. The authors suggested that biological control using polyvalent, broad-host range phages could be a promising strategy to enhance pathogen lethality and thus food safety. Novel methods to protect and deliver viable phages can improve functionality, and recently achieved advances supported the development of phage-based antimicrobials. In particular about bacteriophages against *Listeria monocytogenes* there are some studies, but to be used specifically in dairy industry, this procedure requires further investigations.

The phage P100 is an evaluated virus to combat *Listeria monocytogenes*, showing similarity to the phage A511 and is commercially available as Listex™ P100 preparation, which is increasingly used in (Austrian) dairy processing facilities. Nevertheless, the use of phages for food safety purposes raises some concerns, for example the continuous use of phages may result in resistant *L. monocytogenes* strains. Fister et al. (2015) studied 486 *L. monocytogenes* isolates obtained from 59 dairies, which were screened over 15 years for the presence of P100 insensitive *L. monocytogenes*. The overall number of P100 resistant *L. monocytogenes* isolates for all years and all investigated plants was 2.7%; thus, insensitivities were detected in 5 of 59 dairy plants. Adsorption tests suggested that the detected resistant *L. monocytogenes* isolates had receptor modifications, resulting in the application of bacteriophage P100 being recommended as a complementing treatment in emergency cases. With this in mind, the use of other alternative products still will be necessary to improve the efficiency of this bacteriophage.

8.4 ESSENTIAL OILS AND EXTRACTS

To combat *Listeria monocytogenes*, essential oils (EOs) were also investigated. Das et al. (2013) reviewed the application of berry products which could be developed as natural alternatives to antibiotics to control a wide range of food-borne pathogens. In the study, the American native berries cranberry, blueberries, and strawberry were investigated, and the authors described antibacterial activities and bioactive compounds, such as vitamin C, vitamin A, calcium, iron, folate, magnesium, manganese, and flavonoids, which were the major components amongst over 150 identified individual phytochemicals. It can be hypothesized that phytochemicals present in berry products have different modes of action and synergistic antimicrobial activities that might contribute to a broad spectrum of antimicrobial effectiveness, while decreasing efficacious doses and limiting potential development of resistance. Some characteristics of berry extracts, such as poor solubility in water, could be a limiting factor for their usability in food; therefore, further investigations on bioavailability of these active compounds are required to understand the underlying molecular mechanisms of action.

Vidács et al. (2018) tested the EOs of cinnamon, marjoram, and thyme against immature and mature biofilms of *Listeria monocytogenes* formed on polypropylene (PP) surfaces, because biofilms of pathogenic can develop on food contact surfaces commonly used in food industry, such as stainless steel, plastic, and glass. After optimization of the concentration of EOs, disinfection time, and pH in the EO-based disinfection solutions used for the assay (Response Surface Box-Behnken Design) the results showed that at a concentration of EOs of 1.1–15.8 mg/mL and pH 4.5–7.5 the optimized disinfectants could successfully eliminate 24 and 168 h (immature and mature) old biofilms within 10 min exposure time.

The disinfectant impact of the EO-based natural solutions was in most cases equivalent or better compared to the peracetic acid-based chemical sanitizers or to sodium hypochlorite used in food industry. In literature, innumerable articles can be found about the antibacterial effect of EOs, but only a few studies work on their effects on biofilms. This newly, on EO-basis created alternative, natural disinfectants, usable in households can also be used as a promising alternative against *Listeria monocytogenes* biofilms in the dairy industry.

Ferula assa-foetida oleo-gum-resin also was investigated against some pathogens, including *Listeria monocytogenes.* Zomorodian et al. (2018) evaluated the chemical composition of these species from Larestan in Fars province, Iran. The presence of (E)–1-Propenyl sec-butyl disulfide (21.65%) was identified to be the major constituent of the EO followed by 10-epi-γ-Eudesmol (19.21%) and (Z)–1-Propenyl sec-butyl disulfide (10.20%). The antibiofilm activity was evaluated, and the results showed that the EO was active against all tested Gram-positive and Gram-negative bacteria at a concentration range of 0.5–8 and 16–128 μL/mL, respectively. According to these findings, the authors assumed that the antimicrobial activity of the EO may be due to the presence of (E)–1-Propenyl sec-butyl disulfide and (Z)-1-Propenyl sec-butyl disulfide as the major sulfur compounds of the oil. Thus, the oil can be considered as a potential source for developing novel antimicrobial agents in order to control fungal and bacterial infections or to improve the quality and shelf life of food products.

Furthermore, extracts from Leguminosae family were evaluated against *Listeria monocytogenes.* Pina-Pérez and Ferrús Pérez (2018) revised the worldwide published studies of antimicrobial potential of the most popular legumes against food-borne pathogens. The authors wrote about studies, treating the potential of peptides obtained from legumes, or soy derived isoflavones against biofilm pathogens; and discussed their applicability—in food packaging or films/coatings—to protect industrial surfaces. Legumes are one of the promising plant groups together with grains, which demonstrated antimicrobial potential against some of the most important food-borne pathogens. Despite such promising advances, further tests are required to evaluate microbiological risk assessment before such compounds can be permitted for the control of bacterial proliferation within industrial processes.

Da Silva Dannenberg et al. (2016) proposed the use of EO from pink pepper tree fruit against *Listeria monocytogenes* and evaluated antimicrobial and antioxidant activity in vitro and in situ (applied to Minas-type fresh cheese), amongst minimum inhibitory concentration (MIC) and minimum bactericidal concentration (MBC) during storage for about 30 days at 4°C. The EOs extracted from green and mature fruit were extracted, from which the EO of the mature fruit showed the best results, evidencing its food biopreservative capacity against 83% of the 18 tested *L. monocytogenes* isolates and showing a simultaneous decrease of the concentration of primary and secondary oxidative compounds.

8.5 ANTIMICROBIAL COATING

Another alternative to combat *Listeria monocytogenes* biofilms has been seen in the development of antimicrobial coatings. Huang et al. (2017) evaluated the antilisteria properties of biodegradable polylactide coatings modified with titanium dioxide.

Polylactide is a biodegradable, linear, aliphatic, and thermoplastic polyester with low toxicity and good processability, and titanium dioxide is commonly used as a photocatalyst for preparation of self-cleaning antimicrobial surfaces. The underlying mechanism relies on formation of radicals from the breakdown of water and oxygen following illumination with ultraviolet (UV) light. For the performance of the tests, free-standing films were prepared by casting solutions made of titanium dioxide and previously extruded polylactide. Results showed that polylactide alone could support reduction of 2.84±0.10 log CFU of *Listeria monocytogenes* when incubated at 23°C for 2 h. However, it could be shown that the log reduction for Listeria could be increased to >4 log CFU by the use of titanium dioxide, and when polylactide composites were additionally treated with ultraviolet region A (UV-A). The inactivation kinetics of *L. monocytogenes* followed a diphasic die-off with an initial 30 min lag period and decline in bacterial levels over a further 90 min period. The results concluded that polylactide-titanium dioxide coating shows potential as an antimicrobial strategy, although further work is necessary to assess if such a protective film maintains its function under commercial conditions.

8.6 COLD PLASM

Coutinho et al. (2018) revised cold plasm processing of milk and dairy products as an antimicrobial food preserving strategy. This procedure is a nonthermal technology that has gained attention in recent years as a potential alternative method for chemical and thermal disinfection in foods using ambient or moderate temperatures and short treatment times. The technology uses an electrically energized matter in a gaseous state composed of charged particles, free radicals, and some radiation. The plasma causes cell death and destruction of genetic material, etching of cell surfaces, induced by reactive species and volatilization of compounds, when combined with intrinsic photodesorption of UV photons. The authors described the processing factors involved in this technology as its fundamental parameters and furthermore treat the mechanisms of microbial inactivation. It additionally provides an overview of the effects of nonthermal plasma on quality of dairy products, considering a physicochemical, sensory, and microbiology perspective. Regarding control of food spoilage caused by *Listeria monocytogenes*, some studies in literature could be found, which already tested this technology. Although cold plasma has high microbial inactivation efficiency at low temperatures and a low impact on the product matrix, some studies performed to date have demonstrated certain acceleration of lipid oxidation and a negative impact on the sensory characteristics of processed dairy products. Thus, further investigations are indispensable to elucidate the effects of cold plasma on the quality of these products.

8.7 FINAL CONSIDERATIONS

Listeria monocytogenes is a persistent food-borne pathogen, which poses a risk to human health and can be transmitted to the processing facility by the raw material, the workers, tools, and trucks. Good Manufacturing Practices (GMP), employee training, and planning of powerful preventive antimicrobial strategies can help to eliminate or reduce the content of this food-borne microorganism within the dairy industry. However, the mutating power of this pathogen makes it difficult to combat it. In this way, there is a need to search for new methodologies which have the power of effective listeria food spoilage control without any risk to cause resistance. Thus, with this review article, it is hoped to elucidate new trends to researchers, farmers, and entrepreneurs in the dairy industry.

REFERENCES

Akrami-Mohajeri, F., Derakhshan, Z., Ferrante, M., et al. 2018. The prevalence and antimicrobial resistance of *Listeria* spp in raw milk and traditional dairy products delivered in Yazd, central Iran (2016). *Food Chem. Toxicol*; 114: 141–144.

Amajoud, N., Leclercq, A., Soriano, J.M., et al. 2018. Prevalence of *Listeria* spp. and characterization of *Listeria monocytogenes* isolated from food products in Tetouan, Morocco. *Food Control*; 84: 436–441.

Beresford, M.R., Andrew, P.W., Shama, G. 2001. *Listeria monocytogenes* adheres to many materials found in food-processing environments. *J. Appl. Microbiol*; 90: 1000–1005.

Braga, V., Vázquez, S., Vico, V., et al. 2017. Prevalence and serotype distribution of *Listeria monocytogenes* isolated from foods in Montevideo-Uruguay. *Brazilian J. Microbiol*; 48: 689–694.

Buchanan, R.L., Gorris, L.G.M., Hayman, M.M., Jackson, T.C., Whiting, R.C. 2017. A review of *Listeria monocytogenes*: An update on outbreaks, virulence, dose-response, ecology, and risk assessments. *Food Control*; 75: 1–13.

Burt, S. 2004. Essential oils: Their antibacterial properties and potential applications in foods—A review. *Int. J. Food Microbiol*; 94: 223–253.

Carpentier, B., Cerf, O. 2011. Review—Persistence of *Listeria monocytogenes* in food industry equipment and premises. *Int. J. Food Microbiol*; 145: 1–8.

Castro, H., Ruusunen, M., Lindström, M. 2017. Occurrence and growth of *Listeria monocytogenes* in packaged raw milk. *Int. J. Food Microbiol*; 261: 1–10.

Chambel, L., Sol, M., Fernandes, I., et al. 2007. Occurrence and persistence of *Listeria* spp. in the environment of ewe and cow's milk cheese dairies in Portugal unveiled by an integrated analysis of identification, typing, and spatial-temporal mapping along production cycle. *Int. J. Food Microbiol*; 116: 52–63.

Coffey, A., Ross, R.P. 2002. Bacteriophage-resistance systems in dairy starter strains: Molecular analysis to application. Antonie van Leeuwenhoek. *Int. J. Gen. Mol. Microbiol*; 82: 303–321.

Costa, A., Lourenco, A., Civera, T., Brito, L. 2018. *Listeria innocua* and *Listeria monocytogenes* strains from dairy plants behave similarly in biofilm sanitizer testing. *Lwt*; 92: 477–483.

Coutinho, N.M., Silveira, M.R., Rocha, R.S., et al. 2018. Cold plasma processing of milk and dairy products. *Trends Food Sci. Technol*; 74: 56–68.

Da Silva, E.P., De Martinis, E.C.P. 2013. Current knowledge and perspectives on biofilm formation: The case of *Listeria monocytogenes. Appl. Microbiol. Biotechnol*; 97: 957–968.

Da Silva Dannenberg, G., Funck, G.D., Mattei, F.J., Da Silva, W.P., Fiorentini, Â.M. 2016. Antimicrobial and antioxidant activity of essential oil from pink pepper tree (*Schinus terebinthifolius* Raddi) in vitro and in cheese experimentally contaminated with *Listeria monocytogenes. Innov. Food Sci. Emerg. Technol*; 36: 120–127.

Dalzini, E., Bernini, V., Bertasi, B., Daminelli, P., Losio, M.-N., Varisco, G. 2016. Survey of prevalence and seasonal variability of *Listeria monocytogenes* in raw cow milk from Northern Italy. *Food Control*; 60: 466–470.

Das, Q., Islam, M.R., Marcone, M.F., Warriner, K., Diarra, M.S. 2013. Potential of berry extracts to control foodborne pathogens. *Food Control*; 73: 650–662.

Dutta, V., Elhanaf, D., Kathariou, S., 2017. Conservation and distribution of the benzalkonium chloride resistance cassette bcrABC in *Listeria monocytogenes. Appl. Environ. Microbiol*; 79: 6067–6074.

Dutta, V., Elhanafi, D., Osborne, J., Martinez, M.R., Katharioua, S. 2014. Genetic characterization of plasmid-associated triphenylmethane: Reductase in *Listeria monocytogenes. Appl. Environ. Microbiol*; 80: 5379–5385.

Elhanafi, D., Utta, V., Kathariou, S. 2010. Genetic characterization of plasmid-associated benzalkonium chloride resistance determinants in a *Listeria monocytogenes* train from the 1998–1999 outbreak. *Appl. Environ. Microbiol*; 76: 8231–8238.

Émond, É., Moineau, S. 2007. Bacteriophages and food fermentation 93-124, In S. McGrath and D. van Sinderen (ed.), *Bacteriophage: Genetics and Molecular Biology.* Caister Academic Press, Norfolk, United Kingdom.

Ferreira, V., Wiedmann, M., Teixeira, P., Stasiewicz, M.J. 2014. *Listeria monocytogenes* persistence in food-associated environments: Epidemiology, strain characteristics, and implications for public health. *J. Food Prot*; 77: 150–170.

Fister, S., Fuchs, S., Stessl, B., Schoder, D., Wagner, M., Rossmanith, P. 2015. Screening and characterisation of bacteriophage P100 insensitive *Listeria monocytogenes* isolates in Austrian dairy plants. *Food Control*; 59: 108–117.

Goodridge, L., Fong, K., Wang, S., Delaquis, P. 2018. Bacteriophage-based weapons for the war against foodborne pathogens. *Curr. Opin. Food Sci*; 20: 69–75.

Haubert, L., Cunha, C.E.P. da, Lopes, G.V., Silva, W.P. da. 2018. Food isolate *Listeria monocytogenes* harboring tetM gene plasmid-mediated exchangeable to *Enterococcus faecalis* on the surface of processed cheese. *Food Res. Int*; 107: 503–508.

Ho, V.T.T., Lo, R., Bansal, N., Turner, M.S. 2018. Characterization of *Lactococcus lactis* isolates from herbs, fruits, and vegetables for use as biopreservatives against *Listeria monocytogenes* in cheese. *Food Control*; 85: 472–483.

Huang, S., Guild, B., Neethirajan, S., Therrien, P., Lim, L.T., Warriner, K. 2017. Antimicrobial coatings for controlling *Listeria monocytogenes* based on polylactide modified with titanium dioxide and illuminated with UV-A. *Food Control*; 73: 421–425.

Ibarra-Sánchez, L.A., Van Tassell, M.L., Miller, M.J. 2018. Antimicrobial behavior of phage endolysin PlyP100 and its synergy with nisin to control *Listeria monocytogenes* in Queso Fresco. *Food Microbiol*; 72: 128–134.

Jones, C.E., Shama, G., Jones, D., Roberts, I.S., Andrew, P.W. 1997. Physiological and biochemical studies on psychotolerance in *Listeria monocytogenes. J. Appl. Microbiol*; 83: 31–35.

Kondrotiene, K., Kasnauskyte, N., Serniene, L., et al. 2018. Characterization and application of newly isolated nisin producing *Lactococcus lactis* strains for control of *Listeria monocytogenes* growth in fresh cheese. *LWT - Food Sci. Technol*; 87: 507–514.

Kozak, S.M., Brown, S.R.B., Bobak, Y., D'Amico, D.J. 2018. Control of *Listeria monocytogenes* in whole milk using antimicrobials applied individually and in combination. *J. Dairy Sci*; 101: 1889–1900.

Leong, D., NicAogáin, K., Luque-Sastre, L., et al. 2017. A 3-year multi-food study of the presence and persistence of *Listeria monocytogenes* in 54 small food businesses in Ireland. *Int. J. Food Microbiol*; 249: 18–26.

Lotfi, M., Tajik, H., Moradi, M., Forough, M., Divsalar, E., Kuswandi, B. 2018. Nanostructured chitosan/monolaurin film: Preparation, characterization, and antimicrobial activity against *Listeria monocytogenes* on ultrafiltered white cheese. *Lwt*; 92: 576–583.

Lourenço, A., Kamnetz, M.B., Gadotti, C., Diez-Gonzalez, F. 2017. Antimicrobial treatments to control *Listeria monocytogenes* in queso fresco. *Food Microbiol*; 64: 47–55.

Martin, N.H., Boor, K.J., Wiedmann, M. 2018. Effect of post-pasteurization contamination on fluid milk quality. *J. Dairy Sci*; 101: 861–870.

Moshtaghi, H., Mohamadpour, A.A. 2007. Incidence of *Listeria* spp. in raw milk in Shahrekord, Iran. *Foodborne Pathog. Dis*; 4: 107–110.

Müller, A., Rychli, K., Muhterem-Uyar, M., et al. 2013. Tn6188—A novel transposon in *Listeria monocytogenes* responsible for tolerance to benzalkonium chloride. *PLoS One*; 8: 1–11.

Navratilova, P., Schlegelova, J., Sustackova, A., Napravnikova, E., Lukasova, J., Klimova, E. 2004. Prevalence of *Listeria monocytogenes* in milk, meat, and foodstuff of animal origin and the phenotype of antibiotic resistance of isolated strains. *Vet. Med. (Praha)*; 49: 243–252.

Oxaran, V., Lee, S.H.I., Chaul, L.T., et al. 2017. *Listeria monocytogenes* incidence changes and diversity in some Brazilian dairy industries and retail products. *Food Microbiol*; 68: 16–23.

Pina-Pérez, M.C., Ferrús Pérez, M.A. 2018. Antimicrobial potential of legume extracts against foodborne pathogens: A review. *Trends Food Sci. Technol*; 72: 114–124.

Rankin, S.A., Bradley, R.L., Miller, G., Mildenhall, K.B. 2017. A 100-year review: A century of dairy processing advancements—Pasteurization, cleaning and sanitation, and sanitary equipment design. *J. Dairy Sci*; 100: 9903–9915.

Rørvik, L.M., Caugant, D.A., Yndestad, M. 1995. Contamination pattern of *Listeria monocytogenes* and other *Listeria* spp. in a salmon slaughterhouse and smoked salmon processing plant. *Int. J. Food Microbiol*; 25: 19–27.

Sauders, B.D., Overdevest, J., Fortes, E., et al. 2012. Diversity of *Listeria* species in urban and natural environments. *Appl. Environ. Microbiol*; 78: 4420–4433.

Scallan, E., Hoekstra, R.M., Angulo, F.J., et al. 2011. Foodborne illness acquired in the United States—major pathogens. *Emerg. Infect. Dis*; 17: 7–15.

Simmons, C., Stasiewicz, M.J., Wright, E., et al. 2014. *Listeria monocytogenes* and *Listeria* spp. contamination patterns in retail delicatessen establishments in three U.S. States. *J. Food Prot*; 77: 1929–1939.

Tahoun, A.B.M.B., Abou Elez, R.M.M., Abdelfatah, E.N., Elsohaby, I., El-Gedawy, A.A., Elmoslemany, A.M. 2017. *Listeria monocytogenes* in raw milk, milking equipment, and dairy workers: Molecular characterization and antimicrobial resistance patterns. *J. Glob. Antimicrob. Resist*; 10: 264–270.

Vázquez-boland, J.A., Kuhn, M., Berche, P., et al. 2001. *Listeria* pathogenesis and molecular virulence determinants *Listeria* pathogenesis and molecular virulence determinants. *Clin. Microbiol. Rev*; 14: 584–640.

Vidács, A., Kerekes, E., Rajkó, R., et al. 2018. Optimization of essential oil-based natural disinfectants against *Listeria monocytogenes* and *Escherichia coli* biofilms formed on polypropylene surfaces. *J. Mol. Liq*; 255: 257–262.

Vongkamjan, K., Fuangpaiboon, J., Turner, M.P., Vuddhakul, V. 2016. Various ready-to-eat products from retail stores linked to occurrence of diverse *Listeria monocytogenes* and *Listeria* spp. isolates. *J. Food Prot*; 79: 239–245.

Williams, S.K., Roof, S., Boyle, E.A., et al. 2011. Molecular ecology of *Listeria monocytogenes* and other *Listeria* species in small and very small ready-to-eat meat processing plants. *J. Food Prot*; 74: 63–77.

Yang, J., Wang, J., Yang, K., et al. 2018. Antibacterial activity of selenium-enriched lactic acid bacteria against common food-borne pathogens in vitro. *J. Dairy Sci*; 101: 1930–1942.

Zomorodian, K., Saharkhiz, J., Pakshir, K., Immeripour, Z., Sadatsharifi, A. 2018. The composition, antibiofilm, and antimicrobial activities of essential oil of *Ferula assa-foetida* oleo-gum-resin. *Biocatal. Agric. Biotechnol*; 14: 300–304.

9 Potentialities of Medicinal Plant Extracts Against Biofilm-Forming Bacteria

Muhammad Bilal
Huaiyin Institute of Technology

Hira Munir
University of Gujrat

Hafiz M. N. Iqbal
Tecnologico de Monterrey

CONTENTS

9.1 INTRODUCTION

Two modes of growth are exhibited by bacteria, i.e., sessile aggregate and planktonic cell that is also known as a biofilm. Biofilm is referred to as an association between bacteria during which the cells stick with one another on the surface of extracellular polymeric substance (EPS) that is also formed by bacterial cells themselves (Hall-Stoodley et al., 2004). Leeuwenhoek (Dutch researcher) determined an animalcule present on the surface of the tooth by means of an ordinary microscope, which led to the discovery of microbial biofilms. In 1973, Characklis observed that microbial biofilms showed a high resistance towards disinfectants such as chlorine. Later in 1978, another researcher Costerton coined the word biofilm and highlighted the significance of biofilm. These bacterial biofilms exist everywhere in nature and are also present in wastewater channels, industrial places, bathrooms, hospitals, labs, and hotels. The formation of biofilm can be carried out on both living and nonliving surfaces, and they can also form floating layers on the surface of aqueous solutions (Costerton et al., 1999). It is considered to be the paramount way of bacterial growth and development in the industrial, clinical, and natural environment. Biofilms are composed of heavily crowded inhabitants of multispecies bacterial cells, surrounded by a self-synthesized polymeric matrix, and these cells are attached to the surface or tissue (Romero et al., 2016).

9.2 COMPOSITION OF BIOFILM NETWORK

Biofilms are a group of microorganisms that produce EPSs such as proteins, DNA, RNA, polysaccharides, and water. The major portion that is mainly involved in the flow of the nutrients in the matrix of bacterial biofilm is water (up to 97%). The structure of biofilms consists of two components that are water channels for the transportation of nutrients and a highly packed cell region with no pores in them (Jamal et al., 2015). Biofilms formed by bacteria are beyond the approach of the human immune system and antibiotics. Bacterial cells that form biofilm possess the potential ability to tolerate and neutralize the antibacterial agents that causes a delay in treatment. These microorganisms turn on several specific genes that stimulate the appearance of stress-related genes, which consequently switches to the resilient phenotypes due to some alterations, e.g., temperature, cell density, osmolarity, and pH (Fux et al., 2005). Different constituents of biofilm signify the integrity of biofilm, making them resistant against numerous environmental factors (Vinodkumar et al., 2008).

9.3 MECHANISM OF BIOFILM FORMATION

The development of biofilms is an intricate process during which the microbial transformation occurs from planktonic to the sessile mode of growth (Okada et al., 2005). The formation of biofilm is reliant on the manifestation of some typical genes that lead to biofilm establishment (Sauer et al., 2004; Okada et al., 2005). Biofilm development is a multistep process during which the bacteria undergo few changes after attachment to the surface. It involves the following imperative phases: (i) initial attachment to a surface, (ii) formation and aggregation of microcolony,

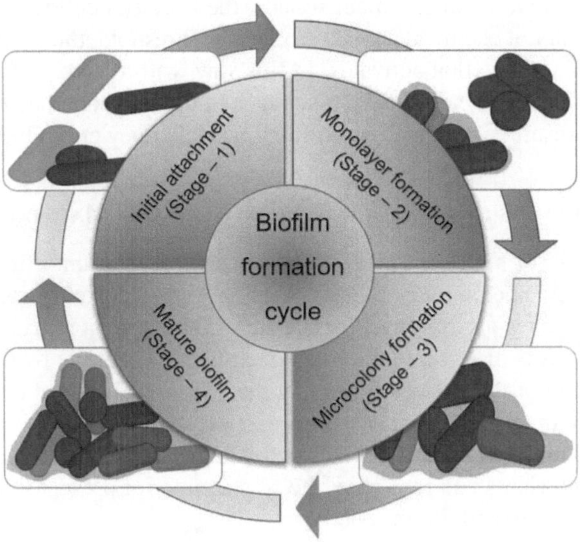

FIGURE 9.1 Four key stages involved in the biofilm formation cycle. Biofilm formation is an environment and species dependent process, which is a complex cycle. Following stage 1 to stage 4, individually occurring cells form clusters and subsequently grow together to form a mature biofilm. The fully developed biofilm disperses and forms new biofilms. It is also worth mentioning that the biofilm formation by motile and nonmotile bacterial species differs from each other.

(iii) formation of three-dimensional structure, and (iv) maturation of biofilm and detachment (Costerton et al., 1999). Figure 9.1 diagrammatically represents the four key stages involved in the biofilm formation cycle.

9.3.1 Attachment to Matrix

As the bacterial cell reaches near to the support/surface, its motion becomes slow, resulting in a reversible relation with the surface and with the already present micro-organisms on the surface. For the establishment of biofilm, a solid-liquid interface provides the best conditions for the bacteria to grow and attach (e.g., water, blood) (Costerton et al., 1999). For better attachment and formation of biofilm, a hydrophilic, rough, and coated surface can provide a better environment. A rise in water temperature, flow velocity, and level of nutrients increases the attachment of bacteria with the surface. Locomotor structures on the surface like pili, flagella, fimbriae, polysaccharides, and proteins are very important and provide an advantage in the formation of biofilm (Donlan and Costerton, 2002).

9.3.2 Formation of Microcolonies

Formation of microcolonies takes place when the bacterial cells adhered physically with the surface and that binding becomes constant and firm resulting in microcolony formation. As an outcome of chemical signals, the bacterial entities start multiplication. As the intensity of chemical signals crosses the threshold, the genetic mechanism for polysaccharide production activates. In this way, with the help of these chemical signals, the divisions of bacterial cells take place in the embedded matrix of exopolysaccharides that result in microcolony formation (Mckenney et al., 1998).

9.3.3 Three-Dimensional Structure Formation and Maturation

During the process of biofilm synthesis and after the formation of microcolony, the expression of a few specific genes takes pace necessary for biofilm establishment. These expressed genes are required for the EPS that is a major structural matrix of biofilm synthesis. It is discovered that the attachment of bacteria by themselves triggers the generation of the extracellular polymeric matrix followed by the formation of water-filled channels for nutrients the transportation within a biofilm structure. Scientists speculated that these channels act as a circulatory system and involve in distributing the nutrients and removing the waste material present in the microcolonies of biofilm (Parsek and Singh, 2003).

9.3.4 Dispersal or Detachment

After the formation, it is observed that the microorganism leaves the biofilm itself on a regular basis. After leaving, the bacterial cells rapidly multiply and disperse. The dispersal/detachment of bacterial cells undergoes a natural pattern (Costerton et al., 1999). The dispersal of bacterial cells occurs due to the dispersion of newly synthesized cells or the liberation of biofilm aggregates because of the flowing effects or quorum sensing (QS) (Baselga et al., 1994; Jamal et al., 2015). It is worth noting that the bacterial entities that are dispersed still have the ability of antibiotic insensitivity. Figure 9.2

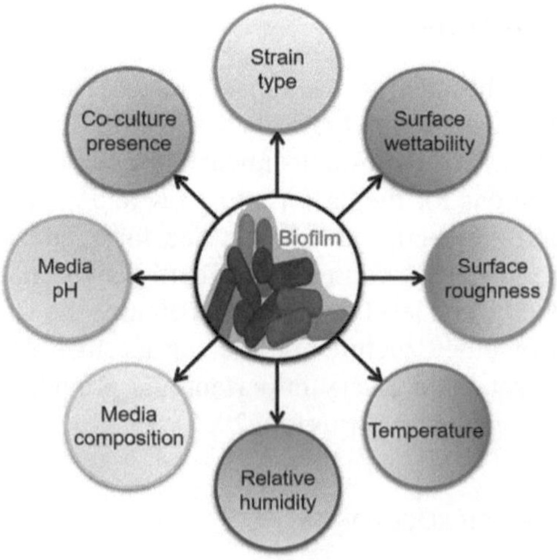

FIGURE 9.2 Influencing factors that play key role in biofilm formation.

depicts the major influencing factors that play a prominent contribution to the formation of biofilm.

9.4 QUORUM SENSING

QS is a mechanism in which several species of bacteria communicate/connect with one another during the biofilm formation (Naves et al., 2010). Signaling molecules produced during QS attaches to the receptors of new bacteria and facilitates the transcription among bacterial species (Figure 9.2) (Miller and Bassler, 2001). This QS system aids the communication between different species of bacteria, and it also helps to form biofilm when there are food shortages and different environmental stress conditions. Several clinically important bacterial strains utilize QS to govern the cooperative formation of virulent factors. In Gram positive bacterial strains, the QS system is completed by a series of steps including synthesis, recognition, and response to autoinducible peptides (Sreenivasan et al., 2013). Exposure to pressure for billions of years has given rise to different strategies in bacterial cell survival, which acclimatized these microorganisms to nearly every ecological niche. One of the most important growth modes for bacteria is the formation of biofilm that is present in approximately 90% of bacteria. QS or cell-to-cell communication of bacterial species is recognized to have a significant contribution to biofilm establishment with its surrounding extracellular matrix (ECM). It is unveiled that the biofilm-forming bacteria (Figure 9.3) possess a highly increased resistance pattern towards disinfectants and antibiotics (Naves et al., 2010). Therefore, biofilm formation is the main challenge in several contexts, such as in industrial corrosion and biofouling and nosocomial and chronic infections (Lopez et al., 2010; Rendueles et al., 2013).

9.5 MEDICINAL PLANTS AS A SOURCE OF NOVEL ANTIBIOFILM ENTITIES

Plants are a good pool of different bioactive substances that are having several medicinal and biological potentialities such as antimicrobial potential (Ribeiro et al., 2018; Oo et al., 2018; Romulo et al., 2018). The World Health Organization (WHO)

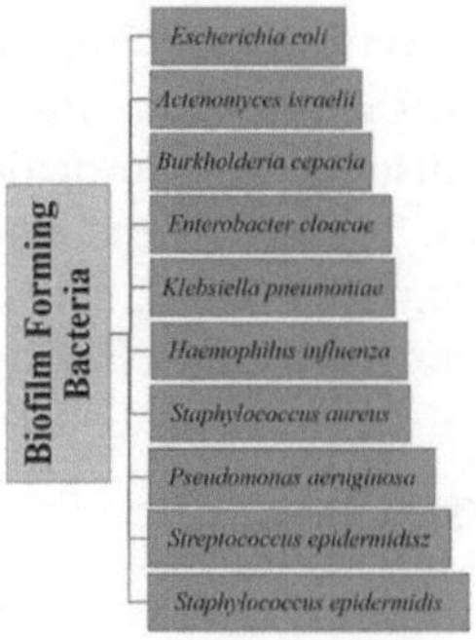

FIGURE 9.3 A list of bacterial species forming biofilm.

also acknowledges the importance of medicinally important plants as a backbone of primary health for approximately half of the population of the world, specifically in those countries having poor resources (Am et al., 2018). Research on medicinally valuable plants as a promising source of new antimicrobials with different action mechanisms is now well recognized. In different medicinal plants, though the active constituents are present in low concentration, still they are acting as a good source of antimicrobial agents as compared to synthetic drugs (Cox et al., 2000). The usage of medicinal plants to treat and cure various ailments has now been a common exercise in medicine in many of the cultures (Erazo et al., 2006; Pérez et al., 2006). Extracts and other active compounds of medicinal plants had gained importance as they are being used in curing illnesses and diseases since ancient times (Maki and Mermel, 1998). Different researchers are working on the development and discovery of new and novel pharmaceuticals having improved therapeutic potential and lesser side effects from medicinal plants (Namasivayam and Roy, 2013).

It is estimated that approximately 80% of the different populations rely on traditional medicine derived from local medicinal plants for their healthcare. For the last many years, the medicinal plants are acting as the backbone of traditional medicine (Karthick Raja Namasivayam and Avimanyu, 2011). Immense research investigation on medicinal plants and their bioactive compounds is currently ongoing; the main emphasis is placed on one of the antimicrobial potentials of plants, that is, against bacteria that form biofilms (Knetsch and Koole, 2011). In research by Namasivayam and Roy (2013), the antibiofilm potential of different plant extracts such as *Ocimum tenuiflorumi*, *Azadirachta indica*, *Tridax procumbens*, and *Vitex negunda* against *Escherichia coli* was determined. Table 9.1 enlists a recent illustration of plant extracts, fractions, and their underlying molecular mechanisms involved in antibiofilm activities against an array of different target bacteria.

9.6 MEDICINAL PLANTS WITH ANIBIOFILM ACTIVITY

9.6.1 *PUNICA GRANATUM* (POMEGRANATE)

It is a commonly consumed fruit that is considered an effective remedy to treat different diseases traditionally. In a study, pomegranate (methanol extract) was employed to explore its inhibitory activity against the biofilm-forming capacity of different microorganisms. The tested fraction markedly inhibited the biofilm formed by several bacteria such as *S. aureus*, MRSA, *Candida albicans*, and *E. coli*. Furthermore, an extract of pomegranate also disturbed the formation of a germ tube with regard to virulence in *Candida albicans*. To discover the main component of pomegranate, high-pressure thin layer chromatography (HP-TLC) was performed that confirms the presence of ellagic acid as a major component (Bakkiyaraj et al., 2013).

9.6.2 *CAMELLIA SINENSIS* (ASSAM TEA)

Kawarai et al. (2016) determined that tea can prevent *Streptococcus mutans* adhesion to surfaces and biofilm development. In contrast to green tea, the Assam tea exhibits a more pronounced antibiofilm activity against *Streptococcus mutans*. The high

TABLE 9.1
Recent Illustration of Plant Extracts, Fractions, and Their Molecular Mechanisms Involved in Antibiofilm Effects Against an Array of Different Target Bacteria

Plant Extract	Fraction Used	Antibiofilm Mechanism Involved	Target Bacteria	References
Licorice roots	Ethanol extract	Disruption of bacterial cell wall Disintegration of the plasma membrane	*Streptococcus pyogenes*	Wijesundara and Rupasinghe (2019)
Purple coneflower stems	Ethanol extract	Disruption of bacterial cell wall Disintegration of the plasma membrane	*Streptococcus pyogenes*	Wijesundara and Rupasinghe (2019)
Purple coneflower flowers	Ethanol extract	Disruption of bacterial cell wall and plasma membrane	*Streptococcus pyogenes*	Wijesundara and Rupasinghe (2019)
Slippery elm inner bark	Ethanol extract	Disintegration of the plasma membrane and disruption of bacterial cell wall	*Streptococcus pyogenes*	Wijesundara and Rupasinghe (2019)
Eucalyptus globulus	Hydromethanolic extract	Substantial reduction in biofilm cells	*Staphylococcus aureus*	Gomes et al. (2019)
Juglans regia	Hydromethanolic extract	Pronounced inhibition of preformed biofilms resulting in colony forming unit reduction	*Staphylococcus aureus*	Gomes et al. (2019)
Eugenia erythrophylla	Acetone leaf extract	Inhibition of the availability of nutrients. Preventing infections by reduction of colonization surfaces and various epithelial of the body.	*Escherichia coli, Enterococcus faecalis, Pseudomonas aeruginosa, Staphylococcus aureus, Salmonella typhimurium, Bacillus cereus*	Famuyide et al. (2019)
Eugenia umtamvunensis	Acetone leaf extract	Inhibition of the availability of nutrients. Preventing infections by reduction of colonization surfaces and various epithelial of the body.	*Enterococcus faecalis, Escherichia coli, Bacillus cereus, Salmonella typhimurium, Staphylococcus aureus, Pseudomonas aeruginosa,*	Famuyide et al. (2019)

(Continued)

TABLE 9.1 (*Continued*)
Recent Illustration of Plant Extracts, Fractions, and Their Molecular Mechanisms Involved in Antibiofilm Effects Against an Array of Different Target Bacteria

Plant Extract	Fraction Used	Antibiofilm Mechanism Involved	Target Bacteria	References
Eugenia natalitia	Acetone leaf extract	Inhibition of the availability of nutrients. Preventing infections by reduction of colonization surfaces and various epithelial of the body.	*Escherichia coli, Pseudomonas aeruginosa, Staphylococcus aureus, Enterococcus faecalis, Bacillus cereus, Salmonella typhimurium*	Famuyide et al. (2019)
Syzygium masukuense	Acetone leaf extract	Inhibition of the availability of nutrients. Preventing infections by reduction of colonization surfaces and various epithelial of the body.	*Escherichia coli, Bacillus cereus, Enterococcus faecalis, Salmonella typhimurium, Staphylococcus aureus, Pseudomonas aeruginosa*	Famuyide et al. (2019)
Syzygium gerrardii	Acetone leaf extract	Inhibition of the availability of nutrients. Preventing infections by reduction of colonization surfaces and various epithelial of the body.	*Escherichia coli, Bacillus cereus, Pseudomonas aeruginosa, Staphylococcus aureus, Enterococcus faecalis, Salmonella typhimurium*	Famuyide et al. (2019)
Argemone mexicana	Acetone extract	Quenching of the biofilm by bioactive constituents	*Staphylococcus aureus, Enterococcus, Calotropis gigantea*	Ramya and Gopinath (2019)
Calotropis gigantea	Methanolic extract	Quenching of the biofilm by bioactive constituents	*Escherichia coli, Proteus vulgaris, Pseudomonas aeruginosa*	Ramya and Gopinath (2019)
Aloe vera	Aqueous extract	Ability of the extract to penetrate and disintegrate the bacterial biofilm	*Staphylococcus aureus*	Saddiq and Al-Ghamdi (2018)
Hymenocallis littoralis leaves	Crude extract	Antibiofilm constituents interact with the active site residues of sortase A, adhesion proteins, and Als3.	*Staphylococcus aureus* NCIM 2654 *and Candida albicans* NCIM 3466	Nadaf et al. (2018)
Juglans regia	Aqueous extract	-	*Pseudomonas aeruginosa*	Dolatabadi et al. (2018)

(*Continued*)

TABLE 9.1 (*Continued*)
Recent Illustration of Plant Extracts, Fractions, and Their Molecular Mechanisms Involved in Antibiofilm Effects Against an Array of Different Target Bacteria

Plant Extract	Fraction Used	Antibiofilm Mechanism Involved	Target Bacteria	References
Juglans regia	Methanolic extract	-	*Pseudomonas aeruginosa*	Dolatabadi et al. (2018)
Acacia arabica	Methanolic extract	Disruption of bacterial cell wall and plasma membrane	*Staphylococcus equorum* JB 359	Khalid et al. (2017)
Acacia arabica	Methanolic extract	Disruption of bacterial cell wall and plasma membrane	*Microbacterium oxydans* M2-2	Khalid et al. (2017)
Acacia arabica	Methanolic extract	Disruption of bacterial cell wall and plasma membrane	*Alcaligenes faecalis* KH-11	Khalid et al. (2017)
Acacia arabica	Methanolic extract	Disruption of bacterial cell wall and plasma membrane	Brevundimonas sp. 266XY4	Khalid et al. (2017)
Acacia arabica	Methanolic extract	Disruption of bacterial cell wall and plasma membrane	*Acinetobacter guillouiae* Phen 8.	Khalid et al. (2017)
Acacia arabica	Methanolic extract	Disruption of bacterial cell wall and plasma membrane	*Bacillus pumilus* HN-30	Khalid et al. (2017)
Melia azadirachta	Ethanol extract	Disruption of bacterial cell wall and plasma membrane	*Staphylococcus equorum* JB 359	Khalid et al. (2017)
Tamarix aphylla	Ethanol extract	Disruption of bacterial cell wall and plasma membrane	*Staphylococcus equorum* JB 359	Khalid et al. (2017)
Piper betel leaf	Ethanolic extract	Inhibition of pyocyanin synthesis and reduction of swimming, swarming, and twitching capability of the bacteria. Anti-QS capabilities.	*Pseudomonas aeruginosa strain* PAO1	Datta et al. (2016)
Zingiber officinale	Methanolic fraction	Inhibition of the virulence genes critical for pathogenesis. Inhibition of glucan synthesis and adherence.	*Streptococcus mutans*	Hasan et al. (2015)
Cocculus trilobus	Ethyl acetate	Sortase activity	Gram-positive bacteria	Kim et al. (2002)
Fritillaria verticillata	Ethyl acetate	Sortase activity	Gram-positive bacteria	Kim et al. (2002)
Liriope platyphylla	Ethyl acetate	Sortase activity	Gram-positive bacteria	Kim et al. (2002)

performance liquid chromatography (HPLC) and ultrafiltration techniques were used for the identification and purification of QS inhibitors in Assam tea. In Assam tea, a compound having a molecular weight lesser than 10 kDa exhibits a high level of galloylated catechins with a potential against biofilm formation than green tea. Whereas polysaccharides like pectin with a molecular mass less than 10 kDa present in green tea was discovered to increase the formation of biofilm (Kawarai et al., 2016).

9.6.3 *Allium sativum* (Garlic)

Garlic is recognized as an abundant source of numerous compounds having strong antimicrobial activities. An extract of garlic shows inhibitory effects on the QS system. Bjarnsholt et al. (2005) determined that extracts of garlic reduce *P. aeruginosa* sensitivity to respiratory burst, tobramycin, and phagocytosis by polymorphonuclear leukocytes in a model rat infected with pulmonary disease (Bjarnsholt et al., 2005). It also decreases the virulence factors and reduces the formation of QS signals in *Pseudomonas aeruginosa* in a rat model (Harjai et al., 2010). In another study conducted by Persson et al. (2005), it was concluded that extracts of garlic exhibited antibiofilm effects on the formation of biofilm against six selected bacterial species. Moreover, biological screening and rational design of different compounds of garlic were also executed, leading to the exploration of a strong QS inhibitor: *N*-(heptyl sulfanyl acetyl)-l-homoserine lactone. This compound interrupts the QS signaling through competitively inhibiting transcriptional regulators LasR and LuxR (Persson et al., 2005).

9.6.4 *Cocculus trilobus* and *Coptis chinensis*

Kim and his team discovered that extracts of *Coptis chinensis* and *Cocculus trilobus* have the potential to block the attachment of bacteria with surfaces having a coat of fibronectin. They showed antiadhesin potential at the attachment phase during the formation of biofilm by interfering with the catalytic activity of a membrane enzyme sortase. This enzyme function is to catalyze the covalent attachment of surface proteins to peptidoglycan in bacterial species. Aqueous and ethyl acetate extracts of these medicinal plants were analyzed, and ethyl acetate extract of *C. trilobus* showed a strong activity while suppressing bacterial attachment through targeting sortase (Kim et al., 2002).

9.6.5 Cranberry Polyphenols

Its fruits have high amounts of polyphenols. Different studies have shown that cranberry extract is a rich source of polyphenols (high molecular weight) that hinders the formation of biofilm and also inhibits the adhesion and settlement of bacteria that are human pathogens (Yamanaka et al., 2004; Duarte et al., 2006; Labrecque et al., 2006; Yamanaka et al., 2007). A component of cranberry affects the enzyme activity that causes the destruction of carbohydrate production, ECM, proteolytic activities, coaggregation, and bacterial cell hydrophobicity. It is suggested that cranberry having bioactive compounds can serve for the treatment and prevention of different oral problems such as periodontitis and dental caries (Bodet et al., 2008).

9.6.6 *Herba patriniae* (Patrinia)

In their research investigation, Fu et al. (2017) constructed a reporter-based system luxCDABE for the detection of expression of genes that are associated with biofilm formation in *P. aeruginosa*. Then, herbal extracts (no. 36) were analyzed for their inhibitory potential against those genes by this reporter-based system. The outcomes revealed that the *Herba patriniae* extract showed a substantial suppressing impact on the genes that were associated with biofilm formation. This herbal extract also reduces the production of exopolysaccharide in *P. aeruginosa*. The results revealed Patrinia as a promising candidate for the identification of novel antimicrobial compounds against *P. aeruginosa* (Fu et al., 2017).

9.6.7 *Ginkgo biloba* (Gingko)

This plant extract has been shown to drastically inhibit the formation of biofilm by *E. coli* on the surfaces of polystyrene, nylon membranes, and glass, with a quantity of 100 µg/mL, without disturbing the growth of bacteria. Ginkgolic acid present in *G. biloba* suppresses prophage and curli genes present in *E. coli*, and this compound causes these inhibitory effects. Curli and prophage genes are responsible for the production of biofilm (He et al., 2013; Lee et al., 2014a). In another research, cinnamaldehyde was reported as it affects the formation and structure of biofilm. It also inhibits the swimming motility of *E. coli* (Niu and Gilbert, 2004).

9.6.8 Phloretin

It acts as an antioxidant compound and is abundantly present in apples. Phloretin discovered that it can reduce the biofilm formation capacity and fimbria in bacterial strain (*E. coli*) without any effect on the planktonic cells. It also impedes the binding of *E. coli* with epithelial cells (Human colon) and suppresses the tumor necrosis factor. Phloretin inhibits different genes such as toxin genes (hlyE and stx 2), curli genes (csgB and csgA), prophage genes, and autoinducer 2 importer genes in biofilm formed by *E. coli* (Lee et al., 2011). Phloretin also possesses the potential of antibiofilm formation against *S. aureus* even at low concentrations having inhibitory efficiency of approximately 70% (Lopes et al., 2017).

9.6.9 Processing Strategies, Separation, and Extraction of Bioactive Compounds

Several natural products derived from medicinal plants possess antibiofilm and antimicrobial properties in vitro. Natural products having the antibiofilm properties mainly inhibit the formation of the polymeric matrix, suppress cell attachment and adhesion, and decrease the production of virulence factors, thus blocking of QS system and biofilm formation. The different mechanisms related to the antibiofilm potential of plants are illustrated in Figure 9.4. Several herbal extracts are used for their antibiofilm potential, and the bioactive components or molecules of these extracts are still unidentified, and there is a need to further investigate them.

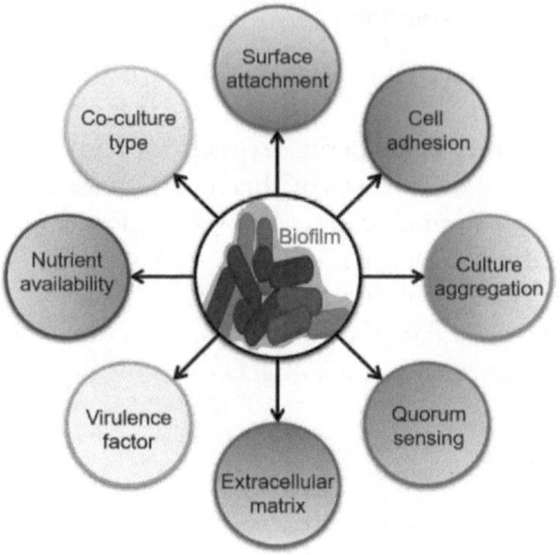

FIGURE 9.4 Main target points, which are considered crucial for the inhibition of biofilm formation. Each representative point could be suppressed or induced subject to the biofilm type and inhibitory action of bioactive agent.

Therefore, the extraction and separation of effective components for antibiofilm are important. In the past few decades, techniques like structure-based virtual screening and chromatographic separation were extensively used for the identification of bioactive components that act as antibiofilm compounds from medicinal plants.

9.7 PHENOLIC COMPOUNDS ASSOCIATED WITH ANTIBIOFILM ACTIVITY

Polyphenols from plants are a huge class of bioactive substances with unique physicochemical attributes and biological functionalities. Typical polyphenolic compounds from plants comprised anthocyanins, coumarins, lignins, tannins, flavonoids, lignans phenolic acids, and stilbenes (Pereira et al., 2009). These biologically active compounds display a remarkable contribution in providing confrontation against a large array of pathogenic microorganisms and safeguard from the toxic effect of toxins and free radicals (Quideau et al., 2011; Daglia, 2012). Currently, plant-based polyphenolic compounds have received a continuously increasing acknowledgment by the general public as well as the scientific community owing to their abundance and wide presence in seeds, fruits, vegetables, and other foodstuffs. The consistent ingestion of these substances has known advantageous impacts on the health of human beings due to the scavenging capacity of free radicals generated during the oxidative reactions (Quideau et al., 2011). The role of plant phenolics as antibiofilm candidates has been revealed and supposed to be suppression of biofilm by interfering with the regulatory mechanisms of bacterial species such as QS or other regulatory networks without affecting their growth. A list of plant polyphenols involved in antibiofilm activities is portrayed in Figure 9.5.

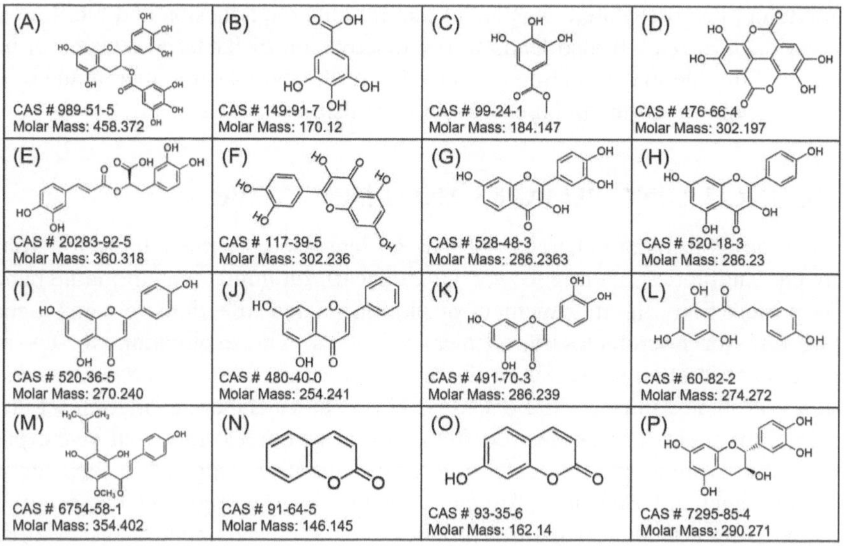

FIGURE 9.5 Biologically active compounds with antibiofilm inhibition potentialities. (A) Epigallocatechin gallate, (B) Gallic acid, (C) Methyl gallate, (D) Ellagic acid, (E) Rosmarinic acid, (F) Quercetin, (G) Fisetin, (H) Kaempferol, (I) Apigenin, (J) Chrysin, (K) Luteolin, (L) Phloretin, (M) Xanthohumol, (N) Coumarin, (O) Umbelliferone, and (P) Catechin. For further details on their pharmacokinetics data, please visit the representative CAS #. Molar mass is given in g/mol.

Among the polyphenols, tannins are represented to be one of the largest groups and are subcategorized into hydrolysable tannins (including ellagitannins and gallotannins) and condensed tannins (including catechins and proanthocyanidins) (Nagy et al., 2011). Flavonoids represent extensively disseminated phenolic compounds in plants and are known to be associated with aroma and color of fruits and flowers to appeal pollinators. As a consequence, they are involved in seeding, propagation, development, and growth of seedlings. They have shown the potential to display antibiofilm potentialities. A large number of flavonoid compounds including luteolin, quercetin, apigenin, chrysin, fisetin, and kaempeferol are present in red wine obtained from *Vitis vinifera* that are able to inhibit the biofilm-forming capacity of *S. aureus* (Cho et al., 2015). Polyphenolic compounds extracted from grape pomace exhibited a pronounced antibiofilm and antibacterial potential against *S. aureus* (Xu et al., 2014).

Coumarins are derivatives that occur naturally in a wide range of biological functionalities such as vasodilatation, photosensitizing, antimicrobial, antibiofilm, and anti-inflammatory activities (Kalkhambkar et al., 2008). Lee et al. (2014b) documented that umbelliferone and coumarin showed potential activity against the development of biofilm formation by *E. coli* O157:H7. The biofilm-forming ability of *E. coli* has been reported to inhibit the use of coladonin (Lee et al., 2014b). Coumarins including esculin, esculetin, nodakenetin, and psoralen also result in the suppression of biofilm development of *P. aeruginosa* (Durig et al., 2010; Ding et al., 2011).

Xanthohumol is a prenylated chalconoid that inhibits the adhesion and biofilm construction of *S. aureus*. It also leads to the inactivation of bacterial isolates in preformed biofilm due to destructing the integrity of the cytoplasmic membrane of the bacteria by inhibiting the lipid metabolism (Rozalski et al., 2013).

9.8 CONCLUDING REMARKS AND PERSPECTIVES

The bacteria biofilms are characterized by an increased tolerance to antimicrobial agents and antibiotics, leading to worrying concern on human health maintenance. At the present time, the management of biofilm-related infections remains a challenging task for microbiologist's clinical scientists. The exploration of new and effective antimicrobial strategies is of great significance to surpass these antibiotic resistance problems in microbial diseases. This chapter elaborates that a vast array of antibiofilm agents can be derived from natural resources. In recent years, plant resources have gained considerable researchers' attention endowed to their numerous health-beneficial attributes. To date, a plethora of investigations has examined the suppressing potential of natural products on the development of bacterial biofilm, representing their perspective as alternative candidates for the cure of bacterial infections. Suppression of the QS network and the interference in each step of biofilm establishment formation constitute the major mechanism of action for biofilm destabilization. Identification and separation of novel antibiofilm agents by chromatography and some other means have unveiled the way for the isolation of effective constituents with broad-spectrum antibiofilm potentialities. Nevertheless, additional investigations are required to identify molecular structures of the plant-derived bioactive compounds with biofilm disruption effects. Moreover, it would have a great significance to enhance the safety, specificity, and effectiveness of natural antibiofilm agents individually or combining with other antibiotics for the appraisal of these antibiofilm entities in clinical trials.

REFERENCES

Am Q, Ma F-A, Md G-G, De La Puerta R. Potential therapeutic applications of the genus *Annona*: Local and traditional uses and pharmacology. *J Ethnopharmacol.* 2018; 225:244–270.

Bakkiyaraj D, Nandhini JR, Malathy B, Pandian SK. The antibiofilm potential of pomegranate (*Punica granatum* L.) extracts against human bacterial and fungal pathogens. *Biofouling.* 2013; 29:929–937.

Baselga R, Albizu I, Amorena B. *Staphylococcus aureus* capsule and slime as virulence factors in ruminant mastitis. A review. *Vet Microbiol.* 1994; 39:195–204.

Bjarnsholt T, Jensen PO, Rasmussen TB, Christophersen L, Calum H, Hentzer M, Hougen HP, Rygaard J, Moser C, Eberl L, Hoiby N, Givskov M. Garlic blocks quorum sensing and promotes rapid clearing of pulmonary *Pseudomonas aeruginosa* infections. *Microbiology.* 2005; 151:3873–3880.

Bodet C, Grenier D, Chandad F, Ofek I, Steinberg D, Weiss EI. Potential oral health benefits of cranberry. *Crit Rev Food Sci Nutr.* 2008; 48:672–680.

Cho HS, Lee JH, Cho MH, Lee J. Red wines and flavonoids diminish *Staphylococcus aureus* virulence with antibiofilm and antihemolytic activities. *Biofouling.* 2015; 31:1–11.

Costerton J, Stewart PS, Greenberg EP. Bacterial biofilms: A common cause of persistent infections. *Science.* 1999; 284:1318–1322.

Cox SD, Mann CM, Markham JL, Bell HC, Gustafson JE, Warmingto JR, Wyllie SG. The mode of antimicrobial action of the essential oil of *Melaleuca alternifolia* (tea tree oil). *J Appl Microbiol.* 2000; 88:170–175.

Daglia M. Polyphenols as antimicrobial agents. *Curr Opin Biotechnol.* 2012; 23:174–181.

Datta S, Jana D, Maity TR, Samanta A, Banerjee R. Piper betel leaf extract affects the quorum sensing and hence virulence of *Pseudomonas aeruginosa* PAO1. 3. *Biotechnology.* 2016; 6:18.

Ding X, Yin B, Qian L, Zeng Z, Yang Z, Li H, Lu Y, Zhou S. Screening for novel quorum-sensing inhibitors to interfere with the formation of *Pseudomonas aeruginosa* biofilm. *J Med Microbiol.* 2011; 60:1827–1834.

Dolatabadi S, Moghadam HN, Mahdavi-Ourtakand M. Evaluating the antibiofilm and antibacterial effects of *Juglans regia* L. extracts against clinical isolates of *Pseudomonas aeruginosa. Microb Pathog.* 2018; 118:285–289.

Donlan RM, Costerton JW. Biofilms: Survival mechanisms of clinically relevant microorganisms. *Clin Microbiol Rev.* 2002; 15:167–193.

Duarte S, Gregoire S, Singh AP, Vorsa N, Schaich K, Bowen WH, Koo H. Inhibitory effects of cranberry polyphenols on formation and acidogenicity of *Streptococcus mutans* biofilms. *FEMS Microbiol Lett.* 2006; 257:50–56.

Durig A, Kouskoumvekaki L, Vejborg RM, Klemm P. Chemoinformatics-assisted development of new antibiofilm compounds. *Appl Microbiol Biotechnol.* 2010, 87, 309–317.

Erazo S, Delporte C, Negrete R, Garcia R, Zaldivar M, Ittura G, Caballero E, López JL, Backhouse N. Constituents and biological activities of *Schinus polygamus. J Ethnopharmacol.* 2006; 107:395–400.

Famuyide IM, Aro AO, Fasina FO, Eloff JN, McGaw LJ. Antibacterial and antibiofilm activity of acetone leaf extracts of nine under-investigated south African Eugenia and Syzygium (Myrtaceae) species and their selectivity indices. *BMC Complement Altern Med.* 2019; 19(1):141.

Fu B, Wu Q, Dang M, Bai D, Guo Q, Shen L, Duan K. Inhibition of *Pseudomonas aeruginosa* biofilm formation by traditional Chinese medicinal herb *Herba patriniae. Biomed Res Int.* 2017; 2017:9584703.

Fux C, Costerton JW, Stewart PS, Stoodley P. Survival strategies of infectious biofilms. *Trends Microbiol.* 2005; 13:34–40.

Gomes F, Martins N, Ferreira IC, Henriques M. Antibiofilm activity of hydromethanolic plant extracts against *Staphylococcus aureus* isolates from bovine mastitis. *Heliyon* 2019; 5(5):e01728.

Hall-Stoodley L, Costerton JW, Stoodley P. Bacterial biofilms: From the natural environment to infectious diseases. *Nat Rev Microbiol.* 2004; 2:95–108.

Harjai K, Kumar R, Singh S. Garlic blocks quorum sensing and attenuates the virulence of *Pseudomonas aeruginosa. FEMS Immunol Med Microbiol.* 2010; 58:161–168.

Hasan S, Danishuddin M, Khan AU. Inhibitory effect of *Zingiber officinale* towards *Streptococcus mutans* virulence and caries development: In vitro and in vivo studies. *BMC Microbiol.* 2015; 15:1.

He J, Wang S, Wu T, Cao Y, Xu X, Zhou X. Effects of ginkgoneolic acid on the growth, acidogenicity, adherence, and biofilm of *Streptococcus mutans* in vitro. *Folia Microbiol.* 2013; 58:147–153.

Jamal M, Tasneem U, Hussain T, Andleeb S. Bacterial biofilm: Its composition, formation, and role in human infections. Research & reviews. *J Microbiol Biotechnol* 2015; 4(3):1–14.

Kalkhambkar RG, Kulkarni GM, Kamanavalli CM, Premkumar N, Asdaq SMB, Sun CM. Synthesis and biological activities of some new fluorinated coumarins and 1-aza coumarins. *Eur J Med Chem.* 2008; 43:2178–2188.

Karthick Raja Namasivayam S, Avimanyu, B. Silver nanoparticle synthesis from *Lecanicillium lecanii* and evaluationary treatment on cotton fabrics by measuring their improved antibacterial activity with antibacterial antibiotics against *Staphylococcus aureus* (ATCC 29213) and *Escherichia coli* (ATCC 25922) strains. *Int J Pharm Pharm Sci.* 2011; 4(3):185–197.

Kawarai T, Narisawa N, Yoneda S, Tsutsumi Y, Ishikawa J, Hoshino Y, Senpuku H. Inhibition of *Streptococcus mutans* biofilm formation using extracts from Assam tea compared to green tea. *Arch Oral Biol.* 2016; 68:73–82.

Khalid M, Hassani D, Bilal M, Butt ZA, Hamayun M, Ahmad A, Huang D, Hussain, A. (2017). Identification of oral cavity biofilm forming bacteria and determination of their growth inhibition by *Acacia arabica*, *Tamarix aphylla* L., and *Melia azedarach* L. medicinal plants. *Arch Oral Biol.* 81:175–185.

Kim SW, Chang IM, Oh KB. Inhibition of the bacterial surface protein anchoring transpeptidase sortase by medicinal plants. *Biosci Biotechnol Biochem.* 2002; 66:2751–2754.

Knetsch MLW, Koole LH. New strategies in the development of antimicrobial coatings: The example of increasing usage of silver and silver nanoparticles. *Polymers.* 2011; 3:340–366.

Labrecque J, Bodet C, Chandad F, Grenier D. Effects of a high-molecular-weight cranberry fraction on growth, biofilm formation, and adherence of *Porphyromonas gingivalis. J Antimicrob Chemother.* 2006; 58:439–443.

Lee JH, Kim YG, Cho HS, Ryu SY, Cho MH, Lee J. Coumarins reduce biofilm formation and the virulence of *Escherichia coli* O157:H7. *Phytomedicine* 2014a; 21:1037–1042.

Lee JH, Kim YG, Ryu SY, Cho MH, Lee J. Ginkgolic acids and *Ginkgo biloba* extract inhibit *Escherichia coli* O157:H7 and *Staphylococcus aureus* biofilm formation. *Int J Food Microbiol.* 2014b; 174:47–55.

Lee JH, Regmi SC, Kim JA, Cho MH, Yun H, Lee CS, Lee J. Apple flavonoid phloretin inhibits *Escherichia coli* O157:H7 biofilm formation and ameliorates colon inflammation in rats. *Infect Immun.* 2011; 79:4819–4827.

Lopes LAA, Dos Santos Rodrigues JB, Magnani M, de Souza EL, de Siqueira-Junior JP. Inhibitory effects of flavonoids on biofilm formation by *Staphylococcus aureus* that overexpresses efflux protein genes. *Microb Pathog.* 2017; 107:193–197.

Lopez D, Vlamakis H, Kolter R. Biofilms. *Cold Spring Harb Perspect Biol* 2010; 2:a000398.

Maki DG, Mermel LA. Infections due to infusion therapy. In: Bennett JV, Brachman PS, editors. *Hospital Infections.* 4th ed. Philadelphia, PA: Lippincott-Raven; 1998. pp. 689–724.

Mckenney D, Hübner J, Muller E, Wang Y, Goldmann DA, Pier GB. The ica locus of *Staphylococcus epidermidis* encodes production of the capsular polysaccharide/adhesin. *Infect Immunity.* 1998; 66:4711–4720.

Miller MB, Bassler BL. Quorum sensing in bacteria. *Ann Rev Microbiol.* 2001; 55:165–199.

Nadaf NH, Parulekar RS, Patil RS, Gade TK, Momin AA, Waghmare SR, Dhanavade MJ, Arvindekar AU, Sonawane KD. Biofilm inhibition mechanism from extract of *Hymenocallis littoralis* leaves. *J Ethnopharmacol.* 2018; 222:121–132.

Nagy M, Granˇcai D, Muˇcaji P. *Farmakognózia Biogenéza Prírodných Látok.* Martin, TN: Osveta; 2011.

Namasivayam SK, Roy EA. Antibiofilm effect of medicinal plant extracts against clinical isolate of biofilm of *Escherichia coli. Int J Pharm Pharm Sci.* 2013; 5(2):486–489

Naves P, Del Prado G, Huelves L, Rodriguez-Cerrato V, Ruiz V, Ponte MC, Soriano F. Effects of human serum albumin, ibuprofen, and N-acetyl-l-cysteine against biofilm formation by pathogenic *Escherichia coli* strains. *J Hospital Infect.* 2010; 76:165–170.

Niu C, Gilbert ES. Colorimetric method for identifying plant essential oil components that affect biofilm formation and structure. *Appl Environ Microbiol.* 2004; 70:6951–6956.

Okada M, Sato I, Cho SJ, Iwata H, Nishio T, Dubnau D, Sakagami Y. Structure of the *Bacillus subtilis* quorum-sensing peptide pheromone ComX. *Nat Chem Biol.* 2005; 1:23–24.

Oo O, Pa S, Ps F. Antimicrobial and antiprotozoal activities of twenty-four Nigerian medicinal plant extracts. *S Afr J Bot.* 2018; 117:240–246.

Parsek MR, Singh PK. Bacterial biofilms: An emerging link to disease pathogenesis. *Ann Rev Microbiol.* 2003; 57:677–701.

Pereira, DM, Valentão P, Pereira JA, Andrade PB. Phenolics: From chemistry to biology. *Molecules.* 2009; 14:2202–2211.

Pérez RM, Hernández H, Hernández S. Antioxidant activity of *Tagetes erecta* essential oil. *J Chil Chem Soc* 2006; 51:883–886.

Persson T, Hansen TH, Rasmussen TB, Skinderso ME, Givskov M, Nielsen J. Rational design and synthesis of new quorum-sensing inhibitors derived from acylated homoserine lactones and natural products from garlic. *Org Biomol Chem.* 2005; 3:253–262.

Quideau S, Deffieux D, Douat-Casassus C, Pouysegu L. Plant polyphenols: Chemical properties, biological activities, and synthesis. *Angew Chem Int Ed Engl.* 2011; 50:586–621.

Ramya R, Gopinath SM. Investigation of antibiofilm potential of *Argemone mexicana* and *Calotropis gigantea* against clinical isolates. *Int J Curr Microbiol App Sci.* 2019; 8(1):3184–3193.

Rendueles O, Kaplan JB, Ghigo JM. Antibiofilm polysaccharides. *Environ Microbiol* 2013; 15:334–346.

Ribeiro IC, Mariano EG, Careli RT, Morais-Costa F, de Sant'Anna FM, Pinto MS, de Souza MR, Duarte ER. Plants of the Cerrado with antimicrobial effects against *Staphylococcus* spp. and *Escherichia coli* from cattle. *BMC Vet Res.* 2018; 14(1):32.

Romer CM, Vivacqua CG, Abdulhamid MB, Baigori MD, Slanis AC, Gaudioso de Allori MC, Tereschuk ML. Biofilm inhibition activity of traditional medicinal plants from Northwestern Argentina against native pathogen and environmental microorganisms. *Rev Soc Bras Med Trop* 2016; 49(6):703–712

Romulo A, Ea Z, Rondevaldova J, Kokoska L. Screening of in vitro antimicrobial activity of plants used in traditional Indonesian medicine. *Pharm Biol.* 2018; 56(1):287–293.

Rozalski M, Micota B, Sadowska B, Stochmal A, Jedrejek D, Wieckowska-Szakiel M, Rozalska B. Antiadherent and antibiofilm activity of *Humulus lupulus* L. derived products: New pharmacological properties. *BioMed Res Int.* 2013; 2013:101089.

Saddiq AA, Al-Ghamdi H. Aloe vera extract: A novel antimicrobial and antibiofilm against methicillin resistant *Staphylococcus aureus* strains. *Pakistan J Pharm Sci.* 2018; 31:2123–2130.

Sauer FG, Remaut H, Hultgren SJ, Waksman G. Fiber assembly by the chaperone-usher pathway. *Biochim Biophys Acta.* 2004; 1694:259–267.

Sreenivasan P, Nujum ZT, Purushothaman KK. Clinical response to antibiotics among children with bloody diarrhea. *Indian Paediatr.* 2013; 50:340–341.

Vinodkumar CS, Kalsurmath S, Neelagund YF. Utility of lytic bacteriophage in the treatment of multidrug-resistant *Pseudomonas aeruginosa* septicemia in mice. *Indian J Pathol Microbiol.* 2008; 51:360–366.

Wijesundara NM., Rupasinghe, HP. Bactericidal and anti-biofilm activity of ethanol extracts derived from selected medicinal plants against *Streptococcus pyogenes. Molecules.* 2019; 24(6):1165.

Xu C, Yagiz Y, Hsu WY, Simonne A, Lu J, Marshall MR. Antioxidant, antibacterial, and antibiofilm properties of polyphenols from muscadine grape (*Vitis rotundifolia* Michx.) pomace against selected foodborne pathogens. *J Agric Food Chem.* 2014; 62:6640–6649.

Yamanaka A, Kimizuka R, Kato T, Okuda K. Inhibitory effects of cranberry juice on attachment of oral streptococci and biofilm formation. *Oral Microbiol Immunol.* 2004; 19:150–154.

Yamanaka A, Kouchi T, Kasai K, Kato T, Ishihara K, Okuda K. Inhibitory effect of cranberry polyphenol on biofilm formation and cysteine proteases of *Porphyromonas gingivalis. J Periodontal Res.* 2007; 42:589–592.

Part IV

Microbial Biofilms: Biomass, Plant Growth, Soil Nutrient, and Wastewater Management

10 Electroactive Biofilms (EAB)

Role in a Bioelectrochemical System for Wastewater Treatment and Bioelectricity Generation

Vijay Kumar Garlapati
Jaypee University of Information Technology

Sunandan Naha
Indian Institute of Technology Guwahati

Swati Sharma
Jaypee University of Information Technology

Pranab Goswami
Indian Institute of Technology Guwahati

Surajbhan Sevda
Indian Institute of Technology Guwahati
National Institute of Technology Warangal

CONTENTS

10.1 INTRODUCTION

Bacteria interact with the different atmospheric conditions to acquire energy through various metabolic pathways and also exist in biofilms at interphases to protect for various environmental stresses. The biofilm formation involves nutrient transport, waste management networks, and communication within the system (Semenec and Franks, 2015; Bogino et al., 2013). Electroactive biofilms (EABs) are biofilms produced by exoelectrogen bacteria on electrode surfaces in bio electrochemical systems (BESs). BESs encompass a wide array of technologies, which include microbial fuel cell (MFC) and different microbial electrolysis cell (MEC) configurations (Aracic et al., 2014; Wang and Ren, 2013). BESs are designed such a way whether it is used for electricity or different industrial commodities production (Table 10.1) (Borole et al., 2012).

In all BESs, the exoelectricigens act as semi catalysts by donating or receiving electrons from the electrodes via extracellular electron transfer (EET). The EABs of BES share some common analog with the traditional biofilms with the constitutional parts of water, microorganisms, and extracellular polymeric substances (Stöckl et al., 2016). This chapter puts forth the dynamics of anodic and cathodic EABs and transfer mechanisms along with the discussion of factors affecting the EAB formation

TABLE 10.1
Different Types of BESs Based on the Cathode Substrates

Configuration of BES	Cathode Substrate	End Product
MFC	Oxygen	Bioelectricity
MEC	Proton	Biohydrogen
BES	Acetate	Ethanol
BES	Oxygen	Hydrogen peroxide

Source: Borole et al., 2012.

and its role in BES-based wastewater treatment and bioelectricity production with the focus on possible prospects.

10.2 ANODIC BIOFILM DYNAMICS AND THEIR TRANSFER MECHANISMS

Anodic (electron acceptor) biofilms help in decreasing the charge transfer resistance in MFC, which enhances the power output of the system (Baranitharan et al., 2015). The anodic EABs are considered as the essential components in the performance of MFCs, and hence, understanding the possible mechanisms and affecting factors of EAB formation is a crucial topic of interest for further step of MFC advancement. The notable characteristics of anodic EABs need to be the presence of an external electron acceptor which selects exoelectrogens that utilize soluble electron acceptors only. The function of exoelectrogens of anodic EABs is not only serving as an attachment entity but also serves as an electron reserve by influencing the structural and functional gradients of the biofilm matrix. The exoelectrogen components, namely *c*-type cytochromes, pili, and endogenous electron mediators, play a significant role in the structure and conductivity of anodic EABs (Richter et al., 2009; Shi et al., 2009; Malvankar and Lovley, 2012).

The interaction of microbes with anodic biofilms is mainly through different EET-based mechanisms such as direct (DEET), electron shuttle (SEET), or pilin (PEET). In DEET and SEET, inner and outer membrane redox proteins and endogenously produced redox mediators, respectively, play a crucial role in electron transfer. In the case of PEET, the periplasmic extensions (-pili structures) help in EET. The different EET mechanisms, coupled with anodic biofilms, were summarized in Figure 10.1 (Saratale et al., 2017). Usually, the electron transfer in shorter distances is accompanied by DEET and SEET, and PEET is utilized in the case of large-distance electron transfer

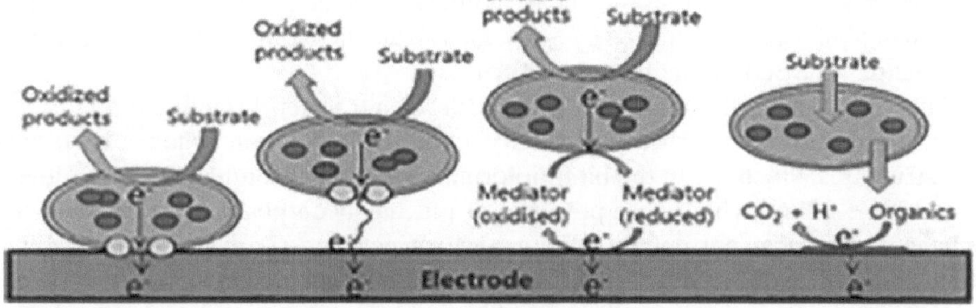

FIGURE 10.1 EET mechanisms coupled with anodic biofilms. (Adapted from Saratale,G.D., et al., *Chemosphere*, 178, 534–547, 2017. With permission.)

(Kalathil and Pant, 2016). The structure and conduction of anodic biofilms vary with the type of EET mechanism, type of culture (mixed or pure), and available substrate for oxidation (Semenec and Franks, 2014). The cultures of *Geobacter sulfurreducens* grown in Fe(III) oxides (insoluble electron acceptor) utilize pili and OmcS for oxidation (PEET), and it is not feasible while grown in Fe(III) citrate (soluble electron acceptor) (DEET or SEET) (Mehta et al., 2005; Reguera et al., 2005). *Geobacter sulfurreducens* utilize PEET and DEET for the formation of anodic biofilms, whereas the pure cultures of *Shewanella oneidensis* adopt DEET and SEET for anodic biofilm formation (Malvankar and Lovley, 2014; Kotloski and Gralnick, 2013).

In DEET, *c*-type cytochromes play an important role in the direct transfer of electrons to anodes. The examples include OmcZ (Semenec and Franks, 2014) and OmcS (Leang et al., 2010) of *G. sulfurreducens* and MtrC *S. oneidensis*(Richter et al., 2012). As a postulation, four stepwise models for DEET have been proposed, which include substrate electron donor oxidation and electron uptake through periplasmic cytochromes, transfer of taken electrons from cytochromes of periplasm to exocytoplasm, transport of electrons from the cytochromes of exocytoplasm to interfacial cytochromes of the anode, and final electron transfer from interfacial cytochromes to the electrode surface cytochromes. Among all four steps, the first step of substrate donor oxidation coupled electron uptake considered as a rate-limiting step (Bonanni et al., 2012; Richter et al., 2009). The studies showcased the importance of enhanced electron transfer from substrate to the anode in further improvement of current production by the MFC. They necessitated the importance of understanding the exact mechanism of DEET.

The pivotal role of outer membrane cytochromes of anodes in the attachment of EABs to the anodic surface was drawn from the study where the anode has been treated with heat/acid before MFC construction resulted in improved current production. The prior treatment of anode with heat/acid facilitates the formation of –OH and C=O groups, which further helps in peptide and hydrogen bonds formed with the $-CONH_2$ of cytochromes of *Desulfovibrio desulfuricans* anode biofilms (Kang et al., 2014).

The redox mediators such as pherazines play an essential role in SEET for transferring electrons to the anode by acting as electron acceptors. Aerobic electrician *P. aeruginosa* biofilms usually produce pherazines for tolerating the anoxic conditions (Pierson and Pierson, 2010). During anoxic conditions, the higher content of phenazines helps in the development of biofilms, which further helps in enhanced current production of MFCs (Shen et al., 2014; Dietrich et al., 2013). Riboflavin and riboflavin-5′-phosphate produced by *S. oneidensis* anodic biofilms also serve as electron shuttles in SEET (Marsili et al., 2008a).

The long-distance-based PEET involves electron transfer through pili (extracellular amino acid-based structures, usually found in many Gram-negative bacteria) like structures which aid in the biofilm formation via adhesion/nanowires (Giltner et al., 2012; 2006). The higher presence of pili has been positively correlated to enhanced electrical current production in exoelectrigen-*Geobacter*-based MFC (Reguera et al., 2006). More research has to be done to understand the primitive role of pili—as the sole organelle in PEET or as conductive material or as working machinery coupling with cytochrome proteins (Malvankar et al., 2015; Liu et al., 2014; Lovley, 2012).

TABLE 10.2
Components Responsible for Electron Transfer in Different Exoelectrigens in MFCs

Compounds Aid in Electron Transfer	Exoelectrigen	References
c-di-GMP, sRNA, and QS	*P.putida,P. fluorescens, P. aeruginosa, B. cenocepacia*	Fazli et al. (2014)
Type IV pili, c-type cytochrome (Pgc A, OmcZ,Omp B)	*G. sulfurreducens*	Smith et al. (2014), Malvankar et al. (2012)
Anthra quinone-2,6-disulfonate	*T. potens strain JR*	Wrighton et al. (2011)
c-Type cytochromes (Omp B, Omp C, and OmcE)	*G. metallireducens*	Smith et al. (2013); Tremblay et al. (2011)
c-Type cytochromes	*Desulfovibrio desulfuricans*	Kang et al. (2014)
Tetraheme cytochrome C3 (Tplc3, cycA), transmembranecomplexes (QrcA)	*D. alaskensis*	Keller et al. (2014)
Flavins, riboflavin; Cytochrome (MtrF)-heme	*S. oneidensis*	Coursolle et al. (2010); Fitzgerald et al. (2012)
c-Type cytochromes (MtrC, OmcA, FAD transporter), cytochrome (MtrF)- heme	*S. putrefaciens*	Pandit et al. (2014)
2,6-Di-tert-butyl-pbenzoquinone	*K. pneumoniae*	Lifang et al. (2010)

Source: Saratale, G.D., et al., *Chemosphere*, 178, 534–547, 2017. With permission.

The different compounds that aid in the electron transfer in different exoelectrogens in MFCs are listed in Table 10.2 (Saratale et al., 2017).

10.3 CATHODE BIOFILM DYNAMICS: CATHODE TO MICROBE ELECTRON TRANSFER

The cathodic biofilm formation plays an important role in the production of biofuel/industrial commodities production and wastewater treatment through BESs by providing external voltage. Biocathodes are usually comprised of electrophilic microorganisms which receive electrons from an electrode. The performance and efficiency of BESs for industrial commodities and wastewater treatment are mainly dictated by cathodic biofilms, which alleviates the need for external precious metal catalysts for the BES performance (Jafary et al., 2015; Fu et al., 2013; Lovley and Nevin, 2013; Nevin et al., 2011). The metagenomic analysis of cathodic biofilms reveals the presence of electrophilic microbes such as *Geobacter* sp., *Thiobacillus* sp., *Proteobacteria*, *Firmicutes*, and *Chloroflexi* (Vilar-Sanz et al., 2013; Kato et al., 2012; Wrighton et al., 2010). The probable mechanism of EET that has been reported from biocathodes includes direct contact through cytochromes/hydrogenases (Rosenbaum et al., 2011; Huang et al., 2011; Strycharz et al., 2011). The cathodic biofilm community switches the mechanism from SEET (−100 mV) to DEET (+200 mV) based on the applied electrode potential (Milner et al., 2014). More research has to be done to characterize the long-distance-based shuttles to understand the cathodic SEET mechanism (Aulenta et al., 2012; Rosenbaum et al., 2011).

10.4 ANALYSIS AND CHARACTERIZATION OF MICROBIAL EABs

The analysis and characterization of cathodic and anodic EABs facilitate the understanding of the probable mechanism of biofilm formation and its maintenance towards attaining the enhanced BES performance towards the production of electricity, biofuels, industrial commodities, and wastewater treatment (Stöckl et al., 2016). Various techniques are available to investigate the microbial attachments to anodic and cathodic chambers of BESs, which range from classical electrochemical techniques to the latest optical and spectroscopic approaches (Saratale et al., 2017). The various techniques used to investigate for a deep understanding of biofilm formation include Cyclic Voltammetry (Jain and Connolly, 2013), spectro-electrochemistry (Jain et al., 2011), and Electrochemical Impedance Spectroscopy (EIS) in BESs (He and Mansfeld, 2009). A vast number of techniques are available to investigate the BES-based microbial biofilm adhesion to electrodes, which include Atomic Force Microscopy (Mangold et al., 2008), Scanning Electron Microscopy (Alhede et al., 2012), Confocal Laser Scanning Microscopy (CLSM) (Neu and Lawrence, 2014), and Raman Microscopy (Lebedev et al., 2014). The understanding and investigation/monitoring of versatile and dynamic biofilms need the exploitation of more than one optical/spectroscopic techniques. The overview of the commonly employed techniques to monitor the EABs is provided in Table 10.3.

10.5 CATHODIC EABs VS. ANODIC EABs

The anodic EABs are considerably thicker than the cathodic EABs in pure culture dual-chambered MFC. However, the thickness of cathodic EABs of environmental samples is 100 times denser than the anodic EABs (Strycharz-Glaven et al., 2013; Yuan et al., 2013; Kiely et al., 2011). Under anaerobic conditions, the growth of biofilms at the cathode is difficult than that at the anode due to the requirement of high energy input to tolerate the negative reduction potential for poising the electrode (Morita et al., 2011). Although cathode can provide an endless supply of electrons to the electrophobic bacteria, due to some limitations, the poising of cathode takes place at relatively low potentials, which places the cathode less attractive as an electron donor. The limitations include first, cathode does not provide the carbon source unlike in case of the anode and second, need to substitute the usually supplied acetate as a carbon source or available CO_2 (which can be used as an energy source by autotrophic electrotrophs) with bicarbonate (Rozendal et al., 2008). The overall conspiracy in the case of cathodic EABs needs more research for a better understanding of the biofilm formation and mechanisms of electron consumption towards enhanced efficiency of cathodic EABs towards BES-based production of industrial commodities and wastewater treatment.

10.6 FACTORS AFFECTING THE BIOFILM FORMATION AND PERFORMANCE

In BES electrically active microbial community form biofilm basically upon the electrode called anode which are grown by oxidation of organic as well as inorganic matter, and therefore they transfer some portions of the produced electrons towards

TABLE 10.3
Commonly Used Techniques to Monitor the EABs

Tool/Technique	Possible Purpose	References
Fluorescence microscopy	EAB's visualization	Rowe et al. (2018); Bose et al. (2014)
Confocal microscopy	Architecture, dimensions of EAB's	Nevin et al. (2009); Strycharz et al. (2008)
Scanning electron microscopy (SEM)	Morphology and nanowire structures of EABs	Patil et al. (2013); Torres et al. (2009); Marsilli et al. (2008)
Atomic force microscopy	Measurement of individual EABs	El-Naggar et al. (2010)
Transmission electron microscopy (TEM)	Spatial arrangement and cellular ultrastructure of EABs	Gorby et al. (2006)
UV/Visspectroscopy	Measurement of absorption spectra/ redox mediators and determination of the spectroelectrochemical properties of cell membrane–cytochromes of EABs	Virdis et al. (2016); Nakamura et al. (2009)
HPLC	Qualitative and quantitative information of redox mediators associated with EABs	Kato et al. (2012); Virdis et al. (2016)
16S rRNA sequencing and phylogenetic analysis	Mixed microbial community composition and characterization associated with the EABs	Franks et al. (2009)
Cyclic voltammetry	EET mechanisms of EABs, rate-limiting steps prediction of EET mechanisms	Harnischand Freguia (2012); Marsili et al. (2008b); Fricke et al. (2008)
Fluorescence in situ hybridization (FISH)	EAB's microbial community analysis	Li et al. (2016)
Electrochemical impedance spectroscopy (EIS)	Capacitance, charge transfer, resistance, and diffusion resistance of EABs	Marsili et al. (2008b); terHeijne et al. (2018); Sanchez-Herrera et al. (2014); Seker and Ramasamy (2013)
Confocal resonance Raman microscopy (CRRM)	Presence and distribution of c-type cytochromes of EABs	Yates et al. (2016)
Surface-enhanced resonance Raman spectroscopy (SERRS)	Characterization of Outer Membrane Cytochromes (OMCs) in a catalytically active microbial biofilm	Millo et al. (2011)

Source: Adapted from Kiran, R., Patil, S.A., *Introduction to Biofilm Engineering*, American Chemical Society, Washington, DC,2019. With permission.

the electrode. The factors which affect such biofilm formation in BES are discussed as follows.

10.6.1 Inoculum Source

To grow the microbial population generally pure as well as mixed type of cultures can be used as anode of BES. Among both types of cultures pure cultures are mainly used for emphasizing on basic ideas of Microbial Electrochemical Technology (MET) along with metabolism-based study of a particular type of microorganism and active declining of few specific compounds like volatile organic carbons, phenol, and some pharmaceuticals as well. In case of mixed cultures which are basically more applicable for maximum industrial- as well as municipal-based applications though they don't need to be sterilized, may be implanted for processing of complex substrates and they are lower sensitivity to the changes regarding environmental situations. Even in many cases mixed cultures produce higher as well as much more stable current densities compared to the pure type of cultures (Kumar et al., 2017). Transfer of electrons to the outside of the cell, towards the anode is a characteristic feature of microbes in nature (Chabert et al., 2015). The basic anode inoculum is typically obtained from anthropogenic-based environments such as aerobic and anaerobic effluent, digestates, soil, and sediments (Miceli et al., 2012). Sometimes previously enriched anode inoculum could have been used. It is also seen that wastewater serves as a very good source of electrically active microbial communities which boosts the bioelectricity generation by enhancing the activation time of the anode electrodes (Madjarov et al., 2016). The presence of methanogenic inoculum in anaerobic sludge imparts an added advantage compared to the aerobic sludge by means of electron transfer initiation which leads to the current production. Microbial community from naturally occurring biofilms like those which are found in marine environments reorganizes an electroactive anodic biofilm faster compared to the plankton populations (Erable et al., 2009). It was seen that biofilm as well as effluents from previously functioned BES makes a faster activation for the presence of electrically activated microbes. The combination of inoculum from various sources may be an efficient strategy to extend the diversity of electrically active microbes. Sometimes bioaugmentation can be used in which pure cultures of already known electrically activated microbes are applied to the mixed cultures which reinforce to revive the system again if any discrepancies occurred in the system process. This method will lead to the highly stable with enhanced electro-producing activity and also the degradation of substrate as a result of interactions among the anode attached microbes and applied pure culture microbes.

10.6.2 Pretreatment of Inoculum

In BES, microbial communities like methanogens, homoacetogens, sulfite, or nitrite decomposers need to contest with electrically active microbes by means of electron transfer strategy for the oxidation of substrates as well as few other kinds of activities rather than the production of current, which basically leads to the declination of coulombic efficiency. So pretreatment of the microorganisms is an option to discourage the growth of unwanted microbes. Methanogenesis plays a major role in MET for development of biofilm by generating electron transferring shuttles (Rotaru et al., 2014). Though long-term or recapitulated treatment of microbes doesn't lead to the effectiveness in large-scale application context, an effective pretreatment of inoculum needs to be developed, which inhibits the methogenic communities permanently and makes no harm on electroactive microbes at the same time. But sometimes after pretreatment regular processing of wastewater and other aquatic bodies in unprotected condition can lead to the competition of microbes to anodes. If further presence of methanogen happened during continuous BES operation then some strategies might help like anode potential regulation, retention time shortening, long-time interaction to oxygen, and starvation. Now we will discuss about the various pretreatment strategies as follows:

10.6.3 Treatment by Heat

Treatment by heat basically depends on the particular ability of the phylum Firmicutes to produce endospores which are heat resistant in nature and enhance the survival rate compared with the microbes containing no endospore when the heat is applied to them beyond their limit of tolerance. Heat is broadly used in dark crude fermentative production of hydrogen to remove the methanogenic microbes which doesn't contain spore and be applied to the BES (Vamshi and Mohan, 2016). But there is a drawback of treatment by heat, i.e., some electrically active microbes like *Geobacter*and other nonspore forming microorganisms can be eliminated.

10.6.4 Pretreatment by Chemical Substances

There are some chemicals, e.g., 2-bromoethanesulfonic acid (BESA), amoxicillin-trihydrate, oxytetracycline, chloramphenicol, chlortetracycline, chloroethanesulfonate, thiamphenicol, lumazine, tetrahydrochloride, hypoxanthine,neomycin, and 2-iodopropane (Catal et al., 2015), which can selectively inhibit the growth of methanogenic microbes. Among these, BESA is widely used though it specifically inhibits the methanogenic microbes in very low concentration without any harm on the electrically active microorganisms. But the regular use of chemicals might have toxic impacts on the environment, so large-scale application in a regular manner might not be entertained. Still chemicals from biological sources can be an efficient alternative for the inhibition of methanogenic organisms.

10.6.5 Interaction with Oxygen

Some well-known electrically active microbes contain obligative anaerobic microbes (e.g., *Desulfovibrio* and *Pelobacter*), aerotolerant anaerobic microorganisms (e.g., *Geobacter*), and facultative anaerobic microorganisms (e.g., *Pseudomonas* and *Shewanella*). However, there might be some aerobic microbes which produce electricity. So basically anode chamber is kept in an anaerobic mode to inhibit the contact with oxygen and to prevent the loss of energy produced by aerobic chemical respiration as well. But strong anaerobic conditions enhance the activity of methanogenic microorganisms. In an analysis, it was found that the treatment of anode biofilm with the dissolved oxygen of concentration 5 mg/L with an effective time of 48–120 h will diminish methanogenic microbes and also influence the metabolism of facultative as well as microaerobic electrically active microorganisms (Ajayi et al., 2010). However, instead of the opposing effects of aerobically controlled metabolisms in BES, intermittent sparging of air may enhance coulombic efficiencies along with the efficiency of removal of COD (Chemical Oxygen Demand) as well with a less effect of the microorganism population.

10.6.6 Electroactive Microbial Community Enrichment

In a typical BES, the anode electrode requires electroactive microbial communities which could make the electron transfer mechanism from the anodic cell chamber towards the anode electrode. There are several studies done to establish the strategy to effectively enrich electrically active microbes during initial startup phase as well as to conduct the MET process. These studies have been done basically using domestic wastewater, aerobic sludge, anaerobic sludge, activated sludge, MFC effluent applied with various types of external resistance, and obtained power density ranging from 1.8 to 1,057 mW/m^2 depending upon various conditions (Zhang et al., 2011; Cho and Ellington, 2007).

10.7 ELECTROCHEMICAL ENRICHMENT

In BES, besides the types of anodic effluent (enrichment media), various other electrochemical factors will affect the power production output, which has been discussed in the below lines.

10.7.1 Imposed Potential of Anode Electrode

The potential of electrode basically enhances the composition structure of the population of microorganism or sometimes it can influence the cellular pathways for transferring electron. Though electrically active microbes obtained energy to sustain their optimum growth by transfer of electrons to the anode, the amount of obtained energy can be increased by enforcing the anode electrode by a higher potential compared with the potential of the donor of the electron (Zhu et al., 2013). Generally the highest limit of the membrane protein potential of maximum microbial strains is almost 0 V with respect to the standard hydrogen electrode, so it needs to be done higher than 0 V. The standard hydrogen electrode may not gain energy for the microbes. Now a strategy to fulfil the objective might have an option by imposing anode potentials in a negative manner so that electrically active microbes can have their respiration metabolism at lower potentials. It has been found that several microorganisms can increase their energy efficiency by applying various proteins situated at the outer part of the membrane of those microorganisms. In maximum cases, direct and mediated transfer of electron has the Nernst-Monod potential which provides half of the highest density of current close to −0.2 V with respect to the standard hydrogen electrode. However there are some exceptions like *Pseudomonas* sp., which contains the redox potential of −0.03 V; thus, it might be expected that it might grow on greater potentials of anode (Torres et al., 2009).

In transfer of electron by mediator, the anode potential is required along with the produced current build upon the potential of oxidation of the mediator. Electrically active microbes that use mediators are basically unable to produce higher densities of current due to limitations of diffusion of the mediators. Higher positive potentials of anode may be required to vanquish the losses of diffusion compared to microbes which perform direct transfer of electron. Regardless of the energy-based advantage of higher positive potentials of anode, there might be a deficiency of consensus wherever an anodic potential either in negative or positive mode should be chosen to backup the growth of an electrically active anode-based biofilm for optimizing power density. Basically an anode potential in a positive mode appears as beneficial in the preliminary steps of the anode-based formation biofilm to enhance the thickness which will lead to the resistance against the shear forces. Afterwards, a lower potential of anode may be applied to enhance the relative abundance of electrically active microbes. However, at the time of earlier phase, the requirement of energy for employing a potential at about extended periods of time may surpass the benefit achieved in output of power, and that's why this technique might be used exclusively for less periods of time.

10.7.2 External Resistance

In BES, if there is a stable cathodic state, the potential of anode can be regulated by employing an appropriate external resistor. Decreasing the resistance externally gives rise to an increase in anode potential which enhances the density of biomass of the biofilm. In some circumstances, a slack structure of biofilm along with a higher proportion of extracellular polymeric substances results in an enhanced amount of void spaces, which generally seems advantageous for the transport of proton and buffer but reduces the biofilm electrical conductivity (Zhang et al., 2011).

10.7.3 Enrichment Media

In an anodic chamber that is inoculated by mixed microorganism culture, it has been found that there is a competition for substrate among various types microbes. To prevent such competitions, pre-enrichment of electrically active microbes by nonfermentable substrates, e.g., acetate, may give rise to the exclusive enrichment of electrically active microbes in the biofilm on anode due to the absence of methanogens. But the growth media consisting of fermentable substrates like sucrose or glucose gives rise to various microorganism populations which also include electrically inactive microbes (Chae et al., 2009). The microorganism population structure basically depends upon the supplied electron donor, and there are several substrates which sustain the growth of electrically active microbes. Electrically active microbes need lower energy than other different groups of microbes and thus can grow in exhausted substrate conditions. Incorporation of simple composition substrates and acceptors of electron at anode may accelerate the establishment of a rigid electrically active microbial population which leads to the enhanced production of power.

10.8 CONDITIONS OF OPERATION

The configuration and function of the electrically active biofilm particularly count on the functional conditions of BES. The various conditions are discussed here as follows.

10.8.1 Temperature

In MET, temperature is a very important factor because it determines the shapes as well as the pathways of metabolism of the microorganism populations. Because of kinetic impediments, lower temperature enhances the startup time. Nevertheless, it was observed that the operating temperature does not impact the final power density, e.g., microbes developing a biofilm on anode at 30°C are also capable to acclimatize to lower temperatures (4°C–15°C) without reducing the power density (Michie et al., 2011). Generally, biofilms developed in lesser temperature are not so responsive to changes in temperature than biofilms developed in high temperature. This phenomenon is exclusively important for wastewater treatment at diverse seasonal temperatures, which need to be ideally started up when there is a lower temperature, though it may result in an extended time for start-up. However, the ideal temperature should always be resolved for the exclusive MET application by remembering the factors, i.e., electrically active microorganism population as well as economical expediency.

10.8.2 pH

The formation of electrically active biofilm is largely influenced by pH, which controls over pathways of metabolism of electrically active microbes. However, production of bioelectricity has been found at pH less than 3 as well as pH greater than 12, but neutral or slightly alkaline pH is generally needed for the formation of an electrically active biofilm from various mixed microorganism population, specially for substrates in complex organic nature because hydrolysis can exclusively proceed at neutral pH (Yuan et al., 2011).

10.9 DESIGN OF BES

The electrically active biofilm basically arises from the primary attachment of microbes to electrode. The material of electrode and the hydro dynamically driven forces regulate the adhesion, detachment, composition, and thickness of the biofilm, which influences the conductivity of the biofilm as well as transfer rate of electron. Now these responsible factors are discussed as follows.

10.9.1 Hydrodynamic Forces

Batch mode operated BES without recirculation may cooperate the growth of suspended microbes which perform mediated transfer of electron; however, continuous mode operated BES with recirculation helps the attachment of electrically active microbes to the electrode. Mixing in a mechanical manner also enhances the hydrodynamic forces in anode, decreasing resistance to transfer of mass, uprising diffusion, and blocking localized gradients of pH in the EAB. Anolyte recirculation at various rates or repeated bubbling with nitrogen can be utilized to enhance the hydrodynamic forces in the anode. But excessive forces cause detachment of biofilm, specifically at the preliminary stage of establishment of biofilm. To solve this problem a strategy has been done by implementing the anode at the bottom level of the reactor so that microorganisms can settle by gravitational activity, obtaining a density of current of 1.86 A/m^2in 1.64 days as contrasted to0.5 A/m^2acquired in an MFC in horizontal mode in 1.87 days (Li et al., 2017).

10.9.2 Modification of Electrode Surface

Structural material of anode and the structure of surface typically affect the attachment of microorganisms and the electroactive conjunctions between microbes and the surface of the electrode. The surface can be modified to enhance microbial adhesion as well as the transfer of electron. Modification procedures basically include treatment of surface chemically or physically, imposition of electroactive or highly conductive coatings, and utilization of composite electrodes of graphite metal (Wei et al., 2011). Preliminary attachment of suspended microorganisms on electrode may be increased by treating conductive electrode surfaces by exclusive chemicals. As we know, most of the known electrically active microbes are charged negatively, so we can add positive charged oxides of metal on electrode to enhance the adhesion. Carbon electrode can be treated by heat to enhance the area of the electrode surface for microorganism adhesion. Treatment by acid not only enhances the surface area of the electrode but also increases the protonation of functional groups, which enhances the total positive charge. So, attachment of microbes to electrodes may be improved by enhancing the area of the surface, roughness, and also modifying the charge of the surface to the positive direction. Nevertheless, to target exclusively the adhesion of electrically active microbes, the modification of electrode needs to be associated with the inhibition of the competitors.

10.9.3 Membranes

Usually in BES, anode and cathode are separated by a membrane. It has been found that the used membrane type may impact on the exuberance as well as the phylogenetically modified distribution of electrically active microbes. In several studies, it was reported that anion exchange membranes generally make 2–5 times more production of power than cation exchange membranes. Transfer of other cations than protons by cation exchange membranes may negatively affect the migration of proton. As a consequence, a gradient of pH forms across membrane, repressing the formation of biofilm which leads to enhancement in start-up time (Bakonyi et al., 2018).

10.10 EABs IN WASTEWATER TREATMENT AND BIOELECTRICITY GENERATION

A huge potential of specified energy is preserved in wastewaters as a rich source of organic matter which is biodegradable by electroactive microorganisms which form the biofilm on the electrode. It might be known to all that the energy retained in wastewater cannot be associated directly with the values of COD. But in MET, the potential for treatment of wastewater can be evidenced by their removal rates of COD, which can achieve up to 90% with coulombic efficiencies higher than 80% (Rahimnejad et al., 2015). It has been calculated that in anaerobic-based digestion 1 kg of COD may be converted to 4.16 kWh of power, so basically if MET is implemented to be efficient as a technological approach for wastewater treatment with continuous generation of energy, it should reach a similar rate of conversion of substrate (Arends and Verstraete, 2012). Analysts are responding to enhanced interest in the BES application of treatment of wastewater to remove the oxidizable toxic matter from the industrial effluents as well as wastewaters from domestic sector. Under the influence of various operating factors as well as the types of BES used, COD removal from 60% to 90% has been reported. When treating wastewater consists of peptone, starch, and fish extract in the single chamber BES in air cathode, COD removal showed a 93%–95% efficiency. With such positive attributes, after exactly implementing scale-up factors, BES might be depicted as a major technology for treatment of wastewater from any source. BES in various modes can be implemented for wastewater treatment and produces energy in the form of bioelectricity.

10.11 CONCLUSION

The manipulation of metabolism of electroactive microorganisms to effectuate or regulate electrochemically driven reactions, which generally happen in the environment or anthropogenically, must lead to huge technology advancements. This will take the lead for the advancement of latest products and procedures in several fields like bioremediation, bioenergy, biosynthesis, biofouling prevention processes, and biosensor design biocorrosion mitigation. A multidimensional approach and comprehensive researches are in promotion, discovering the potentials of applications of EAB and refinement in the design of process and configuration to fit the expected applications. Execution of such BES is not so simple as it needs to address the different undeniable technological, microbiological, and economic challenging factors. Substitute of the membrane with different cheaper material or exclusive design of reactor may enhance the economic attainability of the bioelectrochemical technology processes. Regulating the anode electrode potential on the starting day after development of EABs confines the microorganism population competition on the electrode and enhances electroactive microbial growth. This basically supports in extending highest power generation from the microbially covered electrode without vanquishing the substrate for metabolisms in the competitive manner, e.g., methanogenesis. More experiments and analysis with "true" wastewaters but not "synthetic" wastewaters also need to be evaluated to determine the exact potential of the unique technology. MET holds brilliant promise towards the energy in sustainable approach through hydrogen production (bioelectrolyzer) or direct generation (MFC). The implementation of such novel process of technology as a renewable source of energy will support in lessening the warning to the mankind and fabricating the earth a much better place to live. Also, establishment of this novel technology-based biosensors will enhance faster estimation of several parameters and develop real-time monitoring by online mode, and it is possible practically. Although a huge number of possible implementations of EAB are described in the literature, more unique applications can also be achieved in the near future.

REFERENCES

Ajayi FF, Kim KY, Chae KJ, Choi MJ, Kim IS (2010) Effect of hydrodynamic force and prolonged oxygen exposure on the performance of anodic biofilm in microbial electrolysis cells. *Int J Hydrogen Energy* 35: 3206–3213.

Alhede M, Qvortrup K, Liebrechts R, Høiby N, Givskov M, Bjarnsholt T (2012) Combination of microscopic techniques reveals a comprehensive visual impression of biofilm structure and composition, *FEMS Immunol Med Microbiol* 65: 335–342. doi: 10.1111/j.1574-695X.2012.00956.x.

Aracic S, Semenec L, Franks AE (2014) Investigating microbial activities of electrode-associated microorganisms in real-time. *Front Microbiol* 5: 663.

Arends JBA, Verstraete W (2012) 100 years of microbial electricity production: Three concepts for the future. *Microb Biotechnol* 5: 333–346.

Aulenta F, Catapano L, Snip L, Villano M, Majone M (2012) Linking bacterial metabolism to graphite cathodes: Electrochemical insights into the H_2–producing capability of *Desulfovibrio sp. Chemsuschem* 5: 1080–1085.

Bakonyi P, Koók L, Kumar G, Tóth G, Rózsenberszki T, Nguyen DD, Chang SW, Zhen G, Bélafi-Bakó K, Nemestóthy N (2018) Architectural engineering of bioelectrochemical systems from the perspective of polymeric membrane separators: A comprehensive update on recent progress and future prospects. *J Membr Sci* 564: 508–522.

Baranitharan E, Khan MR, Prasad DMR, Teo WF, Tan GY, Jose R (2015) Effect of biofilm formation on the performance of microbial fuel cell for the treatment of palm oil mill effluent. *Bioproc Biosyst Eng* 38: 15–24.

Bogino PC, Oliva MD, Sorroche FG, Giordano W (2013) The role of bacterial biofilms and surface components in plant-bacterial associations. *Int J Mol Sci* 14: 15838–15859.

Bonanni PS, Schrott GD, Robuschi L, Busalmen JP (2012) Charge accumulation and electron transfer kinetics in *Geobacter sulfurreducens* biofilms. *Energy Environ Sci* 5: 6188–6195.

Borole AP, Reguera G, Ringeisen B, Wang ZW, Feng Y, Kim BH (2012) Electroactive biofilms: Current status and future research needs. *Energy Environ Sci* 5: 9945–9945.

Bose A, Gardel EJ, Vidoudez C, Parra EA, Girguis PR (2014) Electron uptake by iron-oxidizing phototrophic bacteria. *Nat Commun* 5: 3391.

Catal T, Lesnik KL, Liu H (2015) Suppression of methanogenesis for hydrogen production in single-chamber microbial electrolysis cells using various antibiotics. *Bioresour Technol* 187: 77–83.

Chabert N, Amin Ali O, Achouak W (2015) All ecosystems potentially host electrogenic bacteria. *Bioelectrochemistry* 106: 88–96.

Chae K-J, Choi M-J, Lee J-W, Kim K-Y, Kim IS (2009) Effect of different substrates on the performance, bacterial diversity, and bacterial viability in microbial fuel cells. *Bioresour Technol* 100: 3518–3525.

Cho EJ, Ellington AD (2007) Optimization of the biological component of a bioelectrochemical cell. *Bioelectrochemistry* 70: 165–172.

Dietrich LEP, Okegbe C, Price-Whelan A, Sakhtah H, Hunter RC, Newman DK (2013) Bacterial community morphogenesis is intimately linked to the intracellular redox state. *J Bacteriol* 195: 1371–1380.

El-Naggar MY, Wanger G, Man K, Yuzvinsky TD, Southam G, Yang J, Lau WM, Nealson KH, Gorby YA (2010) Electrical transport along bacterial nanowires from *Shewanella oneidensis* MR-1. *PNAS* 107(42): 18127–18131.

Erable B, Roncato M-A, Achouak W, Bergel A (2009) Sampling natural biofilms: A new route to build efficient microbial anodes. *Environ Sci Technol* 43: 3194–3199.

Fazli M, Almblad H, Rybtke ML, Givskov M, Eberl L, Tolker-Nielsen T (2014). Regulation of biofilm formation in *Pseudomonas* and *Burkholderia* species. *Environ Microbio* 16: 1961–1981.

Fitzgerald L, Petersen ER, Ray RI, Little BJ, Cooper CJ., Howard EC, Biffinger JC (2012) *Shewanella oneidensis* MR-1 Mshpilin proteins are involved in extracellular electron transfer in microbial fuel cells. *Process Biochem* 47: 170–174.

Franks AE, Nevin KP, Glaven RH, Lovley DR (2009) Microtomingcoupled to microarray analysis to evaluate the spatial metabolic status of *Geobacter sulfurreducens* biofilms. *ISME J* 4(4): 509–519.

Fricke K, Harnisch F, Schroder U (2008) On the use of cyclic voltammetry for the study of anodic electron transfer in microbial fuel cells. *Energy Environ Sci* 1(1): 144–147.

Fu Q, Kobayashi H, Kuramochi Y, Xu J (2013) Bioelectrochemical analyses of a thermophilic biocathode catalyzing sustainable hydrogen production. *Int J Hydrogen Energy* 38: 15638–15645.

Giltner CL, van Schaik EJ, Audette GF, Kao D, Hodges RS, Hassett DJ, Irvin RT (2006) The *Pseudomonas aeruginosa* type IV pilin receptor binding domain functions as an adhesin for both biotic and abiotic surfaces. *Mol Microbiol* 59: 1083–1096.

Giltner CL, Nguyen Y, Burrows LL (2012) Type IV pilin proteins: Versatile molecular modules. *Microbiol Mol Biol Rev* 76: 740–772.

Gorby YA, Yanina S, Mclean JS, Rosso KM, Moyles D, Dohnalkova A, Beveridge TJ, Chang IS, Kim BH, Kim KS, Culley DE (2006) Electrically conductive bacterial nanowires produced by *Shewanella oneidensis* strain MR-1 and other microorganisms. *PNAS* 30: 11358–11363.

Harnisch F, Freguia S (2012) A basic tutorial on cyclic voltammetry for the investigation of electroactive microbial biofilms. *Chem Asian J* 7(3): 466–475.

Huang LP, Regan JM, Quan X (2011) Electron transfer mechanisms, new applications, and performance of biocathode microbial fuel cells. *Bioresource Technol* 102: 316–323.

Jafary T, Daud WRW, Ghasemi M, Jahim JM, Ismail M, Lim SS (2015) Biocathode in microbial electrolysis cell: present status and future prospects. *Renew Sust Energ Rev* 47: 23–33.

Jain A, Connolly J (2013) Extracellular electron transfer mechanism in *Shewanella loihica* PV-4 biofilms formed at indium tin oxide and graphite electrodes. *Int J Electrochem Sci* 8: 1778–1793.

Jain A, Gazzola G, Panzera A, Zanoni M, Marsili E (2011) Visible spectroelectrochemical characterization of *Geobacter sulfurreducens* biofilms on optically transparent indium tin oxide electrode. *Electrochim Acta* 56: 10776–10785.

Kalathil S, Pant D (2016) Nanotechnology to rescue the bacterial bidirectional extracellular electron transfer in bioelectrochemical systems. *RSC Adv* 6: 30582–30597.

Kang CS, Eaktasang N, Kwon DY, Kim HS (2014) Enhanced current production by *Desulfovibrio desulfuricans* biofilm in a mediator-less microbial fuel cell. *Bioresour Technol* 165: 27–30.

Kato S, Hashimoto K, Watanabe K (2012) Microbial interspecies electron transfer via electric currents through conductive minerals. *Proc Natl Acad Sci USA* 109 (25): 10042–10046.

Keller KL, Rapp-Giles BJ, Semkiw ES, Porat I, Brown SD, Wall JD (2014) New model for electron flow for sulfate reduction in *Desulfovibrio alaskensis* G20. *Appl Environ Microbiol* 80: 855–868.

Kiely PD, Rader G, Regan JM, Logan BE (2011) Long-term cathode performance and the microbial communities that develop in microbial fuel cells fed different fermentation end products. *Bioresour Technol* 102: 361–366.

Kiran R, Patil SA (2019) Microbial electroactive biofilms. In: *Introduction to Biofilm Engineering* (Rathinam and Sani (Eds.)), ACS Symposium Series; American Chemical Society: Washington, DC: pp.159–186.

Kotloski NJ, Gralnick JA (2013) Flavin electron shuttles dominate extracellular electron transfer by *Shewanella oneidensis*. *mBio* 4: e00553–e00612.

Kumar G, Bakonyi P, Zhen G, Sivagurunathan P, Koók L, Kim SH, Tóth G, Nemestóthy N, Bélafi-Bakó K (2017) Microbial electrochemical systems for sustainable biohydrogen production: Surveying the experiences from a start-up viewpoint. *Renew Sustain Energy Rev* 70: 589–597.

Leang C, Qian X, Mester T, Lovley DR (2010) Alignment of the c-type cytochrome OmcS along pili of *Geobacter sulfurreducens*. *Appl Environ Microbiol* 76: 4080–4084.

Lebedev N, Strycharz-Glaven SM, Tender LM (2014) Spatially resolved confocal resonant Raman microscopic analysis of anode-grown *Geobacter sulfurreducens* biofilms, *Chem Phys Chem* 15: 320–327.

Li C, Lesnik KL, Fan Y, Liu H (2016) Redox conductivity of current-producing mixed species biofilms. *PLoS One* 11(5): e0155247.

Li T, Zhou L, Qian Y, Wan L, Du Q, Li N, Wang X (2017) Gravity settling of planktonic bacteria to anodes enhances current production of microbial fuel cells. *Appl Energy* 198: 261–266.

Lifang D, Frang BL, Shungui Z, Yin HD, Jinren NI (2010). A study of electronshuttle mechanism in *Klebsiella pneumoniae* based microbial fuel cells. *Environ Sci Technol* 55: 99–104.

Liu X, Tremblay PL, Malvankar NS, Nevin KP, Lovley DR, Vargas M (2014) A *Geobacter sulfurreducens* strain expressing *Pseudomonas aeruginosa* type IV pili localizes OmcS on pili but is deficient in Fe(III) oxide reduction and current production. *Appl Environ Microbiol* 80: 1219–1224.

Lovley DR (2012) Long-range electron transport to Fe(III) oxide via pili with metallic-like conductivity. *Biochem Soc Trans* 40: 1186–1190.

Lovley DR, Nevin KP (2013) Electrobiocommodities: Powering microbial production of fuels and commodity chemicals from carbon dioxide with electricity. *Curr Opin Biote chnol* 24: 385–390.

Madjarov J, Prokhorova A, Messinger T, Gescher J, Kerzenmacher S (2016) The performance of microbial anodes in municipal wastewater: Pre-grown multispecies biofilm vs. natural inocula. *Bioresour Technol* 221: 165–71.

Malvankar NS, Lovley DR (2012) Microbial nanowires: A new paradigm for biological electron transfer and bioelectronics. *Chemsuschem* 5: 1039–1046.

Malvankar NS, Lovley DR (2014) Microbial nanowires for bioenergy applications. *Curr Opin Biotechnol* 27: 88–95.

Malvankar NS, Tuominen MT, Lovley DR (2012) Lack of cytochrome involvement in long-range electron transport through conductive biofilms and nanowires of *Geobacter sulfurreducens*. *Energy Environ Sci* 5: 8651–8659.

Malvankar NS, Vargas M, Nevin K, Tremblay PL, Evans-Lutterodt K, Nykypanchuk D, Martz E, Tuominen MT, Lovley DR (2015) Structural basis for metallic-like conductivity in microbial nanowires. *mBio* 6: e00084.

Mangold S, Harneit K, Rohwerder T, Claus G, Sand W (2008) Novel combination of atomic force microscopy and epifluorescence microscopy for visualization of leaching bacteria on pyrite. *Appl Environ Microbiol* 74: 410–415,

Marsili E, Baron DB, Shikhare ID, Coursolle D, Gralnick JA, Bond DR (2008a) Shewanella secretes flavins that mediate extracellular electron transfer. *Proc Natl Acad Sci USA* 105: 3968–3973.

Marsili E, Rollefson JB, Baron DB, Hozalski RM, Bond DR (2008b) Microbial biofilm voltammetry: Direct electrochemical characterization of catalytic electrode-attached biofilms. *Appl Environ Microbiol* 74 (23): 7329.

Mehta T, Coppi MV, Childers SE, Lovley DR (2005) Outer membrane C-type cytochromes required for Fe(III) and Mn(IV) oxide reduction in *Geobacter sulfurreducens*. *Appl Environ Microbiol* 71: 8634–8641.

Miceli JF, Parameswaran P, Kang DW, KrajmalnikBrown R, Torres CI (2012) Enrichment and analysis of anode-respiring bacteria from diverse anaerobic inocula. *Environ Sci Technol* 46: 10349–10355.

Michiel S, Kim JR, Dinsdale RM, Guwy AJ, Premier GC (2011) Operational temperature regulates anodic biofilm growth and the development of electrogenic activity. *Appl Microbiol Biotechnol* 92: 419–30.

Millo D, Harnisch F, Patil SA, Ly HK, Schroder U, Hildebrandt P (2011) In situ spectroelectrochemical investigation of electrocatalytic microbial biofilms by surface-enhanced resonance Raman spectroscopy. *Angew Chem Int Ed* 50 (11): 2625–2627.

Milner E, Scott K, Head I, Curtis T, Yu E (2014) Electrochemical investigation of aerobic biocathodes at different poised potentials: Evidence for mediated extracellular electron transfer. *10th ESEE: European Symposium on Electrochemical Engineering*. Sardinia, Italy. 41: pp. 355–360.

Morita M, Malvankar NS, Franks AE, Summers ZM, Giloteaux L, Rotaru AE, Rotaru C, Lovley DR (2011) Potential for direct interspecies electron transfer in methanogenic wastewater digester aggregates. *mBio* 2(4): e00159-11.

Nakamura R, Ishii K, Hashimoto K (2009) Electronic absorption spectra and redox properties of C type cytochromes in living microbes. *Angew Chem Int Ed* 48(9): 1606–1608.

Neu TR, Lawrence JR (2014) Investigation of microbial biofilm structure by laser scanning microscopy. *Adv Biochem Eng Biotechnol* 146: 1–51.

Nevin KP, Kim B, Glaven RH, Johnson JP, Woodard TL, Methe BA, Didonato RJ, Covalla SF, Franks AE, Liu A, Lovley DR (2009) Anode biofilm transcriptomics reveals outer surface components essential for high density current production in *Geobacter sulfurreducens*fuel cells. *PLOS One* 4(5): e5628.

Nevin KP, Hensley SA, Franks AE, Summers ZM, Ou J, Woodard TL, Snoeyenbos-West OL, Lovley DR (2011) Electrosynthesis of organic compounds from carbon dioxide is catalyzed by a diversity of acetogenic microorganisms. *Appl Environ Microbiol* 77: 2882–2886.

Pandit S, Khilari S, Roy S, Pradhan D, Das D (2014) Improvement of power generation using *Shewanella putrefaciens* mediated bioanode in a single chambered microbial fuel cell: Effect of different anodic operating conditions. *Bioresour Technol* 166: 451–457.

Patil SA, Gorecki K, Hagerhall C, Gorton L (2013) Cisplatin-induced elongation of *Shewanella oneidensis* MR-1 cells improves microbe-electrode interactions for use in microbial fuel cells. *Energy Environ Sci*6: 2626–2630.

Pierson LS 3rd, Pierson EA (2010) Metabolism and function of phenazines in bacteria: Impacts on the behavior of bacteria in the environment and biotechnological processes. *Appl Microbiol Biotechnol* 86: 1659–1670.

Rahimnejad M, Adhami A, Darvari S, Zirepour A, Oh SE (2015) Microbial fuel cell as new technology for bioelectricity generation: A review. *Alexandria Eng J* 54: 745–756.

Reguera G, McCarthy KD, Mehta T, Nicoll JS, Tuominen MT, Lovley DR (2005) Extracellular electron transfer via microbial nanowires. *Nature* 435: 1098–1101.

Reguera G, Nevin KP, Nicoll JS, Covalla SF, Woodard TL, Lovley DR (2006) Biofilm and nanowire production leads to increased current in *Geobacter sulfurreducens* fuel cells. *Appl Environ Microbiol* 72: 7345–7348.

Richter H, Nevin H, Lowy DA, Lovley DR, Tender LM (2009) Cyclic voltammetry of biofilms of wild type and mutant *Geobacter sulfurreducens* on fuel cell anodes indicates possible roles of OmcB, OmcZ, type IV pili, and protons in extracellular electron transfer. *Energy Environ Sci* 2: 506, doi: 10.1039/b816647a.

Richter K, Schicklberger M, Gescher J (2012) Dissimilatory reduction of extracellular electron acceptors in anaerobic respiration. *Appl Environ Microbiol* 78: 913–921.

Rosenbaum M, Aulenta F, Villano M, Angenent LT (2011) Cathodes as electron donors for microbial metabolism: Which extracellular electron transfer mechanisms are involved? *Bioresource Technol* 102: 324–333.

Rotaru AE, Shrestha PM, Liu F, Shrestha M, Shrestha D, Embree M, Zengler K, Wardman C, Nevin KP, Lovley DR (2014) A new model for electron flow during anaerobic digestion: Direct interspecies electron transfer to *Methanosaeta* for the reduction of carbon dioxide to methane. *Energy Environ Sci* 7: 408–15.

Rowe AR, Rajeev P, Jain A, Pirbadian S, Okamoto A, Gralnick JA, El-Naggar MY, Nealson KH (2018) Tracking electron uptake from a cathode into *Shewanella* cells: Implications for energy acquisition from solid-substrate. *mBio* 9(1): e02203–e02217.

Rozendal RA, Jeremiasse AW, Hamelers HV, Buisman CJ (2008) Hydrogen production with a microbial biocathode. *Environ Sci Technol* 42: 629–634.

Sanchez-Herrera D, Pacheco-Catalan D, Valdez-Ojeda R, Canto-Canche, B, Dominguez-Benetton X, Domínguez-Maldonado J, Alzate-Gaviria L (2014) Characterization of anode and anolyte community growth and the impact of impedance in a microbial fuel cell. *BMC Biotechnol* 14(1): 1–10.

Saratale GD, Saratale RG, Shahid MK, Zhen G, Kumar G, Shin HS, Choi YG, Kim SH (2017) A comprehensive overview on electro-active biofilms, role of exoelectrogens and their microbial niches in microbial fuel cells (MFCs). *Chemosphere* 178: 534–547.

Seker N, Ramasamy RP (2013) Electrochemical impedance spectroscopy for microbial fuel cell characterization. *J Microb Biochem Technol* 5(S6): S6–004.

Semenec L, Franks AE (2014) The microbiology of microbial electrolysis cells. *Microbiol Aust* 35: 201–206.

Semenec L, Franks AE (2015) Delving through electrogenic biofilms: From anodes to cathodes to microbes. *AIMS Bioengineering* 2(3): 222–248.

Shen HB, Yong XY, Chen YL, Liao ZH, Si RW, Zhou J, Wang SY, Yong YC, Ou Yang PK, Zheng T (2014) Enhanced bioelectricity generation by improving pyocyanin production and membrane permeability through sophorolipid addition in *Pseudomonas aeruginosa*-inoculated microbial fuel cells. *Bioresource Technol* 167: 490–494.

Shi L, Richardson DJ, Wang Z, Kerisit SN, Rosso KM, Zachara JM, Fredrickson JK (2009) The roles of outer membrane cytochromes of Shewanella and Geobacter in extracellular electron transfer. *Environ Microbiol Rep* 1: 220–227.

Smith J, Lovley DR, Tremblay PL (2013) Outer cell surface components essential for Fe(III) oxide reduction by *Geobacter metallireducens. Appl Environ Microbiol* 79: 901–907.

Smith JA, Tremblay PL, Shrestha PM, Snoeyenbos-West OL, Franks AE, Nevin KP, Lovley DR (2014) Going wireless: Fe(III) oxide reduction without pili by *Geobacter sulfurreducens* strain JS-1. *Appl Environ Microbiol* 80: 4331–4340.

Stöckl M, Schlegel C, Sydow A, Holtmann D, Ulber R, Mangold KM (2016) Membrane separated flow cell for parallelized electrochemical impedance spectroscopy and confocal laser scanning microscopy to characterize electro-active microorganisms. *Electrochimica Acta* 220: 444–452.

Strycharz SM, Woodard TL, Johnson JP, Nevin KP, Sanford RA, Lo FE, Lovley DR (2008)Graphite electrode as a sole electron donor for reductive dechlorination of tetrachlorethene by *Geobacter lovleyi. Appl Environ Microbiol* 74 (19): 5943–5947.

Strycharz SM, Glaven RH, Coppi MV, Gannon SM, Perpetua LA, Liu A, Nevin KP, Lovley DR (2011) Gene expression and deletion analysis of mechanisms for electron transfer from electrodes to *Geobacter sulfurreducens. Bio electrochemistry* 80: 142–150.

Strycharz-Glaven SM, Glaven RH, Wang Z, Zhou J, Vora GJ, Tender LM (2013) Electrochemical investigation of a microbial solar cell reveals a nonphotosynthetic biocathode catalyst. *Appl Environ Microbiol* 79: 3933–3942.

ter Heijne A, Liu D, Sulonen M, Sleutels T, Fabregat-Santiago F (2018) Quantification of bio-anode capacitance in bioelectrochemical systems using electrochemical impedance spectroscopy. *J Power Sources* 400: 533–538.

Torres CI, Krajmalnik-Brown R, Parameswaran P, Marcus AK, Wanger G, Gorby YA, Rittmann BE (2009) Selecting anode-respiring bacteria based on anode potential: Phylogenetic, electrochemical, and microscopic characterization. *Environ Sci Technol* 43(24): 9519–9524.

Tremblay PL, Aklujkar M, Leang C, Lovley DR (2011) A genetic system for *Geobacter metallireducens*: Role of flagella and pili in extracellular electron transfer. *Environ Microbiol Rep* 4: 82–88.

Vamshi KK, Mohan SV (2016) Selective enrichment of electrogenic bacteria for fuel cell application: Enumerating microbial dynamics using MiSeq platform. *Bioresour Technol* 213: 146–154.

Vilar-Sanz A, Puig S, Garcia-Lledo A, Trias R, Balaguer MD, Colprim J, Baneras L (2013) Denitrifying bacterial communities affect current production and nitrous oxide accumulation in a microbial fuel cell. *PLoS One* 8: e63460.

Virdis B, Harnisch F, Batstone DJ, Rabaey K, Donose BC (2012) Non-invasive characterization of electrochemically active microbial biofilms using confocal Raman microscopy. *Energy Environ Sci* 5 (10): 7017.

Virdis B, Millo D, Donose BC, Lu Y, Batstone DJ, Krömer JO (2016) Analysis of electron transfer dynamics in mixed community electroactive microbial biofilms. *RSC Adv* 6(5): 3650–3660.

Wang H, Ren ZJ (2013) A comprehensive review of microbial electrochemical systems as aplatform technology. *Biotechnol Adv* 31: 1796–1807.

Wei J, Liang P, Huang X (2011) Recent progress in electrodes for microbial fuel cells. *Bioresour Technol* 102: 9335–9344.

Wrighton KC, Virdis B, Clauwaert P, Read ST, Daly RA, Boon N, Piceno Y, Andersen GL, Coates JD, Rabaey K (2010) Bacterial community structure corresponds to performance during cathodic nitrate reduction. *ISME J* 4: 1443–1455.

Wrighton KC, Thrash JC, Melnyk RA, Bigi JP, Byrne-Bailey KG, Remis JP, Schichnes D, Auer M, Chang CJ, Coates JD (2011) Evidence for direct electron transfer by a gram-positive bacterium isolated from a microbial fuel cell. *Appl Environ Microbiol* 77: 7633–7639.

Yates MD, Eddie BJ, Kotloski NJ, Lebedev N (2016) Toward understanding long-distance extracellular electron transport in an electroautotrophic microbial community. *Energy Environ Sci* 9 (11): 3544–3558.

Yuan Y, Zhao B, Zhou S, Zhong S, Zhuang L (2011) Electrocatalytic activity of anodic biofilm responses to pH changes in microbial fuel cells. *Bioresour Technol* 102: 6887–6891.

Yuan Y, Zhou S, Tang J (2013) In situ investigation of cathode and local biofilm microenvironments reveals important roles of OH^- and oxygen transport in microbial fuel cells. *Environ Sci Technol* 47: 4911–4917.

Zhang L, Zhu X, Li J, Liao Q, Ye D (2011) Biofilm formation and electricity generation of a microbial fuel cell started up under different external resistances. *J Power Sources* 196: 6029–6035.

Zhu X, Tokash JC, Hong Y, Logan BE (2013) Controlling the occurrence of power overshoot by adapting microbial fuel cells to high anode potentials. *Bioelectrochemistry* 90: 30–35.

11 Bioremediation of Agroindustrial Wastewater by Cultivation of *Spirulina* sp. and Biomass Used as Animal Feed Supplement

M. Karthik
Tamil Nadu Agricultural University

K. Ashokkumar
Kerala Agricultural University

N. Arunkumar
Central University of Tamil Nadu

R. Krishnamoorthy
Vanavarayar Institute of Agriculture

P. Arjun
PRIST Deemed University

CONTENTS

11.1 INTRODUCTION

Agro-industries generate a huge amount of waste and wastewater during processing, and the wastewater contains high amounts of organic and inorganic pollutants and heavy metals. The strengthening of agroindustrial production and release of waste and wastewater generated to lands have raised several environmental issues, including eutrophication, surface and ground water pollution, odor pollution, and gas emissions. The source of organic waste is highly collected from animal manure, but there are considerable levels of organic waste generated from agro industries, like sago, sugar, palm oil, olive oil, piggery, poultry, dairy, fish farming, and rice mill wastewater (Bernet and Béline 2009). Agro-industries released greater levels of organic load along with the wastewater when stored, resulting in obnoxious odors, irritating color, lower pH, and higher biological oxygen demand (BOD) and chemical oxygen demand (COD). Physicochemical characteristics of agroindustrial wastewater are shown in Table 11.1.

Today, various physical, chemical, and biological approaches are available to remove pollutants from the agroindustry wastewaters; nevertheless, the most commonly used methods are aerobic and anaerobic digestion (Yaakob et al. 2014). However, these methods achieved only in secondary treatment, and they remove organic pollutants and have almost minimal effect on the control of inorganic pollutants (Westerman and Bicudo 2005). Also, removal of inorganic pollutant requires costliest physiochemical methods (Benemann 1979). However, in biological methods, fungi and bacteria used to eliminate the contaminants by absorbing them, and thus they have long been a stronghold of wastewater treatment in several industries. Since they are efficient and extensively used only for wastewater treatment, various biological treatments are available today, in which, cyanobacteria used wastewater as a substrate for their growth and removal of organic pollutants at a lower cost. Biological treatment is an efficient and sustainable alternative to conventional chemical methods of wastewater treatment and recycling.

In wastewater treatment, microalgae were first used by Oswald and Gotaas (1957). Blue green algae (BGA) are dominated and widely distributed microorganisms that inhabit the different aquatic polluted environment (Gibson and Smith 1982). The benefit of *Spirulina* cultivation is to eliminate the pollutants from agroindustry

TABLE 11.1
Physicochemical Properties of Agroindustrial Wastewater

S. No	Agroindustrial Wastewater	pH	BOD	COD	TS	N	P	Na	Ca	NO_3	References
			mg/L								
1	Beet sugar factory—Vinasse wastewater	6.0	—	—	—	24,000	600	26,000	6,000	—	Coca et al. (2015)
2	Brine (pickle) factory wastewater	4.8	10,800	—	—	200	1,600	64,000	—	—	Duangsri and Satirapipathkul (2011)
3	Coconut milk skim wastewater	7.5	983.33	4,917	—	104.16	0.70	—	—	—	Azimatun Nur et al. (2015)
4	Dairy wastewater	6.3	—	1,300	2,780	—	8.2	141	218	89	Suad and Ahmed (2014)
5	Fish farming wastewater	8.2	—	—	—	—	0.96	—	—	28.33	Nogueira et al. (2018)
6	Olive oil mill wastewater	5.4	—	56,740	43,200	2,900	350	—	—	100	Markou et al. (2012)
7	Palm oil mill wastewater	7.8	1,490	2,830	—	—	—	—	—	—	Parkavi et al. (2011)
8	Paper mill effluent	7.3	83.5	107	20.6	—	—	—	—	—	Setiawan et al. (2019)
9	Prawn hatchery wastewater	8.6	—	—	—	—	2.3	—	360	8.7	Sandeep et al. (2015)
10	Rice mill wastewater	—	—	—	1,459	800	338	263	98.3	—	Usharani et al. (2014)
11	Rice noodle factory wastewater	3.5	2,550	4,400	5.5	—	—	181.5	84	3.49	Vetayasuporn (2004)
12	Sago industry wastewater	7.0	3,592	7,726	2,781	17.18	7.2	42.35	—	—	Karthik and Kumar (2016)
13	Sugar mill wastewater	7.1	816.67	260	—	—	—	—	—	—	Deshmane et al. (2015)
14	Piggery wastewater	7.9	30	155	1,280	150	88	—	—	19.9	Chaiklahan et al. (2010)

Note: mg/L, Milligram per liter; BOD, Biological Oxygen Demand; COD, Chemical Oxygen Demand; TS, Total Solids; N, Nitrogen; P, Phosphorous; Na, Sodium; Ca, Calcium; NO_3, Nitrate.

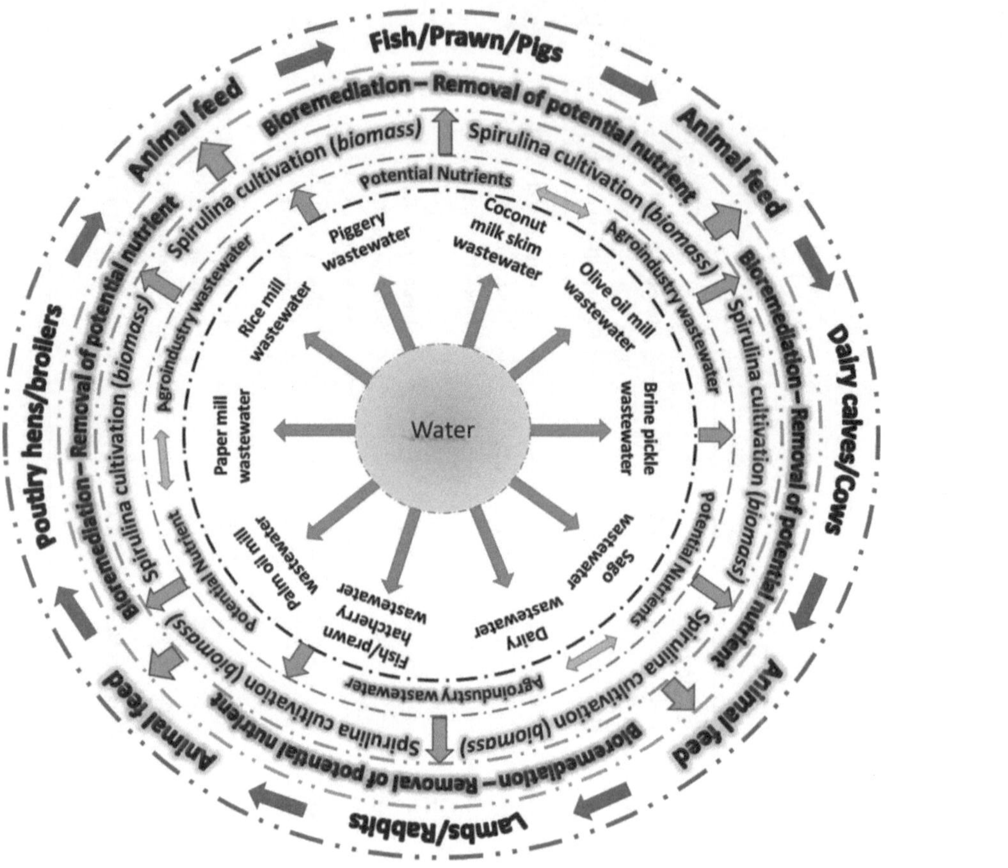

FIGURE 11.1 Outline of bioremediation of agroindustrial wastewater by cultivation of *Spirulina platensis* for removal of environmental pollutants and biomass used as animal feed supplement.

wastewater, and the harvested biomass could be used for manufacturing other useful by-products and treated wastewater, and according to World Health Organization (WHO) and Central Pollution Control Board (CPCB), the permissible limit is used for irrigation purpose. The schematic representation of bioremediation in agroindustrial wastewater by cultivation of *Spirulina platensis* for removal of environmental pollutants is shown in Figure 11.1.

11.2 *SPIRULINA*

The word microalgae is considered as a common term, and it includes cyanobacteria (BGA) used for microalgae-based wastewater treatment. The BGAs are photosynthetic prokaryotes, a microalgae are photosynthetic eukaryotes (Salces et al. 2019). Cyanobacteria gain their carbon and energy sources from photosynthetic mechanisms, which are similar to those found in plants (Holm-Hansen 1968).

Cyanobacterial cells (e.g., *Spirulina*) have a higher digestible potential due to the nonexistence of cellulose, unlike the majority of algae, which helps in their consumption for human and animals. *Spirulina*, due to its faster growth rate, ease of cultivation, harvesting, and processing, offers a substantial scope for the bioremediation of the agroindustry wastewater and concomitant production of biomass used for animal feed for poultry, lambs, rabbits, calves/cows, fish/prawns, and pigs. Furthermore, *Spirulina* is one of the main commercial ventures in the agroindustrial domain.

11.2.1 Strain Selection

The important factor for selection of strain is production of biomass, reduces the pollutant concentration, which is suited to the environmental and cultivation conditions. The genus *Spirulina* belongs to Oscillatoriaceae family, and it consists of filamentous cyanobacteria, characterized by spiral-shaped chains of cells (trichomes) enclosed in a thin sheath. *Spirulina oscillariodes* is the first species of *Spirulina* isolated from freshwater stream (Turpin 1827). Based on the growth and environmental conditions, *Spirulina* characteristics could vary within the species. Presently, various species of *Spirulina* have been reported from relatively different environments such as soil, sand, marshes, brackish water, seawater, freshwater, thermal springs, warm waters from power plants, and fish ponds. Domestic wastewater has been used for first wastewater treatment by using *Spirulina platensis* (Kosaric et al. 1974). However, the filamentous cyanobacteria *Spirulina platensis* is a suitable strain used for cultivation and bioremediation of agroindustrial wastewater treatment by several environmental and operational features to removal of pollutants and biomass production (Table 11.2).

Several *Spirulina* strains have been isolated by Vonshak and Tomaselli (2003), and among them, most of the strains are grown in an optimum temperature between 24°C and 28°C and others can grow up to 40°C –42°C. Isolation of native *Spirulina* strains has been used for increased cell multiplication and biomass production. For example, Sheshardi and Thomas (1979) noted that the cultivation of a locally isolated species of *Spirulina* in Zarrouk medium supplemented with anaerobically digested effluent demonstrated significant greater productivity (12.39 g/m^2day) compared to a standard strain (10.88 g/m^2day).

11.2.2 *Spirulina* Cultivation in Agroindustrial Wastewater

Cultivation of cyanobacteria carried out in photobioreactors and open ponds, namely, raceway ponds, or high-rate algae ponds (HRAPs). The HRAPs are the most commonly used systems for *Spirulina* cultivation and wastewater treatment. The HRAPs consist of rectangular basins/channels, where the wastewater is retained in a constant motion with the help of a powered paddle wheel (Gouveia 2011). In an open raceway pond, reasonably slow stirring rate is affected by a flow rate less than 30 cm/s. The population density of *Spirulina* cell mass increases show havoc on output rate, and this problem occurred in large-scale industries (Richmond and Grobbelaar 1986).

Nowadays, there is an increasing demand of wastewater recycling and sustainable energy consumption. However, a huge volume of water and synthetic nutrients are required for *Spirulina* cultivation, and biomass production is the foremost issue cost

TABLE 11.2
List of Agroindustrial Wastewater with Supplemented of Nutrients for *Spirulina platensis* Cultivation

S. No	Agroindustrial Wastewater	Nutrients Supplemented for *Spirulina platensis* Cultivation	References
1	Beet sugar factory (vinasse) wastewater	Schlosser culture medium	Coca et al. (2015)
2	Brine (pickle) factory wastewater	Zarrouck's medium	Duangsri and Satirapipathkul (2011)
3	Coconut milk skim wastewater	Modified Bangladesh No. 3 synthetic nutrient	Azimatun Nur et al. (2015)
4	Fish farming wastewater	Modified technical fertilizer	Wijayanti et al. (2019)
5	Olive oil mill wastewater	Zarrouck's medium	Ismail et al. (2013)
6	Palm oil mill wastewater	Zarrouck's medium	Azimatun Nur et al. (2019)
7	Paper mill wastewater	Supplement chemicals (urea, KH_2PO_4, $CaCO_3$)	Setiawan et al. (2019)
8	Prawn hatchery wastewater	Nallayam Research Centre medium	Sandeep et al. (2015)
9	Rice mill wastewater	Zarrouck's medium	Usharani et al. (2014)
10	Rice noodle factory wastewater	Zarrouck's medium	Vetayasuporn (2004)
11	Sago factory wastewater	Zarrouck's medium	Karthik and Kumar (2016)
12	Piggery wastewater	Zarrouck's medium	Chaiklahan et al. (2010)

wise (Zhai et al. 2017). *Spirulina* needs a bicarbonate-rich growth medium, and also the most typically used one is the formulation of Zarrouck (1966). Concerning this problem, agroindustrial wastewater contains a high level of macro and micronutrients that are used as a substrate for *Spirulina* growth and proposed a solution to attain economically achievable cultivations (Azimatun Nur et al. 2019). In *Spirulina* cultivation, a number of essential nutrients are needed for the multiplication of cells so as to enable a fast and effective growth. These essential nutrients may be suggested to the particular growing medium for large-scale production (Table 11.2). Various agroindustry wastewaters contain a number of nutrients used for *Spirulina* growth, and the nutrients are nitrates, phosphates, heavy metals, and other nutrients.

In our previous studies (Karthik and Kumar 2016), we tend to find that *Spirulina platensis* cultivated in the anaerobically digested sago wastewater diluted with water at totally different dilution concentrations such as 80:20, 60:40, 50:50, 40:60, 20:80, and 100% (undiluted), nutrient supplemented with various concentrations of $NaHCO_3$ and $NaNO_3$ as sources of carbon and nitrogen, respectively, is based on Zarrouck's broth composition. The laboratory model raceway pond (65 L working volume) with an optimized dilution level of 80:20 and nutrient supplementation levels of $NaHCO_3$ at 0.2 M (16.8 g/L) and $NaNO_3$ at 0.03 M (2.5 g/L) concentrations was used. The *Spirulina* cell multiplication was periodically counted with microscopic observation. Biomass of *Spirulina* was produced at 0.4 g/L on a dry weight basis within a period of 18 days (Figure 11.2).

FIGURE 11.2 Laboratory model raceway pond (LMRP) for *Spirulina* cultivation in sago wastewater and biomass & microphotograph of *Spirulina*. (From Karthik, M. and Kumar, K., Int. J. Agric. Sci., 8, 1763–1767, 2016. With permission.)

11.2.3 Factors Influencing *Spirulina* Growth in Agroindustrial Wastewater

Elimination of potential nutrients in agroindustrial wastewater was highly achieved by *Spirulina* cultivation. The bioremediation of *Spirulina* cultivation in particular operating conditions or systems is needed for cell multiplication. Thus operating systems need to be set properly and monitored regularly. These operating systems are explained in detail in the following sections.

11.2.3.1 Abiotic Factors

11.2.3.1.1 Effect of Light on Spirulina *Growth*

Spirulina cultivation throughout the growth phase needed light intensity, and an optimum range of light intensity is 20–30 K lux (Ogawa and Teuri 1970). The simple way to inhibit algal cultures from light limitation is to lessen the depth of the culture vessel. Oswald (1988) remarked that the productivity in light limited ponds is reciprocally related to the depth. Generally, the recommended culture vessel depths are between 15 and 50 cm (Benemann 1979). The effective stirring that exerts on the output of *Spirulina* cell mass is highlighted as the population density and the extent to increase the light limitation. Although light is the most often limiting factor for growth of cyanobacteria, an excessive amount of light might also cause lowered photosynthetic efficiency, which is called photoinhibition (Oliver and Ganf 2000). According to Carvalho et al. (2004), the ideal light intensity for *Spirulina platensisis* was 72 L mol photon/m^2s. The moderate light intensity for cultivation of *Spirulina* suggested a low light intensity at the beginning to avoid photolysis (Dubey 2006).

11.2.3.1.2 *Effect of pH on* Spirulina *Growth*

The pH involved directly acts on the physiological properties of algae during growth period and is based on the availability of nutrient. The pH of culture growing period is determined directly or indirectly based on the solubility of carbon source and minerals present in the nutrient medium. However, the *Spirulina* strains are grown well at pH 9–11. During mass cultivation the optimum level of pH is 8.4–9.5 for depletion of sodium and bicarbonate ions (Vincent and Silvester 1979). *Spirulina platensis* is taken into account to be an alkalophilic organism naturally (Grant et al. 1990). The pH is maintained by inflow of CO_2, where intermittent addition of inorganic acid to a high alkalinity growth medium is being additionally helpful. The usage of sodium bicarbonate CO_2 has increased the pH (Çelekli et al. 2009). Most of the agroindustry wastewaters contain a low pH, and the addition of sodium bicarbonate helps to increase the pH level. Furthermore, pH concentrations were frequently increased in algal cultures due to photosynthetic CO_2 assimilation (Chevalier et al. 2000).

11.2.3.1.3 *Effect of Temperature on* Spirulina *Growth*

Temperature is a vital physical factor that strongly influences the oxygen evolving activity of the photosystem II (PSII), includes a variety of effects on the cyanobacterial membranes, and influences nutrient accessibility and its uptake (Inoue et al. 2001). An optimum temperature has enhanced the algal growth, and it ranged from 30°C to 35°C (Chevalier et al. 2000). Ogbonda et al. (2007) examined differentiation of *Spirulina* biomass composition due to augmentations of temperature, and the higher crude protein content was recovered at a temperature of 30°C with pH 9.0.

11.2.3.1.4 *Effect of Nutrient Concentration on* Spirulina *Growth*

Spirulina required higher nutrient concentrations, particularly at high salt concentrations compared to other microalgae. This could be the reason for natural *Spirulina* specifically occurring in salt lakes (Ogawa and Teuri 1970). The essential nutrient is carbon required for *Spirulina* cultivation that can uptake both organic and inorganic forms. *Spirulina* has the potential to utilize both CO_2 and HCO^-_3 as an inorganic carbon source. Jaiswal et al. (2005) remarked that the inorganic form of carbon is employed through the mechanism of CO_2 concentrating from extracellular environment. Also, the intercellular carbon is in the kind of HCO_3, and it's converted to CO_2 by carbonic anhydrase enzyme.

Nitrogen is the utmost important nutrient in microalgae (Becker 1994), and it ranged from 1% to 10% in biomass and is reliant on the amount, the availability, and the type of nitrogen sources like NO_3, NO_2 or NH_4^+, and N_2 (Grobbelaar 2004). The source of nitrogen is present in nitrate form, and the microalgae spends cell energy to reduce this ion to nitrite through nitrate-reductase enzyme. After nitrite formation, reduction by nitrite reductase generating ammonia occurs, which is the form of nitrogen used by the alga for its metabolism (Huertas et al. 2000). This might be due to $NaHCO_3$ and $NaNO_3$ serving as a source of carbon and nitrogen, and principally $NaHCO_3$ is a favored carbon source for *Spirulina* growth and development.

Ammonia, CO $(NH_2)_2$ and NO_3, owing to their availability and relatively low-cost, are frequently selected as nitrogen sources for microalgae mass cultivation (Hsieh and Wu 2009). Ayala and Vargas (1987) observed that the biomass of crude protein was recovered based on the availability of nitrogen content present in the medium. Thus, agroindustry wastewater containing high nitrogen may be considered as a suitable growing medium for single cell protein biomass production.

Among the essential macronutrients, carbon and nitrogen are required in high amounts for growth and development of microalgae. Other macronutrients like phosphorus, sulfur, potassium, calcium, sodium, and magnesium are required for microalgae cultivation. Nitrogen is typically required for the synthesis of amino acids in cells (Colla et al. 2007), while phosphorus is used for energy transfer and metabolism (Sari et al. 2012). Micronutrients are required in lesser quantities for the cultivation of microalgae, which are manganese, molybdenum, copper, iron, zinc, boron, chloride, and nickel. Furthermore, chelating agent (e.g., EDTA) is also added in the commercial algal cultures to prevent the growth limitations (Oliver and Ganf 2000).

11.2.3.1.5 Effect of Dissolved Oxygen on Spirulina *Growth*

Oxygen stimulates the bacterial degradation of organic matter in wastewater, liberating more CO_2 and nutrients, and are assimilated by the microalgae, producing surplus algal biomass (Oswald and Gotaas 1957; Oswald 1988). Now, the cultivation and production of *Spirulina* has been completely carried out in the open ponds system (Vonshak 1997); on the other hand it's expected that *Spirulina* and other algae are cultivated in the more controlled and closed systems in the future. They provide vigorous stirring into an open raceway pond, resulting in a significant reduction of dissolved oxygen (DO). Vigorous stirring has a substantial reduction in DO, maintaining a high cell density culture.

11.2.3.1.6 Effect of Water Quality on Spirulina *Growth*

Water quality is an essential factor for *Spirulina* biomass production. Wastewater is used for algae cultivation and has dual influences during growth period, by affecting solubility of added nutrients and accumulation of heavy metals (Argaman and Spivak 1974). When the agroindustry wastewater is diluted *Spirulina platensis* grows very well at any dilution. Because some agroindustry wastewater contains highly concentrated color turbidity, the algae can survive with enhanced light needed for photosynthetic activity and achieve a high cell concentration. Even without dilution of wastewater utilized by *Spirulina* it can grow well, indicating the adaptation of inorganic pollutant concentration. However, the biomass and elimination percentage of environmental pollutants were higher in dilution of agroindustry wastewater compared to without dilution.

11.2.3.2 Biotic Factors

11.2.3.2.1 Effect of Competition between Species on Spirulina *Growth*

In nature, species have to compete with each other for space and nutrients, and this is reflected into algal cultures as well. Some species inhibit the growth of others in mixed culture; e.g., nutrients competing with some group of cyanobacteria can produce inhibitory substances to the growth of eukaryotic algae (Kawaguchi 1980). However, the contaminations eliminated by chosen strains from extreme habitats, like *Spirulina*, is strongly competitive in environments with high pH and relatively high ammonia concentration (Cañizares-Villanueva et al. 1994).

11.2.3.2.2 Effect of Contamination by Other Microorganisms on Spirulina *Growth*

Microalgae cultivation in the open raceway pond is mostly affected by contamination of other microorganisms and other microalgae genera, and the growth rates could obstruct the growth of the cultivated microalgae. The increased pH can cause precipitation of phosphates and removal of hydrogen, sulfur, and ammonia (Larsdotter 2006). Up to 10% of biomass yield was decreased by insect contamination (Venkatraman and Sindhukanya 1981). The fine wire mesh frame is used for the removal of extraneous materials from pond to avoid contamination. The success of bioremediation in agroindustry wastewater is required to follow proper operating conditions and monitor regularly. Also, avoiding contaminants from both external and internal factors can recover successful *Spirulina* growth and harvest pure *Spirulina* biomass.

11.2.3.3 *Spirulina* Biomass Harvesting

The *Spirulina* biomass has been recovered by filtration method by using a low-cost nylon screen to harvest large-sized filamentous microalgae like *Spirulina* (Toyoshima et al. 2015). After harvesting biomass, it is washed to reduce the salts and the completion of the drying process breaks down the trichome by a grinder. Finally the dried *Spirulina* is stored in the sealed plastic bags to avoid hygroscopic action. Also, the dry *Spirulina* is stored in dark, pest-free and hygienic storerooms to avoid *Spirulina* pigments from deteriorating (Ayala 1998).

11.3 BIOREMEDIATION OF AGROINDUSTRIAL WASTEWATER

Bioremediation is a technology that controls pollution, which is used in biological system to catalyze the degradation of various toxic potential nutrients with less harmful effects. Developing the biological-based wastewater treatment system is deliberate, economic, and ecofriendly (Valderrama et al. 2002). The bioremediation of microalgae *Spirulina*-based wastewater treatment has several benefits when compared with conventional methods. The benefits are cost-effective, needs less energy, lessen the sludge formation, filtrate water used for irrigation and recovered high biomass for making useful by-products like, human or animal consumption, polyunsaturated fatty acids (PUFAs), biofuel, and pigment production.

11.3.1 Removal of Potential Nutrients from Agroindustry Wastewater

Spirulina used in the agroindustrial wastewater treatment for a variety of purposes includes elimination of coliform bacteria, reduction percentage of total suspended solids (TSS) and total dissolved solids (TDS), reduction of COD and BOD, removal

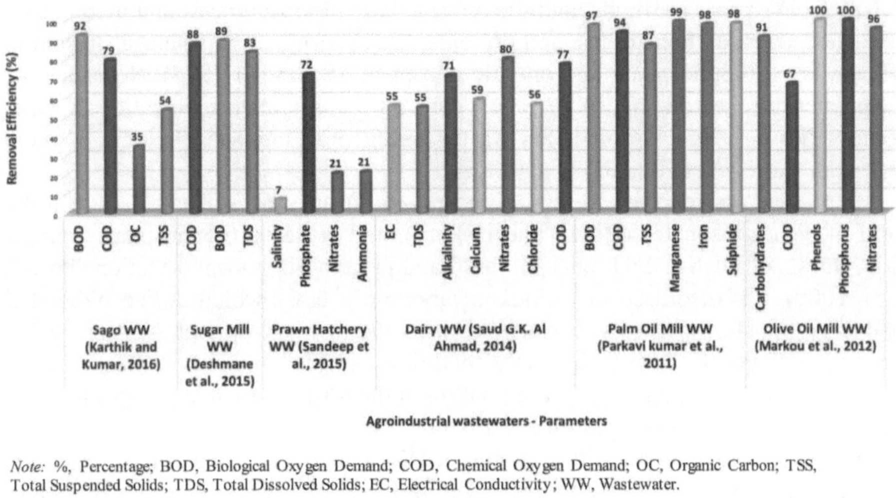

FIGURE 11.3 The percentage reduction efficiency of environmental pollutants due to *Spirulina platensis* cultivation in agroindustrial wastewater. (*Note*: %, Percentage; BOD, Biological Oxygen Demand; COD, Chemical Oxygen Demand; OC, Organic Carbon; TSS, Total Suspended Solids; TDS, Total Dissolved Solids; EC, Electrical Conductivity; WW, Wastewater.)

of nitrogen and phosphorous related nutrients (nitrates, nitrites phosphates, ammonia), calcium, sulfide, phenols, alkalinity (Figure 11.3), and heavy metals. Heavy metals such as lead, nickel, zinc, cadmium, mercury, copper, and arsenic are stated as ecological pollutants (Dias 2002). Inductively coupled plasma mass spectrometry (ICP–MS) based analytical method has been highly used for trace metal analysis in plants (Leiterer et al. 1997).

The phenomenon of ammonia removal in agroindustry wastewater by *Spirulina* cultivation is a major factor. Since gaseous ammonia is also considered as a pollutant, quantifying the rates of ammonia volatilization during growth is required. Due to the presence of sodium chloride in wastewater, ammonia volatilization is reduced, and more nitrogen leftover exists for the assimilation of *Spirulina* growth (Chiu et al. 1980). It might be of interest to study the connection between quantity of sodium chloride added, *Spirulina* growth rate, and degree of ammonia stripping. The eradication of inorganic phosphorus in microalgae culture results from two phenomena which is biological assimilation and chemical precipitation as insoluble phosphate. As with ammonia nitrogen, the environmental conditions that are overcome during *Spirulina* cultivation also help in phosphate elimination by abiotic process (Laliberté et al. 1997).

Nitrogen has been used for synthesis of amino acids, nucleic acids, and pigment production in microalgae (Richmond 2008). The NH_4^+ normally utilized a nitrogen source supply for microalgae, additionally able to use NO_3^- and, in a lesser extent, NO_2^- (Jia and Yuan 2016). Phosphorus is another vital macronutrient that plays a key role in cellular metabolic processes (Tiessen 1995). It is present in nucleic acids

(DNA, RNA), proteins, lipids, and the intermediate of biosynthesis and metabolism of nucleic acids, carbohydrates, and proteins. Major forms of phosphorus present in effluent are orthophosphate and organic phosphate (Schindler 1977). Nevertheless, the greater nitrogen and phosphorus concentration cause the eutrophication in water bodies (Tiessen 1995). Agroindustry wastes are the most important sources of phosphorus in wastewater (Bennett et al. 2001).

Microalgae are reported to utilize phosphorus from wastewaters, mainly in the form of orthophosphates (HPO_4^{2-} and $H_2PO_4^-$), and utilize orthophosphates during biosynthesis of nucleic acid, phospholipid, and protein via phosphorylation (Powell et al. 2009). It is also used in various metabolic processes which utilize ATP/ADP as energy transfer processes, because it forms the primary part of ATP and ADP (Conley et al. 2009). During favorable conditions, surplus amount of phosphorus is preoccupied by microalgae and stored within in the form of organic phosphate granules for future use (Rasoul-Amini et al. 2014).

BOD is broadly defined as the amount of DO needed by aerobic microorganisms to break down organic compounds existing in the known water sample (Singh and Pandey 2018). Therefore, it is a measure of oxygen demand by bacteria to metabolize the organic compound. COD quantifies the organic compounds that can chemically oxidize and not just the level of biodegradable pollutants (Raouf et al. 2012). Colak and Kaya (1988) studied the possibilities of microalgae for declining the BOD and COD levels and reported a net reduction of 68.4% and 67.2% for BOD and COD, respectively, in domestic wastewater. The reduction ranges of 93.2% and 86.3% of BOD and COD level, respectively, were recorded in sago agroindustrial wastewater (Karthik and Kumar 2016).

Photosynthetic microorganism utilizes nutrients existing in the agroindustry effluent for their growth and release of oxygen in water, and then heterotrophic aerobic bacteria then utilize this released oxygen and in turn released CO_2 by bacterial respiration (Munoz and Guieysse 2006). This released CO_2 is further utilized by photosynthetic microorganisms.

Application of bioremediation using indigenous microorganisms like cyanobacteria for decontamination of the agroindustrial wastewater polluted with heavy metals as well as organic contaminants provide a sustainable method for environmental resources. Heavy metals are a major concern in any species of *Spirulina* cultivation, and if heavy metals occurred, cells are absorbing heavy metals from the substrate. Also, *Spirulina* has been used to eliminate heavy metals from the contaminated agroindustry wastewater (Richmond and Hu 2013).

Several algal species have a great potential of sequestrating toxic heavy metals from aqueous environment, and this sequestering process involves various mechanisms (Upadhyay et al. 2019). Basically, it depends upon the algal species, metal ions, and solution condition. The living microalgal cells accumulate microelements (Ca, Mg, Cu, Zn, Co, Mo, Cr, Pb, and Se) inside their cell through active transport (Rajfur et al. 2012). Metals like mercury, cadmium, lead, and arsenic are the most possible adulterates in *Spirulina* products. Nickel, copper, and zinc are contaminants; however, they are much less toxic and have a narrow range of optimum concentrations of *Spirulina* (Pande et al. 1981; Kotangale et al. 1984). The issues seem nonexistent when *Spirulina* is grown in the artificial environment, such as wastewaters, after the values of arsenic (0.06–2.00 ppm), cadmium (0.01–0.10 ppm), mercury (0.01–0.20 ppm), and lead (0.60–5.10 ppm) are suggested by the relevant norms (WHO 1973).

11.4 *SPIRULINA* AS ANIMAL FEED SUPPLEMENT

Spirulina biomass predominantly contains carbohydrates, proteins, essential amino acid, fatty acids, chlorophylls, carotenoids, and vitamins (Velasquez et al. 2016). Currently, *Spirulina*-based products are sold worldwide to dietary or functional food in health food shops. They are widely used for animal feed as fishes (Zhang et al. 2019), poultries (Wikdors and Ohno 2001), rabbits (Gerencsér 2012), lambs (Holman and Malau-Aduli 2014), dairy cows/calves (Riad et al. 2019), and pigs (Nguedia et al. 2019). Also, *Spirulina* is used as a livestock feed supplement with a replacement rate of 5%–10% of the traditional feed. *Spirulina*-fed animals increased the body weight, feed intake, fertility, immune response, antioxidant activity, and external appearance, like healthy skin and fur (Tsiplakou et al. 2017).The edible microalga (*Spirulina*) has rich potential nutritious feed resource for different animal species. Several research findings remarked that *Spirulina* has a potential to improve the animal growth, productivity, fertility, aesthetic quality, and enhanced food security (Holman and Malau-Aduli 2012).

11.4.1 Fish and Prawn

Spirulina has been used as a partial or complete replacement feed supplement of protein in fish feed for diverse fish species such as great sturgeon, rainbow trout, olive flounder, catfish, parrot fish, goldfish, guppy, mrigal carp, and tilapia (Zhang et al. 2019). *Spirulina* feed supplement on fish growth is not only species-specific but is also influenced by the amount of *Spirulina* used for intakes. Now, the effects of *Spirulina*-fed ratio were investigated on a lesser number of commercially available fish species.

11.4.2 Prawn

In China, *Spirulina* is used as a feed supplement for prawns (*Penaeus monodon*), and it can be imported to promote growth and immunity (Abdulrahman et al. 2019). According to Nakagawa and Gomez-Diaz (1975) investigations, *Spirulina* is used as a feed supplement for giant freshwater prawn (*Macrobrachium rosenbergii*), resulting in significant improvement of growth and feed utilization ratio. The feed supplementation of *Spirulina* ranged between 5% and 20% (Nakagawa and Gomez-Diaz 1975). Additionally, *Spirulina* has been used for diets in common carp at different levels, resulting in increased white blood cell (WBC) counts at the ratio of 5 g of *Spirulina*/kg diet supplement (Abdulrahman et al. 2019).

11.4.3 Poultry

Worldwide, *Spirulina* has been used as a feed supplement for broiler and layer diets to improve egg yolk color and egg fertility (Ross and Dominy 1990). *Spirulina* holds high nutritional limits and is used as a natural quality feed additive in poultry diets (Belay et al., 1997). Poultry meat contains a rich source of carotenoids and fatty acids, especially for gamma-linolenic acid (GLA), with infers of health benefits to humans (Guroy et al. 2012). *Spirulina* supplementation has increased the level of yellow and redness in poultry muscle tissues (Venkataraman et al. 1994). Díaz et al. (2017) found that *Spirulina*-fed poultries have improved the quality of meat, nutritional content, and increased the ratio of PUFAs, total saturated fatty acids (SFA), docosahexaenoic acid (DHA), and total n−3 fatty acids.

11.4.4 Rabbits

The quality of rabbit meat improved when rabbits received dietary *Spirulina*. Peiretti and Meineri (2011) noticed that dietary *Spirulina* used rabbit feed supplement as a causal factor, resulting in enhanced GLA and n−6/n−3 PUFA ratios present in the rabbit muscle lipid content. The feed supplementation of *Spirulina* has to improve the meat color by oxidative stability (Dalle Zotte and Szendro 2011).

11.4.5 Lambs

Spirulina is used as a diet supplement for lambs, resulting in increased live weights and average daily gains (ADG) compared with nonleaded lambs (Bezerra et al. 2010). Holman et al. (2012) also reported that diet supplement of *Spirulina* has increased the lamb live weight as well as body condition. In addition *Spirulina* is also used a feed supplement in sheep cross-breeding industries to boost lamb growth rates and gain the lamb live weight.

11.4.6 Dairy Cattle

The dairy cows utilized a feed supplementation of *Spirulina platensis* at the rate of 2 g/day per cow to improving the range of milk's food safety (Simkus et al. 2007). It substantially increased the yield of milk, fat, protein, and lactose, and decreases the somatic cell count. Dietary *Spirulina* fed to cows has increased 25% of milk fat, 9.7% of milk protein (9.7%), and 11.7% of lactose. Thus, *Spirulina*-fed cow's milk increased PUFAs and monosaturated fatty acids and decreased the saturated fatty acid content (Christaki et al. 2012). The dairy cows are fed dietary *Spirulina* to increase milk production and improve body condition at 21%, and 8.5%–11%, respectively (Kulpys et al. 2009).

11.4.7 Pigs

Spirulina is used to increase the body weight, sperm quality, and volume. Hugh et al. (1985) noticed that dietary *Spirulina* is used as a feed supplement of crossbred weanling pigs, resulting in ~9% of increased growth rates when compared to the unsupplemented one and to improve the boar's fertility. According to Granaci (2007),

Spirulina supplemented boars increase the sperm volume by 11%, quality of sperm and increase 5% of poststorage viability when compared to unsupplemented boars.

11.5 CONCLUSION

Bioremediation of *Spirulina* can be effective to eliminate the organic and inorganic nutrients from agroindustrial wastewaters. It is an ecofriendly and cost-effective method, and the production of oxygen (O_2) takes place with less energy requirement, which efficiently removes the nitrates, phosphates, COD, BOD, and heavy metals and reduces the greenhouse gas emission due to consumption of CO_2 by *Spirulina* during the growing period. *Spirulina* biomass has a high nutritional value, and hence it is used as a feed supplement for animals including dairy cows, lambs, fish, prawns, rabbits, pigs, hens, and broilers. The supplement of *Spirulina* to various animal species has to improve the health, productivity, and product quality. Besides, the contamination can be removed by chosen strains from extreme environments, like *Spirulina*, which is strongly competitive in habitat with higher pH and ammonia contents. Nowadays, future studies need to focus on native isolates from the respective agroindustrial wastewater areas they strain can be easily adopting, recycling of wastewater and recovery of high biomass. The selected strains scaling up from laboratory scale to large scale, to identify a higher potential to eliminate the pollutants, process simulation, appropriate design options, benefit costs analyses and optimization of all aspects concerning to performance of the innovative technology so as to be inexpensive compared to traditional methods.

REFERENCES

Abdulrahman, N.M., H.J. Hama, S.R. Hama, B.R. Hassan and P.J. Nader. 2019. Effect of microalgae *Spirulina* spp. as food additive on some biological and blood parameters of common carp *Cyprinus carpio* L. *Iraqi Journal of Veterinary Sciences*, 33(1): 27–31.

Argaman, Y. and E. Spivak. 1974. Wastewater as source of algae. *Water Resources*, 8: 317.

Ayala, F. 1998. Guía sobre el cultivo de *Spirulina*. In: *Biotecnología de Microorganismos Fotoautótrofos*. Motril, Granada: España. p. 3–20.

Ayala, F. and T. Vargas. 1987. Experiments on *Spirulina* culture on waste-effluent media and at the pilot plant. *Hydrobiologia*, 91: 151–152.

Azimatun Nur, M.M., G.M. Garcia, P. Boelen and A.G.J. Buma. 2019. Enhancement of C-phycocyanin productivity by *Arthospira platensis* when growing on palm oil mill effluent in a two-stage semi-continuous cultivation mode. *Journal of Applied Phycology*, 31(5): 2855–2867.

Azimatun Nur, M.M., M.A. Irawan and H. Hadiyanto. 2015. Utilization of coconut milk skim effluent (CMSE) as medium growth for *Spirulina platensis*. *Procedia Environmental Sciences*, 23: 72–77.

Becker, E.W. 1994. *Microalgae, Biotechnology, and Microbiology*. Cambridge: Cambridge University Press.

Belay, A., T. Kato and Y. Ota. 1997. *Spirulina* (*Arthrospira*): Potential application as an animal feed supplement. *Journal of Applied Phycology*, 8: 303–311.

Benemann, J.R. 1979. Production of nitrogen fertilizer with nitrogen-fixing bluegreen algae. *Enzyme and Microbial Technology*, 1: 83–90.

Bennett, E.M., R. Carpenter and N.F. Caraco. 2001. Human impact on erodable phosphorus and eutrophication: A global perspective. *BioScience*, 51: 227–234.
Bernet, N and F. Béline. 2009. Challenges and innovations on biological treatment of livestock effluents. *Bioresource Technology*, 100: 5431–5436.
Bezerra, L.R., A.M.A. Silva, S.A. Azevedo, R.S. Mendes, J.M. Mangueira and A.K.A. Gomes. 2010. Performance of Santa Inés lambs submitted to the use of artificial milk enriched with *Spirulina platensis*. *Ciência Animal Brasileira*, 11: 258–263.
Cañizares-Villanueva, R.O., A. Ramos, AI. Corona, O. Monroy, M. de la Torre and C. Gomez-Lojero. 1994. *Phormidium* treatment of anaerobically treated swine wastewater. *Water Research*, 28: 1891–1895.
Carvalho, J.C.M., F.R. Francisco, K.A. Almeida, S. Sato and A. Converti. 2004. Cultivation of *Arthrospira* (*Spirulina platensis*) (Cyanophyceae) by fed-batch addition of ammonium chloride at exponentially increasing feeding rates. *Journal of Phycology*, 40: 589–597.
Çelekli, A., M. Yavuzatmaca and H. Bozkurt. 2009. Modeling of biomass production by *Spirulina platensis* as function of phosphate concentrations and pH regimes. *Bioresource Technology*, 100(14): 3625–3629.
Chaiklahan, R., C. Nattayaporn, S. Wipawan, P. Kalyanee and B. Boosya. 2010. Cultivation of *Spirulina platensis* using pig wastewater in a semi-continuous process. *Journal of Microbiology and Biotechnology*, 20: 609–614.
Chevalier, P., D. Proulx, P. Lessard, W.F. Vincent and J. de la Noüe. 2000. Nitrogen and phosphorus removal by high latitude mat-forming cyanobacteria for potential use in tertiary waste water treatment. *Journal of Applied Phycology*, 12: 105–112.
Chiu, R.J., H.I. Liu, C.C. Chen, Y.C. Chi, H. Shao, P. Soong and P.L.C. Hao. 1980. The cultivation of *Spirulina platensis* on fermented swine manure. In: *Animal Wastes Treatment and Utilization*, Chang, P.O. (Ed.), *Proceedings of the International Symposium on Biogas, Microalgae and Livestock*, Taiwan, p. 435.
Christaki, E., M. Karatzia, E. Bonos, P. Florou-Paneri and C. Karatzias. 2012. Effect of dietary *Spirulina platensis* on milk fatty acid profile of dairy cows. *Asian Journal of Animal and Veterinary Advances*, 7: 597–604.
Coca, M., V.M. Barrocal, S. Lucas, G. González-Benito and M.T. García-Cubero. 2015. Protein production in *Spirulina platensis* biomass using beet vinasse-supplemented culture media. *Food Bioproducts Processing*, 94: 306–312.
Colak, O. and Z. Kaya. 1988. A study on the possibilities of biological wastewater treatment using algae. *Doğa. Türk biyoloji dergisi*, 12: 18–29.
Colla, L.M., C.O. Reinehr, C. Reichert and J.A.V. Costa. 2007. Production of biomass and nutraceutical compounds by *Spirulina platensis* under different temperature and nitrogen regimes. *Bioresource Technology*, 98(7): 1489–1493.
Conley, D.J., H.W. Paerl, R.W. Howarth, D.F. Boesch, S.P. Seitzinger and K.E. Havens. 2009. Controlling eutrophication: Nitrogen and phosphorus. *Science*, 323: 1014–1015.
Dalle Zotte, A. and Z. Szendro. 2011. The role of rabbit meat as functional food. *Meat Science*, 88: 319–331.
Deshmane, A.B., V.S. Darandale, D.S. Nimbalkar, T.D. Nikam and V.S. Ghole. 2015. Sugar mill effluent treatment using *Spirulina* for recycling of water, saving energy, and producing protein. *International Journal of Science and Technology*. doi: 10.1007/s13762-015-0891-1.
Dias, M.A. 2002. Removal of heavy metals by an *Aspergillus terreus* strain immobilized in polyurethane matrix. *Letters in Applied Microbiology*, 34 (1): 46–50.
Díaz, M., C. Pérez, C. Sánchez, S. Lauzurica, V. Cañeque, C. González and J. De La Fuente. 2017. Feeding microalgae increases omega 3 fatty acids of fat deposits and muscles in light lambs. *Journal of Food Composition and Analysis*, 56: 115–123.

Duangsri, P. and C. Satirapipathkul. 2011. *Spirulina* sp. production in brine wastewater from pickle factory. In: *International Conference on Bioscience, Biochemistry, and Bioinformatics IPCBEE*, Singapore, vol. 5, pp. 415–418.

Dubey, R.C. 2006. *A Textbook of Biotechnology.* Fourth revised and enlarge edition. New Delhi: S. Chand and Company Ltd, pp. 419–421.

Gerencsér, Zs., Zs. Szendrő, Zs. Matics, I. Radnai, M. Kovács, I. Nagy, A. Dal Bosco and A. Dalle Zotte. 2012. Dietary supplementation of *Spirulina* (*Arthrospira platensis*) and thyme (*Thymus vulgaris* L.). Part 1: Effect on productive performance of growing rabbits. *World Rabbit Science Association Proceedings 10th World Rabbit Congress*, Egypt, pp. 657–661.

Gibson, C.E and R.V. Smith. 1982. Freshwater plankton. In: *The Biology of Cyanobacteria*, Carr, N.G., Whitton, B.A. (Eds.). London: Blackwell, pp. 463–489.

Gouveia, L. 2011. Microalgae as a feedstock for biofuels. *Springer Briefs in Microbiology.* doi: 10.1007/978-3-642-17997-6.

Granaci, V. 2007. Achievements in the artificial insemination of swine. *Bulletin of University of Agricultural Sciences and Veterinary Medicine Cluj-Napoca. Animal Science and Biotechnologies*, 63/64: 382–386.

Grant, W.D., W.E. Mwatha and B.E. Jones. 1990. Alkalophiles: Ecology, diversity, and application. *FEMS Microbiology Reviews*, 75: 225–270.

Grobbelaar, J.U. 2004. Algal nutrition: Mineral nutrition. In: *Handbook of Microalgal Culture: Biotechnology and Applied Phycology*, Richmond, A. (Ed.). Oxford: Blackwell Publishing Ltd., pp. 97–115.

Guroy, B., I. Sahin, S. Mantoglu and S. Kayali. 2012. *Spirulina* as a natural carotenoid source on growth, pigmentation, and reproductive performance of yellow tail cichlid *Pseudotropheus acei. Aquaculture International*, 20: 869–878.

Holman, B.W.B and A.E.O. Malau-Aduli. 2012. *Spirulina* as a livestock supplement and animal feed. *Journal of Animal Physiology and Animal Nutrition.* doi: 10.1111/j.1439-0396.2012.01328.x.

Holman, B.W.B., A. Kashani and A.E.O. Malau-Aduli. 2012. Growth and body conformation responses of genetically divergent Australian sheep to *Spirulina* (*Arthrospira platensis*) supplementation. *American Journal of Experimental Agriculture*, 2: 160–173.

Holm-Hansen, O. 1968. Ecology, physiology, and biochemistry of blue-green algae. *Annual Review of Microbiology*, 22: 47–70.

Hsieh, C.H and W.T. Wu. 2009. Cultivation of microalgae for oil production with a cultivation strategy of urea limitation. *Bioresource Technology*, 100: 3921–3926.

Huertas, E., O. Montero and L.M. Lubián. 2000. Effects of dissolved inorganic carbon availability on growth, nutrient uptake, and chlorophyll fluorescence of two species of marine microalgae. *Aquacultural Engineering*, 22: 181–197.

Hugh, W.I., W. Dominy and E. Duerr. 1985. *Evaluation of Dehydrate Spirulina (Spirulina platensis) as a Protein Replacement in Swine Starter Diets.* Honolulu: Hawaii Institute of Tropical Agriculture and Human Resources.

Inoue, N., Y. Taira, T. Emi, Y. Yamane, Y. Kashino and H. Koike. 2001. Acclimation to the growth temperature and the high temperature effects on photo system II and plasma membranes in a mesophilic cyanobacterium *Synechocystis* sp. PCC6803. *Plant and Cell Physiology*, 42: 1140–1148.

Ismail, H., A.M. Azza, A.B.D, El-All and H.A.M. Hassanein. 2013. Biological influence of some microorganisms on olive mill wastewater. *Egyptian Journal of Agricultural Research*, 91(1): 1–9.

Jaiswal, P., R. Prasanna and A.K. Kashyap. 2005. Modulation of carbonic anhydrase activity in two nitrogen fixing cyanobacteria, *Nostoc calcicola* and *Anabaena* sp. *Journal of Plant Physiology*, 162: 1087–1094.

Jia, H and Q. Yuan. 2016. Removal of nitrogen from wastewater using microalgae and microalgae–bacteria consortia. *Cogent Environmental Science*, 2:1275089.

Karthik, M. and K. Kumar. 2016. Optimization of growth conditions and development of a laboratory model race way pond for cultivation of *Spirulina platensis* in anaerobically digested cassava sago factory effluent. *International Journal of Agriculture Sciences*, 8(37): 1763–1767.

Kawaguchi, K. 1980. Microalgae production systems in Asia. In: *Algae Biomass*, Shelef, G., Soeder, C.J. (Eds.). Amsterdam: Elsevier, pp. 25–33.

Kosaric, N., H.T. Nguyen and M.A. Bergougnou. 1974. Growth of *Spirulina maxima* in the effluents from secondary wastewater treatment plants. *Journal of Biotechnology and Bioengineering*, 16: 881–896.

Kotangale, L.R., R. Sarkar and K.P. Krishnamoorthi. 1984. Toxicity of mercury and zinc to *Spirulina platensis. Indian Journal of Environmental Health*, 26: 41–46.

Kulpys, J., E. Paulauskas, V. Pilipavicius and R. Stankevicius. 2009. Influence of cyanobacteria *Arthrospira (Spirulina) platensis* biomass additive towards the body condition of lactation cows and biochemical milk indexes. *Agronomy Research*, 7: 823–835.

Laliberté, G., E.J. Olguin and J. De La Noüe. 1997. Mass cultivation and wastewater treatment using *Spirulina*. In: *Spirulina platensis (Arthospira) Physiology, Cell-biology and Biotechnology*, Vonshak, A. (Ed.). London: Taylor and Francis, p. 165.

Larsdotter, K. 2006. Wastewater treatment with microalgae—A literature review. *Vatten*, 62: 31–38.

Leiterer, M., J.W. Einax, C. Löser and A. Vetter. 1997. Trace analysis of metals in plant samples with inductively coupled plasma–mass spectrometry. *Fresenius Journal of Analytical Chemistry*, 359: 423–426.

Markou, G., I. Chatzipavlidis and D. Georgakakis. 2012. Cultivation of *Arthospira* (*Spirulina*) *platensis* in olive-oil mill wastewater treated with sodium hypochlorite. *Bioresource Technology*, 112: 234–241.

Munoz, R and B. Guieysse. 2008. Algal bacterial processes for the treatment of hazardous contaminants: A review. *Water Research*, 40: 799–815.

Nakagawa, H and G. Gomez-Diaz, 1975. Usefulness of *Spirulina* sp. meal as feed additive for giant freshwater prawn, *Macrobrachium rosenbergii. Suisanzoshuku*, 43: 521–526.

Nguedia, G., E. Miégoué, F. Tendonkeng, C. Sawa, H. Feulefack Defang, J. Fossi, M. Mama, D. A. Ebile, Y. Fongang and E.T. Pamo. 2019. Effect of *Spirulina* level on post-weaning growth of guinea pig *Cavia porcellus* in western Cameroon. *JSM Veterinary Medicine and Research*, 1: 9.

Nogueira, A.M.S., J.S. Junior, H.D. Maia, J.P.S. Saboya and W.R.L. Farias. 2018. Use of *Spirulina platensis* in treatment of fish farming wastewater. *Revista Ciencia Agronomica*, 49(4): 599–606.

Ogawa, T. and G. Teuri. 1970. Blue green alga *Spirulina. Journal of Fermentation Technology*, 48: 361–378.

Ogbonda, K.H., R.E. Aminigo and G.O. Abu. 2007. Influence of temperature and pH on biomass production and protein biosynthesis in a putative *Spirulina* sp. *Bioresource Technology*, 98: 2207–2211.

Oliver, R.L and G.G. Ganf. 2000. Freshwater blooms. In: *The Ecology of Cyanobacteria: Their Diversity in Time and Space*, Whitton, B.A., Potts, M. (Eds.). Dordrecht: Kluwer, pp. 149–194.

Oswald, W.J. 1988. Micro-algae and waste-water treatment. In: *Micro-algal Biotechnology*, Borowitzka, M.A., Borowitzka, L.J. (Eds.). Cambridge: Cambridge University Press.

Oswald, W.J and H.B. Gotaas. 1957. Photosynthesis in sewage treatment. *Transactions of the American Society of Civil Engineers*, 122: 73–105.

Oswald, W.J., H.B. Gotaas, C. Golueke, W. Kellen, E. Gloyna and E. Hermann. 1957. Algae in waste treatment. *Sewage and Industrial Wastes*, 29: 437–457.

Pande, A.S., R. Sarkar and K.P. Krishnamoorthi. 1981. Toxicity of copper sulphate to the alga *Spirulina platensis* and the ciliate *Tetrahymena pyriformis. Indian Journal of Experimental Biology*, 19: 500–502.

Parkavi, K., S. Kuppusamy, H.M. Yusop and S.I.S. Alwi. 2011. POME treatment using *Spirulina platensis* Geitler. *International Journal of Current Science*, 1: 11–13.

Peiretti, P.G and G. Meineri. 2011. Effects of diets with increasing levels of *Spirulina platensis* on the carcass characteristics, meat quality, and fatty acid composition of growing rabbits. *Livestock Science*, 140: 218–224.

Powell, N., A. Shilton, Y. Chisti and S. Pratt. 2009. Towards a luxury uptake process via microalgae—Defining the polyphosphate dynamics. *Water Research*, 43: 4207–4213.

Rajfur, M., A. Klos and M. Waclawek. 2012. Sorption of copper (II) ions in the biomass of alga *Spirogyra* sp. *Bioelectrochemistry*, 87:65–70.

Raouf, N.A., A.A. Al-Homaidan and I.B.M. Ibraheem. 2012. Microalgae and wastewater treatment. *Saudi Journal of Biological Sciences*, 19: 257–275.

Rasoul-amini, S., N. Montazeri-najafabady, S. Shaker, A. Safari, A. Kazemi and P. Mousavi. 2014. Removal of nitrogen and phosphorus from wastewater using microalgae free cells in bath culture system. *Biocatalysis and Agricultural Biotechnology*, 3: 126–131.

Riad, W.A., A.Y. Elsadany and Y.M. EL-diahy. 2019. Effect of *Spirulina platensis* microalga additive on performance of growing Friesian calves. *Journal of Animal and Poultry Sciences*, 10(2): 35–40.

Richmond, A. 2008. *Handbook of Microalgal Culture: Biotechnology and Applied Phycology*. Hoboken, NJ: Wiley.

Richmond, A. and J.U. Grobbelaar. 1986. Factors affecting the output rate of *Spirulina platensis* with reference to mass cultivation. *Biomass*, 10: 253–264.

Richmond, A. and Q. Hu. 2013. *Handbook of Micro Algal Culture, Applied Phycology and Biotechnology*. Chichester: Wiley-Blackwell.

Ross, E. and W. Dominy. 1990. The nutritional value of dehydrated, blue-green algae (*Spirulina plantensis*) for poultry. *Poultry Science*, 69: 794–800.

Salces, B.M., B. Riaño, D. Hernández and M. Cruz García-González. 2019. Microalgae and wastewater treatment: Advantages and disadvantages. In: *Microalgae Biotechnology for Development of Biofuel and Wastewater Treatment*, Alam, M.A., Wang, Z. (Ed.). doi: 10.1007/978-981-13-2264-8_20.

Sandeep, K.P., S.P. Shukla, A. Vennila, C.S. Purushothaman and N. Manjulekshmi. 2015. Cultivation of (*Arthrospira*) *Spirulina platensis* in low cost seawater based medium for extraction of value added pigments. *Indian Journal of Marine Science*, 44(3): 1–10.

Sari, F.Y.A., I.M.A. Suryajaya and H. Hadiyanto. 2012. Kultivasi mikroalga *Spirulina platensis* dalam media POME dengan variasi konsentrasi POME dan komposisi jumlah nutrien. *Jurnal Teknologi Kimia dan Industri*, 1(1): 487–494.

Schindler, D.W. 1977. Evolution of phosphorus limitation in lakes. *Science*, 195: 260–262.

Seshadri, C.V and S. Thomas. 1979. Mass culture of *Spirulina* using low-cost nutrients. *Biotechnology Letters*, 1:287–91.

Setiawan, Y., P.B. Asthary and M. Saepulloh. 2019. CO_2 flue gas capture for cultivation of *Spirulina platensis* in paper mill effluent medium. In: *AIP Conference Proceedings 2120*, 040005. doi: 10.1063/1.5115643.

Simkus, A., V. Oberauskas, J. Laugalis, R. Zelvyte, I. Monkeviciene, A. Sedervicius, A. Simkiene and K. Pauliukas. 2007. The effect of weed *Spirulina Platensis* on the milk production in cows. *Veterinarija ir Zootechnika*, 38: 60.

Singh, A.K and A.K. Pandey. 2018. Microalgae: An eco-friendly tools for the treatment of industrial wastewater and biofuel production. In: *Recent Advances in Phytochemical Management*, Bhargava, R.N. (ed.) Boca Raton, FL: CRC Press Taylor and Francis Group, pp. 167–197.

Suad, G.K and A.L. Ahmad. 2014. Dairy wastewater treatment using microalgae in Karbala city-Iraq. *International Journal of Environment, Ecology, Family and Urban Studies*, 4(2): 13–22.
Tiessen, H. 1995. *Phosphorus in the Global Environment: Transfers, Cycles, and Management*. New York, NY: Wiley.
Toyoshima, M., S. Aikawa, T. Yamagishi, A. Kondo and H. Kawai. 2015. A pilot-scale floating closed culture system for the multicellular cyanobacterium *Arthrospira platensis* NIES-39. *Journal of Applied Phycology*, 27(6): 2191–202.
Tsiplakou, E, M. Abdullah, A. Mavrommatis, M. Chatzikonstantinou, D. Skliros, K. Sotirakoglou, E. Flemetakis, N. Labrou and G. Zervas. 2017. The effect of dietary *Chlorella vulgaris* inclusion on goat's milk chemical composition, fatty acids profile, and enzymes activities related to oxidation. *Journal of Animal Physiology* and *Animal Nutrition*, 102(1): 142–151.
Turpin, P.J.F. 1827. *Spirulina and Oscillarioide*. In: *Dictionnaire des sciences naturelles*, dans lequel, Le Normant, rue de Seine, Paris, FR. 309–310.
Upadhyay, A.K., R. Singh and D.P. Singh. 2019. Phycotechnological approaches toward wastewater management. In: *Emerging and Eco-Friendly Approaches for Waste Management*, Bharagava, R., Chowdhary, P. (Eds.). Singapore: Springer, pp. 423–435.
Usharani, G., G. Srinivasan and S. Sivasakthi. 2014. Analysis of biochemical constituents in *Spirulina platensis* cultivated using rice mill effluent supplementation. *International Journal of Recent Scientific Research*, 5(12): 2183–2187.
Valderrama, L.T., C.M. Del Campo, C.M. Rodriguez, L.E. De Bashan and Y. Bashan. 2002. Treatment of recalcitrant wastewater from ethanol and citric acid production using the microalga *Chlorella vulgaris* and the macrophyte *Lemna minuscula*. *Water Research*, 36 (17): 4185–4192.
Velasquez, S.F., M.A. Chan, R.G. Abisado, R.F.M. Traifalgar, M.M. Tayamen and G.C.G. Maliwat. 2016. Dietary *Spirulina* (*Arthrospira platensis*) replacement enhances performance of juvenile Nile tilapia (*Oreochromis niloticus*). *Journal of Applied Phycology*, 28(2): 1023–1030.
Venkataraman, L.V., T. Somasekaran and E.W. Becker. 1994. Replacement value of blue green algae (*Spirulina platensis*) for fish meal and a vitamin-mineral premix for broiler chicks. *British Poultry Science*, 3: 373–381.
Vetayasuporn, S. 2004. The potential for using wastewater from household scale fermented Thai rice noodle factories for cultivating *Spirulina platensis*. *Pakistan Journal* of *Biological Sciences*, 7(9): 1554–1558.
Vincent, W.F and W.B. Silvester. 1979. Waste water as a source of algae. *Water Research*, 13: 717–719.
Vonshak, A. 1997. *Spirulina platensis* (*Arthospira*). In: *Physiology, Cell Biotechnology*, Vonshak, A. (Ed.). Basingstoke, Hants, London: Taylor and Francis, p. 233.
Vonshak, A. and L. Tomaselli. 2003. *Arthrospira* (*Spirulina*): Systematics and ecophysiology biochemistry. In: *Spirulina platensis* (*Arthrospira*)*: Physiology, Cell-Biology and Biotechnology*, Vonshak, A. (Ed.). London: Taylor & Francis, pp. 505–522.
Westerman, P.W and J.R. Bicudo. 2005. Management considerations for organic waste use in agriculture. *Bioresource Technology*, 96: 215–21.
Wijayanti, M., D. Jubaedah, N. Gofar and D. Anjastari. 2019. Optimization of *Spirulina platensis* culture media as an effort for utilization of *Pangasius* farming waste water. *Sriwijaya Journal of Environment*. 3(3): 108–112.
Wikdors, G.H and M. Ohno. 2001. Impact of algal research in aquaculture. *Journal of Phycology*, 37: 968–74.
World Health Organization. 1973. *Energy and Protein Requirement*. Technical. Report No. 522. Geneva: World Health Organisation.

Yaakob, Z., F.K. Kamrul, R. Rajkumar, M.S. Takriff and S.N. Badar. 2014. The current methods for the biomass production of the microalgae from wastewaters: An overview. *World Applied Sciences Journal*, 31 (10): 1744–1758

Zarrouck, C. 1966. Contribution a` l'e` tude d'une cyanophyce`e. Influence de divers facteurs physiques et chimiques sur la croissance et la photosynthe `se de *Spirulina maxima*. Ph.D. thesis, University of Paris.

Zhai, J., X. Li, W. Li, M.H. Rahaman, Y. Zhao, B. Wei and H. Wei. 2017. Optimization of biomass production and nutrients removal by *Spirulina platensis* from municipal wastewater. *Ecological Engineering*, 108: 83–92

Zhang, F., Y.B. Man, W.Y. Mo and M.H. Wong. 2019. Application of *Spirulina* in aquaculture: A review on wastewater treatment and fish growth. *Reviews in Aquaculture*, 1–18. doi: 10.1111/raq.12341.

12 Role of Microbial Biofilms in Wastewater Management

Anila Fariq, Anum Zulfiqar, Sidra Abbas, and Azra Yasmin
Fatima Jinnah Women University

CONTENTS

12.1 INTRODUCTION

Water, being one of the most important natural resources, is essential for the survival of all life forms on earth. It covers 71% of earth's surface, but the available freshwater is only 2.5%. Enormously growing population requires more quantity of water for utilization leading to water scarcity. The global patterns of urbanization and industrialization are increasing at a tremendous rate which are adversely affecting the natural water quality and quantity (Rajasulochana & Preethy, 2016). Pollution of natural water resources is one of the major problems faced by the global community. This polluted water is disturbing the whole ecosystem and needs to be managed in order to secure the natural resources. Water is polluted by many point and nonpoint sources, which can be broadly classified into three categories:

1. *Domestic waste*: It generally includes the waste produced by households such as partially treated or untreated sewage and road runoff, for example, human waste, garbage, fertilizers for house lawns and gardens, cans, bottles, detergents, and other household items.
2. *Agricultural waste*: It comprises of the waste produced by animal and crop cultivation. For example: cattle waste, vegetable washings, and wastewater from farmhouses and agricultural lands. The use of organic chemicals for agriculture, especially pesticides, is one of the major contributors of water pollution. Pesticides contaminate the surface waters by runoff and the ground water by seepage through the soil.
3. *Industrial waste*: Waste generated by industries is the major source of water pollution. For instance: landfill leachate, paper and pulp waste, chemical processing materials, acid mine water drainage, and metal processing waste (Jones, 2016). Major waste generating industries include mining, power generation, food processing, construction, and manufacturing. Industrial wastewater includes sanitary waste, wastes generated during manufacturing, washing, heating, and cooling operations. These effluents range from biodegradable wastes from paper and pulp industry, tanneries, slaughter houses, sewage, etc. to highly toxic wastes from textile and power generation industries. These effluents pollute the water by increasing Biological Oxygen Demand (BOD), Chemical Oxygen Demand (COD), Total Suspended Solids (TSS), and Total Dissolved Solids (TDS), and high concentrations of heavy metals make the contaminated water unsuitable for irrigation or drinking purpose and also endangering the life of aquatic organisms (Kanu & Achi, 2011).

There are different types of pollutants depending on the composition of waste effluent polluting the water. Pollutants contaminating water may be classified as organic, inorganic, or biological based on the nature of source material. Heavy metals are the major inorganic pollutants present in water, posing serious threats to flora and fauna due to their high toxicity and carcinogenicity. However, nitrates, phosphates, chlorides, sulphates, fluorides, and oxalates are also included in hazardous inorganic pollutants. Majority of organic pollutants are released due to pesticides including fungicides, insecticides, and herbicides; other pollutants in this category are polyaromatic hydrocarbons (PAHs), biphenyls, polychlorinated biphenyls, phenols, halogenated aromatic hydrocarbons, formaldehyde, oils, detergents, greases, etc. Additionally, other hydrocarbons, aldehydes, ketones, alcohols, proteins, lignin, pharmaceuticals, etc. are also present in wastewater. The pollutants can take different forms in water such as colloidal and solvated in suspended forms (Gupta et al., 2012). Some thriving communities of microorganisms in wastewater can cause different diseases. Microbes including bacteria, algae, fungi, planktons, viruses, amoeba, and various worms are harmful for other organisms. Various microbial agents contaminate the drinking water causing water pollution. Water-borne diseases are one the major causes of mortality and morbidity mostly in developing countries. Water related problems usually arise due to various factors, including (1) Ingestion with pathogens like viruses, bacteria, or parasites in water contaminated from animal urine or feces leading to different diseases, e.g., cholera, dysentery, typhoid, diarrhea, and hepatitis. (2) Water scarcity for personal hygiene is another reason for various diseases such as conjunctivitis, skin ulcers, scabies, and trachoma. (3) Diseases due to penetration in human skin or ingestion of infectious agents resulting in clonorchiasis, schistosomiasis, and paragonimias. (4) Diseases associated with biting of pathogen vectors spreading malaria, yellow fever, dengue, and trypanosomiasis (Gambhir et al., 2012).

Water quality is affected by the change in several physical and chemical parameters due to different pollutants and their respective quantity and toxicity. These parameters include pH, chemical oxygen demand, biological oxygen demand, total dissolved solids, total suspended solids, phosphates, nitrates, and other heavy metals (Popa et al., 2012). Wastewater effluents lead to reduced level of dissolved oxygen, bring alteration to physical feature of receiving water bodies, discharge of toxic materials, their bioaccumulation and biomagnifications in aquatic organisms, and high nutrient loads imposing negative impacts on communities and ecosystems. In aquatic ecosystems, low oxygen levels, increase in optimum temperature, inhibition of sunlight due to suspended solids, and food chain contamination as a result of harmful waste effluents threaten the growth and survival of normal biota (Akpor & Muchie, 2011).

Water pollution is generally an impact of anthropogenic activities which not only affects the humans but disturbs the whole ecosystem. Natural aquatic habitats are being polluted to dangerous levels resulting in the loss of biodiversity (Table 12.1). Different pollutants have variable impacts on aquatic flora and fauna endangering their existence and destabilizing the natural ecosystem functioning. Water polluted with excessive nutrients may lead to the overgrowth of algal populations which is known as eutrophication. This situation changes the color, odor, and taste of water ultimately disrupting the balance of aquatic ecosystem. Overgrowth of toxic algae consumed by other species living in water results in eruption of diseases and even death. Chemicals, heavy metals and oils are associated with different diseases including damage to liver, nervous system, kidney, reproductive, and digestive system leading to abnormal body functioning and death of aquatic organisms. Human population is directly influenced by water pollution due to contamination of drinking water. A variety of diseases are associated with water due to the presence of disease-causing agents and hazardous chemicals posing risk to public health.

TABLE 12.1
Various impacts of pollutants. Environmental and health impacts of some pollutants

Pollutants	Environmental impacts	Health impacts	References
Heavy metals	Algal blooms, habitat destruction due to sedimentation, loss of biodiversity, increased water flow and inhibitory effects on plant growth.	Cancer, organ failure, damage to nervous system, improper development and growth, death.	Akpor et al., (2014)
Pharmaceuticals	Altering microbial community, irreversible changes to microbial genome, contamination of food chain and development of drug resistance in organisms.	Endocrine disruption, accumulation in body tissues, damage to reproductive system and inhibitory effect on cell proliferation.	Sirés and Brillas, (2012)., and Patneedi, and Prasadu, (2015).
Pesticides	Persistent in nature leading to bioaccumulation and biomagnification, toxic to non-target biota and food chain contamination	Cancer, diabetes, asthma, neurological diseases, and damage to reproductive system.	Carvalho, (2017), and Kim et al., (2017)
Plastics	Non-biodegradable, deposition in sediments, water column and biota, contamination of food web.	Suffocation, blockage of digestive tract and biofouling, endocrine disruption, change in metabolic processes, and behavioural modifications.	Ogunola et al., (2018)
Microbes	Deterioration of water quality, change in color and odor harming the aesthetics of aquatic systems, increased biological oxygen demand, excessive growth complexing the navigation and channel capacity, blockage of water treatment plants and submerging the growth of beneficial aquatic flora and fauna	Water-borne diseases including stomach ulcers, degenerative heart diseases, gastrointestinal disorders, Pneumonia, liver disorders, and different types of infection etc.	Akpor, and Muchie, (2011).

Liver and kidney damages, hormonal problems, cancer, heart disease, nervous and reproductive system damage, hepatitis, stomach infections, and multiple sclerosis are the major diseases caused by polluted water (Khan & Ghouri, 2011). Some pollutants are particularly dangerous for life due to their persistence resulting in their accumulation in food chain. For example, Dichlorodiphenyltrichloroethane accumulates in the body of planktons. When these planktons are consumed by fish, a higher concentration of DDT gathers in their body. Further intake of these fishes by birds or other animals leads to higher levels of toxic chemical concentrated in them. In this way, the concentration of DDT increases at every successive trophic level and the nontarget species also get exposed to the toxic pollutant. This process is termed as biomagnification that threatens the survival of all life forms at each level of the food chain (Bharathy, 2018).

Wastewater treatment is necessary to avoid the harmful impacts of hazardous pollutants on environment and ensure the safety of public health. Furthermore, making wastewater reusable for various purposes like irrigation and drinking may limit the problem of water scarcity and depletion of natural water resource depletion (Massoud et al., 2009). Pollutants in wastewater can be removed by various physical, chemical, and biological techniques. In order to treat pollutants of variable nature, different waste removal methods are combined together for complete elimination of harmful pollutants. The basic levels of wastewater treatment have three categories in which pollutant treatment takes place with physiochemical and biological methods, i.e., primary, secondary, and tertiary treatments. Primary treatment involves physical and chemical methods for removal of solid particles like sedimentation, flocculation, and sorption. Secondary treatment is done with biological agents for removal of nutrients and other organic matter. Various types of bioreactors are used for this purpose. Tertiary treatment involves removal of all the remaining pollutants after primary and secondary treatments. These technologies are combined in wastewater treatment plants for complete removal of pollutants making water safe for disposal and reuse (Sala-Garrido et al., 2011).

12.1.1 Physical Methods

Some of the physical methods are discussed below:

- *Screening*: It is the initial step in wastewater treatment which is used for the removal of solid particles that can damage the equipment in further processing. Screening devices of different sizes and shapes are available depending on the type of solid particles to be removed. The size of the screens ranges from coarse to microscreens for removing large- to microsized solids. The shape of openings may be rectangular or circular containing bars, racks, rods, or perforated plates.
- *Sedimentation*: It is a wastewater treatment type that involves settling of suspended heavy particles in waste mixture due to the action of gravity. This process is used for the removal of solids under gravitational influence depending on the size of particles and water velocity. Gravity separation is carried out in settling tanks known as clarifiers. Clarifiers may have different designs such as horizontal flow, inclined surface, and solids contact (Bhargava, 2016).
- *Filtration*: Filters are microporous structures for the removal of different contaminants from polluted water. Pollutants get separated from the waster by interaction with the filters involving various processes like microfiltration, ultrafiltration, and nanofiltration. These processes are designed for the removal of specific contaminants including suspending solids, ionic species,

dissolved constituents, and colloidal organic matter. (Igunnu & Chen, 2012; Chollom et al., 2015).

- *Flotation*: Flotation is one of the essential components of wastewater treatment. It is used for the removal of oils, suspended solids, biological solids, and greases by their adherence to air or gas. The solid particles form agglomerates when gathered with gas or air, and these clumps get accumulated on the surface of water which is then skimmed off. Compressed air is helpful in flotation process when it is passed through water. Electroflotation and dissolved air flotation are the common methods of wastewater treatment (Deliyanni et al., 2017).
- *Coagulation*: Addition of certain chemicals for the removal of solid pollutants from wastewater is called coagulation. It is also termed as chemical precipitation and is an important step in wastewater treatment. A variety of coagulants including lime, calcium chloride, and magnesium are used for the coagulation process (Moghaddam et al., 2010; Chowdhury et al., 2013). Coagulation is a physiochemical process due to addition of chemicals and removal of flocs by physical method. It is useful for dewatering of sludge, soda ash and lime, correction of pH, and stabilization of water; pH is adjusted using caustic soda, taste and foul odor can be removed by activated carbon; furthermore, micropollutants like atrazine, etc. can also be removed (Sahu & Chaudhari, 2013).

12.1.2 Chemical Methods

Chemical treatment of wastewater involves the addition of chemicals to bring about changes by different chemical reactions. These methods are usually combined with physical or biological techniques for proper wastewater treatment. Some of the chemical treatment technologies are discussed below:

- *Chemical adsorption*: Adsorption is the method of removing soluble materials from a solution using different adsorptive materials. This is a widely used treatment technology due to its versatility and efficiency as it is effective in removing a variety of pollutants. Many different types of adsorbents are used for chemical treatment of wastewater that chemically reacts with the pollutants and helps in their removal. Widely used sorbents include activated carbon, peat, and fly ash; however, due to the environmental concerns biological materials are preferred for adsorption like coconut, egg shells, peanut hulls, fruit peels, and resins (Renge et al., 2012; he Zhang et al., 2010; Fan et al., 2010). Currently, apart from chemical and biomaterials, the trend of using nanomaterials such as nanotubes and nanosheets is increasing due to high efficiency and effectiveness (Tofighy & Mohammadi, 2011).
- *Oxidation processes*: Chemical wastewater treatment technologies are grouped into two categories, namely chemical oxidation and advanced oxidation processes (AOP) in which oxidation of pollutants take place with any reactive chemical species. Chemical oxidation process leads to either

complete mineralization or conversion of pollutants into nontoxic compounds. Application of these methods can be done individually or in combination known as hybrid advanced oxidation technologies (Holkar et al., 2016).

12.1.3 Chemical Oxidation Processes

These methods involve degradation of wastewater pollutants to less toxic compounds by different reactive species like chlorine, bromine, or other highly oxidizing agents (O_3, H_2O_2).

- *Chlorination*: It is one of the traditional methods of wastewater disinfection intended to remove harmful pathogens from the water by using chlorine. Chlorine reacts with organic matter and produce chlorinated organic compounds. Efficiency of chlorination depends on different parameters including pH, chlorine dose, nature of organic compounds, and other water parameters. Chlorination has been applied successfully to degrade endocrine disruptors and anti-inflammatory drugs including nonylphenol, bisphenol A, and triclosan. Water pH is reported to be most critical parameter that affects performance of the process (Noutsopoulos et al., 2015). Although chlorination is being used due to high efficiency and cost-effectiveness, sometimes the oxidized products are more toxic than the parent compounds. A study reported the presence of highly toxic iodinated products after disinfection. Hypoiodous acid was produced by chlorination of iodine compounds which further react with organic matter forming iodinated byproducts which are more hazardous (Gong & Zhang, 2015).
- *Ozonation*: Ozone is an efficient disinfectant for treating wastewater by their oxidation. This process takes place either by direct reaction of pollutants with ozone or indirectly by reaction with reactive hydroxyl radicals produced by ozone decomposition. Many micropollutants are treated with ozone or the addition of hydroxyl radicals boosts the decomposition rates (Hansen et al., 2016). Oxidation potential of ozone is very high, and it readily reacts with pollutants and oxidizes them to less harmful products. Benefits of ozonation include utilization of ozone in gaseous form, no increase in the volume of wastewater, and no sludge generation. However, the limitations of ozone are high cost, production of toxic by-products, low water solubility and stability, and selective reaction at acidic pH with organic compounds (Gosavi & Sharma, 2014; Mehrjouei et al., 2015; Miralles-Cuevas et al., 2017).

12.1.4 Advanced Oxidation Processes

This method involves a series of processes in which oxidation takes place with the help of highly reactive hydroxyl or sulfate radicals. These methods are applied for the treatment of organic matter that is recalcitrant in nature as well as the inorganic contaminants in wastewater. The inactivation of various pathogens also occurs with such treatments (Deng & Zhao, 2015). It is a two-step procedure in which oxidative species of high reactivity are generated in situ in first step, which further react with the contaminants in the second step. Generation of radicals depends on various process parameters which is ultimately responsible for the treatment efficiency. There are many types of advanced oxidation procedures including catalytic, physical, electrochemical, UV, and ozone-based AOPs (Miklos et al., 2018).

Some of the advanced oxidation methods are explained here:

- *Fenton process*: It is the technique in which hydroxyl ions are produced from hydrogen peroxide when iron is present. Ferrous is converted to ferric ions to produce hydroxyl radicals from hydrogen peroxide which are further utilized to treat a variety of contaminants. The ferric ions can be regenerated to ferrous form of iron in a cyclic manner and can be reutilized. Different parameters affect the efficiency of fenton process, i.e., temperature, pH, buffer solvent, and amount of free iron. This process can be easily carried out at normal atmospheric pressure and room temperature. Many compounds are being treated by this method including phenols, polyvinyl alcohol, and catechol. Other benefits of this methods are availability of reagents, easy storage and handling, and environmental safety of these reagents. However, drawbacks of this technology are related to the waste disposal, its decomposition, loss of iron ions, and sludge formation (Babuponnusami & Muthukumar, 2014).
- *Photolysis*: The technique in which breakdown of contaminants takes place by light/radiation absorption is termed as photolysis. Photodegradation may occur either by direct or indirect ways utilizing different sources of light and radiation. In direct photolysis, molecules are decomposed by direct absorption of solar energy, while in indirect photolysis, natural molecules like nitrates generate highly reactive species such as hydroxyl radicals by consuming solar radiations. Efficiency of photolytic degradation of wastewater depends on various environmental parameters such as turbidity, water column depth, season, weather, and geographic location. It is one of the most important procedures for the treatment of substances that are less reactive and difficult to degrade like pharmaceuticals with aromatic rings and chromophore groups. These compounds may undergo direct or indirect photolysis leading to their degradation (Fatta-Kassinos et al., 2011).

12.1.5 Biological Treatment Technologies

Biological treatment of wastewater includes the use of living organisms for degradation of pollutants known as biodegradation. In this process various microorganisms including bacteria, viruses, and fungi are utilized for removal of contaminants from wastewater. These microbes either mineralize the pollutants completely into CO_2 and H_2O or convert them into innocuous products which are less harmful. This is a cost-effective, efficient, and environmental-friendly approach. Microbes have the ability to consume components of wastewater for their metabolism as an energy source. Biodegradation can take place inside the microbial cell or outside by the action of extracellular enzymes secreted by microbes (Rajasulochana & Preethy, 2016). Biological degradation can be widely distributed into aerobic and anaerobic methods depending on the microbial characters responsible for the decomposition of pollutants.

- *Aerobic treatment*: This treatment involves the organisms that require oxygen for their survival. The reactions occur in the presence of molecular oxygen releasing decomposed products. The end products are usually carbon dioxide, water, and biomass. This treatment is typically used for the pollutants that are difficult to degrade, such as components of refinery wastewater and municipal waste. Various methods are designed for anaerobic treatment of wastewater, e.g., activated sludge and fixed film processes.
- *Anaerobic treatment*: Microbes that do not require oxygen for their growth and survival are involved in this process. In the absence of oxygen, such microbes perform degradation reactions leading to pollutant removal from wastewater. After degradation, carbon dioxide, methane, and extra biomass are generated. This type of treatment is mostly preferred for the wastewater with high organic content such as effluents from food and beverage industries which are rich in alcohol and sugar. Digesters and fluidized bed reactors are the types of anaerobic treatment technologies (Mittal, 2011).

Some of the biological methods of wastewater treatment are mentioned below:

- *Activated sludge process*: It is a conventional way of treating wastewater under controlled conditions. This process depends on the microbes that can utilize the organic matter as their nutrient and energy source by its decomposition. The degradation reactions are allowed to take place in an aeration tank where the waste effluent is mixed with the microbial community. Different mechanisms are involved in this method including biosorption, physical adsorption, and partial or complete degradation of contaminants (Eckenfelder & Cleary, 2013). Haydar et al. (2016) studied the degradation of contaminants from waste effluents of tanning industry. Biological and chemical oxygen demands of wastewater were analyzed before and after treatment. The process resulted in 80% and 90% efficiencies for COD and BOD, respectively (Haydar et al., 2016).
- *Biological trickling filters*: This process involves the immobilization of microbial cells on any supporting matrix. Wastewater containing different pollutants is allowed to pass through the matrix with microbial biomass. The percolating material gets decontaminated as the pollutants in it are decomposed by the immobilized microbes. Different types of these filters are commonly used to treat biological/chemical oxygen demand and other micropollutants (Naz et al., 2015). Aziz and Ali (2017) studied 30 different parameters of municipal and dairy effluents and found that the amount of pollutants exceeds the standard values for safe disposal. These wastewaters need proper treatment before their disposal to environment to minimize the harmful impacts. This study suggested that trickling biofilters are suitable for such wastewaters as these can efficiently remove a variety of pollutants.

- *Biological nitrification and denitrification*: This is a biological process for the treatment of polluted water leading to the degradation of nitrogenous compounds. Nitrification is the method of ammonium oxidation to nitrates or nitrites, while denitrification involves generation of nitrogen gas by nitrates/nitrites reduction. Trace organic contaminants are found to be removed by this process with 90% efficiency (Phan et al., 2014). Different pharmaceuticals including ibuprofen, erythromycin, and galaxolide are also reported to be effectively degraded by the process of nitrification (Suarez et al., 2010).

Furthermore, there are different kinds of bioreactors with variable structure, operating parameters, and pollutant removal efficiencies. A variety of bioreactor designs have been developed to treat different types of contaminants in wastewater. These bioreactors include membrane-based bioreactor systems, moving bed biofilm reactors (MBBR), two phase partitioning bioreactors, and immobilized bioreactors. A combination of these reactors can be used to treat the mixture of hazardous pollutants in wastewater (Kanaujiya et al., 2019).

12.2 BIOFILM MEDIATED WASTEWATER TREATMENT

Biofilm based wastewater treatment involves metabolic activities of biofilm dwelling microorganisms which play a major role in pollutant degradation. Biofilm is a strong and unique ecological niche harboring living as well as dead microorganisms including bacteria, protozoans, algae, and various other microflora. Biofilms are ubiquitous in natural environment such as water and soil, living tissues, medical devices (medical biofilms), and industrial and drinking water distribution systems (Chandki et al., 2011). Biofilms provide a wide range of advantages to its members such as protection and resistance against antimicrobial agents (drugs and disinfectants), harsh and stressful environmental conditions such as dehydration, lack of nutrients, and UV radiations (Santos et al 2018).

In multispecies biofilm, the bacterial population has to compete for the nutrients to survive. However, in a multispecies biofilm, bacteria not only compete for required nutrients but also for their integration in the microbial communities. Hence, biofilm forming bacteria have to seek out for the environmental conditions and bacterial colonies that are suitable for existence as well as growth (Abu Khweek & Amer, 2018). Biofilms have distinct heterogeneous internal structure, i.e., comprise of clusters containing cells, excreted polymeric network, and pores filled with the liquid to occupy the free spaces between the clusters. Each cluster has layers containing diverse microbial species, varied polymer compositions, and different densities of active cells. In aerobic heterotrophic biofilms, filamentous structures "streamers" extend out of the film into the external liquid. Masses of cells extend from a thin basal biofilm attached to the solid surface. As the biofilms get older with time, their physical structure changes, and biopolymers form bridges between clusters and increase the cell density (Melo, 2003).

The major component of biofilm architecture is the extracellular polymeric substances (EPS) also called slime. Microorganisms embedded within the EPS stick to each other and to the surface as well. Biofilms mainly consist of microbial cells and

EPS. EPS might differ in physical and chemical properties, but it is mainly composed of polysaccharides, extracellular DNA, proteins, and lipids. In case of Gram-negative bacteria EPS is neutral or polyanionic. However, in case of some Gram-positive bacteria EPS may be cationic. Due to incorporation of a high amount of water by hydrogen bonding, EPS is very hydrated. There is a five-stage universal growth cycle of a biofilm with common characteristics independent of the phenotype of the organisms (Aparna and Yadav , 2008). The biofilm formation is usually confirmed by the presence of EPS. Staining is a widely used technique for the detection of biofilms (Tanaka et al., 2019). Various studies revealed that overproduction of EPS varies the colony morphology and can be used for the identification of certain species. The polysaccharide constituent of the matrix in the biofilm can provide many benefits to the cells like structure, protection, and adhesion. Aggregative polysaccharides adhere bacterial cells to each other as well as to surfaces. Adhesion promotes the colonization of living and nonliving surfaces by promoting the resistance in bacteria against physical stresses of fluid movement which can detach the cells from nutrient substrates. Polysaccharides can also give protection from different stresses like dehydration, immune effectors, phagocytic cells, and amoebae. Additionally, these compounds may provide definite structure to biofilms, classify the bacterial community, and separate nutrient gradient and waste products. This helps in establishing a heterogeneous population capable of withstanding stress created by the sudden changing environments that many bacteria may confront. The ubiquitous polysaccharide structures, properties, and functions aid in successful adaptation of bacteria to almost every niche (Limoli et al., 2015).

Microorganisms within the biofilms have the ability to acclimatize to the surrounding environment and thus can grow easily. The electrochemical properties of the biofilms can easily corrode the substrate on which it is formed as compared to the biofilm-free substrate (Videla & Herrera, 2005).

12.3 BIOFILM FORMATION

Biofilm cycle comprises of three developmental phases. Biofilm formation is initiated by microbial attachment to a substratum, followed by maturation and formation of the extracellular matrix, then detachment and dispersion of microbes take place (Figure 12.1). During these phases, bacterial biofilms form three-dimensional structures that are separated by water channels, which allow entry of nutrients, oxygen, and discharge of waste products (Abu Khweek & Amer, 2018). The maturation of these aggregated microbial cells is mediated by changes in cell phenotype and physiology regulated by quorum sensing (QS) (Jabra-Rizk et al., 2006). For these reasons, biofilm is an ideal habitat for ecological studies (Barriuso & Martínez, 2018).

12.3.1 STAGE 1 (ATTACHMENT)

In Stage 1, the free moving bacteria attach itself to some surface making a reversible connection with it. A rough surface provides a better environment for the attachment and development of biofilms. During the initial attachment stage, bacterial cells display a logarithmic or exponential growth rate.

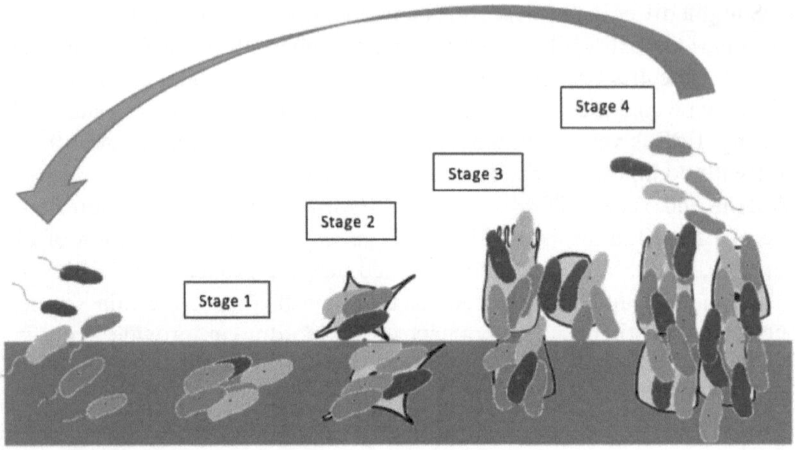

FIGURE 12.1 Different stages of biofilm formation.

12.3.2 Stage 2 (Microcolony Formation)

Stage 2 is characterized by formation of stable microcolonies resulting in irreversible connection with the surface. Release of chemical signal begins for intercommunication. Once the signal reaches a certain intensity, the mechanism for EPS production is activated.

12.3.3 Stage 3 (Three-Dimensional Structure Formation and Maturation)

In Stage 3, certain genes related to biofilm formation are expressed. After the microcolony formation stage of biofilm, expression of certain biofilm related genes take place, necessary for EPS. EPS provides structural integrity to the biofilm matrix.

12.3.4 Stage 4 (Detachment)

Once mature biofilm is developed, bacterial dispersal is observed. Free-floating bacteria are detached and released from the biofilm matrix for further colonization. After getting released from the biofilm, bacterial cells still retain certain biofilm properties, for example antibiotic resistance (Jamal et al., 2015; Aparna & Yadav, 2008).

12.4 ECOLOGICAL SIGNIFICANCE OF BIOFILMS

The durability and dynamic structure of the biofilms along with electrochemical and metabolic characteristics make them attractive for different bioremediation processes as well as monitoring of ecosystems, such as using these microbial communities as biomarkers for monitoring the water quality (Edwards & Kjellerup, 2013).

Biofilms are useful in various industrial processes such as bacterial communities and have been utilized in the past century to neutralize, degrade, and mineralize many xenobiotic compounds in wastewater-activated sludge (Bertin et al., 2007)

However, biofilms can cause severe problems in hospitals, as biofilms can lead to further secondary infections and different other health problems as well as economical problems by prolongling the hospital stays, thus increasing the healthcare cost (Santos et al., 2018).

12.5 CELL TO CELL COMMUNICATION IN BIOFILMS

Biofilm development and QS are interrelated processes in a way that biofilm formation is a mutual group activity involving bacterial populations living in a self-produced extracellular matrix while QS is a cell to cell communication that harmonizes gene expression in response to population cell density. When the population density extends to an optimum concentration, QS might shift to a biofilm lifestyle. In QS, there are interactions within a specie and between species. It depends on the production and detection of extracellular signals. In a fresh culture, bacteria frequently produce signals at very low concentration, and with the increase in population density, signal accumulates in the local environment. The signal interacts with the receptor protein when a threshold concentration is achieved and produce a coordinated as well as modified gene expression in the population. Several types of QS signals include many acyl-homoserine lactone (AHL)-type signals utilize by Proteobacteria, and chemically modified short oligopeptide signals used by Gram-positive species (Solano et al., 2014; Abisado et al., 2018). QS plays a major role in the attachment and EPS production for several bacteria in wastewater treatment (Shrout & Nerenberg, 2012).

12.6 ANALYTICAL APPROACHES TO STUDY BIOFILMS

Different analytical approaches have been employed to qualitatively and quantitatively analyze biofilms. These techniques are used to either quantify biofilm matrix/viable cells of biofilm or can analyze both living and dead cells present in biofilm. Colorimetric methods to assess biofilm matrix include crystal violet assay, 1,9-dimethyl methylene blue, and fluorescein-di-acetate methods. For studying viable cells, BioTimer Assay, LIVE/DEAD BacLight, tetrazolium hydroxide salt, and resazurin methods have been used. Molecular methods to evaluate the bacterial population in biofilms including Polymerase Chain Reaction (PCR) and Flourescence InsituHybridization (FISH) have been reported. Other microscopic techniques used to study microbial biofilms include mass spectrometry, Raman spectroscopy, electron microscopy, and confocal laser scanning microscopy. Owing to the complexity of biofilms, a combination of different techniques is necessary to get better insight of their biochemical, genetic, or physical properties (Pantanella et al., 2013).

12.7 DYNAMICS OF BIOFILM IN WASTEWATER

The biofilm of the wastewater system is comprised of a complex assemblage of microbes, i.e., freshwater diatoms, filamentous bacteria, and their corrosion products. The flow velocity flanking to the substrate is negligible, and the zone of minor flow is termed as the hydrodynamic boundary layer. The attachment of the minute cells on the submerged surface largely alters the low and high velocities of the liquid medium. The cell moves the hydrodynamic boundary layer with low linear velocity, whereas the attachment to the substrate will depend on cell motility and size. Hydrodynamic boundary decreases as the increase in velocity brings more turbulence and mixing to the cell. Higher velocities extend the connection with the surface and exert shear force on the cells attached to the substrate, and as a result disintegration of the cells from the surface occurs.

12.8 ADVANTAGES OF BIOFILM MEDIATED WASTEWATER TREATMENT

Activated sludge process is a widely used technique for wastewater treatment; however, biofilm technologies have been increasingly utilized nowadays. They offer multifarious benefits over suspended microbial cells as they facilitate separation of cells from the rest of treated wastewater during filtration in downstream processing. Biofilms in wastewater either occur as flocs like biofilm, a particle biofilm, and a static biofilm. Particulate and granule biofilms have better settling property than flocs and have better retention time in reactors. The particulate biofilms are used in many reactors like Biofilm Airlift Suspension (BAS), Granular Sludge Blanket (EGSB), Upflow Sludge Blanket (USB), and Internal Circulation (IC) reactors (Rajwar et al., 2018).

In the biofilm-mediated wastewater treatment process, biofilms are attached to substrates, and different nutrients like biochemical oxygen demand, ammonia, nitrate, and dissolved O_2 are supplied from bulk liquid to interface. These nutrients are then utilized to increase the number of microbes for metabolic purposes, and thereby the contaminants in the wastewater are removed. A number of biofilm reactors have been developed so far for the treatment of domestic sewage and a variety of industrial wastewater like biological aerated filter (BAF), biological contact oxidation tank, aerobic inverse fluidized bed biofilm reactor (AIFBBR), biological rotating disc, MBBR, and integrated fixed-film activated sludge reactor (IFAS). In general, biofilm reactors offer several benefits like strong adaptability, high rate of organics and nitrogen removal, low sludge production, and expedient operational management, which result in an increase in demand in the usage of wastewater treatment plants around the world. Recently due to the development of biofilm-based technologies like anaerobic ammonium oxidation, autotrophic denitrification, and gradual improvement of wastewater discharge standards around the world, the biofilm-based treatment systems have become one of the most advanced wastewater treatment methods (Huang et al., 2019).

12.8.1 Biofilm Airlift Suspension Reactor

The biological method for the removal of nitrogen from industrial and municipal wastewater is carried out usually through traditional nitrification and denitrification processes. However, recently the concurrent nitrification and denitrification Simultaneous Nitrification and Denitrification (SND) process has been explored. This process infers that both nitrification and denitrification occur simultaneously in one reactor and under the same operating conditions. A BAS reactor comprises of biofilm-covered biodegradable carriers, and an activated sludge is introduced for the treatment of wastewater containing a low COD:N ratio. A study reported the efficiency of airlift suspension reactor with two biodegradable polymers. A process water from sludge dewatering having ammonium nitrogen 1,150 mg/L concentration was treated in a two-stage system which achieved a nitrogen removal of 75% using biofilm (Walter et al., 2009).

12.8.2 Biological Aerated Filters (BAFs)

BAFs is a type of fixed film reactor that use granular media with a definite surface area and high permeability for secondary and tertiary wastewater treatment. It combines biochemical oxygen demand, solids and ammonia nitrogen (NH_3–N) removal in a single reactor. Nitrification is highly affected by the competition between heterotrophic and autotrophic microorganisms in a single process. High organic loads affect nitrifying organisms in unit. The selection of media plays a major role in the design and operation of BAFs. The granular media helps in maintaining active biomass and a diversity of microbe populations during wastewater treatment. Supporting media in BAFs include clay, schist, phosphorylethanolamine, polyethylene, polystyrene, or zeolite. A study reported two-stage BAFs with modified zeolite for the removal of chemical oxygen demand and ammonia nitrogen at 20°C–26°C and dissolved oxygen above 4.00 mg/L. This study also showed that modified zeolite medium was appropriate for the growth of nitrobacteria biofilms, thereby improving the nitrification process in the reactor.

12.8.3 Aerobic Inverse Fluidized Bed Biofilm Reactor (AIFBBRs)

In AIFBBRs, the solid carrier particles are allowed to move continuously by hydraulic source. Three phase solid-liquid-gas conventional fluidization systems have more density of solid particles than the density of continuous liquid phase. Therefore, fluidization of solid particles is accomplished either by a combination of downward flow of liquid along with the upward flow of the gas or by the upward flow of both gas and liquid. In all abovementioned cases, fluidization of solid particles is in the downward direction which is termed as inverse fluidization. The main advantage of AIFBBRs is that the filter medium does not get clogged. These reactors have high energy efficiency, less pressure drop, elevated gas holdup, and high heat and mass transfer rates from liquid to biofilms. These reactors also hold a relatively high concentration of oxygen in the reacting liquid and significant control on the biofilm thickness. They also show long-term stability and are equally applicable for the treatment of low concentration pollutants. They have been widely used these days for the treatment of industrial wastewaters, e.g., starch, brewery, dairy, steel, textile, sugar, and for the removal of phenol, ferrous iron, aniline, and sulfates (Swain et al., 2018).

12.8.4 Granular Sludge Blanket (EGSB)

An EGSB reactor is used for the removal of high loads of organic carbon, i.e., COD. It comprises of a completely or partially expanded granule bed and has high mechanical stability and settling proficiency which regulates the overall system stability of the reactor. The treatment efficiency of an EGSB reactor is improved under high superficial velocity up to 7–10 m/h, which can be achieved through high recirculation rates and height/diameter ratio. This enhancement assists in high hydraulic mixing, which increases the mass transfer between wastewater and granular sludge. These properties have resulted in the well-known application of EGSB reactors in anaerobic treatment of low-strength as well as medium to high-strength wastewaters. High mixing intensity in an EGSB reactor increases the diffusion of substrate into the biofilm to enhance adequate wastewater–sludge contact (Zheng et al., 2014).

12.8.5 Upflow Anaerobic Sludge Blanket Reactor (UASB)

The upflow anaerobic sludge blanket reactor (UASB) method is mainly used for the treatment of highly concentrated industrial wastewater, but it can also be used for the treatment of municipal wastewater with relatively lower contaminant concentration. UASB reactors have simple design, unsophisticated construction and maintenance, little land requirement, low construction and operational cost, less sludge production, high efficiency of COD removal, capable to withstand fluctuations in temperature, pH, and influent concentration, rapid biomass recovery, and generate bioenergy, i.e., biogas or hydrogen. In UASB, wastewater is introduced from the bottom of the reactor and it flows upward through biologically activated sludge blanket present in the form of granular aggregates. The sludge aggregates are highly stable and show good treatment efficiency upon contact with the granules. The methane and carbon dioxide produced under anaerobic conditions facilitate internal mixing for the formation and maintenance of biological granules (Daud et al., 2018).

12.8.6 Internal Circulation (IC) Reactor

An IC reactor involves an anaerobic digestive process capable to treat high concentrations of COD in industrial wastewater like potato starch, pulp, brewery, and paper. IC reactor requires only a small area and is resistant to shock load, generates biogas, and requires low energy. A case study revealed that the IC reactor COD removal rate is over 80% at COD < 2,300 mg/L (Tran and Bui, 2019).

12.8.7 Biological Contact Oxidation Tank

Biological contact oxidation involves the usage of biofilm and is transitional between activated sludge and the biological filter methods. In a biological contact oxidation tank, the substrate for all microbial habitats submerges, and that is why it is also called as submerged filter. The oxygenation to the wastewater is supplied by mechanical aeration equipment, and it also acts as a substratum for microbial attachment in the aeration tank known as loop-aeration filter or contact aeration tank. It contains about 2%–5% activated sludge suspension in wastewater pool and also purifies wastewater. Therefore, biological contact oxidation has biofilm as well as activated sludge. Moreover, it has high volumetric loading. It is resistant to changes in influent organic loads, require small area, no sludge return and sludge accumulation, require low energy, high purification rate, small treatment time, and easy to operate, manage, and maintain (Li & Zhou, 2011).

12.8.8 Rotating Biological Rotating Contactor (RBC)

A rotating biological contactor (RBC) is a fixed growth bioreactor that provides a suitable alternative to the traditional activated sludge process. An RBC reactor normally comprises of a series of tightly packed large flat or grooved discs mounted on a horizontal shaft that are partially or fully submerged in wastewater. The shaft constantly spins by a mechanical motor or a compressed air drive, and a biofilm is formed onto the whole surface area of the media to metabolize the organic materials present in the wastewater. The movement of media facilitates oxygen transfer and keeps the biomass in aerobic conditions. This movement also gives turbulence in the mixed liquor surface and allows the removal of excess solids from the media. Biomass clarification is done to remove these solids. Microscopic studies show that the outer biofilm layer in RBC reactor is complex and heterogeneous and comprises primarily of filamentous bacteria, protozoa, green eukaryotic algae, and small metazoans while the inner layer is more uniform and compact. The microbial density reduces in the innermost biofilm layer having a higher percentage of nonviable bacteria as compared to outer layer. Biofilm thickness ranges from 0.5 to 4.5 mm in thickness in full-scale disc RBCs used for municipal wastewater treatment. The thickness of the biofilm is controlled to avoid clogging (Cortez et al., 2008).

12.8.9 Moving Bed Biofilm Reactor (MBBR)

The MBBR functions as a 2- (anoxic) or 3-(aerobic) phase system with free floating plastic biofilm carriers. These systems are frequently used for municipal and industrial wastewater treatment, pharmaceutical waste, aquaculture, potable water denitrification, and, in secondary, tertiary, and side-stream applications. The reactor involves a submerged biofilm reactor and a liquid-solid separation unit. The MBBR process has several advantages like capacity to remove carbon-oxidation and nitrogen removal, requires a smaller tank volume, and biomass retention does not require a clarifier, and there is less need of solid loading to the liquid-solid separation unit, and the MBBR does not need a special operational cycle for biofilm thickness as it is a continuous-flow process and liquid-solid separation can be achieved with a variety of processes (McQuarrie & Boltz, 2011).

12.8.10 Integrated Fixed-Film Activated Sludge Reactor (IFAS)

The IFAS reactor consists of biofilm support media in the aerobic condition and is used for biological nitrogen removal. It is a cost-effective option for wastewater treatment plants to support nitrification process throughout the harsh winter conditions without considerably increasing the suspended growth concentration and increasing reactor or clarifier volumes (Sriwiriyarat & Randall, 2005).

12.9 CONCLUSION

This chapter provides a coherent picture of biofilm-based wastewater treatment technology. An in-depth study of microbial community structure, appropriate design, and operation of bioreactors is crucial for the advancement of this treatment. Different factors affecting biofilm formation should be optimized for increased efficiency of bioreactors. More advanced tools like next-generation sequencing for detection of microbial communities in biofilm should be utilized to get a better understanding of their coordinated gene expressions and integrated metabolic activities.

REFERENCES

Abisado, R. G., Benomar, S., Klaus, J. R., Dandekar, A. A., & Chandler, J. R. (2018). Bacterial quorum sensing and microbial community interactions. *MBio*, *9*(3). doi: 10.1128/mBio.02331-17.

Abu Khweek, A., & Amer, A. O. (2018). Factors mediating environmental biofilm formation by *Legionella pneumophila*. *Frontiers in Cellular and Infection Microbiology*, *8*, 38.

Akpor, O. B., & Muchie, B. (2011). Environmental and public health implications of wastewater quality. *African Journal of Biotechnology*, *10*(13), 2379–2387.

Akpor, O. B., Ohiobor, G. O., & Olaolu, D. T. (2014). Heavy metal pollutants in wastewater effluents: sources, effects, and remediation. *Advances in Bioscience and Bioengineering*, *2*(4), 37–43. ISSN 2330-4162.

Aparna, M. S., & Yadav, S. (2008). Biofilms: microbes and disease. *Brazilian Journal of Infectious Diseases*, *12*(6), 526–530.

Aziz, S. Q., & Ali, S. M. (2017). Characterization of municipal and dairy wastewaters with 30 quality parameters and potential wastewater treatment by biological trickling filters. *International Journal of Green Energy*, *14*(13), 1156–1162.

Babuponnusami, A., & Muthukumar, K. (2014). A review on Fenton and improvements to the Fenton process for wastewater treatment. *Journal of Environmental Chemical Engineering*, *2*(1), 557–572.

Barriuso, J., & Martínez, M. J. (2018). In silico analysis of the quorum sensing metagenome in environmental biofilm samples. *Frontiers in Microbiology*, *9*, 1243.

Bertin, L., Capodicasa, S., Occulti, F., Girotti, S., Marchetti, L., & Fava, F. (2007). Microbial processes associated to the decontamination and detoxification of a polluted activated sludge during its anaerobic stabilization. *Water Research*, *41*(11), 2407–2416.

Bharathy, N. (2018). Water pollution and water quality standards for livestock. *International Journal of Science, Environment and Technology*, *7*(6), 1905–1913.

Bhargava, A. (2016). Physico-chemical waste water treatment technologies: an overview. *International Journal of Science Research in Education*, *4*(5), 5308–5319.

Carvalho, F. P. (2017). Pesticides, environment, and food safety. *Food and Energy Security*, *6*(2), 48–60.

Chandki, R., Banthia, P., & Banthia, R., 2011. Biofilms: a microbial home. *Journal of Indian Society of Periodontology*, *15*, 111–114.

Chollom, M. N., Rathilal, S., Pillay, V. L., & Alfa, D. (2015). The applicability of nanofiltration for the treatment and reuse of textile reactive dye effluent. *Water SA*, *41*(3), 398–405.

Chowdhury, M., Mostafa, M. G., Biswas, T. K., & Saha, A. K. (2013). Treatment of leather industrial effluents by filtration and coagulation processes. *Water Resources and Industry*, *3*, 11–22.

Cortez, S., Teixeira, P., Oliveira, R., & Mota, M. (2008). Rotating biological contactors: a review on main factors affecting performance. *Reviews in Environmental Science and Bio/Technology*, *7*(2), 155–172.

Daud, M. K., Rizvi, H., Akram, M. F., Ali, S., Rizwan, M., Nafees, M., & Jin, Z. S. (2018). Review of upflow anaerobic sludge blanket reactor technology: effect of different parameters and developments for domestic wastewater treatment. *Journal of Chemistry*, *2018*, 13.

Deliyanni, E. A., Kyzas, G. Z., & Matis, K. A. (2017). Various flotation techniques for metal ions removal. *Journal of Molecular Liquids*, *225*, 260–264.

Deng, Y., & Zhao, R. (2015). Advanced oxidation processes (AOPs) in wastewater treatment. *Current Pollution Reports*, *1*(3), 167–176.

Eckenfelder, W. W., & Cleary, J. G. (2013). *Activated Sludge Technologies for Treating Industrial Wastewaters*. Lancaster, PA: DEStech Publications, Inc.

Edwards, S. J., & Kjellerup, B. V. (2013). Applications of biofilms in bioremediation and biotransformation of persistent organic pollutants, pharmaceuticals/personal care products, and heavy metals. *Applied Microbiology and Biotechnology*, *97*(23), 9909–9921.

Fan, Y., Wang, B., Yuan, S., Wu, X., Chen, J., & Wang, L. (2010). Adsorptive removal of chloramphenicol from wastewater by NaOH modified bamboo charcoal. *Bioresource Technology*, *101*(19), 7661–7664.

Fatta-Kassinos, D., Vasquez, M. I., & Kümmerer, K. (2011). Transformation products of pharmaceuticals in surface waters and wastewater formed during photolysis and advanced oxidation processes–degradation, elucidation of byproducts, and assessment of their biological potency. *Chemosphere*, *85*(5), 693–709.

Gambhir, R. S., Kapoor, V., Nirola, A., Sohi, R., & Bansal, V. (2012). Water pollution: impact of pollutants and new promising techniques in purification process. *Journal of Human Ecology*, *37*(2), 103–109.

Gong, T., & Zhang, X. (2015). Detection, identification, and formation of new iodinated disinfection byproducts in chlorinated saline wastewater effluents. *Water Research*, *68*, 77–86.

Gosavi, V. D., & Sharma, S. (2014). A general review on various treatment methods for textile wastewater. *Journal of Environmental Science, Computer Science and Engineering & Technology*, *3*, 29–39.

Gupta, V. K., Ali, I., Saleh, T. A., Nayak, A., & Agarwal, S. (2012). Chemical treatment technologies for waste-water recycling—An overview. *RSC Advances*, *2*(16), 6380–6388.

Hansen, K. M., Spiliotopoulou, A., Chhetri, R. K., Casas, M. E., Bester, K., & Andersen, H. R. (2016). Ozonation for source treatment of pharmaceuticals in hospital wastewater–ozone lifetime and required ozone dose. *Chemical Engineering Journal*, *290*, 507–514.

Haydar, S., Aziz, J. A., & Ahmad, M. S. (2016). Biological treatment of tannery wastewater using activated sludge process. *Pakistan Journal of Engineering and Applied Sciences*, *1*, 61–66.

he Zhang, M., lin Zhao, Q., Bai, X., & fang Ye, Z. (2010). Adsorption of organic pollutants from coking wastewater by activated coke. *Colloids and Surfaces A: Physicochemical and Engineering Aspects*, *362*(1–3), 140–146.

Holkar, C. R., Jadhav, A. J., Pinjari, D. V., Mahamuni, N. M., & Pandit, A. B. (2016). A critical review on textile wastewater treatments: possible approaches. *Journal of Environmental Management*, *182*, 351–366.

Huang, H., Peng, C., Peng, P., Lin, Y., Zhang, X., & Ren, H. (2019). Towards the biofilm characterization and regulation in biological wastewater treatment. *Applied Microbiology and Biotechnology*, *103*(3), 1115–1129.

Igunnu, E. T., & Chen, G. Z. (2012). Produced water treatment technologies. *International Journal of Low-Carbon Technologies*, *9*(3), 157–177.

Jabra-Rizk, M. A., Meiller T. F., James C. E., & Shirtliff, M. E. (2006). Effect of farnesol on *Staphylococcus aureus* biofilm formation and antimicrobial susceptibility. *Antimicrobial Agents and Chemotherapy*, *50*, 1463–1469. doi: 10.1128/AAC.50.4.1463-1469.2006.

Jamal, M., Tasneem, U., Hussain, T., & Andleeb, S. (2015). Bacterial biofilm: its composition, formation, and role in human infections. *RRJMB*, *4*(3), 1–15.

Jones, D. L., Freeman, C., & Sánchez-Rodríguez, A. R. (2016). Waste water treatment. In *Encyclopedia of Applied Plant Sciences: Crop Systems* (pp. 352–362). Elsevier-Hanley and Belfus Inc. doi: 10.1016/B978-0-12-394807-6.00019-8.

Kanaujiya, D. K., Paul, T., Sinharoy, A., & Pakshirajan, K. (2019). Biological treatment processes for the removal of organic micropollutants from wastewater: a review. *Current Pollution Reports*, 5(3), 1–17.

Kanu, I., & Achi, O. K. (2011). Industrial effluents and their impact on water quality of receiving rivers in Nigeria. *Journal of Applied Technology in Environmental Sanitation*, *1*(1), 75–86.

Khan, M. A., & Ghouri, A. M. (2011). Environmental pollution: its effects on life and its remedies. *Researcher World: Journal of Arts, Science & Commerce*, *2*(2), 276–285.

Kim, K. H., Kabir, E., & Jahan, S. A. (2017). Exposure to pesticides and the associated human health effects. *Science of the Total Environment*, *575*, 525–535. doi: 10.4028/www.scientific.net/AMM.71-78.4765.

Li, Y., & Zhou, J. (2011). Design of experiment equipment for treating low temperature domestic sewage using biological contact oxidation process. In Sun, D., Sung, W-P & Ran Chen, R. (Eds), *Applied Mechanics and Materials* (Vol. 71, pp. 4765–4769). Trans Tech Publications.

Limoli, D. H., Jones, C. J., & Wozniak, D. J. (2015). Bacterial extracellular polysaccharides in biofilm formation and function. *Microbiology spectrum*, *3*(3). doi: 10.1128/microbiolspec.MB-0011-2014.

Massoud, M. A., Tarhini, A., & Nasr, J. A. (2009). Decentralized approaches to wastewater treatment and management: applicability in developing countries. *Journal of Environmental Management*, *90*(1), 652–659.

McQuarrie, J. P., & Boltz, J. P. (2011). Moving bed biofilm reactor technology: process applications, design, and performance. *Water Environment Research*, 83(6), 560–575.

Mehrjouei, M., Müller, S., & Möller, D. (2015). A review on photocatalytic ozonation used for the treatment of water and wastewater. *Chemical Engineering Journal*, *263*, 209–219.

Melo, L. F. (2003). Biofilm formation and its role in fixed film processes. In Mara, D., & Horan, N. (Eds.), *Handbook of Water and Wastewater Microbiology* (pp. 337–349). London: Academic Press.

Miklos, D. B., Remy, C., Jekel, M., Linden, K. G., Drewes, J. E., & Hübner, U. (2018). Evaluation of advanced oxidation processes for water and wastewater treatment—A critical review. *Water Research*, *139*, 118–131.

Miralles-Cuevas, S., Oller, I., Agüera, A., Llorca, M., Pérez, J. S., & Malato, S. 2017. Combination of nanofiltration and ozonation for the remediation of real municipal wastewater effluents: acute and chronic toxicity assessment. *Journal of hazardous materials*, *323*, pp.442-451.

Mittal, A. (2011). Biological wastewater treatment. *Water Today*, *1*, 32–44.

Moghaddam, S. S., Moghaddam, M. A., & Arami, M. (2010). Coagulation/flocculation process for dye removal using sludge from water treatment plant: optimization through response surface methodology. *Journal of Hazardous Materials*, *175*(1–3), 651–657.

Naz, I., Saroj, D. P., Mumtaz, S., Ali, N., & Ahmed, S. (2015). Assessment of biological trickling filter systems with various packing materials for improved wastewater treatment. *Environmental Technology*, *36*(4), 424–434.

Noutsopoulos, C., Koumaki, E., Mamais, D., Nika, M. C., Bletsou, A. A., & Thomaidis, N. S. (2015). Removal of endocrine disruptors and non-steroidal anti-inflammatory drugs through wastewater chlorination: the effect of pH, total suspended solids and humic acids, and identification of degradation by-products. *Chemosphere*, *119*, S109–S114.

Ogunola, O. S., Onada, O. A., & Falaye, A. E. (2018). Mitigation measures to avert the impacts of plastics and microplastics in the marine environment (a review). *Environmental Science and Pollution Research*, *25*(10), 9293–9310.

Pantanella, F., Valenti, P., Natalizi, T., Passeri, D., & Berlutti, F. (2013). Analytical techniques to study microbial biofilm on abiotic surfaces: pros and cons of the main techniques currently in use. *Annali di igiene*, *25*, 31–42. doi: 10.7416/ai.2013.190341.

Patneedi, C. B., & Prasadu, K. D. (2015). Impact of pharmaceutical wastes on human life and environment. *Rasayan Journal of Chemistry*, *8*(1), 67–70.

Phan, H. V., Hai, F. I., Kang, J., Dam, H. K., Zhang, R., Price, W. E., Broeckmann, A., & Nghiem, L. D. (2014). Simultaneous nitrification/denitrification and trace organic contaminant (TrOC) removal by an anoxic–aerobic membrane bioreactor (MBR). *Bioresource Technology*, *165*, 96–104.

Popa, P., Timofti, M., Voiculescu, M., Dragan, S., Trif, C., & Georgescu, L. P. (2012). Study of physicochemical characteristics of wastewater in an urban agglomeration in Romania. *The Scientific World Journal*, *2012*, 549028.

Rajasulochana, P., & Preethy, V. (2016). Comparison on efficiency of various techniques in treatment of waste and sewage water—A comprehensive review. *Resource-Efficient Technologies*, *2*(4), 175–184.

Rajwar, D., Bisht, M., & Rai, J. P. N. (2018). Wastewater treatment: role of microbial biofilm and their biotechnological advances. In *Microbial Biotechnology in Environmental Monitoring and Cleanup* (pp. 162–174). IGI Global. Edited by Pankaj and Anita Sharma. doi:10.4018/978-1-5225-3126-5.ch010.

Renge, V. C., Khedkar, S. V., & Pande, S. V. (2012). Removal of heavy metals from wastewater using low cost adsorbents: a review. *Scientific Reviews and Chemical Communications*, *2*(4), 580–584.

Sahu, O. P., & Chaudhari, P. K. (2013). Review on chemical treatment of industrial waste water. *Journal of Applied Sciences and Environmental Management*, *17*(2), 241–257.

Sala-Garrido, R., Molinos-Senante, M., & Hernández-Sancho, F. (2011). Comparing the efficiency of wastewater treatment technologies through a DEA metafrontier model. *Chemical Engineering Journal*, *173*(3), 766–772.

Santos, A.L.S.D., Galdino, A.C.M., Mello, T.P.D., Ramos, L.D.S., Branquinha, M.H., Bolognese, A.M., Columbano Neto, J. and Roudbary, M., 2018. What are the advantages of living in a community? A microbial biofilm perspective!. *Memórias do Instituto Oswaldo Cruz*, *113*(9), e180212.

Shrout, J. D., & Nerenberg, R. (2012). Monitoring bacterial twitter: does quorum sensing determine the behavior of water and wastewater treatment biofilms? *Environmental Science & Technology*, 46(4), 1995–2005.

Sirés, I., & Brillas, E. (2012). Remediation of water pollution caused by pharmaceutical residues based on electrochemical separation and degradation technologies: a review. *Environment International*, *40*, 212–229.

Solano, C., Echeverz, M., & Lasa, I. (2014). Biofilm dispersion and quorum sensing. *Current Opinion in Microbiology*, 18, 96–104.

Sriwiriyarat, T., & Randall, C. W. (2005). Evaluation of integrated fixed film activated sludge wastewater treatment processes at high mean cells residence time and low temperatures. *Journal of Environmental Engineering*, 131(11), 1550–1556.

Suarez, S., Lema, J. M., & Omil, F. (2010). Removal of pharmaceutical and personal care products (PPCPs) under nitrifying and denitrifying conditions. *Water Research*, *44*(10), 3214–3224.

Swain, A. K., Sahoo, A., Jena, H. M., & Patra, H. (2018). Industrial wastewater treatment by aerobic inverse fluidized bed biofilm reactors (AIFBBRs): a review. *Journal of Water Process Engineering*, 23, 61–74.

Tanaka, N., Kogo, T., Hirai, N., Ogawa, A., Kanematsu, H., Takahara, J., Awazu, A., Fujita, N., Haruzono, Y., Ichida, S., & Tanaka, Y., 2019. In-situ detection based on the biofilm hydrophilicity for environmental biofilm formation. *Scientific Reports*, *9*(1), p.8070.

Tofighy, M. A., & Mohammadi, T. (2011). Adsorption of divalent heavy metal ions from water using carbon nanotube sheets. *Journal of Hazardous Materials*, *185*(1), 140–147.

Tran, N., & Bui, M. (2019). Applying internal circulation anaerobic reactor for wastewater treatment: a case study in Saigon paper mill wastewater treatment plant. *Architecture Civil Engineering Environment*, *12*(3), 145–151.

Videla, H. A., & Herrera, L. K. (2005). Microbiologically influenced corrosion: looking to the future. *International Microbiology*, *8*, 169–180.

Walters, E., Hille, A., He, M., Ochmann, C., & Horn, H. (2009). Simultaneous nitrification/denitrification in a biofilm airlift suspension (BAS) reactor with biodegradable carrier material. *Water Research*, *43*(18), 4461–4468.

Zheng, M., Yan, Z., Zuo, J., & Wang, K. (2014). Concept and application of anaerobic suspended granular sludge bed (SGSB) reactor for wastewater treatment. *Frontiers of Environmental Science & Engineering*, *8*(5), 797–804.

13 Biosorption and Discolorization of Textile Dye Effluent Using Fungi Isolated From Soil Samples Collected Near Textile Dye Industry

A. Thaminum Ansari
Muthurangam Government Arts College

CONTENTS

13.1 INTRODUCTION

Dyes are natural and xenobiotic compounds that make the world more beautiful through colored substances. However, the release of colored wastewaters represents a serious environmental problem and a public health concern. Color removal, especially from textile wastewaters, has been a big challenge over the last decades, and up to now, there is no single and economically attractive treatment that can effectively decolorize dyes.

Dyes are released into the environment through industrial effluents from three major sources such as textile, dyestuff manufacturing, and paper industries. One of the most pressing environmental problems related to dye effluents is the improper disposal of wastewater from dyeing industry. The conventional treatment methods of color and organic pollutants removal from industrial wastewater lead to severe water pollution, thus developing cost-effective cleanup operations. Microbial degradation seems to be promising compared to other organisms, and the method of application is simpler compared to other available methods. The textile industry accounts for the largest consumption of dyestuffs, at nearly 80% (Lorimer et al., 2001). Industrialization is vital to a nation's economy because it serves as a vehicle for development. Increasing industrialization and urbanization leads to environmental pollution. The discharges of toxic effluents from various industries adversely affect water resources, soil fertility, aquatic organisms, and ecosystem integrity. Among various industries, the textile dyeing industries discharge a large volume of wastewater after the dyeing process. Approximately 10%–15% of the dyes are released into the environment during manufacturing and usage (Keharia et al., 2004).

Since some of the dyes are harmful, dye-containing wastes are important environmental problems (Spadarry et al., 1994). These dyes are poorly biodegradable because of their structures, and treatment of wastewater-containing dyes usually involves physical and chemical methods (Kim and Shoda 1999) such as adsorption, coagulation, flocculation, oxidation, filtration, and electrochemical methods (Calabro et al., 1991). Synthetic dyes are produced to large amounts and are used in different industrial branches including the textile industry. Dye wastewaters discharged from textile and dyestuff industries have to be treated due to their impact on water bodies, and growing public concern over their toxicity and carcinogenicity in particular.

Textile processing industries largely employ azo dyes. They are of environmental interest because of their widespread usage and their potential for forming aromatic hydrocarbons, such as derivatives of benzene, toluene, naphthalene, phenol, and aniline. Azo dyes are the most important group of synthetic colorants that are extensively used in textile, pharmaceutical, and printing industries. It was estimated that 10%–20% of about 0.7 millions of dyestuff being manufactured each year and used in dyeing processes which were found in wastewater (Soares et al., 2001). Several of these dyes are very stable to light temperature and microbial degradation, making them recalcitrant compounds (Nyanhongo et al., 2002). Currently, removal of dyes from effluents is carried out by physicochemical means, including adsorption, precipitation, coagulation, oxidation, filtration, and photodegradation. All these methods have different color removal capabilities, capital costs, and operating rates. Among these methods, coagulation and adsorption are the most commonly used, but they create huge amounts of sludge, which becomes a pollutant creating its own disposal problems. There is a great need to develop an economical and effective way of dealing with textile dyeing waste at the level of the industry itself in the face of ever-increasing production activities.

Treatment of synthetic chemical pollutants using microorganisms, namely fungi and bacteria which produce a variety of enzymes, proved to be efficient in biodegradation of a number of chemical pollutants. Biodegradation constitutes an attractive alternate for physicochemical methods of remediation, mainly due to its reputation as low cost, eco-friendly, and publicly acceptable treatment technology. A large number of microorganisms have been isolated in recent years that are able to degrade dyes previously considered non-degradable. Hence, biodegradation offers an attractive potential for the removal of toxic substances from effluent, thereby resulting in detoxification and safe environmental discharge of the effluents. Textile dyes are classified based on the nature of their chemical structures as azo, diazo, cationic, and basic. Further, depending on their application, dyes are classified as reactive, disperse, mordant, etc., among all reactive, azo dyes are more problematic due to their excess consumption and high water solubility. Furthermore, the sulfonic acid and the azo group are rare among natural products, and thus both confer xenobiotic character on this class of compounds. Anthraquinone-based dyes are more resistant to bacterial degradation due to their fused aromatic structures, which remain colored for long periods of time. A wide variety of azo dyes with anthraquinone, polycyclic, and triphenylmethane groups are being increasingly used in textile dyeing and printing processes. They pose toxicity (lethal effect, genotoxicity, mutagenicity, and carcinogenicity) to aquatic organisms (fish, algae, bacteria, etc.) as well as animals. Chronic effects of dyestuffs, especially of azo dyes, have been studied for several decades. The environmental and subsequent health effects of dyes released through textile industry wastewater are increasingly becoming subject to scientific scrutiny. Wastewater from textile industry is a complex mixture of many polluting substances ranging from organochloride-based waste pesticides to heavy metals associated with dyes and dyeing process (Correia et al., 1994).

Biological discoloration of dyes has been attributed to biodegradation and biosorption. The former is the active way of dye decolorization, and the latter is the passive way of dye decolorization. In view of the fact that biological decoloriation bodes well than the conventional methods, screening of microorganism capable of decolorizing dyes will possibly be useful. In this study, we have screened different fungi and tested their efficiency in dye decolorization.

13.2 PRODUCTION AND USES OF DYES

Synthetic dyes are used extensively in textile dyeing, photography, and as additives in petroleum products. Azo dyes are the largest and most versatile class of dyes and account for more than 50% of the dyes produced annually. Approximately, 40,000 different dyes and pigments are used industrially, presumably more than 2,000 different azo dyes are currently used, and over 7×10^5 tons of these dyes are produced annually worldwide (Zollinger, 1987). Desirable criteria of these dyes are that their high stability in light and washing, and resistant to microbial degradation. Since they are not readily degradable under natural conditions, they are not easily removed from wastewater by conventional wastewater treatment methods. A large amount of azo dyes is used in the dyeing of textiles, and it had been estimated that about 10% of the dyestuff in the dyeing processes do not bind to fibers and are, therefore, released to the environment (Hildenbrand et al., 1999).

Based on the chemical composition, synthetic dyes are classified as azo dyes, nitro dyes, triphenylmethane dyes, phthalocyanine dyes, indigoid dyes, and anthraquinone dyes. Based on the application and usage, dyes are classified into acidic dyes, basic dyes, reactive dyes, polyazo dyes, vat dyes, azoic or naphthol dyes, and disperse dyes. Among the many classes of synthetic dyes used, azo dyes and reactive dyes are the largest and the most versatile. They play prominent roles in all types of application.

Azo dyes are a group of compounds characterized by the presence of one or more azo bonds (–N=N–) in association with one or more aromatic systems. Azo dyes are resistant to biological and chemical degradations. However, studies indicate that most of the azo dyes are toxic. The azo-nitrogen of the dye substrates provides nitrogen requirement of the organism in cultures in the absence of nitrogen. Decolorization of dyes is a reduction process which requires redox equivalents (electron donors) that transfer electrons to the chromatographic group dyes. Nitrate (NO_3^-) has higher oxidation character compared to the chromatographic group dyes. The result obtained was attributed to competition between (NO_3^-) and the chromophoric group for the redox equivalents, which results in preferential reduction of (NO_3^-) relative to the chromophoric group. Decolorization was accompanied by increase in total viable count. Ring opening of the aromatic moiety of the dyes produces the carbon source for the organism.

Basic dyes are the brightest class of dyes and are applied widely in small-scale industries such as textile, carpets, and wool industries whose effluents bearing dyes are released into natural streams. In dying industry, above 30–60 L of water is consumed per kilogram of cloth dyed, and large quantities of effluents are released during the process (Rajeshwari et al., 2001).

13.3 POLLUTION PROBLEMS

Color, being the most discernable indicator of pollution in the aqueous environment, is the first to attract the attention of even an environmentally indifferent person. The ease and cost-effectiveness in synthesis, firmness, and variety in color as compared to natural dyes have replaced natural dyes with synthetic dyes. Synthetic, color-causing substances are toxic and recalcitrant in nature. Many countries have banned the use of several colorants.

The wastewater from the textile industry is known to be strongly colored, presenting a large amount of suspended solids, pH broadly fluctuating, high temperature, and high chemical oxygen demand (COD) and biological oxygen demand (BOD). Color is the first contaminant to be recognized in this wastewater. A very small amount of dye in water (10–50 mg/L) is highly visible and reduces light penetration in water systems, thus causing a negative effect on photosynthesis (Nigam et al., 2000).

The major environmental problem of colorants is the removal of dyes from effluents. The disposal of colored wastes such as dyes and pigments into receiving waters damages the environment, as they are carcinogenic and toxic to humans and aquatic life (Lee et al., 1999). Besides the problem of color, some dyes impart non-visibility and can be modified biologically to toxic or carcinogenic compounds. Nowadays, concern has increased about the long-term toxic effect of water containing these dissolved pollutants. Synthetic dyes have replaced natural dyes and pigments in textile and dye industries.

13.4 TOXICITY OF HEAVY METALS

Most of the heavy metals are well-known toxic and carcinogenic agents, and they represent a serious threat to the human population and the fauna and flora of the receiving water bodies. Heavy metals have a great tendency to bioaccumulate and end up as permanent additions to the environment. When wastewater is discharged to receiving water bodies without removal of heavy metals, heavy metals may be harmful to both human and aquatic life as they are non-degradable and persistent. Heavy metals are considered to be the following elements: Copper, Silver, Zinc, Cadmium, Gold, Mercury, Lead, Chromium, Iron, Nickel, Tin, Arsenic, Selenium, Molybdenum, Cobalt, Manganese, and Aluminum. The removal of heavy metals from wastewater has recently become the subject of considerable interest owing to strict legislations. Industrial wastewater containing heavy metals should be treated before their discharge to the water stream, but its treatment is very costly. There are several techniques to remove heavy metals from wastewater such as filtration and electrocoagulation, but there is some limitation such as long treatment time. Various biological treatments, both aerobic and anaerobic, can be used for heavy metal removal. Heavy metals are the major pollutants in the environment due to their toxicity and threat to creatures and human being at high concentrations. Copper is highly toxic because it is non-biodegradable and carcinogenic. The effects of Ni exposure vary from skin irritation to damage of the lungs, nervous system, and mucous membranes.

13.5 TOXICITY OF DYES

As the concentration of textile dye increases in water bodies, it stops the deoxygenating capacity of the receiving water and cut off sunlight, thereby upsetting biological activity in aquatic life (Zollinger, 1987). Also the photosynthesis process of aquatic plants or stage is affected. It has been observed that the chemicals used to produce dyes today have highly carcinogenic, toxic, and explosive effects on our environment. In addition to being toxic, dye effluents also contain chemicals that are carcinogenic, mutagenic, or teratogenic to various organisms (Hameed et al., 2008). Azo dyes and nitrated polycyclic aromatic hydrocarbons are two groups of chemicals that are abundant in the environment. They cause severe contamination in river and ground water in the vicinity of dyeing industries (Riu et al., 1998). The impact of azo dyes in food industry and their degraded products on human health has caused concerns over a number of years, in spite of legislation controlling their use in several countries. Benzidine (BZ) based azo dyes are widely used in dye manufacturing and carcinogenicity of these dyes on epidemiological studies was carried out by many workers. BZ has long been recognized as a human urinary bladder carcinogen and tumorigenic in a variety of laboratory animals (Haley, 1975). Since BZ is used as a reactant in dye synthesis, workers could be directly exposed to the carcinogen. Experimental studies with rats, dogs, and hamsters have shown that animals administered BZ and BZ-congener-based dyes excrete potentially carcinogenic aromatic amines and their N-acetylated derivatives in their urine (Nony and Bowman, 1980). Oxidative metabolism of polynuclear aromatic hydrocarbon (PAH) is accomplished through highly electrophile oxidases, and some of them may bind covalently to macromolecules such as DNA, RNA, and proteins and induce mutations. PAHs are active inducers of liver enzymes of fish and produce metabolites that are capable of inducing cancer. Even in small concentration, naphthalene is capable of including hepatic histopathological alterations, which may cause severe physico-metabolic dysfunctions leading to death. PAH-contaminated sediments are also a concern for human health because of their potential to contaminate sea food consumed by man (Neff, 1985). Plant growth parameters, namely, germination percentage, seedling survival, and seedling height, have been taken as criteria to assess plant response to specific pollutants. Some of the dyes used in dyeing industries are even carcinogenic and mutagenic, and the effluents even reduce the rate of seed germination and growth of crop plants (Nirmalarani and Janardhanan, 1988).

13.6 DYE DECOLORIZATION

The development of wastewater treatment biotechnology increases interests due to its eco-friendly benefits. Over the past decades, biological decolorization has been investigated as a method to transform, degrade, or mineralize azo dyes (Buna et al., 1996). Moreover, such decolorization and degradation is an environmentally friendly and cost-competitive alternative to chemical decomposition possess (Verma and Madamwar, 2000). Unfortunately, most azo dyes are recalcitrant to aerobic degradation by bacterial cells. However, there are few known microorganisms that have the ability to reductively cleave azo bonds under aerobic conditions. Compared with chemical/physical methods, biological processes have received more interest because of their cost-effectiveness, lower sludge production, and environmental friendliness. Various wood-rotting fungi were able to decolorize azo dyes using peroxidases or laccases. But fungal treatment of effluents is usually time-consuming. Under static or anaerobic conditions, fungi decolorization generally demonstrates good color removal effects. However, aerobic treatment of azo dyes with fungi usually achieved low efficiencies because oxygen is a more efficient electron acceptor than azo dyes.

Recently, dye removal became a research area of increasing interest, as government legislation concerning the release of contaminated effluent becomes more stringent. Various treatment methods for the removal of dyes from industrial effluents like chemical coagulation using alum, lime, ferric chloride, ferric sulfate, and electrocoagulation are very time-consuming and costly with low efficiency. Photo-oxidation by UV/H_2O_2 needs additional chemicals and therefore causes a secondary pollution. Chemical oxidation methods using chlorine and ozone and membrane separation methods are in vogue.

Unfortunately, these methods are quite expensive and show operational problems such as developments of toxic intermediates, lower removal efficiency, and higher specificity for a group of dyes. Special measures therefore are necessary to be taken to remove them from the effluents. Over the past decades, microorganisms have been studied for their ability to degrade recalcitrant organopollutants such as poly aromatic hydrocarbons, chlorophenols, and polychlorinated biphenyls. Several workers have focused their interest on optimization of cultural conditions for dye decolorization.

13.7 BIOSORPTION

Biosorption has been studied for removing heavy metals, dyes, and other organic pollutants by various microorganisms from wastewater. Among these microorganisms, fungal biomass can be produced cheaply and obtained as a waste from various industrial fermentation processes (Kapoor and Viraraghavan, 1995). Compared with live fungal cells, dead fungal biomass possesses various advantages such as absence of nutrient needs and ease of regeneration. Biomass from algae, yeast, filamentous fungi, and bacteria has been used to remove dyes by biosorption. The biosorption capacity of a microorganism is attributed to the heteropolysaccharide and lipid components of the cell wall, which contain different functional groups, including amino, carboxyl, hydroxyl, phosphate, and other charged groups, causing strong attractive forces between the different dyes and the cell wall. The effectiveness of biosorption depends on the following conditions: pH, temperature, ionic strength, time of contact, adsorbent and dye concentration, dye structure, and type of microorganism.

Decolorization by biological means may take place in two ways: either by adsorption (or biosorption) on the microbial biomass or biodegradation by the cells. Biosorption involves the entrapment of dyes in the matrix of the adsorbent (microbial biomass) without destruction of the pollutant, whereas in biodegradation, the original dye structure is fragmented into smaller compounds resulting in the decolorization of synthetic dyes. Over the past few decades, numerous microorganisms have been isolated and characterized for degradation of various synthetic dyes, but most of the reports have dealt mainly with decolorization of azo dyes. There is a dearth of information regarding the degradation and detoxification of basic dyes by microbial systems despite their increased use by the textile industry. Hence, the isolation of a potent species that has the capability of degrading and detoxifying basic dyes is of interest in the biotechnological aspect of dye effluent treatment.

Metal uptake is a combination of a rapid metabolism-independent process, followed by a slower one. Therefore, biosorption of heavy metals can be defined as metal removal by passive linkage in live and dead biomasses from aqueous solutions in a mechanism that is not controlled by metabolic steps. Biosorption process involves solid-phase (sorbent or biosorbent) and liquid-phase (solvent) biomaterials containing dissolved species such as metal ions to be sorbed (sorbate). In biosorption, like every sorption process, binding of sorbate species to biosorbent continues until equilibrium establishes between the sorbate species in liquid and solid phases. Biosorbents contain some molecular groups that have affinity to sorbates such as metal ions. This technology employs various types of biomass to remove heavy metals from contaminated waters. The overall incentives of biosorption development for industrial processes are (a) low cost of biosorbents, (b) great selectivity and efficiency for metal removal at low concentration, (c) potential for biosorbent regeneration and metal valorization, (d) high velocity of sorption and desorption, (e) limited generation of secondary residues, and (f) more environmental-friendly life cycle of the material. So far, an industrially relevant method for removal of toxic metals has not been achieved yet (Volesky and Naja, 2003).

13.7.1 Rhodamine-B

Rhodamine-B (bright reddish-white powder) is a synthetic red to pink colored water and alcohol-soluble dye having brilliant fluorescent qualities with molecular formula $C_{29}H_{29}N_2O_5ClNa_2$. Rhodamine-B is a fluorescent cationic dye in which the dye type becomes more important in textile dyeing because of its more rigid structure than other organic dyes. Due to its cationic structure, it can be applied to anionic fabrics which contain negative charges such as polyester fibers, wool, silk, and acrylic fibers. The dye is brilliant and the most fluorescent among other synthetic dyes. Rhodamine is often used as a tracer dye within water to determine the rate and direction of flow and transport. Rhodamine-B basic dye has wide applications in the dyeing of cotton, silk, paper, bamboo, weed, straw, and leather. Apart from this, it is widely employed in the preparation of carbon paper, ball-point pen, stamp pad inks, paint, and printing ink manufacturing. Rhodamine-B dye was the IARC monographs on the evaluation of carcinogenic risk of chemicals to humans.

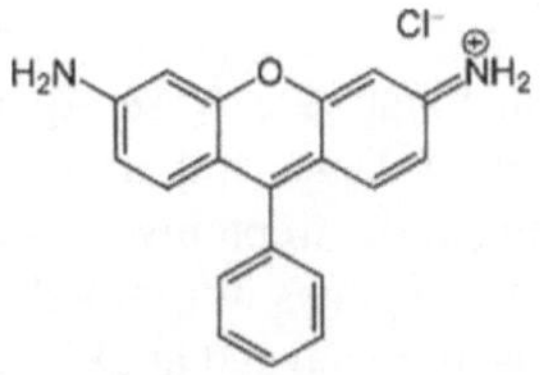

Rhodamine-B
9-(2-carboxyphenyl)-6-diethylamino-3-xanthenylidene]-diethylammonium chloride

13.7.2 Basic Violet-2 Dye

Basic Violet-2 is a synthetic dye of molecular formula $C_{22}H_{24}ClN_3$ and molecular weight 365,899060 [g/mol]. It is used as a non-reactive hair-coloring agent ("direct" dye) in oxidative and semi-permanent hair dye formulations at a maximum concentration of 0.25% and 0.5% in the finished cosmetic product, respectively. It is common practice to apply 35 mL of the formulation over a period of 30 min followed by washing off with water and shampoo (Figure 13.2).

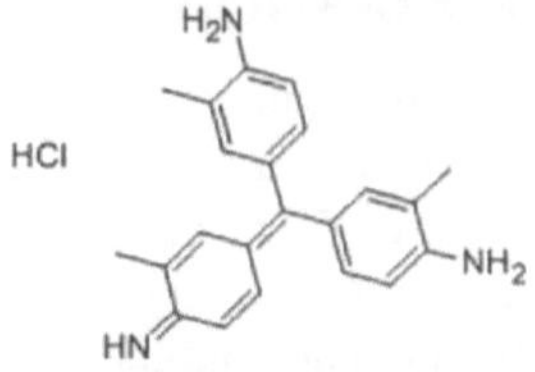

Basic Violet-2
4-[(4-amino-3-methylphenyl)-(4-imino-3-methylcyclohexa-2,5-dien-1-ylidene) methyl]-2-methylaniline hydrochloride

The decolorization of wastewaters and industrial effluents has acquired immense importance because of their threat to mankind due to their mutagenic, carcinogenic, and poisonous natures. There is no simple solution to this problem because the conventional physicochemical methods are either costly or only partially competent to treat these wastes.

13.8 SOIL SAMPLE COLLECTION

Soil samples were collected in sterile plastic bags from soil contaminated by untreated textile wastewater from the vicinity of textile dye manufacturing/textile-processing industries and waste disposal sites from various places in Tamil Nadu, India. The samples were transported to the laboratory in an ice-packed chest box within 1 h of collection. The samples (10 g) were transferred into 500 mL beaker containing

250 mL sterile physiological saline and acclimatized for 10 days at ambient temperature (4°C) for further analysis and stored until use.

13.9 ISOLATION OF FUNGI

Fungi species were isolated on Potato Dextrose Agar (PDA) by the soil dilution method. The media poured in Petri dishes and allowed to solidify for 48 h. To suppress the bacterial growth, 25 mg/L of streptomycin was added to the media (Shazia et al., 2013). After solidification, the plates were filled with diluted soil solution (different proportions). The plates were incubated at 28°C for 72 h. The prominent colonies were picked after incubation periods, and inoculated individually in other PDA plates for further purification.

13.10 IDENTIFICATION OF FUNGI

After an incubation period of 72 h, distinct colonies were counted. The fungal cultures were identified on the basis of macroscopic (colonial morphology, texture, shape, color, diameter, and appearance of the colony) and microscopic (septation in mycelium, presence of specific reproductive structures, shape and structure of conidia, and presence of sterile mycelium) characteristics.

13.11 DYE SOLUTION PREPARATION

Rhodamine-B and Basic Violet-2 dye solutions were prepared by dissolving accurately weighed dyes in distilled water at a concentration of 50 mg/L. To compare dye removal on the same basis, the pH of all the samples were adjusted to 7.5. The maximum absorbance of each dye was measured by scanning the dye forms 900–300 nm in a spectrophotometer. The nanometer at which the maximum absorbance observed was the λ_{max} of the specific dye. The concentrations of dye solution in further experiments were determined at its λ_{max} in the absorbance mode. The calibration curve for the dye was prepared by recording the absorbance values for various concentrations of dye solution at the wavelength of maximum absorbance (λ_{max}). A standard graph was constructed for every dye by measuring the absorbance of known composition of the dyes.

13.12 SCREENING OF FUNGI FOR DYE DECOLORIZATION

Isolation of dye degradation organisms was done by spread-plate technique. To 100 mL of the sterile saline, 1 g of the soil sample was added aseptically and mixed thoroughly. From this, 1 mL was aseptically transferred to 9 mL of sterile saline to obtain a dilution of 10^{-3}. Subsequent serial dilutions were made, and the diluted samples were plated on normal decolorization (NDM) agar incorporated with the dyes (Rhodamine-B and Basic Violet-2 separately) and incubated at 28°C for 48–72 h. After incubation period, the organisms were identified and isolated (Patricia et al., 2002).

13.13 FUNGI AND GROWTH MEDIUM

The fungal strain used was isolated from dye wastewater obtained from the soil of textile dye industries. The principle of sequential selective enrichment batch culture for selection of dye-decolorizing fungi was employed in synthetic wastewater medium (SWM) with Rhodamine-B and Basic Violet-2 as the carbon and nitrogen sources. The basic composition of the SWM was (g/L) $(NH_4)_2SO_4$ 0.28, NH_4Cl 0.23, KH_2PO_4 0.067, $MgSO_4{\cdot}7H_2O$ 0.04, $CaCl_2{\cdot}2H_2O$ 0.022, $FeCl_3{\cdot}6H_2O$ 0.005, NaCl 0.15, $NaHCO_3$ 1.0, and 1mL/L of a trace element solution containing (g/L) $ZnSO_4{\cdot}7H_2O$ 0.01, $MnCl_2{\cdot}4H_2O$ 0.1, $CuSO_4{\cdot}5H_2O$ 0.392, $CoCl_2{\cdot}6H_2O$ 0.248, $NaB_4O_7{\cdot}10H_2O$ 0.177, and $NiCl_2{\cdot}6H_2O$ 0.02. The textile wastewater used for isolation of dye-decolorizing fungi was acclimatized for 8 weeks prior to transfer into 250 mL Erlenmeyer flasks containing 100 mL SWM. After incubation of the flasks, a mixed culture that showed quick and stable decolorization activity was transferred to newly prepared SWM. After five successive transfers, it was plated on SWM agar containing 20 mg/L of each dye and incubated at 30°C for 5 days. Fungal colonies around which clear zones found were quickly picked for further studies and designated as TTW 1–5. To check for the dye-degrading potential of each fungi isolate, preliminary batch experiments were carried out using sterile 250 mL Erlenmeyer flasks containing 100 mL of SWM spiked with dye, after which the solution was inoculated with freshly grown bacterial cells. The final pH was 7.2. The bacterial isolate that showed the highest ability to degrade the basic dyes was selected and used for subsequent investigations. The fungal isolate that showed the highest ability to degrade the basic dyes was selected and used for subsequent investigations. The selected fungi (TTW 4) were characterized and identified as *Aspergillus flavus, Fusarium sp.*, and *Aspergillus niger* using Gram stain, spore test, motility test, and a battery of biochemical and physiological tests (Vanderzannt and Splittstoesser, 1997). The cultural (growth) and microscopic appearances of *A. niger*, *A. flavus*, and *Fusarium* sp. are shown in Figures 13.1 and 13.2.

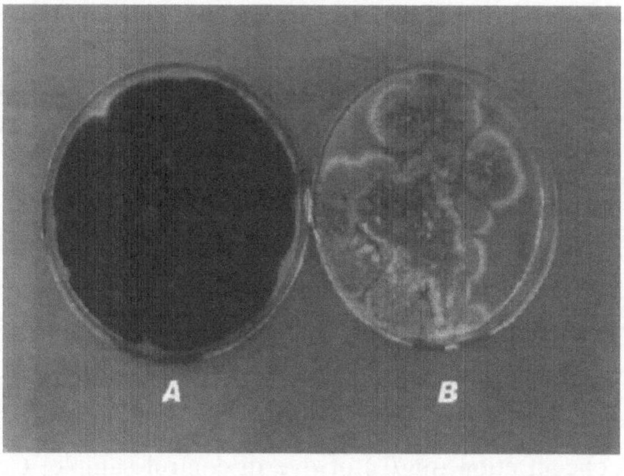

FIGURE 13.1 Growth and appearance of (a) *Aspergillus niger* and (b) *Aspergillus flavus.*

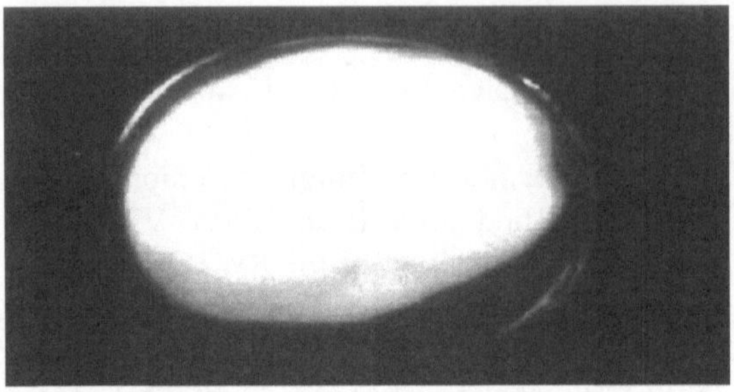

FIGURE 13.2 Growth and appearance of *Fusarium* sp.

13.14 BATCH DECOLORIZATION OPERATION

The decolorization of the basic dyes was studied in 250 mL Erlenmeyer flasks containing 100 mL of SWM and fungi biomass in a batch mode. Each flask was inoculated with 2 mL of freshly grown *A. niger*, *A. flavus*, and *Fusarium* sp. The inoculum size was adjusted at optical density 1.0 at $\lambda = 550$ and 556 nm (1.50×10^7 cells/mL) and incubated under shaking (150 rpm) and static conditions at 30°C. To evaluate the effects of other operational factors on the efficiency of color removal, the batch decolorization experiments were carried out at different initial dye concentrations (1–100 mg/L), temperatures (20°C–40°C), and initial pHs (4–7) under shaking incubation condition. Thereafter, optimal conditions of 40°C, pH 7, and initial dye concentration of 50 mg/L were used in subsequent experiments under shaking condition. Three types of controls were used: uninoculated sterile control, heat-killed control, and sodium azide (0.1% w/v)-amended control. The first indicated the effect of medium components on decolorization, and the latter two showed the adsorption of dyes on cells. After incubation, aliquots (5 mL) were taken and centrifuged at 10,000 rpm for 15 min to separate the fungal cell mass and obtain a clear supernatant, which was used to measure the absorbance of culture samples at the maximum absorption wavelength (λ_{max}) of the respective dyes using a scanning spectrophotometer (Shimadzu UV-2401 PC model Kyoto, Japan). Residual dye concentration of samples was then obtained from a calibration curve of dye concentration versus absorbance prepared for each dye. Decolorization activity was determined by using adsorption (*A*) and dye removal by living biomass (*R*) which was calculated by the following formulae:

$$A(\%) = \frac{C_0 - C_1}{C_0} \times 100\%$$

$$R(\%) = \frac{C_0 - C_{1L}}{C_0} \times 100\%$$

where C_0 is the concentration (mg/L) of dye in control sample, C_1 is the residual concentration (mg/L) of dye in culture samples with killed or sodium azide-treated cells, and C_{1L} is the residual concentration (mg/L) of dye in samples with living biomass. Enumeration of bacterial cells in culture flasks was carried out on a plate count agar (Merck) after tenfold serial dilution of culture samples using the spread-plate method. Viable cell counts obtained after incubation for 24 h at 30°C were expressed as colony-forming units per mL (CFU/mL).

13.15 SUBSTANTIATION OF DYE DEGRADATION

The isolates were inoculated in normal decolorization broth incorporated with the dyes (Rhodamine-B and Basic Violet-2 separately) along with a control (uninoculated). The flasks were incubated at room temperature for 4 days. After incubation, the media along with the cells was centrifuged at 10,000 rpm for 10 min to remove the cells, and the absorbance of the supernatant was analyzed spectrophotometrically (Rhodamine-B (λ_{max}) – 556 nm and Basic Violet-2 (λ_{max}) – 550 nm). The percentage of color removal efficiency was calculated by the following equation:

$$\text{Percentage} = \frac{\text{Absorbance of control} - \text{Absorbance after treatment}}{\text{Absorbance of control}} \times 100$$

13.16 EFFECT OF DIFFERENT GROWTH PARAMETERS

13.16.1 Temperature

To obtain the optimum temperature for the luxuriant growth of the organism, it was grown on normal decolorization broth supplemented with the appropriate dye and incubated at different temperatures, namely 20°C, 25°C, 35°C, and 40°C. The growth was observed after 4 days of incubation. The growth was expressed in dry weight (g/L).

13.16.2 pH

pH is the key parameter that controls the sorption of dyes. The hydrogen ions either act as a bridging moiety between dye molecules and the surface of the biosorbent or may help to produce repulsive forces between the dye molecules. To check the effect of pH, dye solutions at various pHs, namely 3.0, 5.0, and 7.0, were prepared, and the biosorbents were added to the tubes. After the contact time of 60 min, the dye solutions along with the biosorbents were centrifuged at 10,000 rpm for 10 min. The growth was expressed in dry weight (g/L).

13.16.3 Carbon Sources

The effects of carbon sources such as glucose, fructose, maltose, lactose, and starch were studied. The isolates were cultivated in normal decolorization broth amended with different carbon sources and incubated at room temperature for 4 days. After incubation, the growth was measured and expressed in dry weight (g/L).

13.16.4 Nitrogen Sources

The isolates were cultivated in normal decolorization broth amended with different nitrogen sources such as yeast extract, peptone, urea, ammonium sulfate, and sodium nitrate individually incubated at room temperature for 4 days. After incubation, the growth was measured and expressed in dry weight (g/L).

13.17 DECOLORIZATION ASSAYS

Accurately 50 mL of sterile, normal decolorization medium was dispensed into six sterile flasks. The dyes were added in varying concentrations, namely 30, 40, 50, 60, 70, and 80 ppm, respectively. To the flasks, 5% of inocula (5×10^6 spores/mL) were added aseptically and incubated at room temperature for 4 days. Samples were tested at different intervals for dye removal. The optical densities of the samples were analyzed at their λ_{max} by removing the cells present in the samples by centrifuging at 10,000 rpm for 15 min. The percentage of color removal was calculated by comparing it with the standard graph.

13.18 TOXICITY STUDY

Phytotoxicity studies were carried out with 50 mg/L of each dye and its extracted metabolites using seeds of *Triticum aestivum*, *Hordeum vulgare*, and *Lens esculenta* with SWM as control. The degradation metabolites of each dye extracted in ethyl acetate were dried and dissolved in water to form the final concentration of 50 mg/L for phytotoxicity studies (Parshetti et al., 2006). The Petri dishes were kept in dark and observed for germination. Seeds with radicle >1 mm were considered germinated. The germinated seeds were then exposed to a day-and-night cycle length of 10/14 h with a temperature regime of about 28 ± 2°C. The length of plumule (shoot) and radicle (root), and the germination rate (%) were recorded after 7 days.

13.19 BIOSORPTION ASSAY

13.19.1 Microorganisms and Maintenance Conditions

The fungi screened in the above experiments were tested for their efficiency in removing the dye by adsorption.

13.19.2 Fungal Biomass Preparation

Spore suspensions (106 spores/mL) of the isolates were inoculated in 1,000 mL of yeast extract malt and malt extract sucrose liquid medium (YMS) and incubated at room temperature on a rotary shaker at 120 rpm for 48–72 h. The fully grown fungal cells were removed by centrifuging at 10,000 rpm for 15 min. The pellets were removed and washed two times using sterile distilled water and autoclaved in the final wash water at 121°C for 20 min. Such heat treatment is required not only to kill the organisms but also to increase the efficiency of adsorption (Bhole et al., 2004). Water was drained off, and the wet biomass was dried at 37–40 for 4–5 days. After drying, flakes of fungal biomass were crushed in a mortar and pestle, and the powdered biomass was sieved on an electrical sieve. Particles of 140-mesh size were separated for use since they have been shown to give maximum sorption and consistent results.

13.19.3 Biosorption Profile

Biosorption profile was studied by adding 0.1 g of biosorbent in dye solution (200 ppm). The contents were mixed on a rotary shaker at 120 rpm for 10 min and centrifuged at 4,000 rpm for complete sediment of the biosorbent. The unabsorbed dye present in the supernatant was estimated by spectrophotometric analysis. The biosorbent in the sediment (with the sorbed dye) was dried at 37°C for 2–3 days and preserved for further desorption studies. The powdered biomass was used as an adsorbent. The contact time for maximum adsorption was determined by monitoring at 15 min intervals for 60 min.

13.19.4 Effect of pH on Biosorption

pH is the key parameter that controls the sorption of dyes. The hydrogen ions either act as a bridging moiety between dye molecules and the surface of the biosorbent or may help to produce repulsive forces between the dye molecules. To check the effect of pH, dye solutions at various pHs, namely 3.0, 5.0, and 7.0, were prepared, and the biosorbents were added to the tubes. After the contact time of 60 min, the dye solution along with the biosorbents was centrifuged at 10,000 rpm for 10 min. The percentage of dye removal was measured by calculating the initial absorbance and the absorbance after the treatment with the biosorbents (Muraleedharan et al., 1994).

13.19.5 Isolation and Screening

The selective enrichment of soil samples collected from the vicinity of textile-dyeing industries and waste disposal sites led to the isolation of three morphologically different isolates showing clear zones around their colonies on normal decolorization agar containing 200 ppm of the dye. Screening for microbes in the waste disposal sites facilitates the isolation of our interesting microbes. The isolation of different microorganisms from the samples collected from the waste disposal sites indicates the natural adaptation of microorganisms in the presence of toxic dyes.

13.19.6 Isolation of Dye-Degrading Bacteria

Three bacterial isolates that exhibited dye decolorization potentials on SWM agar spiked and screened for their ability to degrade two basic dyes (Rhodamine-B and Basic Violet-2). The fungi isolates decolorized the two dyes albeit to varying degrees within 24 h, and further decolorization of the dyes by the isolates was obtained after 36 h. The variation in decolorization efficiency of the isolates may be attributed to differences in the chemical structure of the dyes and the varying metabolic functions of the different fungi isolates.

Decolorization of the dyes by the isolates was found to be due to degradation to a greater extent than adsorption as % adsorption obtained was quite low compared to % dye reduction by viable cells. Isolate TTW 4 did not show any evidence of adsorption after 24 h. Adsorption and/or degradation are the two mechanisms responsible for dye decolorization by microorganisms. Dye adsorption may be evident from inspection of the fungal growth as those adsorbing dyes will be deeply colored, whereas those causing degradation will remain colorless. While the isolate TTW 4 cells cultured for 8 hours with the dyes were colored, none of the cells was colored with any one of the dyes tested after incubation and decolorization for 24 h. Decolorization assay of a butanol extract of the cell pellets after incubation for 24 h with the dyes showed that the dyes were not adsorbed to the cell (data not shown). This indicates that decolorization of the dyes is mainly due to degradation rather than adsorption to cells. The deeply colored cell mats after 8 h may be explained by the fact that adsorption is frequently the first step of the biodegradation process before transportation of dye into the cytoplasm and its eventual breakdown by viable microbial cells. However, adsorption levels most times are an indication of biotransformation efficiency or its absence as rapid dye biodegraders rarely show high adsorption rates upon decolorization and incubation for an extended time period.

In another experiment carried out using viable, autoclaved (killed), and metabolically poisoned cells of isolate TTW 4, dye decolorization was below 4.5% for the killed and poisoned cells, while the viable cells exhibited color removal ranging from 54% to 78%. Hence, color removal exhibited by viable cells was attributed to the biotransformation of these dyes by the metabolic functions of the bacterium. Prior to now, various authors (e.g. Jang et al., 2005) have reported the isolation of a gene (*tmr*) encoding the enzyme, triphenylmethane reductase (TpmD), from bacteria, and we believe that this enzyme may be responsible for the bioconversion of the dyes tested. The visual change in biomass color of the killed and poisoned cells and their resuspension in methanol shows that the slight decolorization of their culture medium observed is due to the adsorption of dyes on bacterial cells. Decolorization (adsorption) by dead cells may be due to the increase of the cell wall area that ruptured during autoclaving and may also be attributed to the revealing of special sites on cell walls (Table 13.1).

13.20 SUBSTANTIATION OF DYE DEGRADATION

Decolorization percentages of 70%, 64%, and 80% were obtained for Rhodamine-B and 45%, 30%, and 56% were obtained for Basic Violet-2 at normal decolorization medium incorporated with 100 ppm initial dye concentration. It was found that

TABLE 13.1
Table Depicting the λ_{max} of the Dyes

S. No.	Dye Used	λ_{max} (nm)
1	Rhodamine-B (Basic Violet 10)	556
2	Basic Violet-2	550

A. niger was found to effectively degrade the dyes than the other two isolates. Even though all the isolates have been isolated from the soil, the difference between their rates of decolorization may be due to the loss of ecological interactions, which they might be sharing under natural conditions.

13.20.1 Effect of Temperature on the Growth of the Organism

The influence of temperature on the growth of the isolates was detected by cultivating the isolates at various temperatures, and the optimum temperature at which the organism grows effectively was determined. Maximum growth of all the isolates was observed at 30°C. The results show that the temperature effect on decolorization was significant over the examined range as dye decolorization increased as the temperature was elevated to 30°C. On further incubation, the same % dye decolorization was eventually reached in all flasks incubated at different temperatures, suggesting that the test isolate could acclimatize to a broad range of temperatures (20°C–40°C). The optimal temperature for fungal activity was 30°C, and further increase in temperature beyond that resulted in marginal reduction in dye decolorization but essentially, thermal deactivation of decolorization activity under operational temperatures did not occur. Decline in bacterial activity at higher temperatures (>30°C) may be attributed to loss of cell viability or denaturation of the catabolic enzyme. Determination of temperature requirements of microorganisms used for biotechnological applications is paramount, since temperature requirements above ambient ranges may require an energy input and hence not cost-effective.

13.20.2 Effect of pH on the Growth of the Organism

The influence of pH on the growth of the isolates was detected by cultivating the isolates at various pH, and the optimum pH at which the organism grows effectively was determined. The growth of the entire isolates was found to be maximum at 5.5 pH.

13.20.3 Effect of Carbon Sources on the Growth of the Organism

The influence of carbon sources on the growth of the isolates was detected by cultivating the isolates at various carbon sources, and the optimum carbon source for the growth of the organism was identified. The growth of the entire isolates was found to be maximum in the presence of glucose. Among the other substrates (fructose, lactose, maltose, and starch), the isolates grow opulently in the presence of maltose.

13.20.4 Effect of Nitrogen Sources on the Growth of the Organism

The influence of nitrogen sources on the growth of the isolates was detected by cultivating the isolates at various nitrogen sources, and the optimum nitrogen source for the growth of the organism was identified. The maximum growth of all the isolates was observed in the presence of yeast extract and peptone.

13.20.5 Effect of Initial Dye Concentration on Decolorization

The decolorization of dyes was studied at various increasing initial concentrations of each dye (30–80 ppm). The results obtained show complete decolorization of dyes at initial concentrations between 30 and 80 ppm within 24, 48, 72, and 96 h. However, decrease in % decolorization with increase in dye concentration was obtained at concentrations above 30 ppm. Decolorization percentages of 72.2%, 90.5%, and 82.0% were obtained for Rhodamine-B and 50.2%, 30.5% and 52.0% were obtained for Basic Violet-2 at 30 ppm initial dye concentration, and this indicates that an acceptable high color removal can be achieved with *A. niger* in culture broths with dye concentrations below 50 mg/L. For industrial applications, it is important to know whether the microorganisms that decolorize dyes can bear high concentrations of the compound since the dye concentration in a typical industrial effluent can vary between 10 and 50 mg/L. *A.niger* could decolorize the Rhodamine B and Basic Violet-2 dyes at higher concentrations and thus it can be successfully exploited for the treatment of dye – bearing industrial wastewaters. Some authors (e.g. Zablocka-Godlewska et al., 2009) have reported that *Chryseomonas luteola* (42% removal) and *Pseudomonas aeruginosa* (40.5% removal) had the ability to decolorize 50 mg/L concentration of triphenylmethane dyes within 7 days. In comparison, our test isolate *A. niger* showed 65%–90% decolorization of 50 mg/L of the basis dyes tested within 24 h. These results show that *A. niger* has a higher decolorization potential compared to the other fungi reported previously. Decreased % decolorization of dyes obtained at higher concentrations suggests increasing dye toxicity with an increase in dosage. Toxic effect was probably due to inhibition of cellular metabolic activities and cell growth. Several authors have also reported decreasing color removal with increasing dye concentration during decolorization of other dyes by fungi (Tony et al., 2009).

13.20.6 Effect of Agitation on Decolorization

The isolate *A. niger* exhibited effective color removal activity only when incubated under shaking condition, whereas poor decolorization (<30%) for the two dyes was obtained under static condition. Under agitated condition, decolorization percentages of dyes were 72.2%, 90.5%, and 82.0% for Rhodamine-B and 50.2%, 30.5% and 52.0% for Basic Violet-2 within 24, 48, 72, and 96 h of incubation. Incubation under agitated condition was also necessary for better cell growth in contrast to incubation under static condition (data not shown). Poor decolorization of the dyes obtained under static condition could be attributed to the limitation of oxygen needed for the oxidative breakdown of the triarylmethane moiety, since enhanced decolorization was obtained when static cultures were subsequently incubated under agitated condition. *A. niger* also exhibited maximum decolorization of dyes when 0.02% (w/v) yeast extract, starch, or other carbon sources were supplemented in the medium (data not shown). In the absence of a co-substrate, the fungal culture showed reduced decolorization rates, which suggests that the availability of a supplementary carbon source probably for generation of NADH molecules seems to be necessary for growth and decolorization of dyes. A previous report (Ren et al., 2006) have shown that both

NADH/NADPH and molecular oxygen are necessary for the enzyme TpmD to decolorize triphenylmethane dyes, which indicates that the enzyme is an NADH/NADPH-dependent oxygenase. Textile industrial wastewaters usually contain sizing agents such as starch, polyvinyl alcohol (PVA), and carboxymethyl cellulose that provide strength to the fibers and minimize breakage, and these substances may serve as co-substrates for bacteria during effluent treatment for the generation of NADH molecules.

13.20.7 Decolorization Assay

The dye removal percentage was found to increase the reduction of dye concentration. From the results, it was apparent that the fungal isolates degrade Rhodamine-B effectively than Basic Violet-2. The dye (Rhodamine-B) removal percentage was found to be 87.2%, 78.5%, and 90.5% at 30 ppm concentration by *A. flavus*, *Fusarium* sp., and *A. niger*, respectively. From Tables 13.6 and 13.7, it can be seen that *A. flavus* and *A. niger* are able to remove Rhodamine-B to the extent of 87.2% and 90.5%, respectively, after an incubation period of 4 days under stationary condition. The incubation period of 5–7 days is quite long and unsuitable for an industry for actual application process. Hence with a view to the practical application of these studies to develop a treatment process, an incubation period of 1–4 days was selected for the studies on bioremediation under optimal environmental conditions.

When compared to Rhodamine-B, the percentage of dye removal for Basic Violet-2 is not satisfactory. The maximum of dye removal percentage of Basic Violet-2 within the incubation period was found to be 65.3%, 44.8%, and 69.5% at 30 ppm concentration by *A. flavus*, *Fusarium* sp., and *A. niger*, respectively. From the results, *A. niger* was found to be effective among the three isolates.

13.21 BIOSORPTION PROFILE

The fungal isolates were cultivated, and the mycelium was washed and dried. The dried mycelium was ground into pieces using mortar and pestle. The powdered biomass was used as an adsorbent. The percentage of color removal observed in Rhodamine-B dye solution was 75.2%, 62.4%, and 75.0% for *A. flavus*, *Fusarium* sp., and *A. niger*, respectively. The percentage of color removal observed in Basic Violet-2 dye was 50.6%, 22.8%, and 55.4%, respectively.

13.21.1 Effect of pH on Biosorption

The key parameter that controls the sorption of dyes is pH. The dye sorption exhibits saturation kinetics, and for further sorption, fresh biosorbent would have to be added to the dye solution. However, the presence of relatively high concentration of biosorbent in the solution results in reduced distance between the biosorbent particles, thus making many binding sites unoccupied. This would result in low dye adsorption per unit weight of biomass. Increasing the hydrogen ion concentration may act as bridging moiety between dye molecules and the surface of the biosorbent that may increase the sorption Tables (13.2–13. 5, Graphs 13.1–13.2).

TABLE 13.2
Percentage of Color Removal at Different Concentrations of Rhodamine-B

Name of the Organism	Concentration of the Dye (ppm)	24h (%)	48h (%)	72h (%)	96h (%)
Aspergillus flavus	30	58.3	69.3	76.5	87.2
	40	53.2	61.4	73.9	85.3
	50	47.8	56.3	70.1	82.5
	60	42.7	51.4	64.0	76.9
	70	39.2	50.1	61.9	74.1
	80	37.5	49.2	60.7	72.2
Fusarium sp.	30	55.0	62.1	70.9	78.5
	40	52.4	58.2	68.5	71.8
	50	45.2	53.8	62.2	72.3
	60	42.1	49.6	59.8	71.3
	70	38.4	45.0	57.5	68.0
	80	32.9	38.1	47.3	60.7
Aspergillus niger	30	60.0	72.0	79.3	90.5
	40	58.1	68.7	76.7	87.8
	50	54.2	65.3	76.0	87.2
	60	49.6	61.2	72.3	85.3
	70	45.6	56.6	68.9	83.5
	80	43.2	54.4	65.5	82.0

TABLE 13.3
Percentage of Color Removal at Different Concentrations of Basic Violet-2

Organism	Concentration of the Dye (ppm)	24h (%)	48h (%)	72h (%)	96h (%)
Aspergillus flavus	30	34.5	55.2	63.6	65.3
	40	28.3	54.1	61.8	64.1
	50	22.6	53.7	60.2	62.9
	60	20.0	47.3	46.1	58.3
	70	17.3	41.3	49.8	51.9
	80	15.0	35.2	47.2	50.2
Fusarium sp.	30	22.3	34.1	39.7	44.8
	40	20.8	29.6	34.8	41.6
	50	18.6	24.9	30.3	37.3
	60	17.1	22.3	28.7	36.3
	70	15.8	20.7	27.2	34.9
	80	12.1	18.0	25.8	30.5
Aspergillus niger	30	35.7	56.6	66.1	69.5
	40	28.6	55.1	63.3	65.4
	50	24.4	52.6	61.6	63.5
	60	21.3	46.5	56.3	58.2
	70	18.3	40.0	50.1	53.7
	80	16.2	38.0	49.3	52.0

TABLE 13.4
Percentage of Color Removal at Different Intervals of Contact Period of Rhodamine-B

	Percentage of Color Removal at Different Contact Periods			
Organism	**15 min (%)**	**30 min (%)**	**45 min (%)**	**60 min (%)**
Aspergillus flavus	55.6	64.5	69.3	75.2
Fusarium sp.	49.7	55.4	59.0	62.4
Aspergillus niger	55.6	62.1	69.3	75.0

TABLE 13.5
Percentage of Color Removal at Different Intervals of Contact Period of Basic Violet-2

	Percentage of Color Removal at Different Contact Periods			
Organism	**15 min (%)**	**30 min (%)**	**45 min (%)**	**60 min (%)**
Aspergillus flavus	41.3	44.3	47.6	50.6
Fusarium sp.	9.7	12.6	16.8	22.8
Aspergillus niger	38.6	47.3	53.2	55.4

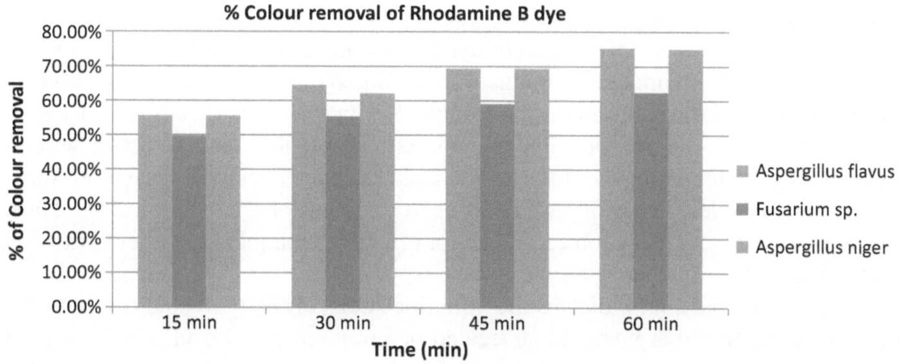

GRAPH 13.1 Percentage of color removal at different intervals of contact period. Dye: Rhodamine-B.

13.22 CONCLUSIONS

There are very few reports on the biodegradation and detoxification of textile and dye-stuff industrial wastes containing basic dyes. In this study, we describe the isolation and characterization of a strain of *A. niger*, *A. flavus*, and *Fusarium* sp. capable of efficiently degrading the basic dyes. The identities of these strains were confirmed using morphological, physiological, and biochemical assays. Degradation of dyes by

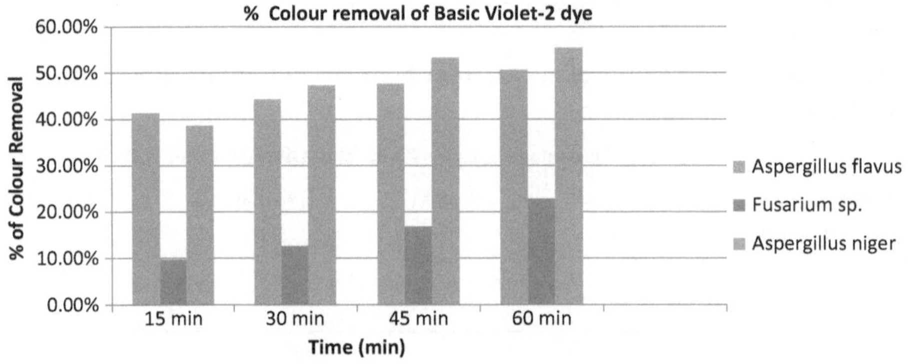

GRAPH 13.2 Percentage of color removal at different intervals of contact period. Dye: Basic Violet-2.

the isolate was found to be dependent on dye concentration, aeration, pH, temperature, and presence of a co-substrate. Phytotoxicity tests carried out on three plant species also indicated detoxification of the dyes after degradation as decolorized samples exhibited lower toxic effects than the raw dyes. Effective dye wastewater treatment using this isolate will demand the optimization of medium components and physicochemical conditions for maximum decolorization and detoxification. The advantages of this biological process are low cost, rapid degradation, and simple handling and, hence, could be applied to treat wastewater from dyeing and printing operations and in bioremediation of dye-contaminated environments. The next focus should be the design and scaling up of efficient tailor-made bioreactors with immobilized fungi for the treatment of dye wastewater and to explore the potentials of producing useful biopolymer products from *A. niger*, *A. flavus*, and *Fusarium* sp. during dye degradation. Based on the results, *A. niger*, identified as isolate TTW 4, was selected as having the best dye decolorization potential, and *A. flavus* was an efficient biosorbent among the three fungal isolates, since it showed little or no sorption of the two dyes and exhibited rapid decolorization of dyes within 24h. Thus, *A. niger* was considered a good candidate for effective biological treatment of textile wastewater and for further research studies.

The presence of fungi indicates the pollution status of the untreated dye industry effluent, suggesting that it should be treated before its disposal using the biological method particularly native and non-native fungi for comparing their degrading efficiencies. Thus, it may be concluded from the study that untreated effluent with high pollutants can be reduced by using native fungus, *A. niger*, and non-native fungus, *A. flavus*. This study emphasized that treated water can be reused for agricultural purpose as evidenced in the present work. Thus, degradation by microbes seems to be the most promising technique for 100% untreated textile dye industry effluent as evidenced in the present investigation. It is well known that water of good quality and free of pollutants is the primary requirement for agricultural and piscicultural practice. After degradation, the treated water could be used for crop cultivation or irrigation and aquaculture purposes.

REFERENCES

Bhole, D.B., Ganguly, B., and Madhuram, A., et al., 2004. Biosorption of methyl violet, basic fuchsin and their mixture using dead fungal biomass. *Current Science*, 86(12): 1641–1645.

Buna, I.M., Nigam, P., and D. Singh, et al., 1996. Microbial decolorization of textile dye containing effluents: a review. *Bioresource Technology*, 58: 217–227.

Calabro, V., Drioli, E., and F. Matera. 1991. Membrane distillation in the textile wastewater treatment. *Destination*, 83: 209–224.

Correia, V.M., Stephenson, T., and S.J. Judd.1994. Characterization of textile wastewater —A review. *Environmental Technology*, 15: 917.

Haley, T.J. 1975. Benzidine revisited; A review of the literature and its congeners. *Clinical Toxicology*, 8: 13.

Hameed, D.K., Mahmoud, A.L., and A. Ahmed. 2008. Equilibrium modeling and kinetic studies on the adsorption of basic dye by a low cost adsorbents: coconut bunch waste. *Journal of Hazardous Materials*, 158: 65–72.

Hildenbrand, S., Schmahl, F.W., and R. Wodarz, et al., 1999. Azo dyes and carcinogenic aromatic amines in cell culture. *International Archives of Occupational and Environmental Health*, 72: 52.

Jang, M.S., Lee, Y.M., and C.H. Kim, et al., 2005. Triphenylmethane reductase from *Citrobacter* sp. strain KCTC 18061P: purification, characterization, gene cloning, and overexpression of a functional protein in *Escherichia coli*. *Applied and Environmental Microbiology*, 71(12): 7955–7960.

Kapoor, A., and T. Viraraghavan. 1995. Fungal biosorption—An alternative treatment option for heavy metal bearing wastewater. A review. *Bioresource Technology*, 53: 195–206.

Keharia, H., Patel, H., and D. Madamwar. 2004. Decololorization screening of synthetic dyes by anaerobic methanogenic sludge using a batch decolorization assay. *World Journal of Microbiology and Biotechnology*, 20: 365–370.

Kim, S.J and M. Shoda. 1999. Purification and characterization of novel peroxidase from *Geotrichum candidum* Dec 1 involved in decolorization of dyes. *Applied and Environmental Microbiology*, 65: 1029–1035.

Lee, C.K., Low, K.S., and P. Gan. 1999. Removal of some organic dyes by acid treated spent bleaching earth. *Environmental Technology*, 29: 99–104.

Lorimer, J.P., Mason, T.J., and Plattes, M., et al., 2001. Degradation of dye effluent. *Pure and Applied Chemistry*, 73(12): 1957–1968.

Muraleedharan, T.R., Iyengar, L., and C. Venkobachar. 1994. Further insight into the mechanism of biosorption of heavy metals by *Ganoderma lucidum*. *Enzyme Technology*, 15: 1015–1027.

Neff, J.M. 1985. The use of biochemical measurements to detect pollutant mediated damage to fish. In: *Aquatic Toxicology and Hazard Assessment, 7th Symposium*. Special Technical Publication No. 854, edited by R.D. Cardwell, R. Purdy, and R.C. Bahner. Philadelphia, PA: American Society for Testing Materials, p. 155.

Nigam, P., Armour, G., and M. Banat., et al., 2000. Physical removal of textile dye and solid-state fermentation of dye adsorbed agricultural residues. *Bioresource Technology*, 72: 219–226.

Nirmalarani, J., and K. Janardhanan. 1988. Effect of South India Viscose factory effluent on seed germination seedling growth and chloroplast pigments content in five varieties of Maize (*Zea mays*). *Madras Agricultural Journal*, 75: 41.

Nony, C.R., and M.C. Bowman.1980. Trace analysis of potentially carcinogenic metabolites of an azo dye and pigment in hamster and human urine as determined by two chromatographic procedures. *Journal of Chromatographic Science*, 18: 64.

Nyanhongo, G.S., Gomes, J, and G.M. Gubitz, et al., 2002. Decolorization of textile dyes by laccases from a newly isolated strain of *Trametes modesta*. *Water Research*, 36: 1449–1456.

Parshetti, G., Kalme, S., and G. Saratale, et al., 2006. Biodegradation of malachite green by *Kocuria rosea* MTCC 1532. *Acta Chimica Slovenica*, 53(4): 492–498.

Patricia, A, Ramalho, H., and M. Scholze, et al., 2002. Improved conditions for the aerobic reductive decolorization of azo dyes by *Candida zeylanoides*. *Enzyme and Microbial Technology*, 31: 848–854.

Rajeshwari, S., Namasivayam, C., and K. Kadiravelu. 2001. Orange peel as an adsorbent in the removal of acid violet 17 (acid dye) from aqueous solution. *Waste Management*, 21: 105–110.

Ren, S.Z., Guo, J., and Y.L. Wang, et al., 2006. Properties of a triphenylmethane dyes decolorization enzyme TpmD from *Aeromonas hydrophila* strain DN322. *Acta Microbiologica Sinica*, 46(3): 385–389.

Riu, J., Schonsee, I., and D. Barcelo. 1998. Determination of sulfonated azo dyes in ground water and industrial effluent by automated solid-phase extraction followed by capillary electrophoresis/mass spectrometry. *Journal of Mass Spectrometry*, 33: 653.

Shazia, I., Kousar, P., and S. Naila, et al., 2013. Tolerance potential of different species of *Aspergillus* as bioremediation tool—Comparative analysis. *Journal of Microbiology Research*, 1: 001–008.

Soares, G.M.B., Costa-Ferreira, M.C., and M.T.P. Amorim. 2001. Decolorization of an anthraquinone-type using a laccase formulation. *Bioresource Technology*, 79: 171–177.

Spadarry, J.T., Isabelle, L., and V. Renganathan. 1994. Hydroxyl radical mediated degradation of azo dyes: evidence for benzene generation. *Environmental Science and Technology*, 28: 1389–1393.

Tony, B.D., Goyal, D., and S. Khanna. 2009. Decolorization of Direct Red 28 by mixed bacterial culture in an up-flow immobilized bioreactor. *Journal of Industrial Microbiology and Biotechnology*, 36(7): 955–960.

Vanderzannt, C., and D.F. Splittstoesser. 1997. *Compendium of Methods for the Microbiological Examination of Foods*. 3rd edition. Washington, DC: American Public Health Association.

Verma, P., and D. Madamwar. 2000. Decolorization of synthetic dyes by a newly isolated strain of *Serratia marcescers*. *World Journal of Microbiology & Biotechnology*, 19: 615–618.

Volesky, B., and G. Naja. 2003. Sorption and biosorption, BV Sorbex, Montreal. *Journal of Technology Transfer and Commercialisation*, 6(2/3/4): 169–211.

Zablocka-Godlewska, E., Przystas, W., and E. Grabinska-Sota. 2009. Decolorization of triphenylmethane dyes and ecotoxicity of their end products. *Environmental Protection Engineering*, 35(1): 61–169.

Zollinger, H. 1987. *Azo dyes and Pigments—Color Chemistry Synthesis, Properties, and Application of Organic Dyes and Pigments*. New York, NY: Wiley VCH, p. 92.

14 Role of Biofilms in Anaerobic Wastewater Treatment Processes for Domestic and Industrial Effluents

Hafiz Muhammad Aamir Shahzad and Sher Jamal Khan
National University of Sciences and Technology (NUST)

CONTENTS

14.1 INTRODUCTION

Water is crucial for survival of human life, but hardly 1% available for human use. Current global water shortages are exacerbated by rapid population growth, climate change, degradation of the environment, industrialization, depletion of potable water, and urbanization (Connor, 2015). The quality of freshwater in streams and rivers is impaired due to untreated water discharged from the industrial sector, agricultural practices, and urban sewer. According to the World Health Organization (Pawari and Gawande, 2015), worldwide 3.1% of deaths are caused by unhygienic and polluted water. Wastewater in a broad term includes water from different sources such as industries, agriculture, stormwater, and the domestic sector. In wastewater, organic material consists of detergents, pesticides, fats, and oils. Furthermore, several forms of microorganisms might exist in wastewater including several types of viruses, bacteria, helminths, and protozoa. Wastewater also contains essential nutrients (ammonia, nitrogen, phosphorus, etc.), various metals, and inorganic materials (cadmium, hydrogen sulfide, lead, mercury, nickel, etc.). In order to mitigate the detrimental effects of sewage, installation of wastewater treatment plants has gained popularity for ensuring safety of surface and groundwater sources in the future.

The choice of wastewater treatment technique is based upon a balance between economical and technical factors as well as the type of pollutants exist in raw water. Different forms of wastewater treatment methods are available depending on the removal of physical and chemical pollutants. However, biological wastewater treatment has gained significant importance in the last few years with characteristics having lower operational cost, comparatively easy handling as well as less deteriorating effects on environment (Rajasulochana and Preethy, 2016). The two broad categories of biological treatment are aerobic and anaerobic wastewater treatment. Aerobic biological wastewater treatment systems have been in use since over a century. Problems associated with aerobic treatments are high operation and maintenance cost, extensive energy needs for air supply, handling of sludge including its settling, and thickening of solids (Acampa et al., 2019). While anaerobic biological treatment has been applied effectively for domestic and high strength industrial wastewater, it can efficiently convert a treatment plant from energy consumption to energy production. Anaerobic treatment has benefits of lower sludge production and lesser nutrient requirements. Energy-rich methane gas is also produced during anaerobic degradation, which is a feasible source of energy when combusted in a combined heat and power plant. Based on structural configuration of biomass, anaerobic wastewater treatment systems can be categorized into two basic configurations: (i) dispersed growth system and (ii) attached growth system. In anaerobic dispersed growth systems, suspended biomass grows in bioreactor without being attached to a carrier material or surface (Usack et al., 2012). In anaerobic attached growth system (also known as biofilm processes), biomass grows while being attached to a support material or surface and forms biofilm (Young and Dahab, 1983). Attached growth biofilm systems have been found to be more effective as compared to suspended growth in terms of biogas production and chemical oxygen demand (COD) removal (Ebrahimi and Najafpour, 2016; Roy et al., 1986).

14.2 DEVELOPMENT OF BIOFILMS

Biomass grows while attached to biofilm support medium. The type of support materials, intracellular interactions, and distribution of polymeric molecules on a support medium affect the attachment of biofilm. Attached growth medium can be submerged in the wastewater where flow to the system is either received intermittently or continuously. The support medium should be long-lasting, insoluble, and chemically resistant. Also the media should provide high specific surface area, high porosity to prevent blockage, and with relatively low cost. Besides this, hydrodynamic conditions of reactor can be affected by geometry and surface area of the media that could ultimately affect the development of biofilms, which in turn affect the treatment of wastewater (Marin et al., 1999). The support medium can be of different kinds, such as natural solid (sand, gravel, stones, soil and rocks), synthetic (rubber or plastic) or biomass agglomerates (granules) itself (Gong et al., 2011; Thaiyalnayaki and Sowmeyan, 2012). Biofilms develop on support medium with flowing wastewater that contains organic matter and nutrients.

The biofilm development is a systematic procedure consisting of series of steps. This starts with adsorption of macromolecules (e.g., polysaccharides, proteins, humic, and nucleic acids) as well as smaller molecules (e.g., lipids, fatty acids, and contaminants such as polychlorinated biphenyls and polyaromatic hydrocarbons) on the surface. These adsorbed molecules can produce films due to multiple effects, which include altering the substrate physicochemical characteristics, serving as an enormous source of nutrients for microorganisms, increasing or suppressing toxic metal ions release from surface, detoxifying bulk solution by adsorbing inhibitory substances, providing trace elements and nutrients needed for a biofilm growth, and triggering biofilm sloughing. The initial steps in biofilm formation are well known because at this point, it is easy to acquire photos of microorganisms.

Bacteria adhere to a substrate by slimy adhesive materials and extracellular polymeric substances (EPS), mainly made of proteins and polysaccharides. Even though literature has well documented the interaction of EPS with attached bacteria, little evidence suggests that EPS is responsible for primary adhesion. However, with the formation of a slimy substance called the biofilm matrix, EPS definitely helps to form mature biofilms. Steps in the formation of mature biofilm in anaerobic biofilm reactor are shown in Figure 14.1. The presence of these three phases (attachment,

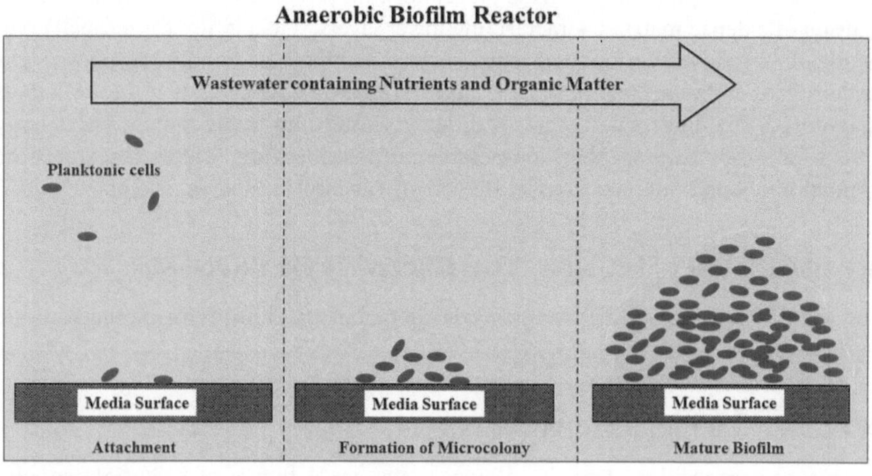

FIGURE 14.1 Steps in formation of mature biofilm.

colonization, and growth) of biofilm formation is generally recognized, although these terms may differ among authors. Once a mature biofilm has been developed, it spreads rapidly and progressively, ultimately covering the whole surface. A mature biofilm varies in thickness depending on the type of reactor, supported medium, and wastewater with applied loading rate. Kennedy et al. (1981) reported that biofilm thickness normally varies from 2 to 4 mm.

14.3 BIOFILMS CHARACTERISTICS

A biofilm can be described as an intricate coherent structure of cells as well as cellular products (e.g., extracellular polymers), that impulsively grow on solid support either attached as dense granules or attached on static solid surface (static biofilms) or on submerged carriers (particle supported biofilms).

Even though biofilms can be seen with naked eye, their structure, microbial composition, and distribution of EPS cannot be seen without the help of microscopes combined with different probes. For instance, fluorescent proteins (FPs) and fluorescent in situ hybridization (FISH) probes can be used as reporter genes. Among biofilm researchers, the most suitable microscope is one that allows the full inspection of living as well as fully hydrated biofilms. Biofilms are generally thin and dense having uneven bumpy surface with different kinds of microorganisms living together in it. Microscopic examination of a biofilm exposes their uneven surface with numerous vents of variable sizes. And it is believed that these vents release biogas and might even serve as conduits for importation of nutrients into the granules (Gulmann et al., 2015).

The biofilm's organic and inorganic composition varies with the composition of wastewater flowing over it. Generally, the accumulation of organic and inorganic nutrients increases with increasing the loading rate of wastewater. Biofilms treating industrial wastewater generally have higher organic content as compared to those treating domestic wastewater. Generally, biofilms present at the bottom of the reactors have higher organic content as compared to those which are at the top. Similarly, a higher concentration of organic materials is present toward the base of the biofilm, with the inclusion of a dense matrix characterizing lower layers. The biofilm's microbial communities break down various nutrients, such as compounds containing phosphorus and nitrogen, carbonaceous materials, and trapped pathogens. When the pollutants are removed, biofilter-treated water is either released into the receiving water body or used for agriculture or other non-potable purposes. Figure 14.2 demonstrates the elimination of contaminants from wastewater by anaerobic biofilm reactor.

14.4 FACTORS AFFECTING THE GROWTH OF BIOFILMS

Some of the key factors affecting the development of biofilms in bioreactors are discussed below.

14.4.1 CHARACTERISTICS OF WASTEWATER

Wastewater composition strongly influences the development of bacterial exopolysaccharides which plays a crucial role in the development of biofilms. Carbohydrated

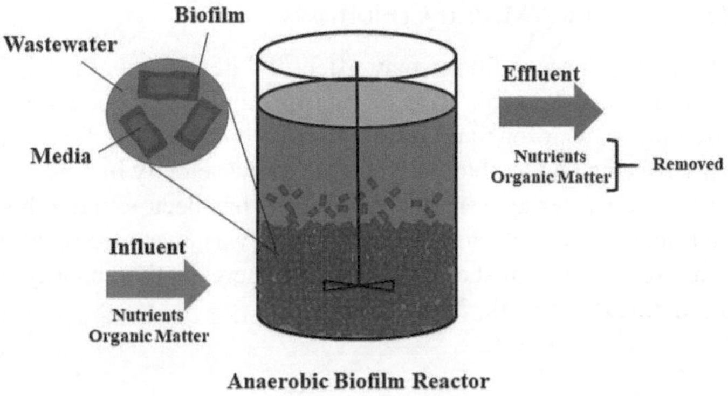

FIGURE 14.2 Elimination of contaminants by anaerobic biofilm reactor.

wastewater promotes biofilms development on media as they facilitate in production of polysaccharides (Fish et al., 2017). Janka et al. (2019) reported that development of biofilm is enhanced or more rapid when reactor is started with synthetic feed and replace it gradually with wastewater to be processed. The growth of methanogenic bacteria that generate polysaccharides can be stimulated by adding a substrate that can be metabolized directly, like methanol that raises the C:N ratio, which allows extracellular polysaccharides to be produced and facilitates bacterial attachment to the supporting material (Miqueleto et al., 2010).

Wastewater, on the other hand, must have all vital nutrients essential for anaerobic growth. Nutrients in wastewater facilitate the transfer of bacteria from planktonic cell to biofilm, while the loss of these nutrients contributes to the elimination of biofilm cells from surfaces of the media. A recommended COD:N:P ratio for biofilm formation is 334:28:5.6 (Thompson et al., 2006). Nutrients that tend to be inadequate might be provided as supplement in the wastewater stream.

14.4.2 pH of the Bioreactor

For the anaerobic systems, pH of the bioreactor should be neutral or close to neutral as much as possible (Garrett et al., 2008). Decrease in pH due to acidogenesis can be avoided with the addition of alkalinity. pH shift greatly affects bacterial activity and development of biofilm by changing mechanisms and ultimately poses harmful effects on microorganisms within the biofilm.

14.4.3 Temperature of the Bioreactor

Microorganisms are very sensitive, even response to trivial changes in temperature. Bacterial population can have healthy growth with optimum temperature while it might be diminished with a slight variation in temperature. Optimum temperature for many bacteria is about 35°C (Daud et al., 2018). In contrast, temperature away from the optimum value reduces the efficiency of bacterial growth due to a decrease in reaction rates of bacterial enzymes.

14.4.4 Hydrodynamics/Mixing Conditions

Hydrodynamic or mixing conditions may affect the development of biofilm, structure, thickness, mass and metabolism of biofilm, and EPS production. Adequate mixing is essential for vigorous and uniform growth of biofilm. Recycling is often used to improve mixing and to obtain sufficient upflow velocity in fluidized bed reactor (FBR). In fact, a greater aggregation of biofilms has been found at higher velocity due to the transportation of more substrate from wastewater to biofilm. It is also reported in literature that high shear stresses also increase the quantity of bacterial attachment on media (Saur et al., 2017).

14.4.5 Type of Media for Attachment of Biofilm

Depending on the type of media used, the ability of microorganisms to adhere to the surfaces can vary significantly. During the initial stages of colonization, nanoscale and microscale surface roughness increases bacteria's adhesion by facilitating more cell attachment to the surface. Media have high adsorption affinity for methanogenic and anaerobic bacteria which is favorable for the growth of biofilms. In addition to these, some other contributors such as elasticity, hydrophobicity, and charge also influence the microbial attachment to media. Table 14.1 summarizes the properties of some different media used for attachment of biofilm.

14.5 BIOFILM ANAEROBIC WASTEWATER TREATMENT SYSTEMS

Biofilm-based wastewater treatments have been considered as an auspicious economical treatment system. For a sustainable environment, biofilm-based bioreactors play an imperative role in biotechnology. Many aspects of their development and technological operations remain poorly understood, and researchers are still conducting extensive investigations for better use of these promising strategies of depollution.

In this section, some commercially available biofilm-based processes are discussed. This includes anaerobic filters (AFs), anaerobic rotating biological contactors (AnRBCs), anaerobic fluidized bed reactors (AnFBRs), upflow anaerobic sludge

TABLE 14.1
Properties of Some Different Attached Growth Media

Type of Media	Normal Size (m)	Bulk Density (kg/m^3)	Specific Surface Area (m^2/m^3)	Void Space (%)
Natural (sand, gravel, stones, soil, and rocks)	0.024–0.128	1442–1600	46–62	50–60
Synthetic (rubber or plastic)	0.61×0.61×1.22	24–45	100–233	95

Source: Daigger and Boltz, 2011.

blanket (UASB) reactors, anaerobic moving bed biofilm reactors (AnMBBRs), and upflow anaerobic hybrid reactors (UAHRs).

14.5.1 Anaerobic Filter (AF)

AF reactors are the anaerobic version of a trickling filter with no air interaction with the biomass (microorganisms are purely anaerobic) and works under flooded conditions. AF also known as a packed bed reactor, having inert support material for the development of biofilm. It was first used in 1957 for the treatment of sewage (Coulter et al., 1957). The influent wastewater passes vertically (upward or downward) in an anaerobic filter through a submerged medium that maintains anaerobic conditions. Soluble organic matter is absorbed as it comes into contact with the biofilm of anaerobic filters, even low concentration of suspended solids can also be removed by retaining them within the interstices of the medium, and eventually biodegrade. The schematic illustration of upflow and downflow anaerobic filters is shown in Figure 14.3.

Although AF are used to treat both domestic and industrial effluent, which have a lower content of suspended solids, in centralized sewerage systems, they can also be used in wastewater treatment system serving individual industries, residential complexes, and housing clusters.

Bodkhe (2008) carried out a study on anaerobic filters for municipal wastewater treatment having COD of 350–450 mg/L. Experiments conducted at various hydraulic retention times (HRTs) indicated that 12 h HRT was most appropriate for AF, resulting in 90% and 95% removal of biological oxygen demand (BOD) and COD, respectively. Specific biogas yield obtained was $0.35\,m^3/kgCOD_{removed}$ with 70% of methane content. Manariotis and Grigoropoulos (2006) evaluated the AF performance for raw municipal wastewater treatment under varied operating conditions as well as hydraulic and organic loadings rates. Bodík et al. (2002) examined the effect of different temperatures and HRTs on the start-up and stable performance of AF for synthetic and real municipal wastewater treatment. They achieved the COD removal of 46%–92% depending on temperature (9°C–23°C) and HRTs (6–46 h). Ladu and Lü (2014) studied the effects of temperature, HRT, and effluent recycling ratios on

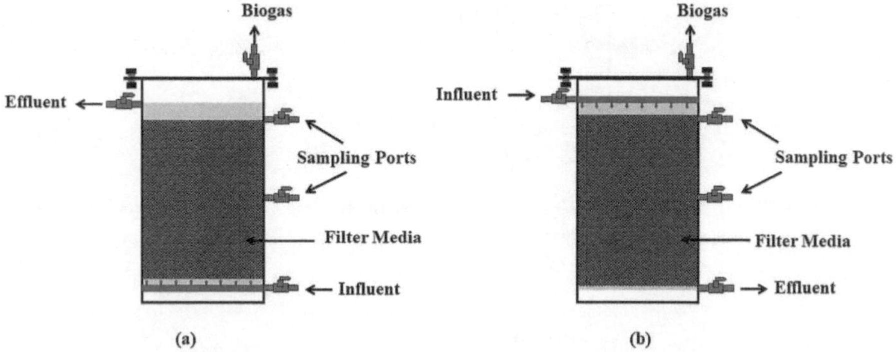

FIGURE 14.3 Schematic of upflow (a) and downflow (b) fixed film reactor.

performance of AF for the treatment of rural domestic wastewater. Results showed that above-mentioned variables have greatly influenced the removal efficacy of AF reactor. Highest removal efficiency and biogas production were achieved at 30°C, with an optimum HRT of 3 days, and with 2:1 effluent recycling ratio.

Guerrero et al. (1997) used upflow anaerobic filter for fish meal processing wastewater treatment. At the maximum applied organic loading rate (OLR) of 5 kgCOD/m^3/day and under mesophilic condition, they achieved the COD removal of above 80% and specific methanogenic yield above 1 kg CH_4-COD/kgVSS/day. Ruiz et al. (1997) examined the performance of AF for slaughterhouse effluent treatment. Under stable operating conditions, decrease in COD removal was observed from 90% to 50% as OLR increased from 1 to 6.5 kgCOD/m^3/day. AF has also been used for the treatment of textile, paper and pulp, and dairy wastewater (Ahn and Forster, 2002; Anderson et al., 1994; Athanasopoulos, 1986).

14.5.2 Upflow Anaerobic Sludge Blanket (UASB) Reactor

UASB reactor was established in Netherlands (Lettinga et al., 1980), depends on the propensity of anaerobic bacteria in order to form biofilm flocks or granules, and kept within the reactor by an effective gas–liquid–solid separator placed at the top of reactor. Schematic illustration of UASB reactor is shown in Figure 14.4.

Within the reactor, retained biomass forms a dense sludge and grows into high granular sludge (1–4 mm) having tremendous settling properties. Wastewater enters

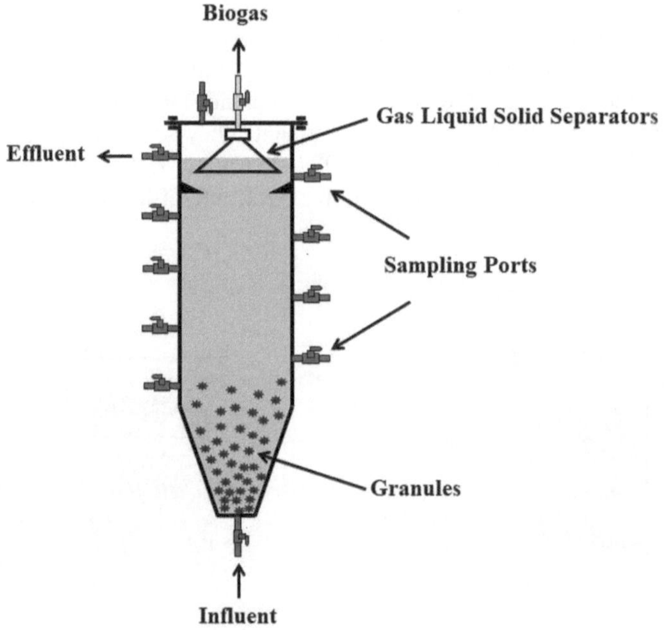

FIGURE 14.4 Schematic of upflow anaerobic sludge blanket (UASB) reactor.

into the reactor from bottom and goes up via thick anaerobic sludge bed (granules). Anaerobic sludge (granules) usually consists of microbial species enriched with methanogens, acetogens, and acidogens. These granules are distributed in the sludge bed by superficial upflow velocity of wastewater and rising biogas bubbles. While biogas and treated effluent are separately collected from top of the reactor and tiny granules are returned back to reactor. Reasonable cost, lower energy requirement, and simple design make UASB reactors superior over other treatment systems.

UASB reactor can be used for low-strength (domestic) and high-strength (industrial) effluents. For optimal bacterial growth and system efficiency, the influence of key process parameters, such as pH, temperature, OLR, HRT, mixing, and granulation, was studied on the performance of UASB (Daud et al., 2018).

Domestic wastewater is usually low in organic matter and high in suspended solids. In domestic wastewater, a huge share of nutrients in black water is responsible for about half of organic matter. UASB reactor showed auspicious results in low-strength domestic effluent treatment.

A pilot scale UASB reactor was used by Zhang et al. (2018) for domestic wastewater treatment having COD concentration of 616 ± 140 mg/L and reported that UASB system maintained COD removal of 76% at 12.5°C–20°C and 6h HRT. Approximately 40% of the removed COD was converted into methane while the biological methane potential of influent COD was 80%. In another study (Elmitwalli and Otterpohl, 2007), for domestic wastewater, UASB reactor attained maximum 84% COD removal at 30°C with 6–8h.

UASB is a commonly used anaerobic reactor for industrial wastewater treatment. Haider et al. (2018) studied the performance of UASB reactor for real textile wastewater treatment under the intermittent operation at HRT of 24h and OLR of 2 kgCOD/m^3/day. They achieved maximum COD and color removal efficacy of 57.5% and 71%, respectively, under intermittent operating condition of 12h feeding with 12h non-feeding. Somasiri et al. (2008) attained maximum 90% COD removal and 92% color removal while treating the textile wastewater in UASB reactor under different operating conditions. UASB has also achieved considerable success among the several anaerobic wastewater treatment technologies for diverse range of effluents generating from textile, sugar, paper and pulp, food processing, soft drinks, slaughterhouse, and coffee-processing industries (Haider et al., 2018; Kalyuzhnyi et al., 1997; Kamali et al., 2016; Mousavian et al., 2019; Narihiro et al., 2009; Ruiz et al., 1997; Selvamurugan et al., 2010).

14.5.3 Anaerobic Fluidized Bed Reactor (AnFBR)

AnFBRs were introduced back in the 1970s (Heijnen et al., 1989). Operating principle of AnFBR is similar to that of UASB, but due to higher upflow velocity, its hydrodynamic conditions change. AnFBR maintains active biomass with biofilm development on the surfaces of fine carrier materials and are fluidized by maintaining a high upflow velocities of wastewater, typically formed by a mixture of influent and recycling flows. In this form, larger surface area is provided by media for the development of biofilm. A schematic of AnFBR is shown in Figure 14.5.

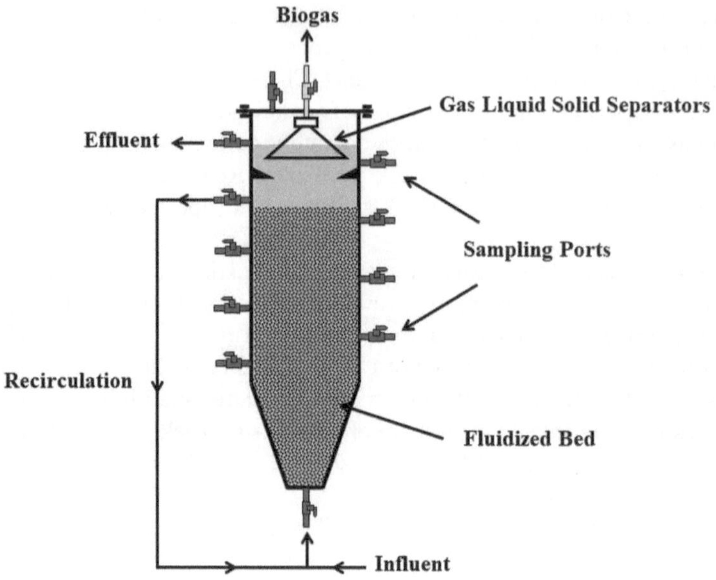

FIGURE 14.5 Schematic of anaerobic fluidized bed reactor (AnFBR).

AnFBRs have been widely used in many biotechnology fields as well as for industrial and domestic effluent treatments (Mendonça et al., 2004; Sanz and Fdz-Polanco, 1990; Souza et al., 2004). Mendonça et al. (2004) described the performance of a full-scale AnFBR for domestic wastewater treatment. Bioreactor accomplished 71% COD removal at upflow velocity of 10.5 m/h, recirculation ratio of 0.85, and HRT of 3.2h. Sanz and Fdz-Polanco (1990) investigated the performance of AnFBR at low temperature for treatment of municipal wastewater. They observed that gradual decrease in temperature from 20°C to 5°C didn't have greater effect on effluent quality. AnFBR operated very stably with 70% and 80% COD and BOD removal, respectively. Souza et al. (2004) conducted an experimental study on AnFBR performance for milk processing wastewater treatment. They observed an average decrease in COD removal from 67% to 48% when COD in feed increased from 462 to 1,472 mg/L. In another study, Şen and Demirer (2003) investigated the performance of AnFBR for COD, BOD, and color removal from textile effluent under different OLR/HRT and effect of glucose concentration as substrate additives. Their study implied that 98%, 95%, and 65% of COD, BOD, and color removal, respectively, were achieved by AnFBR.

14.5.4 Anaerobic Rotating Biological Contactor (AnRBC)

AnRBC is a revised version of aerobic rotating biological contactor and was first introduced by Tait and Friedman in the 1980s (Tait and Friedman, 1980). Typically, RBC unit comprises of narrowly spaced large flat or grooved disks, fixed on a horizontal shaft which is submerged in wastewater (as shown in Figure 14.6).

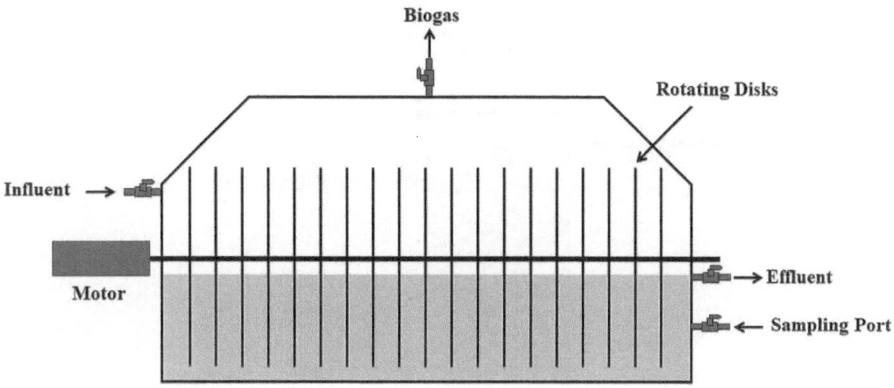

FIGURE 14.6 Schematic of anaerobic rotating biological contractor (AnRBC).

Performance of AnRBC depends on various factors such as rotational speed of disk, HRT, organic and hydraulic loading rates, type of media, temperature, staging, wastewater and biofilm characteristics, effluent recirculation, dissolved oxygen and solids, as well as graduation and media submergence.

Due to the advantages of high specific surface area, simple operation, less energy requirement, able to handle toxic substances, and less accumulation of sloughed biofilm, AnRBC seems to be appropriate treatment process for medium as well as high-strength wastewaters.

Lu et al. (1995) examined the performance of anaerobic RBC for high strength organic wastewater treatment and found AnRBC as an effective and feasible option by achieving 71% and 76% COD and BOD removals, respectively, at OLR of 13.33 kgCOD/m^3/day. In another study, Lu et al. (1997) investigated the performance of AnRBC at different rotational speeds of the disc and submergence. They achieved COD removal of 70%–78% for high-strength organic wastewater treatment at optimum rotational speed of 12 rpm and 100% submergence of disc. Laquidara et al. (1986) studied the performance of AnRBC for medium strength wastewater treatment under loading rate from 10 to 140 g/m^2/day. Also, the AnRBC process found to be an effective process with maximum soluble COD removal of 92% at the lowest loading rate, while relatively 5%–7% decrease in COD removal was observed at the highest loading rate. Maximum methane (20 L/m^2/day) was produced at a loading rate of 90 g/m^2/day COD.

14.5.5 Anaerobic Moving Bed Biofilm Reactor (AnMBBR)

AnMBBR is a novel technology based on microorganisms that grow on the surface of carrier media. Attached media move freely in the bioreactor by mechanical mixing. Fraction of fill media in moving bed biofilm reactor range from 20% to 60% of the total volume of the tank (Azizi et al., 2013). Biofilms grow predominantly within plastic carriers, secured against external abrasion. Implementing moving beds instead of static ones has benefit to avoid clogging and enhance capability to use the entire volume of bioreactor. Besides this, it minimizes or removes the need of

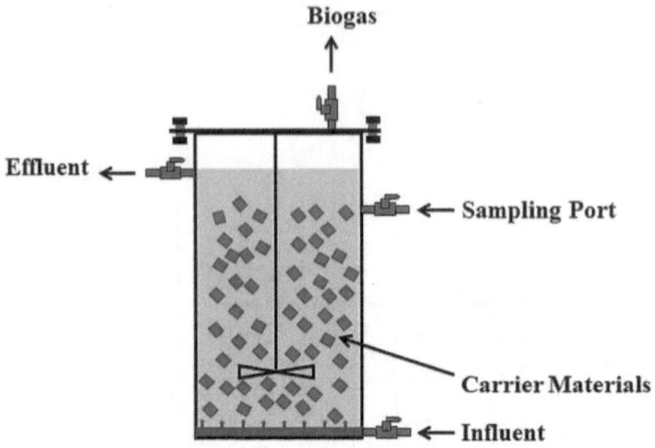

FIGURE 14.7 Schematic of anaerobic moving bed biofilm reactor (AnMBBR).

effluent recirculation, which is required for biofilm processes with fixed beds and UASB systems. Figure 14.7 shows the schematic of AnMBBR.

For the treatment of different kinds of wastewater, AnMBBR process can be an effective strategy (Chai et al., 2014; Sheli and Moletta, 2007). Sheli and Moletta (2007) used AnMBBR for the treatment of wine distillery wastewater at OLR of 1.6 to 29.6 kgCOD/m^3/day (HRT 6.33 to 1.55 days). Almost, at the end of their experimental run, soluble COD removal efficiency still was above 81% at OLR of 29.6 kgCOD/m^3/day or HRT of 1.55 days. Chai et al. (2014) studied the performance of two AnMBBRs for the treatment of winery wastewater filled with different kind of low density carrier materials (R9 and R30). They found that, with carrier material of low specific surface area (R9), more than 80% soluble COD removal was attained at maximum OLR of 29.59 kgCOD/m^3/day, whereas with carrier material having a high specific surface area (R30), a maximum 80% of soluble COD reduction was achieved at OLR of 18.43 kgCOD/m^3/day. Biogas production of both reactors strongly correlated with its OLR and subsequently increased with increase in OLR.

14.5.6 Upflow Anaerobic Hybrid Reactor (UAHR)

UAHR is a combination of UASB and AF. The reactor's lower part serves as a UASB, and upper part contains cross flow media that offers large surface area for microbes to be attached. Figure 14.8 shows the schematic of UAHR. UAHR promotes the benefits of both UASB and AF as well as reduces their limitations (Gupta et al., 2010). UAHR has been studied by many researchers for effective industrial and domestic wastewater treatment (Banu et al., 2007; Bello-Mendoza and Castillo-Rivera, 1998; Gupta et al., 2010). Gupta et al. (2010) investigated and compared the performance of UAHR with UASB for distillery spent wash wastewater treatment. They found that at optimum OLR of 8.7 kgCOD/m^3/day, COD removal was found to be 79% and 75% in UAHR and UASB, respectively. They also found that methane yield in UAHR was 5% more than the UASB reactor and sludge washout was reduced by 25% in UAHR

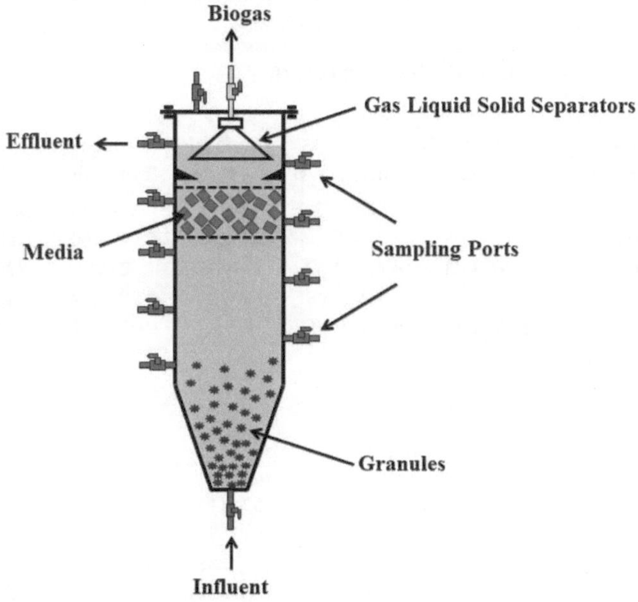

FIGURE 14.8 Schematic of upflow anaerobic hybrid reactor (UAHR).

as compared to UASB. Bello-Mendoza and Castillo-Rivera (1998) checked the ability of UAHR for coffee house wastewater treatment at the OLR of 1.06–2.4 kgCOD/m^3/day. A COD removal of 77% was achieved at the OLR of 1.89 kgCOD/m^3/day and further increase in OLR deteriorated the COD removal efficiency of UAHR. In another study, Banu et al. (2007) examined the performance of hybrid UASB reactor for domestic wastewater treatment. Under different HRTs (7.3–3.3 h), 75%–86% of COD and 70%–91% of BOD removals were achieved. Maximum biogas production was 7,080 mL/day at HRT of 3.3 h, and methane content in the produced biogas was found to be 62% ± 3%.

Performance of the anaerobic reactors discussed above for domestic and industrial wastewater treatment is also summarized in Table 14.2.

14.6 CONCLUSIONS AND A WAY FORWARD

Biofilm-based anaerobic wastewater treatment systems are rapidly extending study areas that have been explored intensively in literature. Due to simple and easy procedure as compared to aerobic treatment, their implementation is considered to be valuable. Studies showed that biofilm reactors have likely potential for anaerobic treatment of domestic and industrial effluent as well as biogas production. Literature also showed that optimization of organic loading rate as well as HRT is predominantly effective for successful biofilm reactor performance. Continued efforts are still required to improve organic removal and energy production from wastewater using current biofilm anaerobic technologies.

TABLE 14.2
Performance of Anaerobic Biofilm Reactors

	Type of Wastewater Treated	OLR (kgCOD/m^3/day)	HRT (Days)	COD Removal (%)	Biogas Production[a]	References
AF	Municipal	—	0.125–12	89–97	0.21–0.35[b]	Bodkhe (2008)
	Domestic	0.19–0.05	1–4	32–81	5.94–8.60[c]	Ladu and Lü
	Industrial	5.26–3.48	4.41–12.22	80–90	2–5[c]	(2014)
	Industrial	6.5–1	1.2–6.5	59–91	0.22–1.34[c]	Guerrero et al.
	Industrial	3.87–1.70	0.49–1.09	71–85[d]	0.17–0.21[b]	(1997)
	Industrial	21–0.15	0.5–3	60–82	0.33–0.38[b]	Ruiz et al. (1997)
						Ahn and Forster (2002)
						Anderson et al. (1994)
UASB	Domestic	—	0.25–0.66	52–64	0.18–0.28[c]	Elmitwalli and
	Industrial	8–2	1	45.92–71	—	Otterpohl (2007)
	Industrial	—	0.25–1	46–70.58	0.26–0.43[b]	Haider et al.
	Industrial	12–2	2–12	60–80	—	(2018)
	Industrial					Selvamurugan et al. (2010)
						Kalyuzhnyi et al. (1997)
AnFBR	Domestic	7–2	—	40–80	4,000–8,000[c]	Mendonça et al.
	Industrial	5–0.38	1–2.08	35–98	—	(2004)
						Şen and Demirer (2003)
AnRBC	High	13.33–0.71	0.22–1.5	42–71	5.8–14.4[c]	Lu et al. (1995)
	Strength	—	0.9–1.3	70–78	0.5–0.7[b]	Lu et al. (1997)
	High Strength					
AnMBBR	Industrial	31.2–1.7	1.55–6.33	39.6–91.7[d]	20–450[b]	Sheli and Moletta
	Industrial	29.59–1.3	1.55–12	39.6–91[d]	0.46–14.6[e]	(2007)
						Chai et al. (2014)
UAHR	Domestic	—	0.055–	75–86	1.8–7.08[c]	Banu et al. (2007)
	Industrial	11.13–4.53	0.122	58–79	23.6–64.33[c]	Gupta et al.
	Industrial	2.59–1.06	4–8	25–77	—	(2010)
			0.41–2.08			Bello-Mendoza and Castillo-Rivera (1998)

[a] The authors reported different units for biogas production.
[b] Biogas production in $m^3/kgCOD_{removed}$.
[c] Biogas production in L/day.
[d] Soluble COD removal.
[e] Volumetric biogas production in L/L/day.

REFERENCES

Acampa, G., Giustra, M.G., Parisi, C.M. 2019. Water treatment emergency: cost evaluation tools. *Sustainability*, 11(9), 2609.

Ahn, J.-H., Forster, C. 2002. A comparison of mesophilic and thermophilic anaerobic upflow filters treating paper–pulp–liquors. *Process. Biochem.*, 38(2), 256–261.

Anderson, G., Kasapgil, B., Ince, O. 1994. Comparison of porous and non-porous media in upflow anaerobic filters when treating dairy wastewater. *Water Res.*, 28(7), 1619–1624.

Athanasopoulos, N. 1986. Cotton fabric desizing and scouring wastewater treatment in upflow anaerobic filter. *Biotechnol. Lett.*, 8(5), 377–378.

Azizi, S., Valipour, A., Sithebe, T. 2013. Evaluation of different wastewater treatment processes and development of a modified attached growth bioreactor as a decentralized approach for small communities. *Sci. World J.*, 2013, 156870.

Banu, J.R., Kaliappan, S., Yeom, I. 2007. Treatment of domestic wastewater using upflow anaerobic sludge blanket reactor. *Int. J. Environ. Sci. Technol.*, 4(3), 363–370.

Bello-Mendoza, R., Castillo-Rivera, M. 1998. Start-up of an anaerobic hybrid (UASB/Filter) reactor treating wastewater from a coffee processing plant. *Anaerobe*, 4(5), 219–225.

Bodík, I., Herdová, B., Drtil, M. 2002. The use of upflow anaerobic filter and AnSBR for wastewater treatment at ambient temperature. *Water Res.*, 36(4), 1084–1088.

Bodkhe, S. 2008. Development of an improved anaerobic filter for municipal wastewater treatment. *Bioresour. Technol.*, 99(1), 222–226.

Chai, S., Guo, J., Chai, Y., Cai, J., Gao, L. 2014. Anaerobic treatment of winery wastewater in moving bed biofilm reactors. *Desalin. Water Treat.*, 52(10–12), 1841–1849.

Connor, R. 2015. *The United Nations World Water Development Report 2015: Water for a Sustainable World.* Paris: UNESCO Publishing.

Coulter, J., Soneda, S., Ettinger, M. 1957. Anaerobic contact process for sewage disposal. *Sewage Ind. Wastes*, 29(4), 468–477.

Daigger, G.T., Boltz, J.P. 2011. Trickling filter and trickling filter-suspended growth process design and operation: a state-of-the-art review. *Water Environ. Res.*, 83(5), 388–404.

Daud, M., Rizvi, H., Akram, M.F., Ali, S., Rizwan, M., Nafees, M., Jin, Z.S. 2018. Review of upflow anaerobic sludge blanket reactor technology: effect of different parameters and developments for domestic wastewater treatment. *J. Chem.*, 2018, 1–13.

Ebrahimi, A., Najafpour, G.D. 2016. Biological treatment processes: suspended growth vs. attached growth. *Iran J. Energy. Environ.*, 7(2), 114–123.

Elmitwalli, T.A., Otterpohl, R. 2007. Anaerobic biodegradability and treatment of grey water in upflow anaerobic sludge blanket (UASB) reactor. *Water Res.*, 41(6), 1379–1387.

Fish, K., Osborn, A., Boxall, J. 2017. Biofilm structures (EPS and bacterial communities) in drinking water distribution systems are conditioned by hydraulics and influence discoloration. *Sci. Tot. Environ.*, 593, 571–580.

Garrett, T.R., Bhakoo, M., Zhang, Z. 2008. Bacterial adhesion and biofilms on surfaces. *Prog. Nat. Sci.*, 18(9), 1049–1056.

Gong, W.-j., Liang, H., Li, W.-z., Wang, Z.-z. 2011. Selection and evaluation of biofilm carrier in anaerobic digestion treatment of cattle manure. *Energy*, 36(5), 3572–3578.

Guerrero, L., Omil, F., Mendez, R., Lema, J. 1997. Treatment of saline wastewaters from fish meal factories in an anaerobic filter under extreme ammonia concentrations. *Bioresour. Technol.*, 61(1), 69–78.

Gulmann, L.K., Beaulieu, S.E., Shank, T.M., Ding, K., Seyfried Jr., W.E., Sievert, S.M. 2015. Bacterial diversity and successional patterns during biofilm formation on freshly exposed basalt surfaces at diffuse-flow deep-sea vents. *Front. Microbiol.*, 6, 901.

Gupta, S.K., Gupta, S., Singh, G. 2010. Anaerobic hybrid reactor: a promising technology for treatment of distillery spent wash. *Int. J. Environ. Pollut.*, 43(1–3), 221–235.

Haider, A., Khan, S.J., Nawaz, M.S., Saleem, M.U. 2018. Effect of intermittent operation of lab-scale upflow anaerobic sludge blanket (UASB) reactor on textile wastewater treatment. *Desalin. Water Treat.*, 136, 120–130.

Heijnen, J., Mulder, A., Enger, W., Hoeks, F. 1989. Review on the application of anaerobic fluidized bed reactors in wastewater treatment. *Chem. Eng. J.*, 41(3), B37–B50.

Janka, E., Carvajal, D., Wang, S., Bakke, R., Dinamarca, C. 2019. Treatment of metformin-containing wastewater by a hybrid vertical anaerobic biofilm-reactor (HyVAB). *Int. J. Environ. Res. Publ. Health*, 16(21), 4125.

Kalyuzhnyi, S., Saucedo, J.V., Martinez, J.R. 1997. The anaerobic treatment of soft drink wastewater in UASB and hybrid reactors. *Appl. Biochem. Biotech.*, 66(3), 291–301.

Kamali, M., Gameiro, T., Costa, M.E.V., Capela, I. 2016. Anaerobic digestion of pulp and paper mill wastes—An overview of the developments and improvement opportunities. *Chem. Eng. J.*, 298, 162–182.

Kennedy, K., Van den Berg, L., Murray, W. 1981. Advanced fixed film reactors for microbial production of methane from waste. *2nd World Congress of Chemical Engineering*, Montréal, Quebec, pp. 317–320.

Ladu, J.L.C., Lü, X.-w. 2014. Effects of hydraulic retention time, temperature, and effluent recycling on efficiency of anaerobic filter in treating rural domestic wastewater. *Water Sci. Eng.*, 7(2), 168–182.

Laquidara, M.J., Blanc, F.C., O'Shaughnessy, J.C. 1986. Development of biofilm, operating characteristics and operational control in the anaerobic rotating biological contactor process. *J. Water Pollut. Control Fed.*, 58, 107–114.

Lettinga, G., Van Velsen, A., Hobma, S.d., De Zeeuw, W., Klapwijk, A. 1980. Use of the upflow sludge blanket (USB) reactor concept for biological wastewater treatment, especially for anaerobic treatment. *Biotechnol. Bioeng.*, 22(4), 699–734.

Lu, C., Li, H.-C., Lee, L.Y., Lin, M.-R. 1997. Effects of disc rotational speed and submergence on the performance of an anaerobic rotating biological contactor. *Environ. Int.*, 23(2), 253–263.

Lu, C., Yeh, A.C., Lin, M.-R. 1995. Treatment of high-strength organic wastewaters using an anaerobic rotating biological contactor. *Environ. Int.*, 21(3), 313–323.

Manariotis, I.D., Grigoropoulos, S.G. 2006. Anaerobic filter treatment of municipal wastewater: biosolids behavior. *J. Environ. Eng.*, 132(1), 23–31.

Marin, P., Alkalay, D., Guerrero, L., Chamy, R., Schiappacasse, M. 1999. Design and startup of an anaerobic fluidized bed reactor. *Water Sci. Technol.*, 40(8), 63–70.

Mendonça, N.M., Niciura, C.L., Gianotti, E.P., Campos, J.R. 2004. Full-scale fluidized bed anaerobic reactor for domestic wastewater treatment: performance, sludge production, and biofilm. *Water Sci. Technol.*, 49(11–12), 319–325.

Miqueleto, A.P., Dolosic, C.C., Pozzi, E., Foresti, E., Zaiat, M. 2010. Influence of carbon sources and C/N ratio on EPS production in anaerobic sequencing batch biofilm reactors for wastewater treatment. *Bioresource Technol.*, 101(4), 1324–1330.

Mousavian, S., Seyedsalehi, M., Paladino, O., Sharifi, P., Kyzas, G.Z., Dionisi, D., Takdastan, A. 2019. Determining biokinetic coefficients for the upflow anaerobic sludge blanket reactor treating sugarcane wastewater in hot climate conditions. *Int. J. Environ. Sci. Technol.*, 16(5), 2231–2238.

Narihiro, T., Terada, T., Kikuchi, K., Iguchi, A., Ikeda, M., Yamauchi, T., Shiraishi, K., Kamagata, Y., Nakamura, K., Sekiguchi, Y. 2009. Comparative analysis of bacterial and archaeal communities in methanogenic sludge granules from upflow anaerobic sludge blanket reactors treating various food-processing, high-strength organic wastewaters. *Microb. Environ.*, 24(2), 88–96.

Pawari, M., Gawande, S. 2015. Ground water pollution and its consequence. *Int. J. Eng. Res. Gen. Sci.*, 3(4), 773–76.

Rajasulochana, P., Preethy, V. 2016. Comparison on efficiency of various techniques in treatment of waste and sewage water—A comprehensive review. *Resour. Efficient Technol.*, 2(4), 175–184.
Roy, D., Gough, R., Jones, L.M. 1986. Comparison of attached and suspended growth methanogenesis in a two phase system. *J. Environ. Sci. Health A.*, 21(8), 769–789.
Ruiz, I., Veiga, M.C., De Santiago, P., Blazquez, R. 1997. Treatment of slaughterhouse wastewater in a UASB reactor and an anaerobic filter. *Bioresour. Technol.*, 60(3), 251–258.
Sanz, I., Fdz-Polanco, F. 1990. Low-temperature treatment of municipal sewage in anaerobic fluidized bed reactors. *Water Res.*, 24(4), 463–469.
Saur, T., Morin, E., Habouzit, F., Bernet, N., Escudié, R. 2017. Impact of wall shear stress on initial bacterial adhesion in rotating annular reactor. *PloS One*, 12(2), e0172113.
Selvamurugan, M., Doraisamy, P., Maheswari, M., Nandakumar, N. 2010. High rate anaerobic treatment of coffee processing wastewater using upflow anaerobic hybrid reactor. *J. Environ. Health Sci.*, 7(2), 129–136.
Şen, S., Demirer, G. 2003. Anaerobic treatment of real textile wastewater with a fluidized bed reactor. *Water Res.*, 37(8), 1868–1878.
Sheli, C., Moletta, R. 2007. Anaerobic treatment of vinasses by a sequentially mixed moving bed biofilm reactor. *Water Sci. Technol.*, 56(2), 1–7.
Somasiri, W., Li, X.-F., Ruan, W.-Q., Jian, C. 2008. Evaluation of the efficacy of upflow anaerobic sludge blanket reactor in removal of color and reduction of COD in real textile wastewater. *Bioresour. Technol.*, 99(9), 3692–3699.
Souza, R., Bresolin, I., Bioni, T., Gimenes, M., Dias-Filho, B. 2004. The performance of a three-phase fluidized bed reactor in treatment of wastewater with high organic load. *Braz. J. Chem. Eng.*, 21(2), 219–227.
Tait, S.J., Friedman, A. 1980. Anaerobic rotating biological contactor for carbonaceous wastewaters. *J. Water Pollut. Control Fed.*, 2257–2269.
Thaiyalnayaki, D., Sowmeyan, R. 2012. Effect of carrier materials in inverse anaerobic fluidized bed reactor for treating high strength organic wastewater. *J. Environ. Anal. Toxicol.*, 2(3). doi: 10.4172/2161-0525.1000134.
Thompson, L., Gray, V., Lindsay, D., Von Holy, A. 2006. Carbon: nitrogen: phosphorus ratios influence biofilm formation by *Enterobacter cloacae* and *Citrobacter freundii*. *J. Appl. Microbiol.*, 101(5), 1105–1113.
Usack, J.G., Spirito, C.M., Angenent, L.T. 2012. Continuously stirred anaerobic digester to convert organic wastes into biogas: system setup and basic operation. *J. Vis. Exp.*, 65, e3978.
Young, J.C., Dahab, M.F. 1983. Effect of media design on the performance of fixed-bed anaerobic reactors. *Water Sci. Technol.*, 15(8–9), 369–383.
Zhang, L., De Vrieze, J., Hendrickx, T.L., Wei, W., Temmink, H., Rijnaarts, H., Zeeman, G. 2018. Anaerobic treatment of raw domestic wastewater in a UASB-digester at 10C and microbial community dynamics. *Chem. Eng. J.*, 334, 2088–2097.

Part V

Application of Microbial Biofilms in Medicine against Chronic Diseases

15 Microbial Biofilms
A New Frontier in Chronic Diseases

Ajla Džanko
International BURCH University

CONTENTS

15.1 INTRODUCTION

In order to come to accurate and concise conclusions on the true nature of bacteria, scientists had to conduct studies on bacteria while they were residing in their natural state and not under artificial conditions. However, this has not always been the case, as the standard system for studying a great number of microorganisms has been established on the experiments and inquiries done in pure cultures and aqueous planktonic states. Nowadays, it has been recognized that planktonic bacterial cells do not provide the required data on the authentic growth and maturation of bacteria present in nature. The fact that the obtained data on bacteria up to now resides on data gathered in their planktonic state has disseminated a nonrealistic representation of their true nature within academic circles and guided further studies where obtained results were satisfying for in vitro setting, however questionable in vivo setting. Subsequently, this has led to many details to arise and clearly indicated that numerous bacterial species exist as part of a greater complex. These complexes are now known to be amalgamations of microorganisms that are capable of forming a protective milieu and are studied by the term biofilms (Donlan 2002). They show finality in their attachment to various surfaces and thus have the repute of providing difficulties when treated with antibiotics and disinfectants. Their ability to attach to surfaces is made possible by the formation of an extracellular polymer substance (EPS). The EPS represents an eminently dehydrated matrix that when observed laboriously is opined to be a product of significant chemical complexity. The EPS made it possible for the biofilms to capture non-cellular components as well as supplementary microbes. However, it is assumed that the main function of the EPS is to enable the accumulation of nutrients. Perhaps, this culminates with the unsurprising fact that cells within the biofilm have complex genetic regulation. When compared to planktonic cells, the structures of biofilms possess nuances in their attributes (phenotypes). Biofilm formations are a direct product of a number of uninterrupted stages that take place as time elapses (Fux et al. 2005, Hall-Stoodley, Costerton and Stoodley 2004). The first stage is rendered by the transport and reversible attachment of bacteria to a particular surface, also allowing the integration of several organic and inorganic substances which are a prominent part of this initial phase. Thereafter, the EPS is secreted, resulting in the formation of bacterial consortia within it. Furthermore, this grants the bacteria the ability to create enduring attachments on surfaces. Lastly, the stage that concludes the process is the colonization and capitulation of the required surface. The continuous process, however, does not end there. Within the biofilm, small variants are a result of constant expansion and maturation phases of these microorganisms.A "primary colonizing unit" is responsible for the secretion of a number of fragments that allure various planktonic bacteria located in that particular domain. This is referred to as secondary colonization. Needless to say, the final organization of any biofilm is a labyrinthine structure that is assembled of bacteria firmly embedded in its EPS. It is also important to note that the structure is also made up of slightly less thick segments of the matrix that contain penetrable water channels. These water channels allow the direct conveyance of the cell's waste results and nutrients. According to studies, biofilms do not exhibit fastidious behavior when attaching to surfaces. Additionally, they have been shown to attach on both natural and synthetic surfaces thus are able to form on most surfaces found in environs (Hall-Stoodley, Costerton, and Stoodley 2004). When contemplating the necessary specimens that will not provide provisional data, scientists have found that the best possible options are to utilize bacteria. However, it would also be superfluous to negate the utilization of other microorganisms such as viruses and protozoa as they have been removed from biofilms and scrutinized in clinics. As previously stated, their ability to form on natural surfaces, for example, plants or even animals and synthetic surfaces such as medical devices grants them an enormous advantage in their formation and steady increase (Fux et al. 2005). Their role in the human body can be deceiving, as they play an important defensive role. Or more accurately, biofilms can accumulate significant purposes which are often judged incorrectly. An exemplary process of this fact is that gut commensal flora that accumulates large amounts of biofilms that are directly linked to epithelial cells, thus creating the environments where pathogens will not gain access to any part of the organism. Astonishingly, as previously noted, the impact that biofilms have on living beings tends to be unfairly misjudged and underestimated in both scientific and academic circles. This results from insufficient data, even with the omnipresent nature of biofilms. Biofilms can be origins for many severe infections, most notably in immunocompromised individuals, as they have certain vulnerabilities to these bacterial states. The aftermath of biofilm-related infections is often serious and deleterious, as several tenacious infections lead to a plethora of medical complications. In order to treat biofilm-related infections, the main fact to understand is that the bacteria which are found embedded in biofilms have certain barriers that enable microorganisms to be guarded against the host organisms defenses and responses. The unwanted outcomes that are produced by biofilms also lead to complications in the treatment options, as the infections tend to reappear in host organisms. These complications are often quite persevering and only possible to remove by surgical interventions in the case of medical device-related infections. Studies have found that the ability of biofilms to execute horizontal gene transfer is significantly greater when compared to their counterparts the planktonic bacteria (Madsen et al. 2012, Stewart and William Costerton 2001, Tormo et al. 2005). This is just one of the initiators of biofilm resistance that is multiple times elevated within a biofilm. Subsequently, this leads to scientists exhibiting great alacrity with regard to the study of biofilms, their intricate structure, and presence as well as the scrupulous ramifications they are capable of producing. Moreover, biofilms could be considered the upcoming scientific endeavor for the field of microbiology, as interest will possibly continue to increase and spread to other scientific fields of study.

One way of understanding biofilms and the infections caused directly by biofilms is to analyze the available literature and note important conclusions. In the next sections, this chapter will analyze and interpret the obtained studies done on biofilms, the infections and diseased which are known to be caused by biofilms. A number of data shown in the next sections will be reviewed and presented, in order to show a clear and concise image of biofilms, their functions, structure, and what their purpose actually represents. The next sections are presented for educational and informative purposes.

15.2 PERIODONTITIS

Biofilms are well-studied formations which have been known to cause a number of tooth-related health issues (Akyıldız et al. 2012). They are often implicated in varying forms of infections which have abysmal consequences (Gu et al. 2013). Periodontitis is a perplexing and troubling infection that arises in the areas of bone where teeth are supported. This condition is a difficult condition that directly affects and mars soft tissue, which repeatedly leads to ramifications to the individual. The causes of this infection are linked to the pernicious relations of particular microorganisms known to bring about various infections and the damaging effects of the natural world and the human immune system. It is also important to note that genetic nuances are also a key factor in the development of the infection, given that genetic predispositions may be prevalent in certain individuals (Slots 2013). Namely, some of the corollaries which occur quite often are the gradual loss of teeth, which is by no means the most dangerous outcome. The resulting infection fosters a myriad of other health complications such as the increase of the possibility of heart attacks and strokes. Nonetheless, periodontitis is quite a common infection that is, fortunately, avoidable with the amelioration of proper oral hygiene (Litin 2003).

If, however, the infection has already transpired, it is of the utmost importance to recognize and ascertain the symptoms. These symptoms will inevitably lead to a plethora of problems which if left untreated may progress into greater issues. This is exactly why discrepancies which may arise in the human body will allocate numerous difficulties. Periodontitis has some significant manifestations which include the inflammation and discoloring of gums, the increased sensitivity of gums, the withdrawal of gums which directly lead to the teeth appearing greater in size, the accumulation of pus between the teeth and gums, putrid breath which leaves an axiomatic unpleasant taste in the mouth.

Biofilms have shown to have ties between their formations and periodontitis. It is not fortuitous that these connections have been drawn through a number of scientific researches as erudite approaches in the genesis of biofilms have had somewhat consistent conclusions. These studies connoted that certain categories of periodontitis should be documented as biofilms that engender infections. A prime example would be the apical peridontitis which has given rise to tiring quandaries (Ricucci and Siqueira 2010). Biofilms have an invariable connection to the surface. This connection is made possible by the production of the EPS. Subsequently, these comminutes of microorganisms are allowed to thrive in such conditions. Dental biofilms, in particular, have numerous labyrinthine divisions as multispecies ecosystems. Namely, oral bacteria are known to engage in interactions that are either competitive or cooperative.

It comes as no surprise that various studies have been able to elucidate and link several bacterial populations to the gradual development of disease as well as some benefits to health. These bacterial inhabitants reside inside dental biofilms and have also displayed some benefits to the health of the individual (Hojo et al. 2009). A prime example of microbial biofilms with scrupulous microbe compositions is dental plaque. Constant studies and research experiments have to be conducted in order to gain insight into the environment of such infectious-causing microorganisms. Oral cavities have allowed the scrutiny of more than 500 peculiar bacterial species to be isolated and studied. This merits scientific inquiry into the fascinating domains of these communities of microorganisms. However, good oral hygiene has been a major factor in preventing the accumulation of dental plaque, therefore never leading to serious health issues. Nevertheless, sporadically maintaining hygiene can grant the increase of bacterial biofilms, thus the development of periodontitis. When the genesis of the infection takes place, the microorganisms have been shown to progressively colonize various areas on the teeth, depending on the specific species. For example, the surface of the teeth is most vulnerable to oral streptococci and *Actinomyces spp*, while the later progression of dental plaque sees an increase of *Fusobacterium nucleatum*, *Treponema denticola*, and *Bacteroides forsythus*. In contrast, oral pathogens like *Porphyromonas gingivalis* have to have established and fully arranged biofilms in order to make further advances toward establishing an association in the gum (Frias, Olle, and Alsina 2001). The inference to be drawn from this is that dental biofilms have a direct role in the origination of the aforementioned infection and that dental plaque is a reasonable factor of concern when contemplating a number of diseases. At the present, there are various tests that enable scientists to disseminate information on the possibility of biofilm formations from specific microorganisms. The tube test method is an exemplar for these types of methodologies. The test requires the inoculation of a specific bacterial population drawn from blood agar in 5 mL of Trypticase soy broth (TSB) in a plastic tube. The culture must then be incubated for the next 18 h at exactly 37°C. After this step in the test, the tubes have to be made vacant in order to color them with safranin or crystal violet. This test results in one of two possible outcomes. If the test is concluded and no biofilm has been detected, the outcome is interpreted as a negative result. Adversely, if slime can be verified by the conspicuously, colored film observed on the walls of the tube, the outcome is interpreted as a positive result (Christensen et al. 1982).

15.3 TOOTH DECAY

Biofilms are unsurprisingly a rather difficult scientific subject for research projects and papers, given their complex and intricate structure. Scientists have found an obvious connection between biofilms and infections (Potera 1999). When biofilms attach on the surfaces of the teeth, they are difficult to eradicate (Potera 1999).

While not the most dangerous sort of ramifications as a direct result of biofilms, dental decay can lead to more serious health issues. The population, as a whole, can be impacted by these issues caused by dental caries as there are no age restrictions or requirements for their formation (Koo and Bowen 2014, Tan et al. 2007). These pathogenic biofilms may be fought against with regular and proper personal hygiene, although the long-term nuances can still be a danger to many individuals. Needless to say that the impoverished part of the population has a greater risk of developing this sort of infection due to the lack of routine and equipment needed in order to maintain salubrious oral hygiene.

Infections that directly lead to tooth decay can have a ranging spectrum of invasion, where, for example, early childhood caries are a rather harmful consequence of infections caused by biofilms (Koo and Bowen 2014). Given the fact that the surface of the teeth can produce the most obvious signs of infection, immediate action has to be shown for the prevention of further development and onset of complications.

The very first indications that linked certain infections and biofilms came in the mid-twentieth century, even though there had been prior knowledge on the existence of bacteria that would attach on the teeth surfaces and have harmful effects. The scientific endeavors which followed had been a step towardtoday's standards of observing microorganisms, their function, and purpose in the ecosystem. However, one of the first known studies done on surface area bacteria had been endeavored by Anton van Leeuwenhoek (Costerton 1999). Moreover, he can be placed as a factor in continues interest arisen further than in the early stages of research.

In today's day and age, many of the symptoms shown to be linked with dental infections can be successfully treated, and constant improvements in the care and health policies are being made so the infections are not a huge cause for concern, given that most of the cases can be preventable. This can be justified by the developments in microbiology and the rise of innovative methodologies being brought upon by these new techniques.

15.4 UTI

Biofilms are persistent communities of bacteria that are more resistant to antibiotics (Tenke et al. 2004). One prevalent infection caused by bacterial biofilms is the urinary tract infection (UTI). UTIs can eventuate because of a number of factors linked to the microbial virulence of their causative agents, which encompass the extracellular facet as well as the surface on which microorganisms form (Delcaru et al. 2016). In order for the UTI to develop, or for that matter in order for any infection to affect an organism, the bacteria must make a link with the host cell (Sparling 1983). Ongoing endourological studies and solutions are being formed by various scientific and academic circles with the utilization of devices on which certain microorganisms form and remain on the surface (Tenke et al. 2004).

Formations of biofilms which are on the surfaces of the host organism are known to have detrimental effects on the duration of the interlude the bacterial communities remain in the urogenital system, thus causing the microbes to exhibit harmful effects (Delcaru et al. 2016, Costerton, Stewart and Greenberg 1999).

UTIs are quite a known type of infection, especially present in the lives of women. Namely, women have a likelihood of suffering from UTIs at least once in their lifetime; the actual percentage of experiencing is approximately 40%–60% ("Urinary Tract Infection (UTI)—Symptoms and Causes" 2019). The chances of creating an all-encompassing treatment strategy for individuals who suffer from complicated UTIs are fairly slim due to the myriad of complications that may or may not arise during the period of infection (Desforges, Stamm, and Hooton 1993). An incessant infection that plagues the medical community isinfections which affect the human urinary system. One way to fight off these infections as mentioned has been with the use of antibiotics (Schmiemann et al. 2010). However, scientific findings have led to the deduction that certain cells of biofilm communities have a greater aversion and resistance to antibiotics, especially when making a comparison with various planktonic bacterial cells (Delcaru et al. 2016, Tabibian et al. 2008)

The ubiquity of these infections is linked to both the gender and the age of an individual (Schmiemann et al. 2010). The consequences of not utilizing the precise instructions when curing UTIs can be deleterious; however, it can be the case for common practice (Schmiemann et al. 2010). Subsequently, when focusing on effective treatment of UTIs, the best possible outcomes can be achieved when resolving early biofilms which are classified as biofilms that have been in existence for less than a day. On the other hand, long-term biofilm amalgamations can be quite difficult to treat, due to the fact that devices would presumably have to be taken out before proper treatment could be implemented (Choong and Whitfield 2001). The complexity of the UTI, also play a major role when contemplating solutions given the fact that complicated UTIs may lead to other health issues, thus a different approach must be taken when dealing with these categories of infections (Schmiemann et al. 2010). The garnered information on UTIs has a correlation with the probability of creating and/or expanding the infection in the population (Schmiemann et al. 2010). Various devices which are purposely situated inside the human body are a direct cause of many biofilm infections (Tenke et al. 2011). The advances which are continuously being made with the aim of improving detection and diagnoses are slowly but surely bringing the spotlight on this important issue. As previously mentioned, a biofilm can be simply elucidated as microorganisms that amass on different surfaces, thereby forming meticulously arranged groups (Tenke et al. 2011). When observing bacterial which form biofilms, it is important to note that in order for these minuscule microorganisms to actually initiate these structures they have to "have awareness" of their propinquity to the specific surface (Tenke et al. 2011).

Some scientific research done on UTIs has been able to ascertain that it could be connoted that bacteria originally attach as sections in disarray which do not experience substantial growth and development when they ramify (Tenke et al. 2011).

One of the most prominent bacteria which cause UTIs is *E. coli*. It is known that about 80% of UTIs iscaused by *E. coli* and also by other bacteria such as *Klebsiella pneumonia*, *Pseudomonas aeruginosa*, *Enterococcus* spp., etc. These bacteria have various factors that provide them with the advantage of causing serious, tenacious infections which foster pernicious complications to the host cell (Delcaru et al. 2016). Suffice to say that before proper care for affected individuals is put to motion a firm grasp on the structure and function of the urinary tract must be first established. An erudite approach must be set in practice for accurate results, which lead to future betterment and understanding of biofilm-caused infections. Namely, the various species of bacteria which form biofilms that affect the urinary tract can show difficulties with accurate identification and thus therapeutic strategies. This persuades scientists to modify their technique, as they have with the modifications implemented in the last couple of years. These findings have been able to draw conclusions on the way of communication between microorganisms.

15.5 KIDNEY STONES

Kidney stones affect the population as a whole, both men and women are not excluded in the possibilities of this scrupulous formation. It would not be an understatement to opine that the unnerving increase of the formation of kidney stones has become a topic of concern due to the implications and problems which might emerge and cause a myriad of complications for the medical community. Data collected over the past few years has indicated that there has been an alarming increase in these formations, due to this studies have been fraught with refutations on the causes. Research has clearly indicated that there are connections between a number of genetic factors of an individual. Not surprisingly the genetic make-up and the subtle nuances in an organism can have rendered palpable reverberations, however, genetic anomalies are slow-changing and thus could be interpreted as a lower-risk cause, given the fact that issues which arise directly from the environment could actually be more detrimental to the formation of kidney stones (Parmar 2004). Kidney stone as noted isa serious health issue which stems from infections of the urinary system and hashad a staggering 10% effect on individuals regardless of their innate identity (Parmar 2004, Hobbs et al. 2018). This health issue such as aforementioned ones merits scrutiny on the issue as the fastidious techniques developed each year help resolve the bacterial communities scattered in the human system.

15.6 OTITIS MEDIA

It has been suggested that almost all bacteria with minor exceptions exist in biofilms which may cause infections (Omar et al 2017, Wolcott, Rhoads and Dowd 2008). Their ubiquity in the environment is thus bound to have many negative influences on a number of living creatures (Fergie et al. 2004). Biofilms are known to cause the inimical disease of otitis media. Communities of bacteria in the form of biofilms are quite a familiar finding in individuals suffering from chronic otitis media (COM). Diverse scientific circles claim that there are correlations between biofilms and the aversion of bacteria to treatment by antibiotics and persistent infections (Akyıldız et al. 2012). It is important to note that biofilms to not function in varying degrees of disarray rather their sizable communities operate in intricately defined structures that may cause damage to the host organism (Gu et al. 2013).

These communities of bacteria are known to cause problems during treatment, noticeably because of their resistance to a number of treatment strategies. In order to form an all-encompassing strategy for treatment, the biofilms must be scrutinized so that a greater understanding of the intricate causes and formations could render cures inevitable. One of the first widely documented studies done on the relationship between otitis media and these microbial communities was achieved by Reyner et al (Akyıldız et al. 2012, Rayner et al. 1998). Due to the fact that much of the underlying causes are still elusive to the scientific community, many other studies must be successfully executed in order to gain useful data. Given the fact that the cause and effect of biofilms are becoming a greater topic of discussion and that inference is being made on their impact factor on otolaryngologic illnesses. When identifying the genesis of biofilm development in individuals that may have shown symptoms, microscopy plays a major role (Akyıldız et al. 2012). Microscopy applies erudite approaches in scrupulous documentation and identification of biofilms, thus garnering the required data. Light microscopes, as well as electronic microscopes, have been used in these specific studies (Akyıldız et al. 2012). Namely, it is not surprising knowing that many diseases might have biofilm formations which allow them to keep their reappearing repute stable and constant in the medical world, otitis media is just one of many examples. COM has incongruous and irrevocable consequences on the person suffering from the formation, and as the name suggests, it is a tenacious illness that is difficult to completely resolve (Brook 1979). Many complications which are oftentimes preventable with proper and timely treatment have a persistent nature and are a periodic cause for medical consultations and queries (Monasta et al. 2012). The severity of the condition differs noticeably when comparing different countries, yet the conundrums arise when the illness affects the younger population as the issues have more significant possibilities for major ramifications (Monasta et al. 2012). The United States Centers for Disease Control and Prevention (CDC) has been able to ascertain that the data implicated biofilms with the cause of over 50% of infections that plague the community (Potera 1999, Gu et al. 2013). Human tissues have shown to be vulnerable to biofilm formations, thus may lead to the genesis of otitis media (Gu et al. 2013). One important aspect of the development of otitis media is the bacteria's impressive ability to communicate with one another. These fascinating forms of communication capabilities foster victorious spread on the surfaces of the body, thereby avoiding maladaptive behavior. Biofilms can interface by the utilization of signaling molecules, and as time elapses, the prowess of their resistance and safeguard becomes difficult to overcome. Given the fact that they are in no danger of environmental disruptions, they are allowed to garner greater areas (Gu et al. 2013). Another prevalent form of otitis media is acute otitis media (AOM) which is found inchildren who are experiencing the symptoms and signs. Otitis media of any type is the reason for a number of medical visits due to children of all ages experiencing the disease (Post 2001, Klein 1994). AOM has also been shown to cause abstruse socioeconomic alternations to numerous countries (Coticchia et al. 2013). Infants have a greater probability of suffering from otitis media when taking into account their unfortified bodies.

Like a plethora of chronic diseases, otitis media has been known to appear more than once in the affected patients which causes greater diligence in the scientific research that often follows (Fergie et al. 2004). The key factors in resolving otitis media are the elucidations and understandings of the human body operating as the host for the biofilms causing agents, furthermore leading to data which is not strewn with gaps of misinformation or inconclusive results. Subsequently, it is quite important that patients are treated with the most accurate strategy when the betterment of their health is a factor and the individuals assigned the duty of treating the patients have the most exact data on the organism in order to circumvent the recalcitrant pathogens (Bluestone, Stephenson, and Martin 1992). In conclusion, bacteria of any form have an obvious influence on diseases related to ear infections and therefore must garner useful information onthe possibilities of treatment (Giebink 1989).

15.7 CHRONIC WOUNDS

Biofilms have been located and studied in over 90% of chronic wounds and cause medical burdens on the population for a plethora of reasons (Attinger and Wolcott 2012). Biofilms in chronic wounds can be quite difficult to scrutinize due to the gaps in knowledge present on the topic (Morgan et al. 2019). The complexity of biofilms never seems to waver, predominantly knowing that they have persistent resistance to the body's immune responses which aim to resolve the initial occurrences (Morgan et al. 2019). Similar infections caused by bacteria tend to have ranging outcomes and are thus difficult to elucidate as the resulting outcomes are often dissimilar and produce more questions than answers. Biofilms are strewn throughout our environment and inside our body so their probability of causing damaging effects is not as surprising when contemplating the real genesis of the conundrum. The significant impact factor of biofilms and chronic wounds has been a continuous topic ofinterest during the last couple of years, and they have gained recognition as being one of the main causes, especially when observing the statistical data of patients suffering from sometimes the deadly complications of not treating the infections (Wolcott et al. 2010, Zhao et al. 2013). The treatment, however, relies on the garnered information on chronic wounds and the key roles of bacterial associations. Without the required information, it can be exceptionally difficult to find an adequate solution and as the infection continues to linger in the host organism rendering treatment futile. When observing environmental factors of these microbial communities, their exemplary domain can be organized on chronic wounds, due to predisposing the unstable tissues to garner infections (Zhao et al 2013). *P.aeruginosa* is often taken as a research tool when studying chronic wounds given that their presence doesn't rapidly spread and affect the individual as it stays at the site of infection (Morgan et al. 2019). This is not a new concept given that most of the information garnered on medical biofilms came directly from the studies of the aforementioned specimen as the specimen has shown favorable tendencies toward chronic wounds (Omar et al. 2017, Lyczak, Cannon, and Pier 2000). The interactions of bacteria on chronic wounds complicate the issues, due to the possibility of causing disease and the severities of the harm caused. It is not tangible. Their fastidious nature acquires them the advantage in the development and allows them to spread. This results in triumphant outcomes for the biofilms, but unfortunate results for the host organism. Many different unions of bacteria are shown to have detrimental consequences on chronic wounds (Percival, Thomas, and Williams 2010). Biofilms often display their perseverance with the combination of a number of implementations, rather than one single cause (Omar et al. 2017). Another exasperating issue that fraught studies of chronic wounds is found in the fact that the signs that indicate the formation and effect of chronic wounds is not immediately observable by the individual and they become abstruse, thus leading to dubious conclusions (Zhao et al. 2013). The recovery process of chronic wounds is hindered by the very presence of bacterial communities and is an increasing issue in the medical field (Attinger and Wolcott 2012). The lucrative values and economic disadvantages have also been impacted by the treatment of chronic wounds (Siddiqui and Bernstein 2010, Scali and Kunimoto 2013), which obviously leads to many drawbacks to the community as a whole.

In conclusion, chronic wounds present a significant problem in medical communities and are seen as a clinical problem that affects the entire population (James et al 2008, Stewart 2008). The scientific researches that are currently doing significant data accumulation have an obligation to obtain relevant and useful information in order for the future of medical applications to see vital progress in the treatment of chronic wounds and similar infections and diseases.

15.8 MEDICAL DEVICE-RELATED INFECTIONS

Medical devices in use today have had a noteworthy influence on the treatment of diverse diseases and are a prevailing implement in most medical institutions (Khardori and Yassien 1995, Desrousseaux et al. 2013). Nonetheless, these tools are not infallible and have been prone to carry microbial associations which are known to cause lasting infections (Khardori and Yassien 1995, Singhai et al. 2012).

A plethora of biofilm formations plays a prominent role in the infections which are caused by impairments of miscellaneous medical devices. Moreover, this can be quite a recurrent incident that is a threat to individuals under medical care, as these proliferations may result in consequential diseases and in the worst-case scenarios moribund patients (Habash and Reid 1999, Jamal et al. 2018, Wi and Patel 2018). These microorganisms are significantly involved with disconcerting biofilm developments. These biofilms are thereby resilient to many treatment methods employed by medical professionals (Khoury et al. 1992, 63).

It is common knowledge that biofilms could be somewhat tended to in order to forestall serious health issues, although the only way to eliminate their presence and industrious reversion in patients was by the withdrawal of the device (Khoury et al. 1992, Costerton et al. 1993, Mack et al. 2006). Given the fact that the host organism's resistance is weakened by tissue deterioration, the bacteria have a greater chance of causing long-lasting harm to the body (Khardori and Yassien 1995). This will inevitably reinforce the bacterial communities and make them remarkably impervious to antibiotic therapy, which may lead to many more medical complications.

Even with the abstruse understandings of biofilm formations, it can still be ascertained that compromised medical devices which are utilized to aid in the curing of illnesses are at greater risk of worsening the overall health conditions and thus escalating the hindrance of therapeutic options (Khardori and Yassien 1995). Medical devices, as previously stated, are often in peril of fostering biofilm milieux, thus prevent their initial purpose for therapeutic and diagnostic efforts (Singhai et al. 2012). This causes problems as the cessation of antibiotic treatment leads to the unwanted return of pernicious biofilms (Costerton et al. 1993).

In order to find a finite solution to device-related infections, continuous research has to be conducted to result in new, applicable solutions for the medical community. However, the expanding technological breakthroughs have made it possible to integrate new medical devices which could possibly be not as vulnerable to acquiring surface-based microorganisms (Reid 1999). The continual pursuit of scientific answers has made it possible to notably lessen the formations of biofilm on numerous medical tools, however, the underlying causes still need to be scrutinized and resolved (Reid 1999). Riveting solutions have to be proposed when dealing with the complex and recalcitrant nature of biofilms, as they have shown to provide a myriad of questions, rather than concrete clues on how to prevent, diagnose, and treat their formations on the host organism. It is also important to note that the increasing interest in medical device-based infections will only further open the possibilities of understanding and grasping the well needed new medical strategies.

15.9 CYSTIC FIBROSIS

Another major conundrum for the medical community lies in the complications caused by specific biofilm formations, which not only have a certain persistence of acclimating onto host organisms but also prolong the duration of chronic infections. Biofilms have the innate ability to fight off a portion of the response of the organisms' immune system, which subsequently leads to severe illnesses to patients. One of the major reasons why biofilms have this ability to persevere and continue to damage the host is the fact that they build up a resistance to antibiotic treatment methods, leading them to quite often fail in their endeavors (Høiby et al. 2010). Biofilms can thus seem unerring and prominent with their irrevocable presence (Høiby, Ciofu, and Bjarnsholt 2010). Specifically in cystic fibrosis, which is a disease that greatly harms both the respiratory and digestive tract, has a distinct correlation to biofilm-related persistence. This persistence is linked directly to the alarming mucous stains formed by biofilms (Høiby, Ciofu, and Bjarnsholt 2010). It can be speculated that biofilms are so seemingly permanent in the host organism due to the fact that the treatment methodologies employ the utilization of superfluous procedures, which have been shown to have limited effects. The genesis of biofilm infections which as noted encompass the regions of the respiratory system can maintain a constant myriad of associations that become ubiquitous for the affected areas. Moreover, biofilms are known to embed themselves onto natural environs (Høiby et al. 2010). What is also important to note is that biofilms can have issues with the prevalence of mutation-related incidences

(Høiby et al. 2010). These occurrences connote the fact that many complications may arise as a result of various factors that affect the organism. Subsequently, many aberrations are a consequence of altered states of the biofilm formations (Høiby et al. 2010). Furthermore, the simple fact that planktonic cells do not exhibit such expressive and distinctive conversions causes many queries. Cystic fibrosis is opined to be a rather complex disease that can only further be complicated and thus made perplexing to medical clinics. Namely, it is a serious illness that still needs new and pioneering studies to be executed on the topic in order to obtain the necessary data that can produce new treatment methodologies. The main issue that needs to be addressed as soon as possible is the underlying factor that enables the protection of biofilms to host responses and the seemingly effortless manner in which they impede the ability of the antibiotic treatment to be successfully executed (Costerton 2001). The exact process of growth and maturation of many biofilms is related to their evolutionary established method that ensures and enhances greater life span and future colonization feasibility (Høiby 2002).

15.10 OSTEOMYELITIS

Osteomyelitis is a curious infection that is not as frequent when considering the fact that it is a subject of interest to many medical professionals. Great deals of questions arise when studying this infection, as it renders abstruse ramifications to the population, even though it affects only a limited amount of patients. Namely, it is an infection that damages the bones of the human body in varying degrees that result from amalgamations of harmful microorganisms (Lew and Waldvogel 2004). Many factors are impactful when dealing with differing levels of osteomyelitis, and it can be restrictive in its phases and segments of infection reaches, as seen both affecting a minor segment of the bone or affecting an alarming consortium of the skeleton (Lew and Waldvogel 2004). It is also important to note that damaged bones have been known to be eminently affected so that their surfaces would become concealed by the biofilm formation (Gristina et al. 1985). One key element that has to be considered when optimizing experiments and studies is that bacterial colonies are the cause of this infection, thus in order to derive a strategy for treatment, a clear and laborious approach has to be endeavored to understand the various stages of growth and development of the microorganisms. Biofilm formations have to be exhaustingly scrutinized, their defenses documented, and their omnipresence described. Osteomyelitis has been observed to damage a distressing portion of bones, as no components of the skeleton have shown victorious defenses and, in turn, are vulnerable to infections, without regard to the age group of the individual suffering from the consequences (Lew and Waldvogel 2004). One of the biggest problems with osteomyelitis is suggested to be the lack of information and knowledge accumulated on the subject, which causes some conclusions and studies to be obfuscated and fraught with deficient data. Biofilm formations are applicable to new solutions and treatment strategies as it is becoming clear that a vast scope of different treatment methods has to be applied to increase the possibility of positive long-term outcomes (Walter et al. 2012, Brady et al. 2008). When discussing the gravity of the infection and whether or not treatment is proceeding successfully, the medical personnel have to be aware of a plethora of factors which indicate concise conclusions; these include the amount of time which has elapsed since the initial formation, the pathogenesis, and the proportion of affected bones (Calhoun, Manring, and Shirtliff 2009). However, the last few decades have resulted in significant progress that introduced new procedures and created many modifications that allowed better control of the infections. These improvements have been tremendously important for future progress in the fields of both microbiology and medicine, yet an infallible approach for treatment has not been established and further progress has to commence in order to acquire answers (Calhoun, Manring, and Shirtliff 2009).

REFERENCES

Akyıldız, İ., Take, G., Uygur, K., Kızıl, Y., &Aydil, U. (2012). Bacterial biofilm formation in the middle-ear mucosa of chronic otitis media patients. *Indian Journal of Otolaryngology and Head & Neck Surgery*, *65*(S3), 557–561. doi: 10.1007/s12070-012-0513-x.

Attinger, C., &Wolcott, R. (2012). Clinically addressing biofilm in chronic wounds. *Advances in Wound Care*, *1*(3), 127–132. doi: 10.1089/wound.2011.0333.

Bluestone, C. D., Stephenson, J. S., & Martin, L. M. (1992). Ten-year review of otitis media pathogens. *The Pediatric Infectious Disease Journal*, *11*(Supplement), S7–S11. doi: 10.1097/00006454-199208001-00002.

Brady, R. A., Leid, J. G., Calhoun, J. H., Costerton, J. W., & Shirtliff, M. E. (2008). Osteomyelitis and the role of biofilms in chronic infection. *FEMS Immunology & Medical Microbiology*, *52*(1), 13–22. doi: 10.1111/j.1574-695x.2007.00357.x.

Brook, I. (1979). Bacteriology of chronic otitis media. *Journal of the American Medical Association*, *241*(5), 487. doi: 10.1001/jama.1979.03290310027008.

Calhoun, J., Manring, M. M., & Shirtliff, M. (2009). Osteomyelitis of the long bones. *Seminars in Plastic Surgery*, *23*(02), 059–072. doi: 10.1055/s-0029-1214158.

Choong, S., & Whitfield, H. (2001). Biofilms and their role in infections in urology. *BJU International*, *86*(8), 935–941. doi: 10.1046/j.1464-410x.2000.00949.x.

Christensen, G., Simpson, W., Bisno, A., & iBeachey, E. (1982). Adherence of slime-producing strains of *Staphylococcusepidermidis* to smooth surfaces. *Infection and Immunity*, *37*, 318–326.

Costerton, J. W. (1999). Bacterial biofilms: a common cause of persistent infections. *Science*, *284*(5418), 1318–1322. doi: 10.1126/science.284.5418.1318.

Costerton, J. W. (2001). Cystic fibrosis pathogenesis and the role of biofilms in persistent infection. *Trends in Microbiology*, *9*(2), 50–52. doi: 10.1016/s0966-842x(00)01918-1.

Costerton, J. W., Khoury, A. E., Ward, K. H., & Anwar, H. (1993). Practical measures to control device-related bacterial infections. *The International Journal of Artificial Organs*, *16*(11), 765–770. doi: 10.1177/039139889301601104.

Costerton, J. W., Stewart, P. S., & Greenberg, E. P. (1999). Bacterial biofilms: a common cause of persistent infections. *Science*, *284*, 1318–1322. doi: 10.1126/science.284.5418.1318.

Coticchia, J. M., Chen, M., Sachdeva, L., & Mutchnick, S. (2013). New paradigms in the pathogenesis of otitis media in children. *Frontiers in Pediatrics*, *1*. doi: 10.3389/fped.2013.00052.

Delcaru, C., Alexandru, I., Podgoreanu, P., Grosu, M., Stavropoulos, E., Chifiriuc, M., & Lazar, V. (2016). Microbial biofilms in urinary tract infections and prostatitis: etiology, pathogenicity, and combating strategies. *Pathogens*, *5*(4), 65. doi: 10.3390/pathogens5040065.

Desforges, J. F., Stamm, W. E., & Hooton, T. M. (1993). Management of urinary tract infections in adults. *New England Journal of Medicine*, *329*(18), 1328–1334. doi: 10.1056/nejm199310283291808.

Desrousseaux, C., Sautou, V., Descamps, S., & Traoré, O. (2013). Modification of the surfaces of medical devices to prevent microbial adhesion and biofilm formation. *Journal of Hospital Infection*, *85*(2), 87–93. doi: 10.1016/j.jhin.2013.06.015.

Donlan, R. M. (2002). Biofilms: microbial life on surfaces. *Emerging Infectious* Diseases, *8*(9), 881–890.

Fergie, N., Bayston, R.,Pearson, J. P., & Birchall, J. P. (2004). Is otitis media with effusion a biofilm infection? *Clinical Otolaryngology and Allied Sciences*, *29*(1), 38–46. doi: 10.1111/j.1365-2273.2004.00767.x.

Frias, J., Olle, E., & Alsina, M. (2001). Periodontal pathogens produce quorum sensing signal molecules. *Infection and Immunity*, *69*(5), 3431–3434.

Fux, C. A., Costerton, J. W., Stewart, P. S., & Stoodley, P. (2005). Survival strategies of infectious biofilms. *Trends in Microbiology*, *13*(1), 34–40.

Giebink, G. S. (1989). The microbiology of otitis media. Retrieved from https://journals.lww.com/pidj/Citation/1989/01001/The_microbiology_of_otitis_media.8.asp.

Gristina, A., Oga, M., Webb, L., & Hobgood, C. (1985). Adherent bacterial colonization in the pathogenesis of osteomyelitis. *Science*, *228*(4702), 990–993. doi: 10.1126/science.4001933.

Gu, X., Keyoumu, Y., Long, L., & Zhang, H. (2013). Detection of bacterial biofilms in different types of chronic otitis media. *European Archives of Oto-Rhino-Laryngology*, *271*(11), 2877–2883. doi: 10.1007/s00405-013-2766-8.

Habash, M., & Reid, G. (1999). Microbial biofilms: their development and significance for medical device-related infections. *The Journal of Clinical Pharmacology*, *39*(9), 887–898. doi: 10.1177/00912709922008506.

Hall-Stoodley, L., Costerton, J. W., & Stoodley, P. (2004). Bacterial biofilms: from the natural environment to infectious diseases. *Nature Reviews Microbiology*, *2*(2), 95–108.

Hobbs, T., Schultz, L. N., Lauchnor, E. G., Gerlach, R., & Lange, D. (2018). Evaluation of biofilm induced urinary infection stone formation in a novel laboratory model system. *The Journal of Urology*, *199*(1), 178–185. doi: 10.1016/j.juro.2017.08.083.

Høiby, N. (2002). Understanding bacterial biofilms in patients with cystic fibrosis: current and innovative approaches to potential therapies. *Journal of Cystic Fibrosis*, *1*(4), 249–254. doi: 10.1016/s1569-1993(02)00104-2.

Høiby, N., Bjarnsholt, T., Givskov, M., Molin, S., & Ciofu, O. (2010). Antibiotic resistance of bacterial biofilms. *International Journal of Antimicrobial Agents*, *35*, 322–332. doi:10.1016/j.ijantimicag.2009.12.011.

Høiby, N., Ciofu, O., & Bjarnsholt, T. (2010). *Pseudomonas aeruginosa* biofilms in cystic fibrosis. *Future Microbiology*, *5*(11), 1663–1674. doi: 10.2217/fmb.10.125.

Hojo, K., Nagaoka, S., Ohshima, T., & Maeda, N. (2009). Bacterial interactions in dental biofilm development. *Journal of Dental Research*, *88*(11), 982–990.

Jamal, M., Ahmad, W., Andleeb, S., Jalil, F., Imran, M., Nawaz, M. A., Hussain, T., Ali, M., Rafiq, M., & Kamil, M. A. (2018). Bacterial biofilm and associated infections. *Journal of the Chinese Medical Association*, *81*(1), 7–11. doi: 10.1016/j.jcma.2017.07.012.

James, G. A., Swogger, E., Wolcott, R., Pulcini, E. D., Secor, P., Sestrich, J., Costerton, J. W., & Stewart, P. S. (2008). Biofilms in chronic wounds. *Wound Repair and Regeneration*, *16*(1), 37–44. doi: 10.1111/j.1524-475x.2007.00321.x.

Khardori, N., & Yassien, M. (1995). Biofilms in device-related infections. *Journal of Industrial Microbiology*, *15*(3), 141–147. doi: 10.1007/bf01569817.

Khoury, A. E., Lam, K., Ellis, B., & Costerton, J. W. (1992). Prevention and control of bacterial infections associated with medical devices. *ASAIO Journal*, *38*(3), M174–M178. doi: 10.1097/00002480-199207000-00013.

Klein, J. O. (1994, November). Otitis media. Retrieved 9 November 2019, from https://www.jstor.org/stable/4458141.

Koo, H., & Bowen, W. H. (2014). *Candida albicans* and *Streptococcus mutans*: a potential synergistic alliance to cause virulent tooth decay in children. *Future Microbiology*, *9*(12), 1295–1297. doi:10.2217/fmb.14.92.

Lew, D. P., & Waldvogel, F. A. (2004). Osteomyelitis. *The Lancet*, *364*(9431), 369–379. doi: 10.1016/s0140-6736(04)16727-5.

Litin, S. C. (2003). *Mayo Clinic Family Health Book*. New York, NY: Harper Resource.

Lyczak, J. B,, Cannon, C. L., & Pier, G. B. (2000). Establishment of *Pseudomonas aeruginosa* infection: lessons from a versatile opportunist. *Microbes and Infections*, 2(9), 1051–1060. doi: 10.1016/S1286-4579(00)01259-4[pii]. PMID: 10967285.

Mack, D., Rohde, H., Harris, L. G., Davies, A. P., Horstkotte, M. A., & Knobloch, J. K.-M. (2006). Biofilm formation in medical device-related infection. *The International Journal of Artificial Organs*, *29*(4), 343–359. doi: 10.1177/039139880602900404.

Madsen, J. S., Burmølle, M., Hansen, L. H., & Sørensen, S. J. (2012). The interconnection between biofilm formation and horizontal gene transfer. *FEMS Immunology & Medical Microbiology*, *65*(2), 183–195.

Monasta, L., Ronfani, L., Marchetti, F., Montico, M., VecchiBrumatti, L., Bavcar, A., Grasso, D., Barbiero, C., & Tamburlini, G. (2012). Burden of disease caused by otitis media: systematic review and global estimates. *PLoS One*, *7*(4), e36226. doi: 10.1371/journal.pone.0036226.

Morgan, S. J., Lippman, S. I., Bautista, G. E., Harrison, J. J., Harding, C. L., Gallagher, L. A., Cheng, A. C., Siehnel, R., Ravishankar, S., Usui, M. L., & Olerud, J. E. (2019). Bacterial fitness in chronic wounds appears to be mediated by the capacity for high-density growth, not virulence or biofilm functions. *PLOS Pathogens*, *15*(3), e1007511. doi: 10.1371/journal.ppat.1007511.

Omar, A., Wright, J., Schultz, G., Burrell, R., & Nadworny, P. (2017). Microbial biofilms and chronic wounds. *Microorganisms*, *5*(1), 9. doi: 10.3390/microorganisms5010009.

Parmar, M. S. (2004). Kidney stones. *British Medical Journal*, *328*(7453), 1420–1424. doi: 10.1136/bmj.328.7453.1420.

Percival, S. L., Thomas, J. G., & Williams, D. W. (2010). Biofilms and bacterial imbalances in chronic wounds: anti-Koch. *International Wound Journal*, *7*(3), 169–175. doi: 10.1111/j.1742-481x.2010.00668.x.

Post, J. C. (2001). Candidate's thesis: direct evidence of bacterial biofilms in otitis media. *The Laryngoscope*, *111*(12), 2083–2094. doi: 10.1097/00005537-200112000-00001.

Potera, C. (1999). Microbiology: forging a link between biofilms and disease. *Science*, *283*(5409), 1837–1839. doi: 10.1126/science.283.5409.1837.

Rayner, M. G., Zhang, Y., Gorry, M. C., Chen, Y., Post, J. C., & Ehrlich, G. D. (1998). Evidence of bacterial metabolic activity in culture-negative otitis media with effusion. *Journal of the American Medical Association*, *279*, 296–299. doi:10.1001/jama.279.4.296.

Reid, G. (1999). Biofilms in infectious disease and on medical devices. *International Journal of Antimicrobial Agents*, *11*(3–4), 223–226. doi: 10.1016/s0924-8579(99)00020-5.

Ricucci, D., & Siqueira Jr., J. F. (2010). Biofilms and apical periodontitis: study of prevalence and association with clinical and histopathologic findings. *Journal of Endodontics*, *36*(8), 1277–1288.

Scali, C., & Kunimoto, B. (2013). An update on chronic wounds and the role of biofilms. *Journal of Cutaneous Medicine and Surgery*, *17*(6), 371–376. doi: 10.2310/7750.2013.12129.

Schmiemann, G., Kniehl, E., Gebhardt, K., Matejczyk, M. M., & Hummers-Pradier, E. (2010). The diagnosis of urinary tract infection. *Deutsches Aerzteblatt* [Online]. doi:10.3238/arztebl.2010.0361.

Siddiqui, A. R., & Bernstein, J. M. (2010). Chronic wound infection: facts and controversies. *Clinics in Dermatology*, *28*(5), 519–526. doi: 10.1016/j.clindermatol.2010.03.009.

Singhai, M., Malik, A., Shahid, M., Malik, M., & Goyal, R. (2012). A study on device-related infections with special reference to biofilm production and antibiotic resistance. *Journal of Global Infectious Diseases*, *4*(4), 193. doi: 10.4103/0974-777x.103896.

Slots, J. (2013). Periodontology: past, present, perspectives. *Periodontology 2000*, *62*(1), 7–19.

Sparling, P. (1983). Bacterial virulence and pathogenesis: an overview. *Clinical Infectious Diseases*, *5*(Supplement 4), S637–S646. doi: 10.1093/clinids/5.supplement_4.s637.

Stewart, P. S. (2008). Biofilms in chronic wounds. *Wound Repair and Regeneration*, *16*(1), 37–44. doi: 10.1111/j.1524-475x.2007.00321.x.

Stewart, P. S., & William Costerton, J. (2001). Antibiotic resistance of bacteria in biofilms. *The Lancet*, *358*(9276), 135–138.

Tabibian, J. H., Gornbein, J., Heidari, A., Dien, S. L., Lau, V. H., Chahal, P., Churchill, B. M., & Haake, D. A. (2008). Uropathogens and host characteristics. *Journal of Clinical Microbiology*, *46*, 3980–3986. doi:10.1128/JCM.00339-08.

Tan,S., Smith, V., Barker, L. K., Thornton-Evans, G., Eke, P. I., & Beltrán-Aguilar, E. D. (2007). Trends in oral health status: United States, 1988–1994 and 1999–2004. *Vital Health Statistics*, *11*(248), 1–92.

Tenke, P., Köves, B., Nagy, K., Hultgren, S. J., Mendling, W., Wullt, B., Grabe, M., Wagenlehner, F. M., Cek, M., Pickard, R., & Botto, H. (2011). Update on biofilm infections in the urinary tract. *World Journal of Urology*, *30*(1), 51–57. doi: 10.1007/s00345-011-0689-9.

Tenke, P., Riedl, C. R., Jones, G. L., Williams, G. J., Stickler, D., & Nagy, E. (2004). Bacterial biofilm formation on urologic devices and heparin coating as preventive strategy. *International Journal of Antimicrobial Agents*, *23*, 67–74. doi: 10.1016/j.ijantimicag.2003.12.007.

Tormo, M. Á., Knecht, E., Götz, F., Lasa, I., & Penadés, J. R. (2005). Bap-dependent biofilm formation by pathogenic species of *Staphylococcus*: evidence of horizontal gene transfer? *Microbiology*, *151*(7), 2465–2475.

Urinary Tract Infection (UTI)—Symptoms and Causes. (2019). Retrieved 11 November 2019, from https://www.mayoclinic.org/diseases-conditions/urinary-tract-infection/symptoms-causes/syc-20353447.

Walter, G., Kemmerer, M., Kappler, C., & Hoffmann, R. (2012). Treatment algorithms for chronic osteomyelitis. *Deutsches Aerzteblatt* [Online]. doi: 10.3238/arztebl.2012.0257.

Wi, Y. M., & Patel, R. (2018). Understanding biofilms and novel approaches to the diagnosis, prevention, and treatment of medical device-associated infections. *Infectious Disease Clinics of North America*. doi: 10.1016/j.idc.2018.06.009.

Wolcott, R. D., Rhoads, D. D., Bennett, M. E., Wolcott, B. M., Gogokhia, L., Costerton, J. W., & Dowd, S. E. (2010). Chronic wounds and the medical biofilm paradigm. *Journal of Wound Care*, *19*(2), 45–53. doi: 10.12968/jowc.2010.19.2.46966.

Wolcott, R. D., Rhoads, D. D., & Dowd, S. E. (2008). Biofilms and chronic wound inflammation. *Journal of Wound Care*, *17*(8), 333–341. doi: 10.12968/jowc.2008.17.8.30796.

Zhao, G., Usui, M. L., Lippman, S. I., James, G. A., Stewart, P. S., Fleckman, P., & Olerud, J. E. (2013). Biofilms and inflammation in chronic wounds. *Advances in Wound Care*, *2*(7), 389–399. doi: 10.1089/wound.2012.0381.

16 Microbial Biofilms-Aided Resistance and Remedies to Overcome It

Ayaz Ahmed and Anum Khalid Khan
University of Karachi

CONTENTS

16.1 INTRODUCTION

Biofilms represent the sessile niche of microbial consortia which colonize biotic or abiotic surfaces embedded with self-excreted polymeric substances (polysaccharide, extracellular DNA (eDNA), proteins, etc.). Microbes within biofilm differ from their planktonic counterparts phenotypically as well as genotypically. Bacterial biofilms due to its resilient nature to antibiotics, chemicals, and host immune disease became a global concern as it enhances bacterial resistance by 1,000 times. The biofilm modes of bacterial growth are the key to persistent infections and responsible for 80% hospital-acquired chronic infections such as urinary tract infections, endocarditis, otitis media, and cystic fibrosis, etc. Biofilm formation starts when few bacterial cells adhere to surface having nutrients or pellicle, and then this adherence becomes irreversible followed by multiplication of cell, production of exopolymeric substances, and altered gene makeup which change actively metabolizing cells to least active cells to further mature biofilm. Mature biofilm has water channels within them to provide nutrients and resembles primitive multicellular organisms in terms of tactics to interact with their environment. There are a number of factors important for antimicrobial resistance posed by biofilms such as:

1. Structure of biofilms
2. State of microbial cells within the biofilm
3. Synergistic approaches of microbial community within biofilms
4. Rate of mutation
5. Efflux pump
6. Quorum sensing (QS).

So, in this chapter, each of the enlisted components will be discussed followed by remedies to overcome this barrier to fight biofilm-related infections:

16.1.1 Biofilm Structure and State of Growth

Mature biofilm looks like a multistory building of different phenotypes of microbial cells embedded within a self-produced matrix of different polysaccharides. The glycocalyx is an important constituent of biofilm, and it serves as a resistant factor to harsh host and environmental factors. Its thickness varies between 0.2–1.0 μm which by virtue of weak chemical interactions such as van Der Waals' and hydrogen bonds provides adhesion of the biofilm with any kind of support (Flemming & Wingender 2001). The composition of different sugars and protein of glycocalyx is also directly proportional to the impermeable nature of biofilms toward antimicrobial substances.

This glycocalyx serves as a gatekeeper and checks the transport of antimicrobial substance, and interestingly, it traps antibacterial molecule up to 25% of its weight and exposed the cell to this low concentration to enhance their resistance (Sugano et al. 2016). The antibiotic degrading exo-enzymes and nature of glycocalyx further strengthen biofilms by degrading most of the antibacterial products thus slowing down the activity of susceptible drugs (Algburi et al. 2017). The type of exopolymeric (polysaccharides) substance varies between different species of bacteria. For instance, *Pseudomonas aeruginosa* produced non-mucoid phenotypes at the early phase of biofilm formation followed by mucoid-type phenotype persister cells that cause problems in cystic fibrosis patients (Ghafoor et al. 2011, Jackson et al. 2004, Owlia et al. 2014). The reason behind these two types might be that non-mucoid types favor enhanced biofilm formation and cell-to-cell communication. Poly-β(1-6)-N-acetylglucosamine (PNAG) is a type of polysaccharide which is shared by many Gram-positive and Gram-negative bacteria that provide structural stability to biofilms by promoting intracellular adhesion (Cywes-Bentley et al. 2013).

Microbes at the surface are exposed to high oxygen concentration and nutrients whereas, their concentration reduced to unavailable for cells in the middle or at the bottom, respectively. This clearly explains the distribution of metabolically active cells at the surface to least or non-metabolically active cells (persistent cells) at the bottom of biofilms. These persistent cells are metabolically inactive cells (doubling time > 1–3 h) that are responsible for enhanced resistance toward antibiotics than their free-living counterpart. The actively growing cells are exposed to antimicrobial in the first phase and get easily killed followed by the dominance of persister cells within the biofilms which, in turn, resulted in antimicrobial-resistant strains to cause chronic infections (Mirani et al. 2015, Wood et al. 2013). The metabolic heterogeneity within biofilm due to oxygen is a key for antimicrobial tolerance. In the presence of pure oxygen, *P. aeruginosa* biofilm gets killed by antibiotics such as ciprofloxacin but reduced oxygen level increased antibiotic tolerance (Walters et al. 2003). A study of *Staphylococcus* biofilm showed the presence of three different phenotypic variants, i.e., planktonic, metabolic inactive, and small colony variants (highly resistant) within the biofilm after treatment with a sub-inhibitory concentration of oxacillin and vancomycin (Mirani et al. 2015, Mirani & Jamil 2013). They further demonstrated that multi drug resistant (MDR) *Staphylococcus aureus* strains have significantly enhanced tendency to form biofilms which with the age, dominated by small colony variant type instead of planktonic cells with reduced gene expression analysis (Mirani et al. 2015). The ATP release within the biofilm persister cells reduced, thus reducing drug targets activity, another mechanism of drug tolerance. Virginio et al. analyzed the relationship between antibiotic resistance and enhanced biofilm formation, they showed gentamicin, ceftazidime, colistin, and ciprofloxacin resistance was due to biofilm mode of growth among *Escherichia coli*, *Klebsiella pneumonia*, and *P. aeruginosa*, respectively (Cepas et al. 2019). They further showed the non-*marcescens* drug-resistant strains have better biofilm-related resistance toward antibiotics than the MDR strains (Cepas et al. 2019). So, biofilm in terms of its structure and nature of metabolic active cells provide resistance to different antibiotic and promotes chronic infections.

16.1.2 Gene Transfer

Biofilm is usually a consortium of synergistic microbes that benefit each other (Stalder & Top 2016). It provides a suitable ground for cell-to-cell contact, thus becoming a hotspot for gene swapping by horizontal gene transfer (HGT) to adapt the environment (Balcazar et al. 2015, Olsen et al. 2013, Skippington & Ragan 2011). Different environmental factors such as surface structure, temperature, pH, nutrients, water content, metabolic rate, cell densities, etc. play a crucial role in the establishment of mature biofilm and HGT (Madsen et al. 2012). Pili and type IV secretion systems are the two main key initiators of cell-to-cell contact within biofilms (Guglielmini et al. 2013, Zechner et al. 2012). One of the components of mature biofilms is eDNA which can be acquired by different microbial cells with the community. Nadzeya et al. showed that under different selective and non-selective conditions, horizontal transfer of antibiotic-resistant genes among co-cultured species was occurred in the early biofilm that the mature one in *Neisseria gonorrhoeae* (Kouzel et al. 2015). They further summarized that these transfers occurred mainly in a thin biofilm structure than in the densely packed biofilms (Kouzel et al. 2015). Another study showed that HGTs with biofilm-coated cells are higher than their planktonic counterparts in aqueous phase (Angles et al. 1993). Besides eDNA, plasmid or mobile genetic elements in biofilms are the major culprits responsible to introduce antibiotic resistance to sensitive strains. HGT is directly proportional to the structure and composition of biofilms and its transfer among different species is required to transfer resistant genes which then transfer to the same species quite easily and frequently (Li et al. 2019; 2018). Different studies showed plasmid transfer within biofilm mass is significantly correlated and responsible for metallo-beta-lactamases (MBLs) or fluoroquinolone-resistant among Gram-negative pathogens (Ciofu & Tolker-Nielsen 2019, Dumaru et al. 2019). Krol et al. showed that different drug-resistant plasmids efficiently entered the biofilm and adopted by recipient *E. coli* strain to become resistant pathogens (Krol et al. 2013). Aguilla-Arco et al. showed *staphylococcal* isolates from biofilm-related hospital-acquired infections showed the presence of plasmid-carrying resistance toward gentamicin, erythromycin, and tetracyclines which can further cause infections to other individuals in the hospital setting (Aguila-Arcos et al. 2017). So, the plasmid and eDNA-encoding antimicrobial genes are the main culprits in the emergence of resistant superbugs within biofilms.

16.1.3 Mutation Rate

Few of the strains within biofilms become hyper-mutable specially ESKAPE (*Enterococcus faecalis*, *Staphylococcus aureus*, *Klebsiella pneumoniae*, *Acinetobacter baumannii*, *P. aeruginosa*, and *Enterobacter species*) pathogens. The presence of these hyper-mutable strains is a major concern that develops antimicrobial resistance. Understanding of mutagenesis among biofilm slow-growing cells suggests adaptive mutation is higher as compared to their planktonic counterpart (Kivisaar 2010, Maharjan & Ferenci 2017). Oxidative stress, SOS response to DNA damage, and Rpo-S-dependent responses are the few factors responsible for high mutability within biofilms (Melnyk & Coates 2015).

Studies showed different antioxidants disperse biofilms and downregulate antibiotic resistance. A study showed *P. aeruginosa* strains were susceptible to oxidative stress-related mutagenesis because within biofilm mode of growth they suppress the expression of antioxidant enzymes (*katA*, *sodB*, *ahpC*, and *PA3529*) were downregulated due to which these cells showed 10- to 100-fold more resistance to rifampicin and ciprofloxacin, respectively (Boles & Singh 2008, Driffield et al. 2008). The expression of *gltB* and *gltC* in *Listeria monocytogenes* 4b G are very important in the maturation of biofilm and adaptive mutation toward antibiotics (Huang et al. 2013). In a study, *P. aeruginosa* strain was tagged with green fluorescence protein which upon mutation gives florescence. They showed a majority of the mutation has happened in small colon variants or microcolonies and then planktonic cells, suggesting that these microcolonies are the hub of mutation under oxidative stress or less DNA repair damage system (Conibear et al. 2009). A group of researchers showed that when *A. baumannii* biofilm was exposed to sublethal dose of antibiotics, biofilm formation and enhanced resistance to the particular antibiotic were shown by the dispersed cells (Santos-Lopez et al. 2019). After genome sequence analysis, they further showed that the mutation rate was higher in the biofilm mode of growth and new mutation is only present in the biofilm-dispersed cells as compared to their original planktonic phenotypes (Santos-Lopez et al. 2019). Within biofilms, *Streptococcus agalactiae* (Group B) express *bceR* gene which is responsible for bacitracin and cathelicidin LL-37 resistance, biofilm maturation, and oxidative stress tolerance (Yang et al. 2019). Wang et al. showed *Vibrio cholera* during biofilm-related intestinal infections model showed the increased expression of *mutS* to overcome the host oxidative stress to further mature biofilm and enhanced antimicrobial resistance (Wang et al. 2018). Densely populated biofilms give bacteria a barrier to overcome host oxidative stress; however, its exposure to sublethal doses resulted in hypermutable strains within biofilms, which further hinders the recovery process (Rodriguez-Rojas et al. 2012).

16.1.4 Microbial Synergism within Biofilm

Biofilms beside barriers to antimicrobials provide a place for bacterial metabolic cooperation. Microbes within biofilms, after adapting to their environment displayed synergistic behaviors (Lee et al. 2014b). The majority of the environmental or clinical biofilm showed the presence of diversified complicated niche of microbial species complimenting each other (Madsen et al. 2016, Pathak et al. 2012, Schwering et al. 2013, Wen et al. 2010). With the advent of sophisticated sequencing and bioinformatic tools, it is now known that various body sites possess a diverse species of microbes including bacteria, fungi, protozoa, etc. For example, an oral cavity is a home of 700 diverse microbial species that exist in equilibrium once destroyed lead to disease (Grice & Segre 2011, Lloyd-Price et al. 2016). Similarly, the gut microbiome is a home of more than 1,000 species and also considered as the second brain of the body (Calvani et al. 2018, Lloyd-Price et al. 2016). The synergistic cooperation of *Candida albicans* and *S. aureus* is a serious concern for a variety of indwelling devices, oral and vaginal infections. Their synergism within a biofilm is usually life-threatening and very harsh in terms of treatment. The hyphea meshwork and polysaccharide β-1,3-glucan of *C. albicans* anchor *S. aureus* further in the biofilm matrix which makes them 1,000-fold tolerant to vancomycin (Harriott & Noverr 2010, Kong et al. 2016). Madushika et al. showed the biofilm of *Bacillus cereus* and *Aspergillus flavus* significantly degrade hexadecane synergistically (Perera et al. 2019). Biofilm synergism was also found dishwasher rubbers. *P. aeruginosa* and *Acinetobacter junii* are the most isolated pathogens which favor bacterial synergism accommodating opportunistic yeast pathogens, *E. dermatitidis* which can be a source of infections in immune-compromised individuals (Zupancic et al. 2018). Another study showed that the virulence factor (type IV pili) in *Haemophilus influenza* only expressed in the presence of *Streptococcus pneumoniae*, thus facilitate adherence of both organisms as a polymicrobial community to cause chronic rhinosinusitis (Cope et al. 2011). In oral biofilms, *Streptococcus gordonii* ferment sugars and facilitate *Veillonella* species which are unable to ferment sugars in return, and they support *S. gordonii* by produced QS molecules (autoinducers) to form mature biofilms which cause oral disease (Mashima & Nakazawa 2015). The association of *S. gordonii*, *Fusobacterium nucleatum*, and *Porphyromonas gingivalis* facilitates each other in polymicrobial biofilms to escape dendritic cell insults and host immune system to establish periodontitis (El-Awady et al. 2019). Within the biofilm, negative correlation also found, for example, *Streptococcus sanguinis* while growth produces H_2O_2 which inhibits the growth of *Streptococcus Mutans*, thus favoring oral health (Giacaman et al. 2015). The QS molecules N-acyl homoserine lactone (AHL) produced by *P. aeruginosa* was reported to inhibit *C. albicans* within the biofilms (Fourie & Pohl 2019). The synergism behaviors are responsible for various health complications such as dental caries, periodontitis, cystic fibrosis, etc.

16.1.5 Efflux Pump

Efflux pumps are membrane-bound proteins that give particular bacteria a status of multidrug resistance by throwing antimicrobial substances such as antibiotics out of the cells. They can be divided into substrate-specific pumps and non-substrate-specific pumps that throwback different substances out of the cells (Webber & Piddock 2003). The genes for these pumps are present on chromosomes as well as mobile genetic elements and plasmids (Marquez 2005). Till date, five superfamilies of these efflux pumps have been reported which are multidrug and toxin extrusion (MATE), small multidrug resistance (SMR), major facilitator superfamily (MFS), ATP-binding cassette (ABC), and resistance-nodulation division (RND) (Jack et al. 2001, Kuroda & Tsuchiya 2009, Lubelski et al. 2007, Nikaido & Takatsuka 2009, Pao et al. 1998). Except for ABC, all other pumps utilize energy from sodium/potassium motive force. Besides making cells harsh toward antibiotics, they also involved in different bacterial phenotypes such as QS, biofilm formation, and virulence expression.

Ilyas et al. in their review suggested four different contributions of these efflux pumps toward biofilm formation, i.e., efflux of quorum quenching, indirect regulation of biofilm-related genes, efflux of antibacterial molecules, and by influencing bacterial aggregations (Alav et al. 2018). Yamasaki et al. showed that efflux pumps AcrB and MdtABC were necessary for the maturation of biofilm as mutant of both the genes form initial biofilm but failed to maintain with the passage of time (Yamasaki et al. 2015). By using in silico studies, Bindu et al. showed AcrR the regulator of the AcrAB efflux pump is important for various phenotypes of Acinetobacter *nosocomialis* such as motility and biofilm formation (Subhadra et al. 2018). Deletion of efflux genes acrB, acrE, and tolC in *E. coli* resulted in immature biofilm and enhanced sensitivity to antibiotics (Bay et al. 2017). The efflux system Bcr/CflA is responsible for *Proteus mirabilis* crystalline biofilm formation, down-regulation of this efflux system by using fluoxetine will result in reduced biofilm mass (Nzakizwanayo et al. 2017). A study was conducted to evaluate 12 natural compounds as efflux pump inhibitors, out of 12, Reserpine was found to be better inhibitors which enhance antimicrobial sensitivity with reduced biofilm formation in *K. pneumonia* clinical isolates (Magesh et al. 2013). Plant-derived products such as plumbagin, nordihydroguaiaretic acid (NDGA), and shikonin significantly inhibit the AcrB efflux system in *E. coli* thus making it vulnerable to various antibiotics and reduction in biofilm formation (Ohene-Agyei et al. 2014). Ketoconazole an antifungal drug has now been shown to possess efflux pump inhibitor activities thus enhances the efficacy of fluoroquinolones and disrupt biofilm formation among MDR *S. aureus* strains (Abd El-Baky et al. 2019). However, several efflux pump inhibitor confirmed its involvement in the biofilm formation and resistance to various antibiotics among nosocomial pathogens.

16.1.6 Quorum Sensing (QS)

QS is a mechanism in which microbes interact via cell-to-cell communications by production and detection of a signal molecule such as autoinducers which will define the fate of bacterial population in terms of virulence expression. The phenomenon of QS was first discovered in *Vibrio fischeri*, a marine bacterium that produced acylated homoserine lactones (AHL) by LuxI operon and sensed by LuxR operon to express virulence genes (LaSarre & Federle 2013). The majority of the Gram-negative strains produce similar AHL with different acylation chain with oxo/hydroxyl substitution at 3rd position (LaSarre & Federle 2013). On the other hand, Gram-positive strains secrete QS peptides, which were sensed by receptor (membrane-bound histidine kinase) and intracellular regulator to express required genes (Platt & Fuqua 2010). Autoinducer 2 is another class of QS molecules which is shared by both members of Gram-positive and Gram-negative strains (Rutherford & Bassler 2012).

Burkholderia cenocepacia was reported to become resilient cells within biofilm due to QS (autoinducer 2; QS gene cepIR, and cciIR) controlled virulence factors (Ganesh et al. 2019, Niu et al. 2008). Muras et al. showed externally added QS molecule (C6-HSL) resulted in the dysbiosis in the oral cavity with the prevalence of *Peptostreptococcus* and *Prevotella* species making condition favorable for periodontal disease (Muras et al. 2019). The presence of Lqs QS system (LqsA) is crucial for *Legionella pneumophila* to completely express their virulence, tolerance to antibiotics, and biofilm mode of growth (Personnic et al. 2019). A combination of citral and phloretin synergistically suppresses the genes responsible for QS (LuxS) which, in turn, reduces the biofilm formation and *Streptococcus pyogenes* and its associated virulence (Adil et al. 2019). The secretion and expression of secretary proteases (protease IV, elastase A and B) are QS dependent in *P. aeruginosa* (Li & Lee 2019).

The QS-dependent transcriptional regulator (VjbR) is important for *Brucella canis* pathogenicity; upon mutation, it significantly makes *B. canis* susceptible to macrophages in mice suggesting the role of QS in developing full-blown infections (Liu et al. 2019). The QS inhibitors such as ciprofloxacin significantly downregulate the virulence expression of *P. aeruginosa* (Hoffmann et al. 2007).

Various factors are responsible to give microbes the status of resilient within the biofilm. First of all, the composite structure of microbes lives in cross proximity shares their signals via QS mechanism, which further matures the biofilm. When the biofilm gets mature, the impermeability barrier posed by glycocalyx or extra-polymeric substance restricts the penetration of antimicrobial to the depth of biofilms. The presence of sub-inhibitory concentration of antibiotics within biofilm exerts positive pressure which develops resistance to antibiotics at chromosomal levels. The mutual understanding among microbes and the sharing of plasmids also result in the development of resistance among strains that are sensitive to particular antibiotics without the introduction of plasmid. Last but not least, the MDR efflux pump among bacteria within biofilm further makes the biofilm more resilient.

Besides all the resilient showed by biofilm, there is a need to treat or cure biofilm formation, especially on prosthetic devices and catheter-related infection. There are numerous studies that have been done suggesting their importance to tackle the biofilm-related infection. The following section will elaborate on the different options to handle biofilm-related infections.

16.2 NATURAL PRODUCT AND ANTIBIOFILM AGENTS

Historically, nature has come to rescue human beings for their ailments. Nature in terms of medicinal plants is a blessing to cure different diseases. Ancients records recovered from different parts of the globe clearly indicated the use of medicinal plants to cure various health conditions such as Egyptian medicine records (Ebers Papyrus). It was written in 1550 BC and enlisted more than 700 natural species mainly from plant origin for healing purposes. Another plant medical record (Mesopotamia) written in 2600 BC has documented nearly 1,000 drugs of plant origin.

The use of traditional Chinese medicines and Indian Ayurveda systems dates back over millennia. The history of both of them enlisted a number of drugs from natural plants and spices such as jimson weed, cinnamon, nutmeg, pepper, clove, etc. However, in the early nineteenth century, the knowledge regarding medicinal plants improved by scientific community by providing in-depth mechanism of natural isolated extract or pure compounds for the cure of various ailments. The alternative plant-derived safe drugs are used even to date (Atanasov et al. 2015, Petrovska 2012).

16.2.1 Plant Extracts

The plant extracts composed of various active components at different ratios were also used to challenge bacterial biofilm mode of growth. These products interfere with bacterial adhesion, matrix formation, decreasing virulence factors or interfering with QS networks, and biofilm formation (Lu et al. 2019a). Jahan et al. reported the antibiofilm activity of methanolic and ethanolic extracts of *Centella asiatica* (Thankuni), *Mentha spicata* (Mentha), *Azadirachta indica* (Neem), *Psidium guajava* (Guava) and spices *Syzygium aromaticum* (Cloves), *Cinnamomum zeylanicum* (Cinnamon) against multidrug-resistant *P. aeruginosa* strains (Jahan et al. 2018). The antibiofilm potential of *Allium sativum* (garlic) extract against *S. pneumoniae*, *B. cereus*, *S. aureus*, *K. pneumoniae*, *P. aeruginosa*, and *E. coli* has been reported. The extract not only prevents more than 50% biofilms but can also disperse biofilms (Mohsenipour & Hassanshahian 2015). Garlic extracts are also responsible for blocking the QS signaling system by interfering with transcriptional regulators LuxR and LasR (Lu et al. 2019). Abraham et al. reported that methanolic extract of the dried fruits of Capparis *spinosa* (caper) showed antibiofilm and anti-QS activities at 0.5–2 mg/mL concentration against *Serratiamarcescens*, *P. aeruginosa*, *E. coli*, and *P. mirabilis* (Issac Abraham et al. 2011)

16.2.2 Essential Oil

Over centuries, essential oils have also been used as medicines to combat bacterial infection. They are volatile hydrophobic compounds present in many plants (Lahiri et al. 2019). Studies were shown the ability of essential oils to inhibit biofilms. For example, Safoura et al. reported the biofilm inhibition *K. pneumoniae* by cumin seed essential oil. The antibiofilm activity of thymol, oregano, and cinnamon essential oils against biofilm-forming *Acinetobacter*, *Sphingomonas*, and *Stenotrophomonas* has also been reported. They disrupt bacterial cell wall or membrane to exert their inhibitory effect (Lahiri et al. 2019).

16.2.3 Pure Compounds

These crude extracts have been further explored for active ingredient that interferes with biofilm formation. They usually referred to as secondary metabolites and differ from primary metabolites (nucleic acids, amino acids, sugars, and fatty acids) which are required for basic cellular functions. Approximately, more than 200,000 plant-derived structures belonging to different classes with diversifying biological activities are known (Springob & Kutchan 2009). The following section will highlight the antibiofilm potential of different classes of phytochemicals:

16.2.3.1 Alkaloids

They are amino acid-derived heterocyclic nitrogenous compounds with more than 12,000 known structures. They are sub-classified into pyridine, tropane, quinoline, isoquinoline, indole, imidazole, steroid, alkaloidal amine, purine (Iriti & Faoro 2009).

The majority of the alkaloids such as quinoline or their derivatives depolarize cell membranes resulting in cytoplasm leakage, thus inhibit bacterial cells. Alkaloids belonging to class isoquinoline act against biofilm-embedded bacteria by inhibiting their nucleic acid synthesis and type-1 topoisomerases. The 3-carboxyl group present in the fluoroquinolones inhibits Type-II topoisomerase. Whereas, methyl quinolone restricts oxygen consumption in the bacterial cells (Lahiri et al. 2019).

Jain et al. showed that alkaloids and flavonoids from curcuma species *C. longa* (Turmeric), *C. caesia* (Black turmeric), and *C. aromatica* (Wild turmeric) hinder biofilm formation by *S. aureus* and *Bacillus subtilis* (Jain & Parihar 2018). The synthetic derivatives of cinchona alkaloids, i.e., 11-triphenylsilyl-10,11-dihydrocinchonidine (11-TPSCD) was found to be active against inhibiting and dispersing mature biofilms (Skogman et al. 2012). Li et al. reported biofilm inhibition of *Staphylococcus epidermidis* by total alkaloids of *Sophora alopecuroides* (Li et al. 2016a).

16.2.3.2 Phenylpropanoids

They are usually known as plant phenolics and are derived from amino acid phenylalanine and tyrosine. They possess a phenyl ring with a propane side chain. They are classified into hydroxycinnamic acid, coumarins, lignans, and flavonoids. Plant extracts are found to be rich in flavonoids. These flavonoids themselves are subclassified into flavanol, flavanone, isoflavone, flavone, flavan-3-ols or catechin, and anthocyanin. The mode of action of flavonoids includes interaction with bacterial proteins and cell wall structures thus inhibiting nucleic acid, cell wall synthesis, or energy metabolism (Lahiri et al. 2019). Moreover, they also interfere with bacterial signaling within the biofilm by inhibiting N-acyl homoserine lactones-mediated QS (Górniak et al. 2019). Table 16.1 highlights the antibiofilm potentials of different flavonoids.

16.2.3.3 Terpenoids

Terpenoids also named as isoprenoids are derived from five-carbon monomer units called isoprene. More than 40,000 different terpenoids are known. On the basis of isoprene molecules, they are divided into hemiterpenes, monoterpenes, sesquiterpenes, diterpenes, triterpenes, tetraterpenes, and polyterpenes having different carbon units, respectively (Springob & Kutchan 2009).

Terpenoids possess an anti-cell adhesion property which makes them potential antibiofilm agents (Lahiri et al. 2019). A phenolic monoterpene, carvacrol is reported to inhibit biofilms of *S. aureus* and *Salmonella enterica* (Jose et al. 2017). Dalleau et al. reported the antibiofilm activity of 10 terpenes against *Candida* species. Of them, carvacrol, geraniol, or thymol inhibits more than 80% biofilms of *C. albicans* and more than 75% *Candida parapsilosis* biofilms (Dalleau et al. 2008). Vetiveria *zizanioides* root extract containing sesquiterpenes a major constituent downregulates adhesin genes like fnbA, fnbB, clfA, thus inhibiting biofilms formed by methicillin-resistant *S. aureus* (MRSA) (Kannappan et al. 2017). The casbane diterpene, isolated from the ethanolic extract of *Croton nepetaefolius*, can inhibit in vitro biofilm formed by oral pathogens *S. mutans*, *Streptococcus salivarius*, *Streptococcus sobrinus*, *Streptococcus mitis*, *S. sanguinis*, and *Streptococcus oralis*. The antibacterial effect of this compound is due to the hydrophobic moiety, and a hydrophilic region having two hydrogen bond donor groups which enable the insertion of the compound in the cell membranes, causing the destabilization of the lipid bilayer, which ultimately effects cellular development. However, the biofilm inhibition is presumably related to the growth inhibition only (Sá et al. 2012). Betulinic acid (BA), a triterpene, isolated from the *Platanus*

TABLE 16.1
Antibiofilm Potential of Isolated Flavonoids from Natural Sources

Compounds	Mode of Action	Microorganisms	References
	Flavanols		
Quercetin	Inhibit • Quorum sensing (violacein production) • Pyocyanin production • Proteolytic and elastolytic activities • Swarming motility • Biofilm formation	• *Chromobacterium violaceum* 124721 • *Pseudomonas aeruginosa* PAO1	Vasavi et al. (2014)
Kaempferol	Inhibits initial attachment	*S. aureus*	Ming et al. (2017)
Rutin	Reduce EPS production	Mono and multi-species of *Escherichia coli* and *Staphylococcus*	Al-Shabib et al. (2017)
Resveratrol	• Inhibit biofilm formation by targeting *AphB* protein. • Inhibit biofilm formation by downregulating virulence genes responsible for stress-induced proteins, hemolysin, ABC transporter, butyrate acetoacetate CoA transferase • Inhibit biofilms by downregulating fimbriae and proteinases.	• *Vibrio cholera* • *Fusobacterium nucleatum* • *Porphyromonas gingivalis*	Augustine et al. (2014) He et al. (2016) Kugaji et al. (2019)
	Flavones		
Luteolin	• Reduce adhesion, invasion, hydrophobicity, and motility downregulate type 1 fimbriae *fimH* gene. • Inhibit QS, biofilm formation, and swarming motility.	• *Escherichia coli* • *P. aeruginosa* • *C. violaceum*	Shen et al. (2014) Bali et al. (2019)
Apigenin	Reduces exopolysaccharide production with reduced expression of fructosyltransferase and glucosyl transferase.	*Streptococcus mutans*	Koo et al. (2003) Koo et al. (2006)
	Isoflavone		
Genistein	Reduced biofilm formation as evident by Crystal violet assay	*Staphylococcus aureus*	Moran et al. (2014)
	Flavanones		
Naringin	Inhibit QS as evident by decreased violacein production, reduced AI-2 and *VanS* gene.	• *C. violaceum* • *Vibrio anguillarum*	Liu et al. (2017)

(*Continued*)

TABLE 16.1 (*Continued*)
Antibiofilm Potential of Isolated Flavonoids from Natural Sources

Compounds	Mode of Action	Microorganisms	References
	Flavan-3-ol		
Epigallocatechin gallate	• Damage the cell membrane and cell wall. • Suppress biofilm formation and inhibit microbial infection in the lungs.	• *Porphyromonas gingivalis* • *Stenotrophomonas maltophilia*	Asahi et al. (2014) Vidigal et al. (2014) Hengge (2019)
	Anthocyanin		
Malvidin	Inhibit QS, i.e., reduced violacein production, biofilm formation, and EPS production.	*Klebsiella pneumoniae*	Gopu et al. (2015)
Petunidin	Inhibits QS, EPS and biofilm	*Klebsiella pneumoniae*	Venkadesaperumal et al. (2015)
Emodin	Deplete QS signal TraR and reduces biofilm thickness	*Escherichia coli*	
	Coumarins		
Umbelliferone	Downregulated curli and motility genes expression to reduce biofilm formation	*Escherichia coli*	Lee et al. (2014a)
Esculetin	Reduces Shiga-like toxin gene *stx2 to destroy virulence within biofilm*	*Escherichia coli*	Lee et al. (2014a)
Nodakenetin, Fraxin	Reduced QS thus reduced biofilm formation	*P. aeruginosa*	Reen et al. (2018)

acerifolia bark is reported to be effective against biofilms of *S. aureus* (Silva et al. 2019). Khan et al. reported clerodane diterpenoids 16-oxo-cleroda-3, 13(14) E-diene-15 oic acid and kolavenic acid isolated from *Polyalthia longifolia* var. pendula (Linn.) can significantly inhibit the biofilms of *S. mutans*, MRSA, *K. pneumoniae*, and *P. mirabilis* (Khan et al. 2017)

16.2.3.4 Polyketides

They originate from two-carbon units derived from acyl precursors such as acetyl-CoA and malonyl-CoA. More than 10,000 polyketides are known, they are classified into chalcones, stilbenes, phloroglucinols, resorcinol, benzophenones, biphenyl, curcuminoids, archidones, pyrones, chromones (Shimizu et al. 2017). Using the core structure of these natural compounds, many synthetic compounds with improved antibiofilm potential have been synthesized. Two chalcones isolated from the flowers of *Butea monosperma* inhibits biofilms of nontypeable *H. influenzae* at concentration ranging from 42 to 70 μg/mL, and 3-hydroxychalcone synthesized using the Claisen–Schmidt condensation reaction inhibited biofilms at 16 μg/mL (Kunthalert et al. 2014).

16.3 BIOFILMS DISRUPTING ENZYME

As stated previously, bacteria in biofilms are embedded in exopolymeric substances that protect cells and increase their resistance to antibiotics. Studies now involve targeting the individual components of extra polymeric substances (EPS) (DNA, proteins, and polysaccharide) by enzymes to dismantle the three-dimensional structure of biofilms.

16.3.1 Deoxyribonuclease I

eDNA is an important part of the biofilm matrix which is released as a result of a process called autolysis. DNase I can nonspecifically degrade single-stranded and double-stranded DNA yielding 5′-phosphorlyated polynucleotides and a free 3′-hydroxyl group (Varela-Ramirez et al. 2017). Tetz et al. reported 53.85%, 52.83%, 50.24%, 53.61%, 51.64%, 47.65%, and 49.52% degradation of 24 h established biofilm of *A. baumannii*, *H. influenzae*, *K. pneumoniae*, *E. coli*, *P. aeruginosa*, *S. aureus*, and *S. pyogenes*, respectively, using DNase I at concentration of 10 μg/mL. In addition, they also reported an increase in bacterial susceptibility to antibiotics such as azithromycin, rifampin, levofloxacin, ampicillin, and cefotaxime in presence of 5 μg/mL of DNase I (Tetz et al. 2009). Moreover, Hall-Stoodley et al. reported DNase I in a dose-dependent manner significantly reduces >90% biofilm mass of five out of six isolates of *Streptococcus pneumonia* (Hall-Stoodley et al. 2008). Another study reported strong antibiofilm activity of DNase I and DNase 1L2 against *S. aureus* and *P. aeruginosa* (Taraszkiewicz et al. 2013). Kaplan et al. reported the use of rhDNase at concentrations of 1–4 μg/L and 10 mg/L to inhibit biofilms of *S. aureus* and *S. epidermidis*, respectively. Besides inhibition rhDNase at 1 mg/mL significantly disperse the already established biofilms and facilitate antibiotic efficacy (Kaplan et al. 2012). DNase I was also reported to inhibit and eradicate to some extent the polymicrobial biofilms composed of *Klebsiella spp.*, *P. aeruginosa*, *S. aureus*, *E. faecalis*, and *Salmonella Typhimurium* at 10 μg/mL (Sharma & Pagedar Singh 2018). DNase was also reported to act synergistically with antibiotics and was reported to enhance the activity of amphotericin B against *C. albicans* biofilms (Martins et al. 2012). DNase I was also used as a coating on polymethylmethacrylate using dopamine as an intermediate which suppresses initial adherence of *S. aureus* and *P. aeruginosa* by 95% and 99%, respectively, without affecting mammalian cells. This can be used to devise biomaterial which can be used in catheters of prosthetic devices to overcome biofilm-related complications (Swartjes et al. 2013).

16.3.2 Dispersin B

Polysaccharides are the main component of biofilm (90% of total EPS) which provide structural integrity, strong adhesion, and halt antimicrobial diffusion. Inhibiting or cleaving polysaccharides is yet another approach to dissolve biofilms using enzymes. Polysaccharide intracellular adhesin that was later described as linear poly-β-1,6-N-acetyl-D-glucosamine (PNAG) are the major constituents of biofilms formed by many phylogenetically diverse bacteria (Matthysse et al. 2008). Dispersin B is a soluble glycoside hydrolase from *Actinobacillus actinomycetemcomitans* which cleaves N-acetyl-D-glucosamine (Kerrigan et al. 2008, Ramasubbu et al. 2005). Antibiofilm activity of dispersin B has been reported against *S. aureus*, *Burkholderia species*, *E. coli*, *S. epidermidis*, *Yersinia pestis*, and *Pseudomonas fluorescens* (Fleming & Rumbaugh 2017). Other studies have highlighted the synergistic potential of this enzyme to enhance the antimicrobial efficacy of tobramycin, cefamandole nafate against *Staphylococcal* biofilms (Donelli et al. 2007) and triclosan against *S. aureus*, *S. epidermidis*, and *E. coli* (Darouiche et al. 2009). This combinatorial efficacy can be used with other antimicrobial agents as cleansing agents, wound gel, and polymeric matrices to avoid biofilm-related infections.

16.3.3 α-Amylases

α-Amylases are known for their wide industrial applications, and they also possess biofilm-inhibiting and eradicating potentials. It cleaves internal α-1,4-glycosidic bond in starch yielding varying length oligosaccharides (de Souza & de Oliveira Magalhães 2010). α-Amylases inhibit *S. aureus* biofilms up to 72%, 89%, and 90% at doses of 10, 20, and 100 μg/mL, respectively, however, it also eradicated the preformed biofilms up to 79%–89% after exposure of 5 and 30 min, respectively (Craigen et al. 2011). Moreover, Craigen et al. also compare α-amylase from *Aspergillus oryzae*, *B. subtilis*, human saliva, and β-amylase from sweet potato, all of them inhibited the *S. aureus* biofilms but amylases from human saliva and sweet potato failed to eradicate biofilms (Craigen et al. 2011). Another study reported biofilm inhibiting the potential of α-amylases extracted from *B. subtilis* S8-18 of marine origin against the clinical strain of MRSA, *Vibrio cholera* and *P. aeruginosa* (Kalpana et al. 2012). Fleming et al. reported reduced biomass, termination of biofilms, and increase in gentamycin susceptibility by cellulase and amylase against monocultured and cocultured biofilms of *S. aureus and P. aeruginosa* (Fleming et al. 2017)

16.3.4 Lyase

Lyase is the enzymes that cause eliminative cleavage to produce double bonds or add groups to double bonds (Nithya & Jayachitra 2016). They degrade polysaccharides and facilitate antimicrobial entrance in the biofilm matrix. Bugli et al. have reported the synergistic effect of clarithromycin and alginate lyase against biofilms of Helicobacter *pylori* (Bugli et al. 2016). Alginate lyase (20 μg/mL), when co-administered with gentamycin (64 μg/mL), liquefies the biofilm matrix and reduces the viable count to 2–3 $\log_{10}$ of two mucoid *P. aeruginosa* strains (Taraszkiewicz et al. 2013). Furthermore, significant positive interaction of alginate lyase with two polyene antifungal (amphotericin B (AMB) and Liposomal AMB) was also reported against two strains of Aspergillus *fumigatus* (Bugli et al. 2013). Hatch and Schiller (1998) suggested the use of alginate lyase to facilitate the diffusion of aminoglycosides in *P. aeruginosa* biofilms.

16.3.5 Lactonase

Lactonases were reported to interfere with bacterial QS thus facilitating removal of biofilms or the development of less resistant biofilm. Lactonases hydrolyze acyl-homoserine lactones to interrupt bacterial communications (Dong et al. 2001). Rajesh et al. isolated AHL-lactonase from an endophytic bacterium of *B. cereus* VT96 which significantly downregulates QS-associated virulence factors in *P. aeruginosa* (Rajesh & Rai 2016). Lactonases at 1 U/mL significantly inhibited the 65%–70% biofilms of four clinical strains of *P. aeruginosa*; furthermore, disperse pre-formed biofilm to make it vulnerable to gentamycin and ciprofloxacin at 0.3 U/mL. It downregulated virulence factors such as pyocyanin, protease activity, elastase activity, and pyochelin secretion up to 90% (Taraszkiewicz et al. 2013).

16.3.6 Lysostaphin

Lysostaphin is a zinc metalloprotease derived from *Staphylococcus simulans* and has endopeptidase activity against penta-glycine crosslink in peptidoglycans of *S. aureus* bacterial cell wall. Lysostaphin significantly reduced *S. aureus* biofilm by killing them as well as eradicated pre-formed biofilm on different surfaces such as polycarbonate, polystyrene, or glass. It also disrupts the matrix of three strains of *S. epidermidis* at 200 μg/mL of lysostaphin for 3 h (Wu et al. 2003). Similarly, Walencka et al. also confirmed the biofilm inhibition potential of lysostaphin against 13 *S. aureus* and 12 *S. epidermidis* clinical strains at different concentrations (Walencka et al. 2005). Lysostaphin also showed synergism with nafcillin to eradicate biofilm of MRSA in an implanted catheter in mice. Furthermore, the synergistic effect of lysostaphin (20 μg/mL) with 10 other antibiotics was reported against MRSA and methicillin sensitive *Staphylococcus aureus*. The highest synergism was found with doxycycline, i.e., its minimum biofilm eradication concentration was reduced from 4 to 0.5 mg/mL (Kokai-Kun et al. 2009). The ability of lysostaphin and other enzymes to disperse biofilms alone or in combination with rifampicin and vancomycin against MRSA and MSSA has also been documented (Hogan et al. 2017).

Still, the enzyme-based antibiofilm therapy is at preliminary stages. There is a chance that these dispersing enzymes facilitate cells within biofilm to reach other sites to cause infection, so these enzymes must be used in combination with antibacterial agents to kill the dispersing microbes. Another hurdle in enzyme therapy is they are expensive as compared to conventional chemical disinfectants and antibiotics, however, further research is required to facilitate enzyme-based antibiofilm therapy (Sadekuzzaman et al. 2015).

16.4 MODIFICATION OF C-DI-GMP SIGNALING SYSTEM

Twenty-five years ago, an important universal bacterial second messenger called Bis-(3′-5′)-cyclic dimeric guanosine monophosphate (c-di-GMP) was discovered. The messenger has been known to regulate bacterial behaviors such as motility, sporulation, resistance to antibiotics and stress, cell cycle, differentiation, virulence, and biofilm formation (Römling et al. 2013). Notably, c-di-GMP plays an important role in switching bacterial lifestyle from motile state to sessile state and vice versa (i.e., dispersion; the fifth stage of biofilm formation) (Valentini & Filloux 2016). This transition causes dispersal of bacteria which results in the spread of infection within host or between hosts. Therefore, targeting c-di-GMP is an effective strategy to combat biofilm-associated infections in clinical and industrial settings.

Cyclic-di-GMP is a small molecule that diffuses freely in the bacterial cytoplasm. The two molecules of GTP combined in a reaction catalyzed by enzyme diguanylate cyclases (DGCs) resulting in the synthesis of c-di-GMP, which is degraded by phosphodiesterases (PDEs) into 5-phosphoguanylyl-(3-5)-guanosine (pGpG) and/or GMP. The catalytic domain of DGC is GGDEF and that of PDEs is either EAL or HD-GYP. In response to internal or external stimuli, enzymatic activity is regulated by these domains. They can exist alone or in association with the receiver or transmission domain (Valentini & Filloux 2016).

High content of c-di-GMP in the cell results in increased biofilm formation (Lin Chua et al. 2017). Although its intracellular levels are allosterically regulated by feedback inhibition (Römling & Balsalobre 2012), an antagonist of DGCs could be used to manage dispersion and formation of biofilms. Sambanthamoorthy et al. reported four small molecules LP 3134, LP 3145, LP 4010, and LP 1062 that prevent biofilm formation by *P. aeruginosa* and *A. baumannii* by inhibiting DGCs. Among them, only two were reported to be non-toxic to eukaryotic cells, therefore, it can be suggested as potential candidates to control biofilm infections (Sambanthamoorthy et al. 2014).

Interestingly, the levels of c-di-GMP are regulated by QS either directly or indirectly. In direct control, the activity of enzymes synthesizing and degrading c-di-GMP is modulated by autoinducer independent of the environmental conditions; however, indirect control is sensitive to the environment (Srivastava & Waters 2012). Terrein, isolated *from Aspergillis terreus*, was the first dual inhibitor of QS and c-di-GMP signaling in *P. aeruginosa*. Terrein inhibits the activity of DGC lowering the intracellular levels of c-di-GMP. The inhibition by terrin was reversed by the introduction of exogenous QS ligands, conforming regulation of c-di-GMP levels by QS (Kim et al. 2018).

Other compounds such as ebselen and the selenone analog of ebselen, i.e., ebselen oxide were discovered using differential radial capillary action of ligand assay. Both molecules reduced DGC activity, altering c-di-GMP signaling which inhibits biofilm formation by *P. aeruginosa* (Wu et al. 2015). In addition, seven small compounds that inhibit DGCs in *V. cholerae* have also been reported elsewhere. Since eukaryotes lack enzymes related to c-di-GMP, small molecules inhibiting this secondary messenger might be an efficient approach to prevent biofilms without affecting the infected host (Sambanthamoorthy et al. 2012).

16.5 BIOFILM-INHIBITING PEPTIDES

The emergence of antibiotic resistance has led to the exploitation of antimicrobial peptides as antibiofilm agents. They serve as defense molecules of host innate immune response, having broad-spectrum activity against pathogens. At present,

3,137 antimicrobial peptides obtained from six kingdoms have been discovered, with 2,642 being antibacterial and 54 being antibiofilm peptides (http://aps.unmc.edu/AP/main.php).

The number of amino acid residues in anti microbial peptides (AMPs) varies from five to over a hundred, mostly present in L-form. Their molecular weight ranges from 1 to 5 kDa. Generally, AMPs are cationic in nature; however, a few anionic antimicrobial peptides with net charge varying from –1 to –7 have also been reported (Yasir et al. 2018). For example, a small anionic antibacterial peptide XLAsp-P1 was isolated from amphibia *Xenopus laevis*. The peptide was found to be effective against various Gram-positive and Gram-negative bacteria (Galdiero et al. 2019).

The use of antimicrobial peptides as antibiofilm agents begin with the discovery of human cathelicidin, LL-37, the peptide has shown strong biofilm inhibition and eradicating potential against *P. aeruginosa* biofilms (Fleming & Rumbaugh 2017). Later, peptides from different sources were isolated, modified, and tested for their antibiofilm potential.

Yasir et al. in their review proposed that AMPs can interfere with biofilm formation through five major mechanisms. These include (i) alteration in membrane potential of biofilm embedded cells; (ii) interruption of bacterial cell signaling systems; (iii) degradation of EPS and biofilm matrix; (iv) inhibition bacterial stringent response; and (v) downregulation of biofilm-associated genes (Yasir et al. 2018). Few antimicrobial peptides from human origin and their derivatives are mentioned in Table 16.2.

Bacteriocins such as nisin A, lacticin Q, and nukacin ISK-1 also inhibit MRSA biofilms by modulating bacterial membrane potential and release of ATP (Okuda et al. 2013). Bovine cyclic bactenecin derivate, Bac8c, alters the bacterial structure by producing holes in cell walls causing the generation of extracellular debris and nucleoid condensation. Importantly, it inhibits the biofilms of *S. mutans* downregulating biofilm-associated genes (gtfB, C, Lux S, comD, and comE) (Ding et al. 2014). Moreover, another peptide RN3 (5-17P22-36) is reported to downregulate the biofilm-forming genes and eradicates the established biofilms of *P. aeruginosa* (Pulido et al. 2016).

The interference of peptides with QS signaling is well documented. BmKn-2 isolated from scorpion venom was modified into peptide BmK-22 and was reported to inhibit and disrupt biofilms of *P. aeruginosa* by inhibiting virulence factor pyocyanin. The peptide also interferes with the QS signaling system by downregulating lasI and rhlR expression (Teerapo et al. 2019). Important QS-inhibiting peptide known as RNA III-inhibiting peptides is reported to interfere with *Staphylococcal* TRAP/agr systems responsible for biofilm synthesis (Wu et al. 2015). KBI-3221, an analog of natural QS peptide, Competence Stimulating Peptide, produced by *S. mutans* can regulate the biofilm formation of oral pathogens (LoVetri & Madhyastha 2010).

It has also been viewed that these antimicrobial peptides not only themselves interfere with biofilm formation but also restores antibiotic susceptibility toward biofilm formers. For example, a study by Field et al. showed that nisin increases the efficacy of polymyxin and colistin toward resistant biofilm of *P. aeruginosa* (Field et al. 2016). Another study reported the 16-fold increase of β-lactams against carbapenem-resistant *K. pneumonia* isolates (Ribeiro et al. 2015). Similarly, other

TABLE 16.2
Natural and Synthetic Antibiofilm Peptides from Humans

Peptide	Source	Pathogen	Mode of Action	References
LL-37	Human	*P. aeruginosa*	Inhibits initial adherence, stimulates twitching motility and regulates two QS system; Las and the Rhl systems	Overhage et al. (2008)
17BIPHE2	Derived from LL-37	*S. aureus*	Eradicate and inhibits biofilm by inhibiting bacterial attachment	Mishra et al. (2015)
GF-17	Derived from LL-37	*S. aureus*	Eradicate and inhibits biofilm by inhibiting bacterial attachment	Mishra et al. (2015)
1037	Derivative of LL-37	*P. aeruginosa*	Restrain biofilm formation by reducing swimming and swarming motilities, promoting twitching motility, and suppressing gene expression.	de la Fuente-Núñez et al. (2012)
KE-18	Truncated memetic of LL-37	*C. albicans* and *S. aureus*	Binds to LPS and LTA and prevent biofilm formation	Luo et al. (2017)
KS- 30, 12 and 20	Derivative of LL-37	*Acinetobacter baumannii*	Biofilm inhibition and eradication	Feng et al. (2013)
LL-31	Truncated variant of LL-37 lacking the six C-terminus residues	*Burkholderia pseudomallei*	Kills bacterial cells by disrupting membrane leading to intercellular leakage	Kanthawong et al. (2012)
P60.4Ac	Derivate of LL-37	MRSA	Degrades biofilms	Haisma et al. (2014)
P10	Derivate of P60.4Ac	MRSA	Degrades biofilms	Haisma et al. (2014)
Hepcidin	Isolated from Human liver	*S. epidermidis*	Acts on PIA and prevents the accumulation of biofilm extracellular matrix.	Brancatisano et al. (2014)
Human beta-defensin 3 (hβd-3)	Epithelial skin cells and cells within apocrine sweat glands	*S. epidermidis*, MRSA	downregulates expression of icaA, icaD, and mec A gene, upregulates expression of icaR	Sutton & Pritts (2014)
HBD3-C15	Derivate of hβd-3 consisting of 15 amino acids from C terminal of hβd-3	*Streptococcus mutans*	Bactericidal, biofilm inhibitory	Ahn et al. (2017)

synthetic peptides IDR-1018, DJK-5, and DJK-6 which can degrade stress-related signaling nucleotide (p)ppGpp can increase the antibiotic susceptibility to inhibit and treat the biofilms of multidrug-resistant pathogens (Grassi et al. 2017). Furthermore, HBD3-C15 is also reported to potentiate the antibiofilm activity of calcium hydroxide and chlorhexidine digluconate against *S. mutans* (Ahn et al. 2017).

Although AMPs have significant in vitro and in vivo efficacy, their clinical use is still hampered. The peptides are expensive drugs, requiring expensive intermediates and high therapeutic doses. Additionally, their stability in physiological conditions is challenging. Host proteolytic enzymes, salts, proteins, and ions can degrade these peptides or can alter their efficacy. Therefore, to address these issues researchers have focused on appropriate designs, combination therapies, and the use of recombinant systems (Dostert et al. 2019).

16.6 BACTERIOPHAGE THERAPY

Bacteriophages are bacterial viruses that reside within bacteria as lysogeny or infect to kill the host (Wu et al. 2015). They found ubiquitously wherever bacteria are present. They are used before the discovery of Fleming's magic bullet, i.e., antibiotics (Kutter et al. 2015). Skin, wounds, and burn infections a house for ESKAPE pathogens used to be treated with bacteriophages. In 2006, the U.S. FDA has approved the use of phages in the packaging of meat and cheese to target foodborne pathogen *Listeria*. Other enteric pathogens such as *E. coli*, *Klebsiella*, *Salmonella*, *Shigella*, and *Vibrio cholerae* have also been targeted using phages (Kutter et al. 2015). There is now paucity in using phage therapy due to the lack of double-blind clinical trials.

The activity of phages is not only limited against planktonic bacteria but also against bacteria in biofilms. They are being investigated as a potential weapon to penetrate biofilms and kill bacterial cells. They can be utilized either as an entity or by delivering phage-encoded enzymes that degrade bacterial cell walls and EPS (Chan & Abedon 2015). The interaction between phage and biofilm is a dynamic process. Adsorption of phages to the receptors of the target bacteria is the key step in phage infection; however, the EPS matrix is the main hurdle in doing so. Phage overcomes this barrier by diffusion or with help of phage-encoded EPS degrading enzymes like polysaccharide depolymerase. These enzymes destroy biofilm architecture allowing the phages to infect bacteria.

Phages offer the advantage of being inexpensive and host-specific keeping nearby microflora intact. There is a large body of in vitro and animal evidence supporting the efficacy of phages in treating biofilm-associated infections (Yang et al. 2012). Ahiwale et al. reported a T7-like lytic phage isolated from Pavana river water, prevent and disperse biofilms of *P. aeruginosa* (Ahiwale et al. 2011). Loss of 99.9% biofilms of *P. mirabilis* and *E. coli* using lytic bacteriophages and 90% reduced biofilm formation by both strains using bacteriophage impregnated hydrogel-coated catheter sections have been reported (Carson et al. 2010). A new podoviridae *S. aureus* bacteriophage (SAP-2) by virtue of cell wall degrading enzymes showed significant antibiofilm activity (Yang et al. 2012). Another bacteriophage PT-6 that produces alginase was isolated which not only depolymerizes the alginate capsule but also enhanced engulfment by phagocytes against cystic fibrosis isolates of *P. aeruginosa* (Carson et al. 2010).

Some phage treatments such as KPO1K2 phage (encoding depolymerases) also enhances ciprofloxacin sensitivity against *K. pneumonia* biofilms and restricts them to the form of resistant variants (Yang et al. 2012). Rahman et al. reported SAP-26 belonging to siphoviridae showed enhanced biofilm removal effect in the presence of antimicrobial agents particularly rifampicin against MRSA and MSSA (Rahman et al. 2011). Kumaran et al. studied the therapeutic outcome of a sequence of phage-antibiotic therapy and concluded that biofilms of *S. aureus* were reduced up to 3 log CFU/mL when phage (SATA-8505) treatment preceded antibiotics (Kumaran et al. 2018).

Moreover, phage cocktails have also been employed against biofilms. Maszewska et al. tested the sensitivity of thirteen phages against 50 uropathogenic *P. mirabilis* strains and found that phages 39APmC32, 65APm2833, and 72APm5211 alone and in cocktail strongly prevent and disrupt biofilms (Maszewska et al. 2018). Also in vitro biofilms of 44 strains of *P. aeruginosa* isolated from chronic rhinosinusitis patients (with and without cystic fibrosis) were treated with a cocktail of four phages for 24 and 48 h. A significant reduction in biofilm biomass was observed regardless of antibiotic resistance (Fong et al. 2017).

Although phages are generally regarded as safe (Kutter et al. 2015) and offer enormous advantages in chronic or antibiotic-resistant infections, they are still subjected to certain drawbacks. These include occasional membrane-bound endotoxins, lack of pharmacokinetic data, and incorporation of phage virulence genes in a bacterial genome, bacterial resistance to phages, narrow host range, and importantly conversion of lytic phages to lysogenic phages. These issues, however, are being addressed using phage cocktails, recombinant phages, or combining antibiotics with phages. Further work in the context of pure phage isolation, optimizing formulation, and clinical trials are needed to adapt them for defined conditions (Sadekuzzaman et al. 2015).

16.7 NANOTECHNOLOGY

In recent years, nanotechnology has provided immense advancement in the upgradation of biomedical devices and targeted-drug delivery. Nanotechnology has been explored in different ways to address antimicrobial resistance. It is considered a promising field to prevent, eliminate, and control the stubborn biofilms.

16.7.1 Nanoparticles as Antibiofilm Agents

Nanotechnology makes use of nanomaterial that has at least one dimension less than 100 nm. The limited size is responsible for an increase in surface area to volume ratio making them highly reactive, also their nanoscale mimics the scale at which biological reactions occur at the cellular level (Neethirajan et al. 2014). The use of metallic nanoparticles (MNPs) such as silver, iron, gold, zinc, copper, and their oxides as antimicrobial agents has already been explored. Now, attention has been diverted toward their ability to overcome biofilms. Table 16.3 enlists some data published on the antibiofilm potential of nanoparticles.

TABLE 16.3
Nanoparticles as Antibiofilm Agents

Method of preparation	Concentration	Pathogen	Mode of Action	Reference
		AgNPs		
UV photo reduction	< 1 μg/mL	*P. aeruginosa*	Generation of ROS and cell membrane damage	Kora & Arunachalam (2011)
Enzymatic reduction of $AgNO_3$ using two lignin-degrading fungus *Aspergillus flavus* and *Emericella nidulans*	2 μg/mL 4 μg/mL 8 μg/mL	*S. aureus* *P. aeruginos* *E. coli*	Reduce Biofilm Matrix	Barapatre et al. (2016)
Treatment Ag+ with leaf extract of *Allophylus cobbe*	0.1–1.0 μg/mL	*P. aeruginosa*, *Shigella flexneri*, *S. aureus*, and *S. pneumoniae*	Reduce Biofilm Matrix	Gurunathan et al. (2014)
Reduction of 1 mM silver nitrate using *Bacillus cereus* supernatant	NA	*S. aureus*	Quorum quenching neutralization of adhesive substances	Chaudhari et al. (2012)
Biosynthesis using *Pseudomonas aeruginosa* (JQ989348) isolated from deep sea water	NA	*P. aeruginosa* *S. aureus*	Quorum quenching	Ramalingam et al. (2014)
Microwave-assisted synthesis using aqueous leaf extract of *Eucalyptus globulus*	30 μg/mL	*P. aeruginosa* and MRSA	Reduction of EPS	Ali et al. (2015)
		AuNPs		
Reduction and capping by Baicalein	100 g/mL	*P. aeruginosa*	Reduction of EPS, swimming, and swarming motility	Rajkumari et al. (2017)
Reduction by aqueous leaf extract of *Garcinia combogia*	20–40 μg/mL	*B. licheniformis*	Reduced biofilm matrix	Nithya & Jayachitra (2016)
Bacteriophage	0.2 mM	*P. aeruginosa*	Reduced biofilm matrix	Ahiwale et al. (2017)
Cyclic peptide (P2)-dependent extracellular biosynthesis using *B. niabensis* 45	32–1024 μg/mL	*Pseudomonas aeruginosa* PAO1 and *Staphylococcus aureus* ATCC25923	Reduced biofilm matrix	Li et al. (2016b)
Adapted coprecipitation method	0.01–0.625 mg/mL	*P. aeruginosa*	Reduced biofilm matrix	Iconaru et al. (2013)

(Continued)

TABLE 16.3 (*Continued*)

Method of preparation	Concentration	Pathogen	Mode of Action	Reference
		CuNP		
One-pot method	100 ng/mL	*P. aeruginosa*	Target cell surface hydrophobicity and exopolysaccharide	LewisOscar et al. (2015)
One-pot synthesis	100 ng/mL	*Vibrio alginolyticus* (ATCC 17749), *Vibrio parahaemolyticus* (ATCC 17802), and *Aeromonas hydrophila* (ATCC 7966)	Reduce cell surface hydrophobicity and 85% of extracellular polysaccharide (EPS) production	Chari et al. (2017)
		MgNP		
Water-based synthesis using sonochemistry	N/A	*E. coli* and *S. aureus*	Reduced biofilm matrix	Lellouche et al. (2012)

16.7.2 Surface Modification Using Nanoparticles

As stated previously, biomedical devices inherent biofilms that are resistant to conventional antibiotics and are a major cause of hospital-acquired infections. Nanomaterials are frequently used to modify traditional surfaces to make them anti-adhesive to prevent bacterial colonization. They can be used alone or in conjugation with other bioactive substances. Taglietti et al. have reported that modifying glass surfaces with silver nanoparticles strongly prevent biofilm formation by *S. epidermidis* (Taglietti et al. 2014). Abdulkareem et al. reported reduced *Streptococcal* biofilms after the coating of titanium surface with zinc oxide (nZnO), hydroxyapatite and a composite (nZnO+nHA) using electrohydrodynamic deposition (Abdulkareem et al. 2015). Modification of titanium alloy implants using dual-layered silver-HA nanocoating has also been reported (Besinis et al. 2017). Coating of iron oxide nanoparticles on catheters inhibits 30.97% biofilm of *S. aureus* and 16.92% biofilm of *E. coli*; however, it did not inhibit biofilms of *S. marcescens* (Salman et al. 2015). There are many other pieces of research that prove surface modification of biomaterials with metallic, polymeric, and composite nanoparticles strongly oppose biofilm formation (Neethirajan et al. 2014).

16.7.3 Targeted Drug Delivery Using Nanoparticles

Nanotechnology also offers an advantage of slow, controlled release of bioactive substances at a specific target. Ramos et al. in their review had reported a list of nano-carriers with their formulations and their effect on biofilm formers. These include liposomes (LIPs), microemulsions, nanoemulsions, cyclodextrins (CDs), solid lipid nanoparticles (SLNs), polymeric nanoparticles (PNs), and MNPs (Dos Santos Ramos et al. 2018). In addition, Ahmed at al. reported that the efficacy of antibiotic chlorhexidine increased when coated on gold nanoparticles against biofilms of *K. pneumonia* 13882 and its clinical strains. The gold conjugated chlorhexidine was able to penetrate biofilms and disperse the matrix. The coupling of nanoparticles not only enhanced the efficacy of antibiotics but also different polymers and enzymes (Ahmed et al. 2016). Shi et al. reported that biopolymer chitosan-coated iron oxide nanoparticles inhibit biofilms of *S. aureus* (Shi et al. 2016). Vinoj et al. showed that coating of gold nanoparticles by (quorum quenching) N-acylated homoserine lactonase obtained from *Bacillus licheniformis* at a concentration of 2–8 μM enhanced reduction in biofilms of clinical strains of *Proteus* without effecting macrophages (Vinoj et al. 2015). The ability of nanoparticles to synergize the active molecules for biofilm inhibition is an attractive feature allowing the efficient use of clinically available drugs.

The ability of nanoparticles to penetrate the living system because of their limited size is of much concern. The intensive use of nanomaterials can result in their accumulation in an ecological and living system, therefore, investigations regarding their short-term and long-term uses are required (Dos Santos Ramos et al. 2018). Still, the emerging field of nanotechnology is addressing many complex issues in biomedicines.

16.8 PHOTODYNAMIC THERAPY

Photodynamic therapy (PDT) is defined as an elegant treatment modality that involves the use of photosensitizing compounds, harmless visible light, and molecular oxygen. The phenomenon was observed over 100 years ago by Oskar Raab that *Paramecium* spp. stained with acridine orange was destructed when exposed to intense light. Since then, the technique was used for the treatment of different types of cancers mainly skin cancers (Allison & Moghissi 2013). In 1960, antimicrobial PDT was first introduced. Macmillan found that 99% of microorganisms stained with toluidine blue were killed within 30 min when irradiated with light of 632 nm wavelengths (Ghorbani et al. 2018). PDT has widely been employed for treatment for cancers, ophthalmological and dermatological disorders (Rosa & da Silva 2014). However, in recent years, the rapid emergence of antibiotic resistance among pathogens led to the exploration of PDT as an alternative treatment of microbial infection.

The basic principle of therapy involves the absorption of energy via photosensitization and its transfer to the molecular oxygen and free radicals causing the oxidative damage to the plasma membrane and nuclear material of microbial population without harming the host cells. It is a two-step process that requires the photosensitizer (PS) and light of an appropriate wavelength. PS is set at ground level and has a stable electronic configuration. Upon exposure to light with a specific wavelength, the electrons from low energy levels move to a higher energy level (excited state). This intermediate state is highly unstable and possesses a half-life ranging from 10–6 to 10–9. This causes a shift in electron spin and converts it to a longstanding triple state. In this excited state (triple-state), PS reacts with an oxygen molecule in two different mechanisms. Type 1 mechanism results due to transfer of electrons from PS in triplet state to organic molecule producing free radicals; and these free radical, in turn, produces ROS such as O_2^-, H_2O_2, and hydroxyl radicals, etc. on reacting with O_2 molecule. Type 2 mechanism involves the transfer of energy to ground-state molecular oxygen resulting in production of singlet oxygen. The oxidative products generated through either of reaction irreversibly damage bacterial biomolecules resulting in death. The two reactions can occur simultaneously or individually depending on the type of PS and conditions of PDT (Hu et al. 2018; Rosa & da Silva 2014; Ghorbani et al. 2018).

PSs are defined as agents that transfer light. Ideally, they should be hydrophilic, stable, and non-toxic till illuminated. Different natural and synthetic compounds can be used as PSs. The commonly used PS includes natural compounds such as hypericin and curcumin or dyes such as methylene blue, toluidine blue, Rose Bengal. Importantly, tetrapyrrole compounds (pigments of life such as porphyrins, phthalocyanines, chlorins, and bacteriochlorins) are also used as PSs. Each of these is reported to excite at a different wavelength and kill bacteria (Ghorbani et al. 2018).

Moreover, the growing field of nanotechnology has also improved the use of PS. The nanoparticles deliver PS, accompany them, or sometimes themselves act as PSs (Ghorbani et al. 2018). AuNR gold nanorods activated with laser were effective against biofilm of oral pathogens (*E. faecalis*, *S. aureus*, *S. mutans*, *S. sobrinus*, *S. oralis*, *S. salivarius*, and *E. coli*) (Castillo-Mart et al. 2015). Misba et al. observed that conjugation of PS toluidine blue O with silver nanoparticles resulted in the production of OH^- ions; indicating type I phototoxicity against biofilms of *S. mutans*. The conjugate TBO-AgNP resulted in loss of bacterial viability by 4 $\log_{10}$ (Misba et al. 2016). Khan et al. reported that biofilm inhibition by PS methylene blue was enhanced when conjugated with gold nanoparticles. The 660-nm diode laser activated the gold nanoparticle–methylene blue conjugate, which effectively reduces the biofilms of *C. albicans* via type I phototoxicity mechanism (Khan et al. 2012).

Fullerenes are the clinically approved nanoparticles that are being upgraded using different cationic compounds (Ghorbani et al. 2018). Other clinically approved PS include Photofrin (hematoporphyrin derivative), PPIX (Protoporphyrin IX), Verteporfin, Radachlorin (now Bremachlorin), Fullerenes, Temoporfin, Foscan (m-tetrahydroxyphenylchlorin) (Hu et al. 2018), or Fotolon (Mono-L-aspartyl chlorine e6 (NPe6) (Allison & Moghissi 2013).

Each PS requires illumination at a certain wavelength to get excited. A light source is a key component in PDT. The first light source used was a conventional bulb, which due to its incoherency, polychromaticity and strong thermal component did not yield appreciable results. Later, laser with extremely fine coherency and spectral bands was employed. These lasers were operated at a low power of ≤100 mV to produce energy in visible (400–700 nm wavelength), ultraviolet (200–400 nm) or near-infrared (700–1,500 nm) regions. Though highly reliable, lasers are expensive. Alternatively, light-emitting diodes (Dos Santos Ramos et al.) that produce high intensity of a broad-spectral band can be employed. They are cost-effective and versatile (Rosa & da Silva 2014).

Diversity of PDT and independence on antibiotic resistance patterns make it a highly attractive approach to get rid of clinical biofilms. PDT not only kills the microbial cells but also degrades the matrix by targeting components of EPS with ROS. PS initially binds the biofilm matrix, sometimes they penetrate the matrix reaching to the cell surface or penetrate further into the cytoplasm or even organelle. The localization of PS depends on the type of PS used. Wherever the PS is localized, exposure to light causes the excessive generation of ROS which suppresses microbial antioxidant defense mechanism. This marks the initiation of oxidative damage of matrix, cell surface, and intracellular components, thus collapsing the biofilms (Hu et al. 2018)

The disruption of *P. aeruginosa* and MRSA biofilms on prosthetic devices was observed when exposed to novel PS RLP068/Cl (Vassena et al. 2014). Type 2 phytotoxicity was observed in *E. faecalis* biofilm when treated with a PS (Rose Bengal with chitosan) and exposed to 5–60 J/cm^2 green lights (Shrestha et al. 2012). Wood et al. reported that seven-day oral plaque biofilms formed on natural enamel surfaces were significantly reduced when incubated with cationic Zn(II) phthalocyanine PS and irradiated with white light (Wood et al. 1999). O'Neill et al. reported toluidine blue O (25 µg/mL) irradiated with helium/neon laser light killed substantial numbers of oral bacteria in multi-species biofilms (O'Neill et al. 2002). 5-Aminolevulinic acid (ALA), precursor of the PS protoporphyrin IX, irradiated with light-emitting diode of 410 nm wavelength was found to be effective against biofilms of MRSA in vitro and in vivo. Intraperitoneal administration of PS on ulcer surfaces in mice decreased bacterial counts on ulcer surfaces (Morimoto et al. 2014). Another study reported the loss of *S. aureus* biofilms from rat tibial bone using ALA-mediated PDT (Bisland et al. 2006).

The use of antibiotics as PS has also been reported. In a study by Wood et al., photodynamic killing of *S. mutans* biofilms by erythromycin and PS "photofrin" and methylene blue was compared. The result suggested that erythromycin was 1–2 $\log_{10}$ more effective than photofrin and 0.5–1 $\log_{10}$ more effective than methylene blue in killing bacteria in biofilms (Wood et al. 2006). Loss in architecture of biofilms of *C. albicans* and Candida *dubliniensis* has also been reported using erythrosine- and LED-mediated PDT (Costa et al. 2011). Moreover, natural compounds such as curcumin are also used as PS activated by LED against *C. albicans* biofilms (Dovigo et al. 2011).

Although PDT offers several advantages, the targeted delivery of PS is still a hurdle. The PS must only target selective microbe without affecting the tissues (Ghorbani et al. 2018). Furthermore, the thick slimy matrix of biofilms shelters the pathogens from photodynamic inactivation (Sadekuzzaman et al. 2015). Functionalization with cationic molecules and the use of nanostructures for targeted delivery are likely to solve these problems (Ghorbani et al. 2018).

16.9 CRISPR INTERFERENCE

Biofilm formation is a complex phenomenon requiring different sets of genes. CRISPR/Cas is a unique system that can inhibit bacterial biofilm through genome modification. The phenomenon has been an integral part of the bacterial adaptive immune system against invading viruses. It has wide applications in genome engineering and has revolutionized many biological fields.

Currently, the CRISPR/Cas system id divided into three types. Out of them. Type II CRISPR/Cas system is the most studied one which differs phylogenetically and only found in bacteria (Hedge et al. 2019). This system makes use of trans-activating RNA (tracrRNA), CRISPR RNA (crRNA), and an endonuclease Cas 9, obtained from the type II system. TracrRNA has sequences complementary to crRNA, this chimeric form is known as single guide RNA (sgRNA). The sgRNA directs Cas9 to genomic DNA at 5′ end and binds to specific sequences called protospacer adjacent motif (PAM) typically NGG sequences. Five nucleotides upstream of PAM sequences mark the target site and is required for the double-stranded endonucleases activity of Cas 9 (Chaterji et al. 2017).

Kang et al. have used CRISPR-Cas9 to knockout the luxS gene in *E. coli*. The gene product is responsible for AI-2 synthesis, which is an important QS signaling molecule and involved in the initial stages of biofilm formation (Kang et al. 2017).

Beyond gene editing, the system is also modified to regulate gene expression. In this regard, the variants of CRISPR-Cas9 tool kits; CRISPRa for transcriptional activation and CRISPRi for transcriptional interference along with catalytic activity deficient dCas 9 have been extensively studied (Yang & Huang, 2019). Zuberi et al. have reported the suppression of luxS in *E. coli* using CRISPRi. The reduced expression of luxS resulted in biofilm inhibition (Zuberi et al. 2017). In another study, The FimH gene of *E. coli* responsible for initial bacterial attachment was suppressed using CRISPRi (Zuberi et al. 2017). Noirot-Gros et al. used the CRISPRi to study the genes associated with biofilm formation in three different strains of *P. fluorescens* (SBW25, WH6, and Pf0-1). Moreover, Zhang and Poh (2018) adapted CRISPRi/dCas9 to regulate the expression of the wcaF gene in *E. coli*. The repressed expression resulted in decreased production of colanic acid, which is a major component of polysaccharides in EPS matrix, eventually inhibiting the biofilms of *E. coli*.

16.10 OTHER REMEDIES

In recent years, there has been an explosion of many small molecules that serve as antibiofilm agents. Most of these agents are obtained directly from nature or modified versions of natural products.

Fatty acids such as oleic acids prevent biofilm formation of *S. aureus* by limiting bacterial adhesion. Another fatty acid, cis-2-decenoic acid, represses and disperses biofilms of *P. aeruginosa*, *E. coli*, *K. pneumoniae*, *P. mirabilis*, *S. pyogenes*, *B. subtilis*, *S. aureus*, and the yeast *C. albicans*. Nitric oxide (NO) is an important biological messenger. It is a signal for lifestyle switching and biofilm dispersion, NO binding to HNOX, a protein harboring heme nitric oxide/oxygen-binding domain inhibits cyclic-di-GMP production which causes biofilm dispersal (Rabin et al. 2015).

Since metal cations are known to stabilize the integrity of the biofilm matrix, the use of chelating agents is yet another effective strategy to overcome biofilms. Sodium citrate inhibits the biofilms of *Staphylococcus* species. Biofilm inhibition using minocycline-ethylenediaminetetraacetic acid (EDTA), tetrasodium EDTA, and disodium EDTA has also been reported (Saleemi et al. 2018). Milk protein lactoferrin also possesses the chelating properties and therefore can inhibit the biofilm formation (Sadekuzzaman et al. 2015). Moreover, ionic liquids such as 1-alkylquinolinium bromide are designed to be biofilm inhibitors (Rabin et al. 2015).

16.11 CONCLUSION

Biofilm formation is an effective strategy employed by microorganisms for their survival in severe environmental conditions. Antibiotics alone are insufficient to fight chronic biofilm-associated infections. Therefore, researches have come up with multiple remedies, to be used alone or in combination, to fight these resilient biofilms. There are several evidences that ensure promising results of antibiofilm strategies such as natural substances, PDT, bacteriophages, bacterial peptides, quorum quenchers, nanotechnology, modification of signaling molecules, and CRISPRi. However, these remedies are in the nascent phase of their development and require further work to validate them. Currently, advancement in multidisciplinary collaboration predominantly of basic sciences, pharmacology, and clinical microbiology is the basic need for the treatment of biofilms.

REFERENCES

Abd El-Baky, R. M., Sandle, T., John, J., Abuo-Rahma, G. E. A. & Hetta, H. F. 2019. A novel mechanism of action of ketoconazole: Inhibition of the NorA efflux pump system and biofilm formation in multidrug-resistant *Staphylococcus aureus*. *Infection and Drug Resistence* 12, 1703–1718.

Abdulkareem, E. H., Memarzadeh, K., Allaker, R. P., Huang, J., Pratten, J. & Spratt, D. 2015. Antibiofilm activity of zinc oxide and hydroxyapatite nanoparticles as dental implant coating materials. *Journal of Dentistry* 43(12), 1462–1469.

Adil, M., Baig, M. H. & Rupasinghe, H. P. V. 2019. Impact of Citral and Phloretin, Alone and in Combination, on Major Virulence Traits of *Streptococcus pyogenes*. *Molecules* 24(23), 4237.

Aguila-Arcos, S., Alvarez-Rodriguez, I., Garaiyurrebaso, O., Garbisu, C., Grohmann, E. & Alkorta, I. 2017. Biofilm-forming clinical *Staphylococcus* isolates harbor horizontal transfer and antibiotic resistance genes. *Frontiers in Microbiology* 8, 2018.

Ahiwale, S., Tamboli, N., Thorat, K., Kulkarni, R., Ackermann, H. & Kapadnis, B. 2011. In vitro management of hospital *Pseudomonas aeruginosa* biofilm using indigenous T7-like lytic phage. *Current Microbiology* 62(2), 335–340.

Ahiwale, S. S., Bankar, A. V., Tagunde, S. & Kapadnis, B. P. 2017. A bacteriophage mediated gold nanoparticles synthesis and their antibiofilm activity. *Indian Journal of Microbiology* 57(2), 188–194.

Ahmed, A., Khan, A. K., Anwar, A., Ali, S. A. & Shah, M. R. 2016. Biofilm inhibitory effect of chlorhexidine conjugated gold nanoparticles against *Klebsiella pneumoniae*. *Microbial Pathogenesis* 98, 50–56.

Ahn, K. B., Kim, A. R., Kum, K.-Y., Yun, C.-H. & Han, S. H. 2017. The synthetic human beta-defensin-3 C15 peptide exhibits antimicrobial activity against *Streptococcus mutans*, both alone and in combination with dental disinfectants. *Journal of Microbiology* 55(10), 830–836.

Al-Shabib, N. A., Husain, F. M., Ahmad, I., Khan, M. S., Khan, R. A. & Khan, J. M. 2017. Rutin inhibits mono and multi-species biofilm formation by foodborne drug resistant *Escherichia coli* and *Staphylococcus aureus*. *Food Control* 79, 325–332.

Alav, I., Sutton, J. M. & Rahman, K. M. 2018. Role of bacterial efflux pumps in biofilm formation. *J Antimicrobial Chemotherapy* 73(8), 2003–2020.

Algburi, A., Comito, N., Kashtanov, D., Dicks, L. M. T. & Chikindas, M. L. 2017. Control of biofilm formation: Antibiotics and beyond. *Applied and Environmental Microbiology* 83(3), 1–16.

Ali, K., Ahmed, B., Dwivedi, S., Saquib, Q., Al-Khedhairy, A. A. & Musarrat, J. 2015. Microwave accelerated green synthesis of stable silver nanoparticles with eucalyptus globulus leaf extract and their antibacterial and antibiofilm activity on clinical isolates. *PLoS One* 10(7), e0131178.

Allison, R. R. & Moghissi, K. 2013. Photodynamic therapy (PDT): PDT mechanisms. *Clinical Endoscopy* 46(1), 24–29.

Angles, M. L., Marshall, K. C. & Goodman, A. E. 1993. Plasmid transfer between marine bacteria in the aqueous phase and biofilms in reactor microcosms. *Applied and Environmental Microbiology* 59(3), 843–850.

Asahi, Y., Noiri, Y., Miura, J., Maezono, H., Yamaguchi, M., Yamamoto, R., Azakami, H., Hayashi, M. & Ebisu, S. 2014. Effects of the tea catechin epigallocatechin gallate on *Porphyromonas gingivalis* biofilms. *Journal of Applied Microbiology* 116(5), 1164–1171.

Atanasov, A. G., Waltenberger, B., Pferschy-Wenzig, E.-M., Linder, T., Wawrosch, C., Uhrin, P., Temml, V., Wang, L., Schwaiger, S., Heiss, E. H., Rollinger, J. M., Schuster, D., Breuss, J. M., Bochkov, V., Mihovilovic, M. D., Kopp, B., Bauer, R., Dirsch, V. M. & Stuppner, H. 2015. Discovery and resupply of pharmacologically active plant-derived natural products: A review. *Biotechnology Advances* 33(8), 1582–1614.

Augustine, N., Goel, A. K., Sivakumar, K. C., Ajay Kumar, R. & Thomas, S. 2014. Resveratrol—A potential inhibitor of biofilm formation in *Vibrio cholerae*. *Phytomedicine* 21(3), 286–289.

Balcazar, J. L., Subirats, J. & Borrego, C. M. 2015. The role of biofilms as environmental reservoirs of antibiotic resistance. *Frontiers in Microbiology* 6, 1216.

Bali, E. B., Turkmen, K. B., Erdonmez, D. & Saglam, N. 2019. Comparative study of inhibitory potential of dietary phytochemicals against quorum sensing activity of and biofilm formation by *Chromobacterium violaceum* 12472, and swimming and swarming behaviour of *Pseudomonas aeruginosa* PAO1. *Food Technology Biotechnology* 57(2), 212–221.

Barapatre, A., Aadil, K. R. & Jha, H. 2016. Synergistic antibacterial and antibiofilm activity of silver nanoparticles biosynthesized by lignin-degrading fungus. *Bioresources and Bioprocessing* 3(1), 8.

Bay, D. C., Stremick, C. A., Slipski, C. J. & Turner, R. J. 2017. Secondary multidrug efflux pump mutants alter *Escherichia coli* biofilm growth in the presence of cationic antimicrobial compounds. *Research in Microbiology* 168(3), 208–221.

Besinis, A., Hadi, S. D., Le, H. R., Tredwin, C. & Handy, R. D. 2017. Antibacterial activity and biofilm inhibition by surface modified titanium alloy medical implants following application of silver, titanium dioxide and hydroxyapatite nanocoatings. *Nanotoxicology* 11(3), 327–338.

Bisland, S. K., Chien, C., Wilson, B. C. & Burch, S. 2006. Pre-clinical in vitro and in vivo studies to examine the potential use of photodynamic therapy in the treatment of osteomyelitis. *Photochemical and Photobiological Sciences* 5(1), 31–38.

Boles, B. R. & Singh, P. K. 2008. Endogenous oxidative stress produces diversity and adaptability in biofilm communities. *Proceedings of the National Academy of Sciences of the United States of America* 105(34), 12503–12508.

Brancatisano, F. L., Maisetta, G., Di Luca, M., Esin, S., Bottai, D., Bizzarri, R., Campa, M. & Batoni, G. 2014. Inhibitory effect of the human liver-derived antimicrobial peptide hepcidin 20 on biofilms of polysaccharide intercellular adhesin (PIA)-positive and PIA-negative strains of *Staphylococcus epidermidis*. *Biofouling* 30(4), 435–446.

Bugli, F., Palmieri, V., Torelli, R., Papi, M., De Spirito, M., Cacaci, M., Galgano, S., Masucci, L., Paroni Sterbini, F., Vella, A., Graffeo, R., Posteraro, B. & Sanguinetti, M. 2016. In vitro effect of clarithromycin and alginate lyase against *Helicobacter pylori* biofilm. *Biotechnology Progress* 32(6), 1584–1591.

Bugli, F., Posteraro, B., Papi, M., Torelli, R., Maiorana, A., Sterbini, F. P., Posteraro, P., Sanguinetti, M. & De Spirito, M. 2013. In vitro interaction between alginate lyase and amphotericin B against *Aspergillus fumigatus* biofilm determined by different methods. *Antimicrobial Agents and Chemotherapy* 57(3), 1275–1282.

Calvani, R., Picca, A., Lo Monaco, M. R., Landi, F., Bernabei, R. & Marzetti, E. 2018. Of microbes and minds: A narrative review on the second brain aging. *Front Med (Lausanne)* 5, 53.

Carson, L., Gorman, S. P. & Gilmore, B. F. 2010. The use of lytic bacteriophages in the prevention and eradication of biofilms of *Proteus mirabilis* and *Escherichia coli*. *FEMS Immunology & Medical Microbiology* 59(3), 447–455.

Castillo-Mart, J. C., Martínez-Castañón, G. A., Martínez-Gutierrez, F., Zavala-Alonso, N. V., Patiño-Marín, N., Niño-Martinez, N., Zaragoza-Magaña, V. & Cabral-Romero, C. 2015. Antibacterial and antibiofilm activities of the photothermal therapy using gold nanorods against seven different bacterial strains. *Journal of Nanomaterials* 16(1), 177–177.

Cepas, V., Lopez, Y., Munoz, E., Rolo, D., Ardanuy, C., Marti, S., Xercavins, M., Horcajada, J. P., Bosch, J. & Soto, S. M. 2019. Relationship between biofilm formation and antimicrobial resistance in gram-negative bacteria. *Microbial Drug Resistance* 25(1), 72–79.

Chan, B. K. & Abedon, S. T. 2015. Bacteriophages and their enzymes in biofilm control. *Current Pharmaceutical Design* 21(1), 85–99.

Chari, N., Felix, L., Davoodbasha, M., Sulaiman Ali, A. & Nooruddin, T. 2017. In vitro and in vivo antibiofilm effect of copper nanoparticles against aquaculture pathogens. *Biocatalysis and Agricultural Biotechnology* 10, 336–341.

Chaterji, S., Ahn, E. H. & Kim, D. H. 2017. CRISPR genome engineering for human pluripotent stem cell research. *Theranostics*, 7(18), 4445–4469.

Chaudhari, P. R., Masurkar, S. A., Shidore, V. B. & Kamble, S. P. 2012. Effect of biosynthesized silver nanoparticles on *Staphylococcus aureus* biofilm quenching and prevention of biofilm formation. *Nano-Micro Letters* 4(1), 34–39.

Ciofu, O. & Tolker-Nielsen, T. 2019. Tolerance and resistance of *Pseudomonas aeruginosa* biofilms to antimicrobial agents-how *P. aeruginosa* can escape antibiotics. *Frontiers in Microbiology* 10, 913.

Conibear, T. C., Collins, S. L. & Webb, J. S. 2009. Role of mutation in *Pseudomonas aeruginosa* biofilm development. *PLoS One* 4(7), e6289.

Cope, E. K., Goldstein-Daruech, N., Kofonow, J. M., Christensen, L., McDermott, B., Monroy, F., Palmer, J. N., Chiu, A. G., Shirtliff, M. E., Cohen, N. A. & Leid, J. G. 2011. Regulation of virulence gene expression resulting from *Streptococcus pneumoniae* and nontypeable *Haemophilus influenzae* interactions in chronic disease. *PLoS One* 6(12), e28523.

Costa, A. C. B. P., de Campos Rasteiro, V. M., Pereira, C. A., da Silva Hashimoto, E. S. H., Beltrame, M., Junqueira, J. C. & Jorge, A. O. C. 2011. Susceptibility of *Candida albicans* and *Candida dubliniensis* to erythrosine- and LED-mediated photodynamic therapy. *Archives of Oral Biology* 56(11), 1299–1305.

Craigen, B., Dashiff, A. & Kadouri, D. E. 2011. The use of commercially available alpha-amylase compounds to inhibit and remove *Staphylococcus aureus* biofilms. *The Open Microbiology Journal* 5, 21–31.

Cywes-Bentley, C., Skurnik, D., Zaidi, T., Roux, D., Deoliveira, R. B., Garrett, W. S., Lu, X., O'Malley, J., Kinzel, K., Zaidi, T., Rey, A., Perrin, C., Fichorova, R. N., Kayatani, A. K., Maira-Litran, T., Gening, M. L., Tsvetkov, Y. E., Nifantiev, N. E., Bakaletz, L. O., Pelton, S. I., Golenbock, D. T. & Pier, G. B. 2013. Antibody to a conserved antigenic target is protective against diverse prokaryotic and eukaryotic pathogens. *Proceedings of the National Academy of Sciences of the United States*110(24), E2209–E2218.

Dalleau, S., Cateau, E., Bergès, T., Berjeaud, J.-M. & Imbert, C. 2008. In vitro activity of terpenes against *Candida* biofilm. *International Journal of Antimicrobial Agents* 31, 572–576.

Darouiche, R. O., Mansouri, M. D., Gawande, P. V. & Madhyastha, S. 2009. Antimicrobial and antibiofilm efficacy of triclosan and DispersinB® combination. *Journal of Antimicrobial Chemotherapy* 64(1), 88–93.

de la Fuente-Núñez, C., Korolik, V., Bains, M., Nguyen, U., Breidenstein, E. B. M., Horsman, S., Lewenza, S., Burrows, L. & Hancock, R. E. W. 2012. Inhibition of bacterial biofilm formation and swarming motility by a small synthetic cationic peptide. *Antimicrobial Agents and Chemotherapy* 56(5), 2696–2704.

de Souza, P. M. & de Oliveira Magalhães, P. 2010. Application of microbial α-amylase in industry - A review. *Brazilian Journal of Microbiology*: [publication of the Brazilian Society for Microbiology] 41(4), 850–861.

Ding, Y., Wang, W., Fan, M., Tong, Z., Kuang, R., Jiang, W. & Ni, L. 2014. Antimicrobial and antibiofilm effect of Bac8c on major bacteria associated with dental caries and *Streptococcus mutans* biofilms. *Peptides* 52, 61–67.

Donelli, G., Francolini, I., Romoli, D., Guaglianone, E., Piozzi, A., Ragunath, C. & Kaplan, J. B. 2007. Synergistic activity of dispersin B and cefamandole nafate in inhibition of staphylococcal biofilm growth on polyurethanes. *Antimicrobial Agents and Chemotherapy* 51(8), 2733–2740.

Dong, Y.-H., Wang, L.-H., Xu, J.-L., Zhang, H.-B., Zhang, X.-F. & Zhang, L.-H. 2001. Quenching quorum-sensing-dependent bacterial infection by an N-acyl homoserine lactonase. *Nature* 411(6839), 813–817.

Dos Santos Ramos, M. A., Da Silva, P. B., Spósito, L., De Toledo, L. G., Bonifácio, B. V., Rodero, C. F., Dos Santos, K. C., Chorilli, M. & Bauab, T. M. 2018. Nanotechnology-based drug delivery systems for control of microbial biofilms: A review. *International Journal of Nanomedicine* 13, 1179–1213.

Dostert, M., Belanger, C. R. & Hancock, R. E. W. 2019. Design and assessment of antibiofilm peptides: Steps toward clinical application. *Journal of Innate Immunity* 11(3), 193–204.

Dovigo, L. N., Pavarina, A. C., Ribeiro, A. P., Brunetti, I. L., Costa, C. A., Jacomassi, D. P., Bagnato, V. S. & Kurachi, C. 2011. Investigation of the photodynamic effects of curcumin against *Candida albicans*. *Photochemistry and Photobiology* 87(4), 895–903.

Driffield, K., Miller, K., Bostock, J. M., O'Neill, A. J. & Chopra, I. 2008. Increased mutability of *Pseudomonas aeruginosa* in biofilms. *Journal of Antimicrobial Chemotherapy* 61(5), 1053–1056.

Dumaru, R., Baral, R. & Shrestha, L. B. 2019. Study of biofilm formation and antibiotic resistance pattern of gram-negative *Bacilli* among the clinical isolates at BPKIHS, Dharan. *BMC Research Notes* 12(1), 38.

El-Awady, A., de Sousa Rabelo, M., Meghil, M. M., Rajendran, M., Elashiry, M., Stadler, A. F., Foz, A. M., Susin, C., Romito, G. A., Arce, R. M. & Cutler, C. W. 2019. Polymicrobial synergy within oral biofilm promotes invasion of dendritic cells and survival of consortia members. *NPJ Biofilms Microbiomes* 5, 11.

Feng, X., Sambanthamoorthy, K., Palys, T. & Paranavitana, C. 2013. The human antimicrobial peptide LL-37 and its fragments possess both antimicrobial and antibiofilm activities against multidrug-resistant *Acinetobacter baumannii*. *Peptides* 49, 131–137.

Field, D., Seisling, N., Cotter, P. D., Ross, R. P. & Hill, C. 2016. Synergistic nisin-polymyxin combinations for the control of *Pseudomonas* biofilm formation. *Frontiers in Microbiology* 7, 1713–1713.

Fleming, D., Chahin, L. & Rumbaugh, K. 2017. Glycoside hydrolases degrade polymicrobial bacterial biofilms in wounds. *Antimicrobial Agents and Chemotherapy* 61(2), e01998–e02016.

Fleming, D. & Rumbaugh, K. P. 2017. Approaches to dispersing medical biofilms. *Microorganisms* 5(2), 15.

Flemming, H. C. & Wingender, J. 2001. Relevance of microbial extracellular polymeric substances (EPSs)—Part I: Structural and ecological aspects. *Water Science and Technology* 43(6), 1–8.

Fong, S. A., Drilling, A., Morales, S., Cornet, M. E., Woodworth, B. A., Fokkens, W. J., Psaltis, A. J., Vreugde, S. & Wormald, P.-J. 2017. Activity of bacteriophages in removing biofilms of *Pseudomonas aeruginosa* isolates from chronic rhinosinusitis patients. *Frontiers in Cellular and Infection Microbiology* 7, 418–418.

Fourie, R. & Pohl, C. H. 2019. Beyond antagonism: The interaction between *Candida* species and *Pseudomonas aeruginosa*. *Journal of Fungi (Basel)* 5(2), 34.

Galdiero, E., Lombardi, L., Falanga, A., Libralato, G., Guida, M. & Carotenuto, R. 2019. Biofilms: Novel strategies based on antimicrobial peptides. *Pharmaceutics* 11(7), 322.

Ganesh, P. S., Vishnupriya, S., Vadivelu, J., Mariappan, V., Vellasamy, K. M. & Shankar, E. M. 2019. Intracellular survival and innate immune evasion of *Burkholderia cepacia*: Improved understanding of quorum sensing-controlled virulence factors, biofilm and inhibitors. *Microbiology and Immunology*, 64(2), 87–98.

Ghafoor, A., Hay, I. D. & Rehm, B. H. 2011. Role of exopolysaccharides in *Pseudomonas aeruginosa* biofilm formation and architecture. *Applied and Environmental Microbiology* 77(15), 5238–5246.

Ghorbani, J., Rahban, D., Aghamiri, S., Teymouri, A. & Bahador, A. 2018. Photosensitizers in antibacterial photodynamic therapy: An overview. *Laser Therapy* 27(4), 293–302.

Giacaman, R. A., Torres, S., Gomez, Y., Munoz-Sandoval, C. & Kreth, J. 2015. Correlation of *Streptococcus mutans* and *Streptococcus sanguinis* colonization and ex vivo hydrogen peroxide production in carious lesion-free and high caries adults. *Archives of Oral Biology* 60(1), 154–159.

Gopu, V., Kothandapani, S. & Shetty, P. H. 2015. Quorum-quenching activity of *Syzygium cumini* (L.) Skeels and its anthocyanin malvidin against *Klebsiella pneumoniae*. *Microbial Pathogenesis* 79, 61–69.

Górniak, I., Bartoszewski, R. & Króliczewski, J. 2019. Comprehensive review of antimicrobial activities of plant flavonoids. *Phytochemistry Reviews* 18(1), 241–272.

Grassi, L., Maisetta, G., Esin, S. & Batoni, G. 2017. Combination strategies to enhance the efficacy of antimicrobial peptides against bacterial biofilms. *Frontiers in Microbiology* 8, 2409–2409.

Grice, E. A. & Segre, J. A. 2011. The skin microbiome. *Nature Reviews Microbiology* 9(4), 244–253.

Guglielmini, J., de la Cruz, F. & Rocha, E. P. 2013. Evolution of conjugation and type IV secretion systems. *Molecular Biology and Evolution* 30(2), 315–331.

Gurunathan, S., Han, J. W., Kwon, D.-N. & Kim, J.-H. 2014. Enhanced antibacterial and antibiofilm activities of silver nanoparticles against Gram-negative and Gram-positive bacteria. *Nanoscale Research Letters* 9(1), 373.

Haisma, E. M., de Breij, A., Chan, H., van Dissel, J. T., Drijfhout, J. W., Hiemstra, P. S., El Ghalbzouri, A. & Nibbering, P. H. 2014.LL-37-derived peptides eradicate multidrug-resistant *Staphylococcus aureus* from thermally wounded human skin equivalents. *Antimicrobial Agents and Chemotherapy* 58(8), 4411–4419.

Hall-Stoodley, L., Nistico, L., Sambanthamoorthy, K., Dice, B., Nguyen, D., Mershon, W. J., Johnson, C., Hu, F. Z., Stoodley, P., Ehrlich, G. D. & Post, J. C. 2008. Characterization of biofilm matrix, degradation by DNase treatment and evidence of capsule downregulation in *Streptococcus pneumoniae* clinical isolates. *BMC Microbiology* 8, 173.

Harriott, M. M. & Noverr, M. C. 2010. Ability of *Candida albicans* mutants to induce *Staphylococcus aureus* vancomycin resistance during polymicrobial biofilm formation. *Antimicrob Agents Chemother* 54(9), 3746–3755.

Hatch, R. A. & Schiller, N. L. 1998. Alginate lyase promotes diffusion of aminoglycosides through the extracellular polysaccharide of mucoid *Pseudomonas aeruginosa. Antimicrobial Agents and Chemotherapy* 42(4), 974–977.

He, Z., Huang, Z., Zhou, W., Tang, Z., Ma, R., & Liang, J. 2016. Antibiofilm activities from Resveratrol against *Fusobacterium nucleatum. Frontiers in Microbiology* 2016, 7, 1065.

Hegde, S., Nilyanimit, P., Kozlova, E., Narra, H. P., Sahni, S. K. & Hughes, G. L. 2019. CRISPR/Cas9-mediated gene deletion of the ompA gene in an *Enterobacter* gut symbiont impairs biofilm formation and reduces gut colonization of *Aedes aegypti* mosquitoes. *bioRxiv*, 389957.

Hengge, R. 2019. Targeting bacterial biofilms by the green tea polyphenol EGCG. *Molecules* 24(13), 2403.

Hoffmann, N., Lee, B., Hentzer, M., Rasmussen, T. B., Song, Z., Johansen, H. K., Givskov, M. & Hoiby, N. 2007. Azithromycin blocks quorum sensing and alginate polymer formation and increases the sensitivity to serum and stationary-growth-phase killing of *Pseudomonas aeruginosa* and attenuates chronic *P. aeruginosa* lung infection in Cftr(-/-) mice. *Antimicrobial Agents Chemotherapy* 51(10), 3677–3687.

Hogan, S., Zapotoczna, M., Stevens, N. T., Humphreys, H., O'Gara, J. P. & O'Neill, E. 2017. Potential use of targeted enzymatic agents in the treatment of *Staphylococcus aureus* biofilm-related infections. *Journal of Hospital Infection* 96(2), 177–182.

Hu, X., Huang, Y.-Y., Wang, Y., Wang, X. & Hamblin, M. R. 2018. Antimicrobial photodynamic therapy to control clinically relevant biofilm infections. *Frontiers in Microbiology* 9, 1299.

Huang, Y., Suo, Y., Shi, C., Szlavik, J., Shi, X. M. & Knochel, S. 2013. Mutations in gltB and gltC reduce oxidative stress tolerance and biofilm formation in *Listeria monocytogenes* 4b G. *International Journal of Food Microbiology* 163(2–3), 223–230.

Iconaru, S. L., Prodan, A. M., Le Coustumer, P. & Predoi, D. 2013. Synthesis and antibacterial and antibiofilm activity of iron oxide glycerol nanoparticles obtained by coprecipitation method. *Journal of Chemistry* 2013, 412079. doi: 10.1155/2013/412079.

Iriti, M. & Faoro, F. 2009. Chemical diversity and defense metabolism: How plants cope with pathogens and ozone pollution. *International Journal of Molecular Sciences* 10, 3371–3399.

Issac Abraham, S. V. P., Palani, A., Ramaswamy, B. R., Shunmugiah, K. P. & Arumugam, V. R. 2011. Antiquorum sensing and antibiofilm potential of *Capparis spinosa. Archives of Medical Research* 42(8), 658–668.

Jack, D. L., Yang, N. M. & Saier, M. H., Jr. 2001. The drug/metabolite transporter superfamily. *European Journal of Biochemistry* 268(13), 3620–39.

Jackson, K. D., Starkey, M., Kremer, S., Parsek, M. R. & Wozniak, D. J. 2004. Identification of psl, a locus encoding a potential exopolysaccharide that is essential for *Pseudomonas aeruginosa* PAO1 biofilm formation. *Journal of Bacteriology* 186(14), 4466–4475.

Jahan, M., Abuhena, M., Azad, A. K. & Karim, M. M. 2018. In vitro antibacterial and antibiofilm activity of selected medicinal plants and spices extracts against multidrug resistant *Pseudomonas aeruginosa. Journal of Pharmacognosy and Phytochemistry* 7(3), 2114–2121.

Jain, A. & Parihar, D. 2018. Antibacterial, biofilm dispersal, and antibiofilm potential of alkaloids and flavonoids of curcuma. *Biocatalysis and Agricultural Biotechnology.*

Jose, W., Costa, R., Magda, T. & Theodora, T. 2017. Antibiofilm activity of natural substances derived from plants. *African Journal of Microbiology Research* 11, 1051–1060.

Kalpana, B. J., Aarthy, S. & Pandian, S. K. 2012. Antibiofilm activity of α-amylase from *Bacillus subtilis* S8–18 against biofilm forming human bacterial pathogens. *Applied Biochemistry and Biotechnology* 167(6), 1778–1794.

Kang, S., Kim, J., Hur, J. K. & Lee, S. S. 2017. CRISPR-based genome editing of clinically important *Escherichia coli* SE15 isolated from indwelling urinary catheters of patients. *Journal of Medical Microbiology*, 66(1), 18–25.

Kannappan, A., Gowrishankar, S., Srinivasan, R., Pandian, S. K. & Ravi, A. V. 2017. Antibiofilm activity of *Vetiveria zizanioides* root extract against methicillin-resistant *Staphylococcus aureus. Microbial Pathogenesis* 110, 313–324.

Kanthawong, S., Bolscher, J., Veerman, E., Marle, J., Soet, J., Nazmi, K., Wongratanacheewin, S. & Taweechaisupapong, S. 2012. Antimicrobial and antibiofilm activity of LL-37 and its truncated variants against *Burkholderia pseudomallei. International Journal of Antimicrobial Agents* 39, 39–44.

Kaplan, J. B., LoVetri, K., Cardona, S. T., Madhyastha, S., Sadovskaya, I., Jabbouri, S. & Izano, E. A. 2012. Recombinant human DNase I decreases biofilm and increases antimicrobial susceptibility in s*taphylococci. J Antibiot (Tokyo)* 65(2), 73–77.

Kerrigan, J., Ragunath, C., Kandra, L., Gyémánt, G., Liptak, A., Janossy, L., Kaplan, J. & Ramasubbu, N. 2008. Modeling and biochemical analysis of the activity of antibiofilm agent Dispersin B. *Acta Biologica Hungarica* 59(4), 439–451.

Khan, A. K., Ahmed, A., Hussain, M., Khan, I. A., Ali, S. A., Farooq, A. D. & Faizi, S. 2017. Antibiofilm potential of 16-oxo-cleroda-3, 13(14) E-diene-15 oic acid and its five new gamma-amino gamma-lactone derivatives against methicillin-resistant *Staphylococcus aureus* and *Streptococcus mutans. European Journal of Medicinal Chemistry* 138, 480–490.

Khan, S., Alam, F., Azam, A. & Khan, A. U. 2012. Gold nanoparticles enhance methylene blue-induced photodynamic therapy: A novel therapeutic approach to inhibit *Candida albicans* biofilm. *International Journal of Nanomedicine* 7, 3245–3257.

Kim, B., Park, J.-S., Choi, H.-Y., Yoon, S. S. & Kim, W. G. 2018. Terrein is an inhibitor of quorum sensing and c-di-GMP in *Pseudomonas aeruginosa*: A connection between quorum sensing and c-di-GMP. *Scientific Reports* 8(1), 8617–8617.

Kivisaar, M. 2010. Mechanisms of stationary-phase mutagenesis in bacteria: Mutational processes in *pseudomonads. FEMS Microbiol Lett* 312(1), 1–14.

Kokai-Kun, J. F., Chanturiya, T. & Mond, J. J. 2009. Lysostaphin eradicates established *Staphylococcus aureus* biofilms in jugular vein catheterized mice. *Journal of Antimicrobial Chemotherapy* 64(1), 94–100.

Kong, E. F., Tsui, C., Kucharikova, S., Andes, D., Van Dijck, P. & Jabra-Rizk, M. A. 2016. Commensal protection of *Staphylococcus aureus* against antimicrobials by *Candida albicans* biofilm matrix. *MBio* 7(5), 1–12..

Koo, H., Hayacibara, M. F., Schobel, B. D., Cury, J. A., Rosalen, P. L., Park, Y. K., Vacca-Smith, A. M. & Bowen, W. H. 2003. Inhibition of *Streptococcus mutans* biofilm accumulation and polysaccharide production by apigenin and tt-farnesol. *Journal of Antimicrobial Chemotherapy* 52(5), 782–789.

Koo, H., Seils, J., Abranches, J., Burne, R. A., Bowen, W. H. & Quivey Jr, R. G. 2006. Influence of apigenin on gtf gene expression in *Streptococcus mutans* UA159. *Antimicrobial Agents and Chemotherapy* 50(2), 542–546.

Kora, A. J. & Arunachalam, J. 2011. Assessment of antibacterial activity of silver nanoparticles on *Pseudomonas aeruginosa* and its mechanism of action. *World Journal of Microbiology and Biotechnology* 27(5), 1209–1216.

Kouzel, N., Oldewurtel, E. R. & Maier, B. 2015. Gene transfer efficiency in gonococcal biofilms: Role of biofilm age, architecture, and pilin antigenic variation. *Journal of Bacteriology* 197(14), 2422–2431.

Krol, J. E., Wojtowicz, A. J., Rogers, L. M., Heuer, H., Smalla, K., Krone, S. M. & Top, E. M. 2013. Invasion of *E. coli* biofilms by antibiotic resistance plasmids. *Plasmid* 70(1), 110–119.

Kugaji, M.S., Kumbar, V.M., Peram, M.R., Patil, S., Bhat, K.G. & Diwan, P.V. 2019. Effect of Resveratrol on biofilm formation and virulence factor gene expression of *Porphyromonas gingivalis* in periodontal disease. *APMIS* 127(4):187–195. doi: 10.1111/apm.12930.

Kumaran, D., Taha, M., Yi, Q., Ramirez-Arcos, S., Diallo, J.-S., Carli, A. & Abdelbary, H. 2018. Does treatment order matter? Investigating the ability of bacteriophage to augment antibiotic activity against *Staphylococcus aureus* biofilms. *Frontiers in Microbiology* 9, 127.

Kunthalert, D., Baothong, S., Khetkam, P., Chokchaisiri, S. & Suksamrarn, A. 2014. A chalcone with potent inhibiting activity against biofilm formation by non-typeable *Haemophilus influenzae. Microbiology and Immunology* 58(10), 581–589.

Kuroda, T. & Tsuchiya, T. 2009. Multidrug efflux transporters in the MATE family. *Biochim Biophys Acta* 1794(5), 763–768.

Kutter, E. M., Kuhl, S. J. & Abedon, S. T. 2015. Re-establishing a place for phage therapy in western medicine. *Future Microbiology* 10(5), 685–688.

Lahiri, D., Dash, S., Dutta, R. & Nag, M. 2019. Elucidating the effect of antibiofilm activity of bioactive compounds extracted from plants. *Journal of Bioscience* 44(2), 52.

LaSarre, B. & Federle, M. J. 2013. Exploiting quorum sensing to confuse bacterial pathogens. *Microbiology and Molecular Biology Reviews* 77(1), 73–111.

Lee, J.-H., Kim, Y.-G., Cho, H. S., Ryu, S. Y., Cho, M. H. & Lee, J. 2014a. Coumarins reduce biofilm formation and the virulence of *Escherichia coli* O157:H7. *Phytomedicine* 21(8), 1037–1042.

Lee, K. W., Periasamy, S., Mukherjee, M., Xie, C., Kjelleberg, S. & Rice, S. A. 2014b. Biofilm development and enhanced stress resistance of a model, mixed-species community biofilm. *ISME Journal* 8(4), 894–907.

Lellouche, J., Friedman, A., Lellouche, J.-P., Gedanken, A. & Banin, E. 2012. Improved antibacterial and antibiofilm activity of magnesium fluoride nanoparticles obtained by water-based ultrasound chemistry. *Nanomedicine: Nanotechnology, Biology and Medicine* 8(5), 702–711.

LewisOscar, F., MubarakAli, D., Nithya, C., Priyanka, R., Gopinath, V., Alharbi, N. S. & Thajuddin, N. 2015. One pot synthesis and antibiofilm potential of copper nanoparticles (CuNPs) against clinical strains of *Pseudomonas aeruginosa. Biofouling* 31(4), 379–391.

Li, B., Qiu, Y., Song, Y., Lin, H. & Yin, H. 2019. Dissecting horizontal and vertical gene transfer of antibiotic resistance plasmid in bacterial community using microfluidics. *Environment International* 131, 105007.

Li, B., Qiu, Y., Zhang, J., Huang, X., Shi, H. & Yin, H. 2018. Real-time study of rapid spread of antibiotic resistance plasmid in biofilm using microfluidics. *Environmental Science and Technology* 52(19), 11132–11141.

Li, X., Guan, C., He, Y., Wang, Y., Liu, X. & Zhou, X. 2016a. Effects of total alkaloids of *sophora alopecuroides* on biofilm formation in *Staphylococcus epidermidis. BioMed Research International* 2016, 4020715. doi: 10.1155/2016/4020715.

Li, X. H. & Lee, J. H. 2019. Quorum sensing-dependent post-secretional activation of extracellular proteases in *Pseudomonas aeruginosa. Journal of Biological Chemistry* 294(51), 19635–19644.

Li, Y., Li, Y., Li, Q., Fan, X., Gao, J. & Luo, Y. 2016b. Rapid biosynthesis of gold nanoparticles by the extracellular secretion of *Bacillus niabensis* 45: Characterization and antibiofilm activity. *Journal of Chemistry* 2016.

Lin Chua, S., Liu, Y., Li, Y., Jun Ting, H., Kohli, G. S., Cai, Z., Suwanchaikasem, P., Kau Kit Goh, K., Pin Ng, S., Tolker-Nielsen, T., Yang, L. & Givskov, M. 2017. Reduced intracellular c-di-GMP content increases expression of quorum sensing-regulated genes in *Pseudomonas aeruginosa. Frontiers in Cellular and Infection Microbiology* 7, 451.

Liu, Y., Sun, J., Peng, X., Dong, H., Qin, Y., Shen, Q., Jiang, H., Xu, G., Feng, Y., Sun, S., Ding, J. & Chen, R. 2019. Deletion of the LuxR-type regulator VjbR in *Brucella canis* affects expression of type IV secretion system and bacterial virulence, and the mutant strain confers protection against *Brucella canis* challenge in mice. *Microbial Pathogenesis* 139, 103865.

Liu, Z., Pan, Y., Li, X., Jie, J. & Zeng, M. 2017. Chemical composition, antimicrobial, and anti-quorum-sensing activities of pummelo peel flavonoid extract. *Industrial Crops and Products* 109, 862–868.

Lloyd-Price, J., Abu-Ali, G. & Huttenhower, C. 2016. The healthy human microbiome. *Genome Medicine* 8(1), 51.

LoVetri, K. & Madhyastha, S. 2010. Antimicrobial and antibiofilm activity of quorum sensing peptides and peptide analogues against oral biofilm bacteria. *Methods in Molecular Biology* 618, 383–392.

Lu, L., Hu, W., Tian, Z., Yuan, D., Yi, G., Zhou, Y., Cheng, Q., Zhu, J. & Li, M. 2019a. Developing natural products as potential antibiofilm agents. *Chinese Medicine* 14(1), 11.

Lubelski, J., Konings, W. N. & Driessen, A. J. 2007. Distribution and physiology of ABC-type transporters contributing to multidrug resistance in bacteria. *Microbiology and Molecular Biology Reviews* 71(3), 463–476.

Luo, Y., McLean, D. T. F., Linden, G. J., McAuley, D. F., McMullan, R. & Lundy, F. T. 2017. The naturally occurring host defense peptide, LL-37, and its truncated mimetics KE-18 and KR-12 have selected biocidal and antibiofilm activities against *Candida albicans*, *Staphylococcus aureus*, and *Escherichia coli* in vitro. *Frontiers in Microbiology* 8, 544.

Madsen, J. S., Burmolle, M., Hansen, L. H. & Sorensen, S. J. 2012. The interconnection between biofilm formation and horizontal gene transfer. *FEMS Immunology and Medical Microbiology* 65(2), 183–195.

Madsen, J. S., Roder, H. L., Russel, J., Sorensen, H., Burmolle, M. & Sorensen, S. J. 2016. Coexistence facilitates interspecific biofilm formation in complex microbial communities. *Environmental Microbiology* 18(8), 2565–2574.

Magesh, H., Kumar, A., Alam, A., Priyam, Sekar, U., Sumantran, V. N. & Vaidyanathan, R. 2013. Identification of natural compounds which inhibit biofilm formation in clinical isolates of *Klebsiella pneumoniae*. *Indian Journal of Experimental Biology* 51(9), 764–772.

Maharjan, R. & Ferenci, T. 2017. The fitness costs and benefits of antibiotic resistance in drug-free microenvironments encountered in the human body. *Environmental Microbiology Reports* 9(5), 635–641.

Marquez, B. 2005. Bacterial efflux systems and efflux pumps inhibitors. *Biochimie* 87(12), 1137–1147.

Martins, M., Henriques, M., Lopez-Ribot, J. L. & Oliveira, R. 2012. Addition of DNase improves the in vitro activity of antifungal drugs against *Candida albicans* biofilms. *Mycoses* 55(1), 80–85.

Mashima, I. & Nakazawa, F. 2015. The interaction between *Streptococcus spp.* and *Veillonella tobetsuensis* in the early stages of oral biofilm formation. *Journal of Bacteriology* 197(3), 2104–2111.

Maszewska, A., Zygmunt, M., Grzejdziak, I. & Rozalski, A. 2018. Use of polyvalent bacteriophages to combat biofilm of *Proteus mirabilis* causing catheter-associated urinary tract infections. *Journal of Applied Microbiology* 125(5), 1253–1265.

Matthysse, A. G., Deora, R., Mishra, M. & Torres, A. G. 2008. Polysaccharides cellulose, poly-beta-1,6-n-acetyl-D-glucosamine, and colanic acid are required for optimal binding of *Escherichia coli* O157:H7 strains to alfalfa sprouts and K-12 strains to plastic but not for binding to epithelial cells. *Applied Environmental Microbiology* 74(8), 2384–2390.

Melnyk, R. A. & Coates, J. D. 2015. The perchlorate reduction genomic island: Mechanisms and pathways of evolution by horizontal gene transfer. *BMC Genomics* 16, 862.

Ming, D., Wang, D., Cao, F., Xiang, H., Mu, D., Cao, J., Li, B., Zhong, L., Dong, X., Zhong, X., Wang, L. & Wang, T. 2017. Kaempferol inhibits the primary attachment phase of biofilm formation in *Staphylococcus aureus*. *Frontiers in Microbiology* 8, 2263.

Mirani, Z. A., Aziz, M. & Khan, S. I. 2015. Small colony variants have a major role in stability and persistence of *Staphylococcus aureus* biofilms. *Journal of Antibiotics (Tokyo)* 68(2), 98–105.

Mirani, Z. A. & Jamil, N. 2013. Role of extra-cellular fatty acids in vancomycin induced biofilm formation by vancomycin resistant *Staphylococcus aureus. Pakistan Journal of Pharmaceutical Science* 26(2), 383–389.

Misba, L., Kulshrestha, S. & Khan, A. 2016. Antibiofilm action of a toluidine blue O-silver nanoparticle conjugate on *Streptococcus mutans*: A mechanism of type I photodynamic therapy. *Biofouling* 32, 313–328.

Mishra, B., Golla, R. M., Lau, K., Lushnikova, T. & Wang, G. 2015. Anti-staphylococcal biofilm effects of human cathelicidin peptides. *ACS Medicinal Chemistry Letters* 7(1), 117–121.

Mohsenipour, Z. & Hassanshahian, M. 2015. The effects of *allium sativum* extracts on biofilm formation and activities of six pathogenic bacteria. *Jundishapur Journal of Microbiology* 8(8), e18971–e18971.

Moran, A., Gutierrez, S., Martinez-Blanco, H., Ferrero, M. A., Monteagudo-Mera, A. & Rodriguez-Aparicio, L. B. 2014. Non-toxic plant metabolites regulate *Staphylococcus* viability and biofilm formation: A natural therapeutic strategy useful in the treatment and prevention of skin infections. *Biofouling* 30(10), 1175–1182.

Morimoto, K., Ozawa, T., Awazu, K., Ito, N., Honda, N., Matsumoto, S. & Tsuruta, D. 2014. Photodynamic therapy using systemic administration of 5-aminolevulinic acid and a 410-nm wavelength light-emitting diode for methicillin-resistant *Staphylococcus aureus*-infected ulcers in mice. *PLoS One* 9(8), e105173.

Muras, A., Mayer, C., Otero-Casal, P., Exterkate, R. A. M., Brandt, B. W., Crielaard, W., Otero, A. & Krom, B. P. 2019. Short chain N-acylhomoserine lactone quorum-sensing molecules promote periodontal pathogens in in vitro oral biofilms. *Applied and Environmental Microbiology* 86, e01941–19. doi: 10.1128/AEM .01941-19.

Neethirajan, S., Clond, M. A. & Vogt, A. 2014. Medical biofilms—Nanotechnology approaches. *Journal of Biomedical Nanotechnology* 10(10), 2806–2827.

Nikaido, H. & Takatsuka, Y. 2009. Mechanisms of RND multidrug efflux pumps. *Biochim Biophys Acta* 1794(5), 769–781.

Nithya, B. & Jayachitra, A. 2016. Improved antibacterial and antibiofilm activity of plant mediated gold nanoparticles using *Garcinia cambogia. International Journal of Pure and Applied Bioscience* 4(2), 201–210.

Niu, C., Clemmer, K. M., Bonomo, R. A. & Rather, P. N. 2008. Isolation and characterization of an autoinducer synthase from *Acinetobacter baumannii. Journal of Bacteriology* 190(9), 3386–3392.

Nzakizwanayo, J., Scavone, P., Jamshidi, S., Hawthorne, J. A., Pelling, H., Dedi, C., Salvage, J. P., Hind, C. K., Guppy, F. M., Barnes, L. M., Patel, B. A., Rahman, K. M., Sutton, M. J. & Jones, B. V. 2017. Fluoxetine and thioridazine inhibit efflux and attenuate crystalline biofilm formation by *Proteus mirabilis. Science Reports* 7(1), 12222.

O'Neill, J. F., Hope, C. K. & Wilson, M. 2002. Oral bacteria in multi-species biofilms can be killed by red light in the presence of toluidine blue. *Lasers in Surgery and Medicine* 31(2), 86–90.

Ohene-Agyei, T., Mowla, R., Rahman, T. & Venter, H. 2014. Phytochemicals increase the antibacterial activity of antibiotics by acting on a drug efflux pump. *Microbiologyopen* 3(6), 885–96.

Okuda, K. I., Zendo, T., Sugimoto, S., Iwase, T., Tajima, A., Yamada, S., Sonomoto, K. & Mizunoe, Y. 2013. Effects of bacteriocins on methicillin-resistant *Staphylococcus aureus* biofilm. *Antimicrobial Agents and Chemotherapy* 57(11), 5572–5579.

Olsen, I., Tribble, G. D., Fiehn, N. E. & Wang, B. Y. 2013. Bacterial sex in dental plaque. *Journal of Oral Microbiology* 5.

Overhage, J., Campisano, A., Bains, M., Torfs, E. C. W., Rehm, B. H. A. & Hancock, R. E. W. 2008. Human host defense peptide LL-37 prevents bacterial biofilm formation. *Infection and Immunity* 76(9), 4176–4182.

Owlia, P., Nosrati, R., Alaghehbandan, R. & Lari, A. R. 2014. Antimicrobial susceptibility differences among mucoid and non-mucoid *Pseudomonas aeruginosa* isolates. *GMS Hygiene and Infection Control* 9(2).

Pao, S. S., Paulsen, I. T. & Saier, M. H., Jr. 1998. Major facilitator superfamily. *Microbiology and Molecular Biology Reviews* 62(1), 1–34.

Pathak, A. K., Sharma, S. & Shrivastva, P. 2012. Multi-species biofilm of *Candida albicans* and non-*Candida albicans* Candida species on acrylic substrate. *Journal of Applied and Oral Science* 20(1), 70–75.

Perera, M., Wijayarathna, D., Wijesundera, S., Chinthaka, M., Seneviratne, G. & Jayasena, S. 2019. Biofilm mediated synergistic degradation of hexadecane by a naturally formed community comprising *Aspergillus flavus* complex and *Bacillus cereus* group. *BMC Microbiology* 19(1), 84.

Personnic, N., Striednig, B., Lezan, E., Manske, C., Welin, A., Schmidt, A. & Hilbi, H. 2019. Quorum sensing modulates the formation of virulent *Legionella persisters* within infected cells. *Nature Communications* 10(1), 5216.

Petrovska, B. B. 2012. Historical review of medicinal plants' usage. *Pharmacognosy Reviews* 6(11), 1–5.

Platt, T. G. & Fuqua, C. 2010. What's in a name? The semantics of quorum sensing. *Trends in Microbiology* 18(9), 383–387.

Pulido, D., Prats-Ejarque, G., Villalba, C., Albacar, M., González-López, J. J., Torrent, M., Moussaoui, M. & Boix, E. 2016. A novel RNase 3/ECP peptide for *Pseudomonas aeruginosa* biofilm eradication that combines antimicrobial, lipopolysaccharide binding, and cell-agglutinating activities. *Antimicrobial Agents and Chemotherapy* 60(10), 6313–6325.

Rabin, N., Zheng, Y., Opoku-Temeng, C., Du, Y., Bonsu, E. & Sintim, H. O. 2015. Agents that inhibit bacterial biofilm formation. *Future Medicinal Chemistry* 7(5), 647–671.

Rahman, M., Kim, S., Kim, S. M., Seol, S. Y. & Kim, J. 2011. Characterization of induced *Staphylococcus aureus* bacteriophage SAP-26 and its antibiofilm activity with rifampicin. *Biofouling* 27(10), 1087–93.

Rajesh, P. S. & Rai, V. R. 2016. Inhibition of QS-regulated virulence factors in *Pseudomonas aeruginosa* PAO1 and *Pectobacterium carotovorum* by AHL-lactonase of endophytic bacterium *Bacillus cereus* VT96. *Bio-Catalysis and Agricultural Biotechnology* 7, 154–163.

Rajkumari, J., Busi, S., Vasu, A. C. & Reddy, P. 2017. Facile green synthesis of baicalein fabricated gold nanoparticles and their antibiofilm activity against *Pseudomonas aeruginosa* PAO1. *Microbial Pathogenesis* 107, 261–269.

Ramalingam, V., Rajaram, R., PremKumar, C., Santhanam, P., Dhinesh, P., Vinothkumar, S. & Kaleshkumar, K. 2014. Biosynthesis of silver nanoparticles from deep sea bacterium *Pseudomonas aeruginosa* JQ989348 for antimicrobial, antibiofilm, and cytotoxic activity. *Journal of Basic Microbiology* 54(9), 928–936.

Ramasubbu, N., Thomas, L. M., Ragunath, C. & Kaplan, J. B. 2005. Structural analysis of dispersin B, a biofilm-releasing glycoside hydrolase from the periodontopathogen *Actinobacillus actinomycetemcomitans*. *Journal of Molecular Biology* 349(3), 475–486.

Reen, F. J., Gutiérrez-Barranquero, J. A., Parages, M. L. & O´Gara, F. 2018. Coumarin: A novel player in microbial quorum sensing and biofilm formation inhibition. *Applied Microbiology and Biotechnology* 102(5), 2063–2073.

Ribeiro, S. M., de la Fuente-Núñez, C., Baquir, B., Faria-Junior, C., Franco, O. L. & Hancock, R. E. W. 2015. Antibiofilm peptides increase the susceptibility of carbapenemase-producing *Klebsiella pneumoniae* clinical isolates to β-lactam antibiotics. *Antimicrobial Agents and Chemotherapy* 59(7), 3906–3912.
Rodriguez-Rojas, A., Oliver, A. & Blazquez, J. 2012. Intrinsic and environmental mutagenesis drive diversification and persistence of *Pseudomonas aeruginosa* in chronic lung infections. *Journal of Infectious Disseases* 205(1), 121–127.
Römling, U. & Balsalobre, C. 2012. Biofilm infections, their resilience to therapy and innovative treatment strategies. *Journal of Internal Medicine* 272(6), 541–561.
Römling, U., Galperin, M. Y. & Gomelsky, M. 2013. Cyclic di-GMP: The first 25 years of a universal bacterial second messenger. *Microbiology and Molecular Biology Reviews* 77(1), 1–52.
Rosa, L. P. & da Silva, F. C. 2014. Antimicrobial photodynamic therapy: A new therapeutic option to combat infections. *Journal of Medical Microbiology & Diagnosis* 3(4), 1.
Rutherford, S. T. & Bassler, B. L. 2012. Bacterial quorum sensing: Its role in virulence and possibilities for its control. *Cold Spring Harbor Perspectives in Medicine* 2(11), a012427.
Sá, N. C., Cavalcante, T. T. A., Araújo, A. X., Santos, H. S. D., Albuquerque, M. R. J. R., Bandeira, P. N., Cunha, R. M. S. D., Cavada, B. S. & Teixeira, E. H. 2012. Antimicrobial and antibiofilm action of casbane diterpene from *Croton nepetaefolius* against oral bacteria. *Archives of Oral Biology* 57(5), 550–555.
Sadekuzzaman, M., Yang, S., Mizan, M. & Ha, S.-D. 2015. Current and recent advanced strategies for combating biofilm. *Comprehensive Reviews in Food Science and Food Safety* 14, 491–509.
Saleemi, M. A., Palanisamy, N. K. & Wong, E. H. 2018. Alternative approaches to combat medicinally important biofilm-forming pathogens. In: *Antimicrobials, Antibiotic Resistance, Antibiofilm Strategies, and Activity Methods* (edited by Kırmusaoğlu, S.). IntechOpen. https://www.intechopen.com/books/antimicrobials-antibiotic-resistance-antibiofilm-strategies-and-activity-methods/alternative-approaches-to-combat-medicinally-important-biofilm-forming-pathogens
Salman, J. A., Mohammed, F., Raghad, A. A. & Inas, A. 2015. Antibiofilm effect of iron oxide nanoparticles synthesized by *Lactobacillus fermentum* on catheter. *World Journal of Pharmaceutical Research* 4(8), 317–328.
Sambanthamoorthy, K., Luo, C., Pattabiraman, N., Feng, X., Koestler, B., Waters, C. M. & Palys, T. J. 2014. Identification of small molecules inhibiting diguanylate cyclases to control bacterial biofilm development. *Biofouling* 30(1), 17–28.
Sambanthamoorthy, K., Sloup, R. E., Parashar, V., Smith, J. M., Kim, E. E., Semmelhack, M. F., Neiditch, M. B. & Waters, C. M. 2012. Identification of small molecules that antagonize diguanylate cyclase enzymes to inhibit biofilm formation. *Antimicrobial Agents and Chemotherapy* 56(10), 5202–5211.
Santos-Lopez, A., Marshall, C. W., Scribner, M. R., Snyder, D. J. & Cooper, V. S. 2019. Evolutionary pathways to antibiotic resistance are dependent upon environmental structure and bacterial lifestyle. *Elife* 8, e47612.
Schwering, M., Song, J., Louie, M., Turner, R. J. & Ceri, H. 2013. Multi-species biofilms defined from drinking water microorganisms provide increased protection against chlorine disinfection. *Biofouling* 29(8), 917–928.
Sharma, K. & Pagedar Singh, A. 2018. Antibiofilm effect of DNase against single and mixed species biofilm. *Foods (Basel, Switzerland)* 7(3), 42.
Shen, X. F., Ren, L. B., Teng, Y., Zheng, S., Yang, X. L., Guo, X. J., Wang, X. Y., Sha, K. H., Li, N., Xu, G. Y., Tian, H. W., Wang, X. Y., Liu, X. K., Li, J. & Huang, N. 2014. Luteolin decreases the attachment, invasion and cytotoxicity of UPEC in bladder epithelial cells and inhibits UPEC biofilm formation. *Food and Chemical Toxicology* 72, 204–211.

Shi, S.-F., Jia, J.-F., Guo, X.-K., Zhao, Y.-P., Chen, D.-S., Guo, Y.-Y. & Zhang, X.-L. 2016. Reduced Staphylococcus aureus biofilm formation in the presence of chitosan-coated iron oxide nanoparticles. *International Journal of Nanomedicine* 11, 6499–6506.

Shimizu, Y., Ogata, H. & Goto, S. 2017. Type III polyketide synthases: Functional classification and phylogenomics. *Chembiochem* 18(1), 50–65.

Shrestha, A., Hamblin, M. R. & Kishen, A. 2012. Characterization of a conjugate between rose bengal and chitosan for targeted antibiofilm and tissue stabilization effects as a potential treatment of infected dentin. *Antimicrobial Agents and Chemotherapy* 56(9), 4876–4884.

Silva, G., Primon-Barros, M., Macedo, A. J. & Gnoatto, S. C. B. 2019. Triterpene derivatives as relevant scaffold for new antibiofilm drugs. *Biomolecules* 9(2), 58.

Skippington, E. & Ragan, M. A. 2011. Lateral genetic transfer and the construction of genetic exchange communities. *FEMS Microbiology Reviews* 35(5), 707–735.

Skogman, M., Kujala, J., Busygin, I., Leinob, R., Vuorela, P. & Fallarero, A. 2012. Evaluation of antibacterial and antibiofilm activities of cinchona alkaloid derivatives against *Staphylococcus aureus. Natural Product Communications* 7, 1173–1176.

Springob, K. & Kutchan, T. M. 2009. Introduction to the Different Classes of Natural Products. In: *Plant-Derived Natural Products: Synthesis, Function, and Application* (edited by Osbourn, A. E. & Lanzotti, V.). Springer US, New York, pp. 3–50.

Srivastava, D. & Waters, C. M. 2012. A tangled web: Regulatory connections between quorum sensing and cyclic Di-GMP. *Journal of Bacteriology* 194(17), 4485–4493.

Stalder, T. & Top, E. 2016. Plasmid transfer in biofilms: A perspective on limitations and opportunities. *NPJ Biofilms Microbiomes* 2, 16022.

Subhadra, B., Kim, J., Kim, D. H., Woo, K., Oh, M. H. & Choi, C. H. 2018. Local repressor AcrR regulates AcrAB efflux pump required for biofilm formation and virulence in *Acinetobacter nosocomialis. Frontiers in Cellular and Infection Microbiology* 8, 270.

Sugano, M., Morisaki, H., Negishi, Y., Endo-Takahashi, Y., Kuwata, H., Miyazaki, T. & Yamamoto, M. 2016. Potential effect of cationic liposomes on interactions with oral bacterial cells and biofilms. *Journal of Liposome Research* 26(2), 156–162.

Sutton, J. M. & Pritts, T. A. 2014. Human beta-defensin 3: A novel inhibitor of *Staphylococcus*-produced biofilm production. Commentary on "Human β-defensin 3 inhibits antibiotic-resistant *Staphylococcus* biofilm formation." *The Journal of Surgical Research* 186(1), 99–100.

Swartjes, J. J., Das, T., Sharifi, S., Subbiahdoss, G., Sharma, P. K., Krom, B. P., Busscher, H. J. & van der Mei, H. C. 2013. A functional DNase I coating to prevent adhesion of bacteria and the formation of biofilm. *Advanced Functional Materials* 23(22), 2843–2849.

Taglietti, A., Arciola, C. R., D'Agostino, A., Dacarro, G., Montanaro, L., Campoccia, D., Cucca, L., Vercellino, M., Poggi, A., Pallavicini, P. & Visai, L. 2014. Antibiofilm activity of a monolayer of silver nanoparticles anchored to an amino-silanized glass surface. *Biomaterials* 35(6), 1779–1788.

Taraszkiewicz, A., Fila, G., Grinholc, M. & Nakonieczna, J. 2013. Innovative strategies to overcome biofilm resistance. *BioMed Research International* 2013, 150653–150653.

Teerapo, K., Roytrakul, S., Sistayanarain, A. & Kunthalert, D. 2019. A scorpion venom peptide derivative BmKn22 with potent antibiofilm activity against *Pseudomonas aeruginosa. PLoS One* 14(6), e0218479.

Tetz, G. V., Artemenko, N. K. & Tetz, V. V. 2009. Effect of DNase and antibiotics on biofilm characteristics. *Antimicrobial Agents and Chemotherapy* 53(3), 1204–1209.

Valentini, M. & Filloux, A. 2016. Biofilms and cyclic di-GMP (c-di-GMP) signaling lessons from *Pseudomonas aeruginosa* and other bacteria. *Journal of Biological Chemistry* 291(24), 12547–12555.

Varela-Ramirez, A., Abendroth, J., Mejia, A. A., Phan, I. Q., Lorimer, D. D., Edwards, T. E. & Aguilera, R. J. 2017. Structure of acid deoxyribonuclease. *Nucleic Acids Research* 45(10), 6217–6227.

Vasavi, H. S., Arun, A. B. & Rekha, P. D. 2014. Anti-quorum-sensing activity of *Psidium guajava* L. flavonoids against *Chromobacterium violaceum* and *Pseudomonas aeruginosa* PAO1. *Microbiology and Immunology* 58(5), 286–293.

Vassena, C., Fenu, S., Giuliani, F., Fantetti, L., Roncucci, G., Simonutti, G., Romanò, C. L., De Francesco, R. & Drago, L. 2014. Photodynamic antibacterial and antibiofilm activity of RLP068/Cl against *Staphylococcus aureus* and *Pseudomonas aeruginosa* forming biofilms on prosthetic material. *International Journal of Antimicrobial Agents* 44(1), 47–55.

Venkadesaperumal, G., Meena, C., Murali, A. & Shetty Halady, P. K. 2015. Petunidin as a competitive inhibitor of acylated homoserine lactones in *Klebsiella pneumoniae*. *RSC Advances* 6, 2592–2601.

Vidigal, P. G., Musken, M., Becker, K. A., Haussler, S., Wingender, J., Steinmann, E., Kehrmann, J., Gulbins, E., Buer, J., Rath, P. M., Steinmann, J. 2014. Effects of green tea compound epigallocatechin-3-gallate against *Stenotrophomonas maltophilia* infection and biofilm. *PLoS One* 9, e92876.

Vinoj, G., Pati, R., Sonawane, A. & Vaseeharan, B. 2015. In vitro cytotoxic effects of gold nanoparticles coated with functional acyl homoserine lactone lactonase protein from *Bacillus licheniformis* and their antibiofilm activity against *Proteus* species. *Antimicrobial Agents and Chemotherapy* 59(2), 763–771.

Walencka, E., Sadowska, B., Rozalska, S., Hryniewicz, W. & Rozalska, B. 2005. Lysostaphin as a potential therapeutic agent for staphylococcal biofilm eradication. *Polish Journal of Microbiology* 54(3), 191–200.

Walters, M. C., 3rd, Roe, F., Bugnicourt, A., Franklin, M. J. & Stewart, P. S. 2003. Contributions of antibiotic penetration, oxygen limitation, and low metabolic activity to tolerance of *Pseudomonas aeruginosa* biofilms to ciprofloxacin and tobramycin. *Antimicrob Agents Chemother* 47(1), 317–323.

Wang, H., Xing, X., Wang, J., Pang, B., Liu, M., Larios-Valencia, J., Liu, T., Liu, G., Xie, S., Hao, G., Liu, Z., Kan, B. & Zhu, J. 2018. Hypermutation-induced in vivo oxidative stress resistance enhances *Vibrio cholerae* host adaptation. *PLoS Pathogenes* 14(10), e1007413.

Webber, M. A. & Piddock, L. J. 2003. The importance of efflux pumps in bacterial antibiotic resistance. *Journal of Antimicrobial Chemotherapy* 51(1), 9–11.

Wen, Z. T., Yates, D., Ahn, S. J. & Burne, R. A. 2010. Biofilm formation and virulence expression by *Streptococcus mutans* are altered when grown in dual-species model. *BMC Microbiology* 10, 111.

Wood, S., Metcalf, D., Devine, D. & Robinson, C. 2006. Erythrosine is a potential photosensitizer for the photodynamic therapy of oral plaque biofilms. *Journal of Antimicrobial Chemotherapy* 57(4), 680–684.

Wood, S., Nattress, B., Kirkham, J., Shore, R., Brookes, S., Griffiths, J. & Robinson, C. 1999. An in vitro study of the use of photodynamic therapy for the treatment of natural oral plaque biofilms formed in vivo. *Journal of Photochemistry and Photobiology B: Biology* 50(1), 1–7.

Wood, T. K., Knabel, S. J. & Kwan, B. W. 2013. Bacterial persister cell formation and dormancy. *Applied and Environmental Microbiology* 79(23), 7116–7121.

Wu, H., Moser, C., Wang, H.-Z., Høiby, N. & Song, Z.-J. 2015. Strategies for combating bacterial biofilm infections. *International Journal of Oral Science* 7(1), 1–7.

Wu, J. A., Kusuma, C., Mond, J. J. & Kokai-Kun, J. F. 2003. Lysostaphin disrupts *Staphylococcus aureus* and *Staphylococcus epidermidis* biofilms on artificial surfaces. *Antimicrobial Agents and Chemotherapy* 47(11), 3407–3414.

Yamasaki, S., Wang, L. Y., Hirata, T., Hayashi-Nishino, M. & Nishino, K. 2015. Multidrug efflux pumps contribute to *Escherichia coli* biofilm maintenance. *International Journal of Antimicrobial Agents* 45(4), 439–41.

Yang, L., Liu, Y., Wu, H., Song, Z., Høiby, N., Molin, S. & Givskov, M. 2012. Combating biofilms. *FEMS Immunology & Medical Microbiology* 65(2), 146–157.

Yang, Y. & Huang, Y. 2019. The CRISPR/Cas gene-editing system—an immature but useful toolkit for experimental and clinical medicine. *Animal Models and Experimental Medicine* 2(1), 5–8.

Yang, Y., Luo, M., Zhou, H., Li, C., Luk, A., Zhao, G., Fung, K. & Ip, M. 2019. Role of two-component system response regulator bceR in the antimicrobial resistance, virulence, biofilm formation, and stress response of group B *Streptococcus. Frontiers in Microbiology* 10, 10.

Yasir, M., Willcox, M. D. P. & Dutta, D. 2018. Action of antimicrobial peptides against bacterial biofilms. *Materials (Basel, Switzerland)* 11(12), 2468.

Zechner, E. L., Lang, S. & Schildbach, J. F. 2012. Assembly and mechanisms of bacterial type IV secretion machines. *Philosophical Transactions of the Royal Society B: Biological Sciences*367(1592), 1073–87.

Zhang, J. & Poh, C. L. 2018. Regulating exopolysaccharide gene wcaF allows control of *Escherichia coli* biofilm formation. *Scientific Reports* 8(1), 13127.

Zuberi, A., Misba, L. & Khan, A. U. (2017). CRISPR interference (CRISPRi) inhibition of luxS gene expression in *E. coli*: An approach to inhibit biofilm. *Frontiers in Cellular and Infection Microbiology* 7, 214.

Zupancic, J., Raghupathi, P. K., Houf, K., Burmolle, M., Sorensen, S. J. & Gunde-Cimerman, N. 2018. Synergistic interactions in microbial biofilms facilitate the establishment of opportunistic pathogenic fungi in household dishwashers. *Frontiers in Microbiology* 9, 21.

17 Biofilms-Associated Infections

Continuous Challenges in Human and Animal Health

Nor Fadhilah Kamaruzzaman, Tan Li Peng, and Ruhil Hayati Hamdan
Universiti Malaysia Kelantan

CONTENTS

17.1 DEFINITION, BIOFILMS CHARACTERISTICS, AND CHALLENGES IN ANTIMICROBIAL THERAPY

Biofilms can be defined as a structured consortium of microbial cells surrounded by the self-produced matrix (Figure 17.1) (Høiby 2017). Biofilms is known to be produced by many of bacteria including the important pathogen that cause life-threatening infections in humans and animals such as *Staphylococcus aureus*, *Pseudomonas aeruginosa*, *Klebsiella pneumoniae*, *Enterobacter* spp., *Salmonella* spp., *Escherichia coli*, etc. (Tasneem et al. 2018). These bacteria are known to cause a serious problem in human and animal health (Jamal et al. 2018; Abdullahi et al. 2016).

Structurally, the biofilms is built of individual (planktonic) bacterial cells attached with the self-released lipopolysaccharides, proteins, lipids, glycolipids, and nucleic acids. These components are recognized as extra-polymeric substances (EPSs). The EPS is responsible to promote adhesion and aggregation of bacteria to the surfaces and provides stability to the biofilms structure (Kamaruzzaman et al. 2018). The lipopolysaccharide produced by the bacteria is different from the bacteria species. For example, *P. aeruginosa* produces Pel (a cationic exopolysaccharide composed of 1–4 linked galactosamine and glucosamine sugars) and Psl (a pentasaccharide composed of D-glucose, D-mannose, and L-rhamnose) (Billings et al. 2013; Jennings et al. 2015) while *S. aureus* and *Staphylococcus epidermidis* produce poly-ß(1,6)-N-acetyl-D-glucosamine (PNAG) (Izano et al. 2008). The nucleic acid is known as extracellular DNA (eDNA) that interacts with extracellular calcium (Ca^{2+}) within the biofilms structure to induce bacterial aggregation via cationic bridging. The positive charge of Ca^{2+} repulses the negative charge of the biofilms's component, thus assisting the adherence of the biofilms to the material and tissue surface. The negatively charged eDNA chelates the action of cationic antimicrobial peptides of the immune system, thus acting as the defense mechanism to the structure (Okshevsky, Regina, and Meyer 2015). Due to the fragility of the structure, the characterization biofilms is often performed *in vitro*. The thickness of the biofilms grown in vitro can vary between the bacterial species, for example, *K. pneumoniae* 231 μm, *P. aeruginosa* 209 μm, and *S. aureus* 8 μm (Singla, Harjai, and Chhibber 2014; Werner et al. 2004; Kamaruzzaman et al. 2017). Figure 17.2 shows the structure of *S. aureus* biofilms grown *in vitro* and visualized by confocal microscope with a thickness of approximately 8.0 μm.

Formation of biofilms can be considered as the survival mechanism for the bacteria. However, in this form, they are inherently resistant to antibiotic action, thus provides additional challenges for the treatment of related infection. Bacteria in the biofilms form can be 10–1,000 times more resistant to antibiotics compared to their

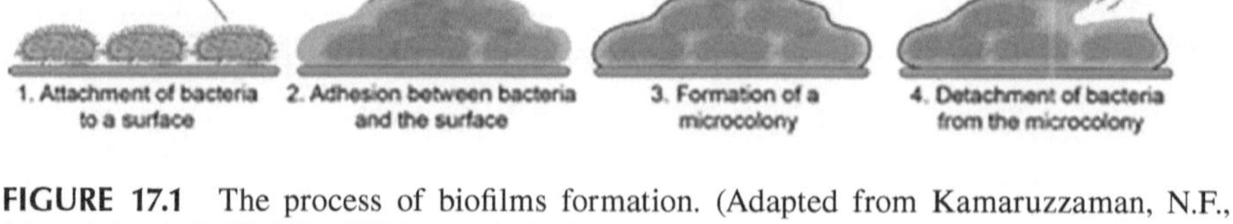

FIGURE 17.1 The process of biofilms formation. (Adapted from Kamaruzzaman, N.F., *Materials*, 11, 1–27, 2018. With permission.)

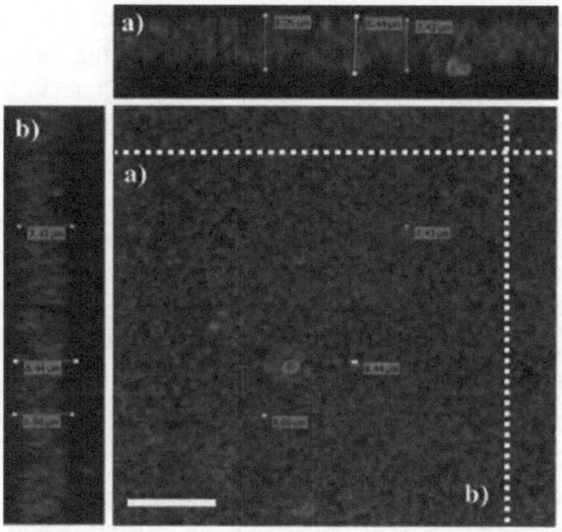

FIGURE 17.2 The three-dimensional structure of *S. aureus* biofilmss visualized with a confocal microscope. *S. aureus* biofilmss were cultured in tryptic soy broth for 48 h, fixed with 4% paraformaldehyde and treated with wheat germ agglutinin to stain n-acetylglucoseamine component of polysaccharide and DAPI to stain the bacteria nuclear material, followed by confocal microscopy z-stack projection that moved through 111 slices across the cell. (a) Horizontal cross-section of biofilms and (b) vertical cross-section of biofilms. White scale bar is 7.5 μm. The approximate thickness of the biofilms was 7.9±0.5 μm. (Adapted from Kamaruzzaman, N.F., *Front. Microbiol.*, 8, 1–10, 2017. With permission.)

susceptibilities as individual (planktonic cells) (Mah and Toole 2001; La et al. 1987; Nickel et al. 1985). This could be due to the thick biofilms matrix that reduces permeation of antibiotics across the biofilms structure (Mah and Toole 2001; Nickel et al. 1985; Singh et al. 2016; Nguyen et al. 2011). Additionally, the eDNA and polysaccharide components of the biofilms can interact with the antimicrobials and prevent further penetration of the antibiotics across the structure (Billings et al. 2013; Johnson et al. 2013). Additionally, the physiological condition which is the lack of oxygen reduces the outer membrane potential of the bacteria within the biofilms and reduces uptake of antibiotics into the cells (Walters III et al. 2003; Borriello et al. 2004; P. S. Stewart et al. 2000). The physiology of the bacteria itself within the biofilms structure is another challenge as it is reported that the bacteria in the biofilms exist as small colony variants and thus present different phenotypes compared to the wild types (Waters et al. 2016). This variant was reported to have a better tolerance toward antimicrobials. Therefore, they have greater tolerance toward antimicrobials. Thus, all these characteristics of biofilms suggest the reason why persistent infections occur in the mentioned bacteria.

17.2 BIOFILMS-RELATED INFECTIONS IN HUMAN

Aggregation of one or more species of microbes particularly bacteria has recently gained more concern in medical history as it is rapidly becoming clear that the formation of a biofilms is the root cause of development of many persistent infections.

Biofilms can be found as the non-attached form as floating mats on the liquid surface or in a submerged state (Bjarnsholt et al. 2013). Biofilms can be formed on both biotic and abiotic surfaces. The growth and activity of bacteria in attached form are enhanced when they are attached to a surface (Heukelekian and Heller 1940) and thus responsible for causing persistent infections of the patients (Costerton, Stewart, and Greenberg 1999).

A few well-known examples of biofilms-associated infections can be collectively classified into device-related biofilms diseases such as catheter-associated urinary tract infection and prosthetic joint infection; non-device related chronic biofilms diseases viz. cystic fibrosis pneumonia, periodontitis, and chronic dermal wounds (P. S. Stewart 2014). Besides, biofilms-related device malfunction has arisen as a new problem in which chemical degradation and physical damage occur as the consequences of the growth of biofilms on the surface. This condition required device removal and introduce further complications to the patient (del Pozo and Patel 2007). All of the mentioned biofilms-related diseases contribute to patient morbidity and increased mortality and represents a considerable economic burden to both individual and country.

It is thus essential to extend our knowledge on the different biofilms-related diseases, the mechanisms involved in biofilms antimicrobial resistance in order to develop new and effective diagnostic, treatment and prevention strategies for this biofilms disease war. The novel strategies for the treatment of biofilms-related infections have been covered in our previous article (Kamaruzzaman et al. 2018). In this chapter, we will focus on the challenges, solutions, and future implications on the biofilms diseases mainly for device-related and non-device-related chronic biofilms diseases.

17.2.1 Device-Related Biofilms Disease

There are about 1 million cases with an estimated 60% of hospital-associated infections are due to biofilms that have formed on indwelling devices (Darouiche 2004). It was estimated that about 40% of the infections are associated with ventricular-assisted devices, 10% for ventricular shunts, 4% for pacemakers and defibrillator, 4% for mechanical heart valves, 2% for breast implants, and another 2% for joint prostheses (Figure 17.3) (Darouiche 2004). The composition of biofilms depends on the devices, and their duration of action may be composed of only a single or of different types of microbial species.

The process of biofilms formation is illustrated in Figure 17.1. The process is generally universal for many of the known bacteria forming bacteria, but the location and the extent of biofilms formation depend on the duration of implantation in the patient. It was reported that bacterial colonization of catheter can occur within the first 24h of implantation, with the formation of the biofilms on the external surfaces continues within the next 10days and in 30days, biofilms could extend until inside the catheter lumen (Jamal et al. 2017). At the time of surgical implantation of medical devices, tissue damage may occur resulted in accumulation of platelets and fibrin at the location of suture and on the devices. Microbial cells have better ability to colonize these locations and eventually produce biofilms on these sites (Donlan and Costerton 2002).

The nature of an indwelling foreign body is a double-edged sword, since, apart from their outstanding benefits, infectious complications are regularly observed

FIGURE 17.3 Medical devices that are associated with the device-related biofilms diseases.

(Fux et al. 2003). Soon after a foreign body enters into a host, the host–pathogen dynamic started to be profoundly influenced (Gristina et al. 1988). The mechanisms of this host–pathogen dynamic begin when the mere presence of the foreign body (i) enables an invasion of bacterial inoculum that can lead to infection, (ii) permits the non-pathogenic organism to opportunistically colonize the foreign body and infect, (iii) allows a pathogen to persist undetected at the site of the infection, (iv) induces a chronic local inflammatory response, and (v) limits the induction and effectiveness of a humoral response in the presence of chronic, persistent infection (Nickel et al. 1985).

Stable communities of bacteria in biofilms often walled off the human immune system, resulting in chronic low-grade inflammation. Due to the physiological heterogeneity of the bacteria within the biofilms, they are also extremely difficult to eradicate using existing antimicrobials (Kamaruzzaman et al. 2018). One of the major challenges in reducing the device-related biofilms diseases is the design and construction of implanted medical devices that able to control the deposition of these host molecules that form host-conditioned surfaces to limit opportunities for adherence to free-floating (planktonic) bacteria via their matrix attachment adhesions.

17.2.1.1 Strategies to Reduce Bacterial Colonization on Indwelling Devices

To overcome bacteria colonization and biofilms formation on the indwelling devices, researchers tried to understand the underneath factors that promote attachment of bacteria on the surface of the device. It was reported that physicochemical properties of the devices' surface play a role to reduce attachment of microbes, for example, hydrophobicity can increases attachment of the bacteria to surface and low hydrophobicity increases repulsion force between the bacteria and the surface (Tribedi and Sil 2014). Many bacteria are more likely to attach to the hydrophobic and non-polar surfaces like Teflon silicon, stainless steel, and other plastics; however, some of the bacteria including the human-associated bacterium *S. epidermidis* prefer polar and hydrophilic substrates (Ista, Baca, and Lbpez 1996).

By understanding the tendencies of the microbes in forming the biofilms on different hydrophobicity and topographical surface, various studies have been conducted to modify the medical device surface as a method to reduce and to avoid bacteria adhesion and biofilms formation. The research trends are now directed toward addressing the development of preventive strategies, rather than treatment approaches, as it is well known that bacteria in the biofilms form are inherently difficult to be killed (Catt and Cappitelli 2019). Currently, there are three major methods for improving the anti-biofilms properties of the medical devices (Table 17.1).

17.2.2 Non-device-Related Chronic Biofilms Disease

Other than causing indirect complications to the device implanted in humans, biofilms-forming bacteria have been revealed to be associated with a wide array of chronic infections and complicate the majority of bacterial infections in humans.

TABLE 17.1
Methods of for Improving the Antibiofilms Properties of Medical Devices

Methods	Description	References
1) Incorporation of novel materials	Novel materials such as ceramics and composites with antimicrobial infused can be incorporated to reduce attachment of biofilms to the surface of devices	Brentel et al. (2011); Du et al. (2012); Zhang et al. (2013); Wang, Shen, and Haapasalo (2013)
2) Physical surface modification	Smoother surface on the device can be introduced, for example, by mechanical modification of the surface as non or nano-porous to reduce attachment of the bacteria	Feng et al. (2015); Desrousseaux et al. (2013); Lagree et al. (2018)
3) Chemical surface modification	The device surfaces can be coated with additional material, for example, surfactants and antimicrobials to reduce attachment and survivability of the bacteria on the surface	Prijck, Smet, and Coenye (2010); Merchan et al. (2010); Lopez et al. (2011); Armentano et al. (2014)

These include chronic rhinosinusitis (CRS) (Fastenberg et al. 2016), airway infections in cystic fibrosis (Høiby et al. 2017), chronic obstructive pulmonary disease (COPD) (Hassett, Borchers, and Panos 2014), endocarditis (Jung et al. 2012), periodontitis (Lasserre, Brecx, and Toma 2018), conjunctivitis (Bispo, Haas, and Gilmore 2015), otitis media (Kaya et al. 2013), decubitus and diabetic ulcers (Kunimitsu et al. 2019), urinary tract infections and prostatitis (Soto 2014; Delcaru et al. 2016). The bacteria not only able to form biofilms on the foreign devices, they are also capable of forming bacteria on tissue inside the bodies. Within the period of development, the bacterial physiology, including the generation of genetic and phenotypic variability will be influenced, as well as the ability of resisting antibiotics after being regularly exposed to sub-minimum inhibitory concentrations molecules. Collectively, these effects would accelerate the emergence and spread of antibiotic-resistant bacteria and thus the biofilms (Andersson and Hughes 2014). The following section describes in detail regarding two important biofilms-related infections, periodontal disease, chronic wound infections, and sinusitis.

17.2.2.1 Periodontal Disease

One example of biofilms-associated infection is periodontal diseases, the most common infectious diseases in oral cavity associated with the establishment of pathogenic biofilms that trigger an immune and inflammatory host response, leading to the destruction of supporting periodontal tissues and eventual tooth loss (Eke et al. 2012). These diseases have also been indicated as potential risk factors for several systemic diseases (Cullinan and Seymour 2013; Li et al. 2000). Oral cavity is an optimum environment for the commensal bacteria, and in conditions where the oral health is not well maintained or in immunosuppressed patient, the pathogenic species may colonize and initiate infection (Zuanazzi et al. 2010). High levels of medically important pathogens in these periodontitis-associated microbiotas may pose a risk for systemic dissemination and development of infections at distant body sites due to the anatomical proximity of the periodontal biofilms to the gingival blood stream.

Periodontitis is characterized by irreversible and progressive degradation of periodontal tissues. The warm moist dental pocket between teeth and gingival tissues provides with an ideal hatchery for microbial growth and proliferation. Therapeutic regimens such as restorations, non-surgical or surgical periodontal therapies, root canal therapy, and dental implants are well accepted; however, secondary biofilms infections still cannot be completely eliminated. Dental materials used and the location of biofilms play the major role on the consequences of these secondary infections (Allaker 2010). There are few major strategies summarized by Cloutier, Mantovani, and Rosei (2015) on the development of devices with antibiofilms activities, for example, antimicrobial agent release, contact killing, and multifunction (Cloutier, Mantovani, and Rosei 2015). The advantages and disadvantages on each of the methods are also summarized by Jiao et al. (2019) as described in Table 17.2.

17.2.2.2 Chronic Wound Infections

Another most reported non-device-related chronic biofilms disease is the chronic wounds in which biofilms appeared on almost 60% of the specimens in comparison with only 6% of biopsies from acute wounds (James et al. 2007). Biofilms were

TABLE 17.2
Advantages and Disadvantages of the Antimicrobials Approaches on Dental Materials

Approaches	Advantages	Disadvantages
Antimicrobial agent release	• High local doses of antimicrobial agents at a specific site	• Limited antimicrobial agents reservoirs (lack of long-term effect)
Contact killing	• Broad-spectrum and strong contact-killing activity • Low risks of antimicrobial resistance development	• Exhibit only bacteriostatic effects • No effects on planktonic bacteria • Problem on "surface biofouling" • Potential cytotoxic
Multifunctional	• Able to activate microbicidal activity in response to the microenvironment • Other non-antimicrobial benefits (remineralization)	• Selection on the combination for synergistic antimicrobial and beneficial properties

Source: Adapted from Jiao, Y., *Int. J. Oral Sci.*, 11, 1–11. With permission.

suspected as one contributing factor that delayed healing or contribute to the recurring infections. The mechanism could be due to the constant stimulation of the inflammatory response that is released by the host cells as the signal for removal of the biofilms. This resulted in damage of normal and healing tissues, proteins and immune cells on the surrounding areas, and contributing to the impairment of healing process (Lawrence et al. 2007). Besides, the chronic inflammatory response does not guarantee for biofilms removal, and it has been hypothesized that, contradictory, this response favors the formation of biofilms. Lawrence et al. (2007) suggested the inflammatory response may induce exudate release from the biofilms that consequently serve as the source of nutrition and helps perpetuate the biofilms (Lawrence et al. 2007).

Similar to biofilms in periodontitis, for chronic wound biofilms, the microenvironment of the wound provides an ideal milieu for the microbes to sustain (Wolcott, Rhoads, and Dowd 2008). This has been partly the challenge of chronic wound management, with increased resilience and complexity to standard approaches of care. With only relatively recent recognition of the existence of biofilms in wounds and their role in delayed healing and chronicity, the development of effective therapeutic strategies to date has been very limited (Metcalf, Parsons, and Bowler 2016). However, wound care researchers can still benefit from the knowledge gained in other industries and in related healthcare areas such as dentistry and indwelling medical devices, because the treatment strategy options are well developed and broadly similar. Although the intention to prevent, remove, and kill bacterial biofilms is the same, challenge in selecting appropriate wound treatments that can be acutely sensitive and fragile must be balanced of safety versus efficacy.

17.2.2.3 Chronic Rhinosinusitis

There is also increasing evidence that chronic inflammations caused by the biofilms are critical to the pathophysiology of CRS (Fastenberg et al. 2016). The common bacteria associated with CRS are *P. aeruginosa* and *S. aureus* (Cryer et al. 2004; Boase et al. 2013; Cryer et. al., 2004; Foreman and Wormald, 2010). The bacteria were reported to form biofilms on silicon elastic devices removed from patients (Ferguson and Stolz 2005). Additionally, biofilms were also found in all sinonasal mucosal samples collected from 16 patients undergoing sinus surgery (Cryer et al. 2004). Thus, biofilms does play a role in causing persistency of the disease in humans despite surgery intervention and targeted long-term antibiotic therapy (Palmer 2006).

17.3 BIOFILMS-ASSOCIATED INFECTIONS IN ANIMAL HEALTH AND THEIR POTENTIAL FOR ZOONOSIS TRANSMISSION

The general impact of biofilms on animal health has been covered extensively by Abdullahi et al. (2016). The following part of this chapter will focus on the burden of biofilms-related infections in the livestock animal that causes economic losses to the farmers, as well as the potential transmission of biofilms-forming bacteria to human as zoonosis diseases. Zoonosis is infectious diseases that can be transmitted from animals to humans. Sixty-one percent of the pathogens known to infect humans are zoonotic (Percival and Garcı 2011). Most emerging infectious diseases considered to be serious public health problems have zoonotic origins, and approximately three-quarters have originated from wild animals (Shin and Park 2018). Close contact with the animal via inhalation, ingestion, contaminated mucous membranes, and damage of intact skin are possible transmission routes of zoonotic pathogens (Shin and Park 2018). The risk is particularly high for personnel that work closely with the animal, for example, veterinarian and farmers. Transmission of zoonotic pathogens in foodborne diseases includes undercooked meat or other animal tissues, seafood, and invertebrates, as well as unpasteurized milk and dairy products and contaminated vegetables (Shin and Park 2018). Additionally, insects serve as important biological or mechanical vectors in transmitting some organisms (Iannino et al. 2018). Pathogens that associated with zoonotic infections are known to form biofilmss (Percival and Garcı 2011). Biofilms are one of the bacterial mechanisms to survive and thrive in the environment. Table 17.3 summarizes pathogens that can form biofilms with the potential of zoonosis transmission to humans.

17.3.1 Mastitis

Mastitis is a disease affecting ruminant specifically large and small ruminant (cows, sheep, and goats). Mastitis is caused by several pathogens including *S. aureus*, *E. coli*, and *Streptococcus agalactiae* (Dogan et al. 2006). Globally, the disease causes economic losses between 16 and 26 billion Euros annually (Gonçalves et al. 2018). Infections by these bacteria cause inflammation in the udder, and the toxin released by the bacteria can cause necrosis of the mammary udder cell, reducing milk production, and thus affecting economic outcome by the farmer (Henriques et al. 2016). The treatment that involves administration of antibiotics has only been partially successful. The disease often recurs and persistent in the animal, causing irreversible

TABLE 17.3
Diseases Associated with the Biofilms-Forming Pathogen with the Potential for Zoonoses Transmission

Disease	Causative Agent	Host	References
Mastitis	*Staphylococcus aureus*, *E. coli*, *S. agalactiea*	Human, small and large ruminant	El-Mahallawy et al. (2017); Vishnupriya et al. (2014)
Wound infection	*Acinetobacter baumanii*	Human, dogs, cats, horse	Tomaras et al. (2003); Maisch et al. (2012); Yang et al. (2019)
Pneumonia	*Mannheimia hemolytica*	Human, cattle	Morck et al. (1990); Takeda et al. (2003); Boukahil and Czuprynski (2015)
Bite wound infection	• *Actinobacillus lignieresii*, *A. equuli*, and *A. suis* • *Staphylococcus aureus* (MRSA), *S. intermedius*	• Human, cattle, horses, pigs • Human, dog, pigeon	Weyant et al. (1996); Raad et al. (2007); Neill et al. (2007)
Cat scratch disease	• *Bartonella henselae* and *B. Quintana*	Human, dogs, cats	Shin and Park (2018)
Fish tank granuloma	*Mycobacterium marinum*	Human, fish	Hashish et al. (2018)
Gastric ulcers, gastritis	*Helicobacter pylori*	Human Oral cavity of dogs	Kandulski, Selgrad, and Malfertheiner (2008); Hathroubi et al. (2018)
Meningitis	*Streptococcus suis* type 2	Human, pig	T. Bjarnsholt (2013); Lun et al. (2007)
Gasteroenteritis	*Aeromonas hydrophila*	Human, reptiles, amphibians, fish	Lynch et al. (2002)
Salmonellosis	*Salmonella gastroenteritis*, *Salmonella enterica* subspecies I, *Salmonella enteritidis*	Human, reptiles, birds, dogs, cats	Shin and Park (2018)
Diarrhea	• *Vibrio cholera and V. parahaemolyticus* • *Escherichia coli 0157*	Human, aquatic organism, birds, reptiles, mammals	Alam et al. (2007); Cantas and Suer (2014)

damage to the udder cells (Zhao and Lacasse 2008). To avoid further losses in the farm productions, animal will be culled.

Persistency of the disease in an animal is believed to be due to several factors. This includes continues development of antimicrobial resistance causing the pathogen to be non-responsive to the antibiotic therapy (Beuron et al. 2014). Additionally, the ability of *S. aureus* to invade and survive within the bovine mammary epithelial cells can cause it to escape the antibiotic therapy, as due to their physicochemical properties, not all antibiotics are able to cross the mammalian cells to exert its activities (Kamaruzzaman et al. 2017). Additionally, the ability of the infecting pathogens to form biofilms on the surface of the mammary udders are other factors that cause ineffectivity of the antibiotics to completely kill the bacteria during the treatment (Henriques, Gomes, and Jos 2016; Fox, Zadoks, and Gaskins 2005; Bardiau et al. 2013). It is well known that bacteria susceptibility toward antimicrobials is reduced when in the form of biofilms. Surprisingly, it was reported that the antibiotic which is commonly used for the mastitis treatment, enrofloxacin has been shown to promote biofilms formation of *E. coli* (Costa et al. 2012). Biofilms formation has also been influenced by milk components and the acidic pH of the environment. This is particularly important as the antibiotics commonly used for mastitis treatment are acidic antibiotics (Costa et al. 2012). Thus, all these evidence directed the role of biofilms to cause persistent mastitis infections in animals.

The problem with mastitis does not stop only in animals. The bacteria associated with mastitis can also be transmitted to humans via milk-borne zoonoses. For example, a study has reported transmission of methicillin-resistant *S. aureus* and *E. coli* isolated from milk in bovine mastitis cases to the animal handlers (Costa et al. 2012; El-Mahallawy et al. 2017). Thus, this implication showed the importance of proper control of mastitis cases, not only to reduce the disease burden in animals but to control the possible transmission of the bacteria to humans.

17.3.2 Pneumonia

Mannheimia hemolytica is the pathogen that is associated with pneumonia and hemorrhagic septicemia in sheep, buffalo, and cattle, and various diseases in poultry and other domesticated animals (Confer 2017). It is part of the bovine nasal flora and colonizes the nasal cavity and tonsillar crypts in cattle and sheep, yet this bacterium causes a devastating disease in cattle which is pneumonic mannheimiosis (shipping fever) (Boukahil and Czuprynski 2015; Lopez and Martinson 2017). The bacteria have been reported to form biofilms on the surfaces of bovine respiratory epithelial cells (Boukahil and Czuprynski 2016). Mannheimia hemolytica has occasionally been isolated from human cases of septicemia, upper respiratory tract infections, and animal bite wounds. Unpasteurized milk can be a risk of infection to dairy consumers (Confer 2017).

17.3.3 Fish Tank Granuloma

Both pathogenic and non-pathogenic species of mycobacteria are capable of forming biofilms, and this capability is not essentially a virulence mechanism (Hashish et al. 2018). *Mycobacterium marinum* is a non-tuberculous mycobacterium (NTM) (Chakraborty and Kumar 2019). It grows in macrophage and aquatic environments. It grows at 28°C–35°C and causes infection in fish (Sunil et al. 2018). It also can infect humans resulting in granulomatous skin lesions. The transmission of disease occurs particularly with personnel handling fish, thus named as "fish tank granuloma." The signs of infection such as reddish bumps (papules) that enlarge over time, along with swollen on the site of the infection (Mason et al. 2016). The international incidence and prevalence of *M. marinum* infection are unknown due to the lack of international surveillance. However, a study conducted in France demonstrated the incidence of

M. marinum infection to be 0.04 per 100,000 inhabitants per year (Mazumder and Gelfand 2019). The treatment requires the usage of multiple anti-mycobacterial for a long period of time. Such drug-tolerant chronic infection is often associated with *in vivo* biofilms (Chakraborty and Kumar 2019). The study of environmental and pathogenic NTM biofilms is especially important because of the industrial and medical implications for infection by pathogenic NTM species (Primm, Lucero, and Falkinham 2004). Specifically, *M. marinum* is a concern in re-circulating water systems in intensive aquaculture farming because once *M. marinum* bacteria are established, they are difficult to eradicate (Sunil et al. 2018).

17.3.4 Salmonellosis

Centers for Disease Control and Prevention (CDC) reported that Salmonella infection in humans caused 450 deaths in the United States annually (CDC 2018). The importance of food-producing animals as the reservoirs for non-typhoidal serovars affecting humans is well-established (Percival and Garcı 2011). *Salmonella enterica* subspecies I is the causative agent for Salmonellosis in humans and other warm-blooded animals (Stevens, Humphrey, and Maskell 2009). Biofilms-forming abilities of Salmonella are correlated with its persistence in fishmeal and feed factories. Studies on Salmonella in fish factories suggest that biofilms-forming ability may be an important factor for the persistence of Salmonella in the environment (Vestby et al. 2009). *Salmonella enteritidis* is the most common serotype isolated in poultry farm and is responsible for many cases of food poisoning in human beings worldwide. Almost 50% of them are able to produce biofilms (Marin, Hernandiz, and Lainez 2007).

17.3.5 Gasteroenteritis

Most gastroenteritis cases (>85%) are attributed to *Aeromonas hydrophila* (Daskalov 2017). *Aeromonas* sp. are Gram-negative and rod shaped. They are motile aquatic bacteria considered important pathogens in reptiles, amphibians, and fish. They are known to be a major problem in fish farming. Fish are thought to act as a reservoir of *Aeromonas hydrophila* possibly leading to infection in mammals (Percival and Garcı 2011). In humans, *Aeromonas* sp. are known to cause gastroenteritis (from mild to cholera-like symptoms) and other infections such as endocarditis, septicemia, hemolytic uremic syndrome, peritonitis, respiratory infections, myonecrosis, osteomyelitis, ocular infections, and meningitis. *Aeromonas hydrophila* has been reported to grow well in biofilms detected in drinking water systems (State et al. 2015).

17.3.6 Diarrhea

Diarrheal diseases are a major cause of morbidity and mortality worldwide with a particular impact on children (Guzman-Otazo et al. 2019). Cholera is an acute, diarrheal illness caused by infection of the intestine with the toxigenic bacterium *Vibrio cholerae serogroup* O1 or O139. An estimated 2.9 million cases and 95,000 deaths occur each year around the world (CDC 2018). Enterohemorrhagic *E. coli* (EHEC) occurs largely as a single serotype (O157:H7) causing sporadic cases and outbreaks of hemorrhagic colitis characterized by bloody diarrhea (Guentzel 1996).

17.3.7 Meningitis

Meningitis is a disease in piglet and human caused by the bacteria *Streptococcus suis* type 2. The bacteria can be isolated from carrier adult pigs' upper respiratory tract, tonsils, and feces of infected herds, known to form biofilms and inherently more resistant to penicillin G and ampicillin (Percival and Garcı 2011; Huong et al. 2014). It has been isolated from human patients worked in pig industry in several Asian and European countries, Canada, New Zealand, Australia, and Argentina (Lynskey, Lawrenson, and Sriskandan 2011; Lun et al. 2007). The prevalence of *S. suis* infection is the highest in Asia with the primary risk factors are thought due to occupational exposure and eating of contaminated food (Huong et al. 2014). The pooled proportions of case-patients with pig-related occupations and history of eating high-risk food were 38.1% and 37.3%, respectively (Huong et al. 2014).

17.3.8 Bite Wound Infection

Actinobacillus genus is a Gram-negative bacterium commensal of equine oral cavity and upper respiratory tract found to be responsible for sleepy foal disease (Stewart et al. 2002). *Actinobacillus lignieresii*, *Actinobacillus equuli*, and *Actinobacillus suis* can be present in the oropharyngeal flora of cattle, horses, and pigs, respectively. Human bitten by these animals may be exposed to these pathogens and potentially developed bite wound infections (Percival and Garcı 2011). It has been roughly estimated that the horse bites account for as high as 20% of overall animal bites in Turkey, which comes after dog bites which are 70%. More extensive muscle damage may develop in most of the horse attacks, which is different from small animal bites (Cantas and Suer 2014). A report stated that a 53-year-old butcher affected with septicemia and acute septic shock caused by *A. equuli* (Ashhurst-smith, Norton, and Thoreau 1998).

17.3.9 Cat Scratch Disease

Cat scratch disease (CSD) is a zoonosis caused by *Bartonella henselae*, a fastidious, hemotropic, Gram-negative bacterium. Cats and dogs are the principal mammal reservoir of the pathogen, and transmission to human can occur through animal scratches and bites. Worldwide, 55% of cases were reported in children younger than 18 years of age and 60% of these are males (Damborg et al. 2015; Nelson, Saha, and Mead 2016). In the United States, CSD is not a notifiable condition. Thus, information on the epidemiology of this disease has been limited to clinical case series and analyses of hospital discharge databases (Nelson, Saha, and Mead 2016). This disease affected immunocompromised patients either through acute or chronic infection with vascular proliferative or suppurative manifestations (Iannino et al. 2018). Previous reports stated that Bartonella infection can be treated using azithromycin, penicillin, tetracyclines, cephalosporins, and aminoglycosides (Breitschwerdt 2014).

To reduce the level of bacteremia in an infected cat or dog effectively, doxycycline, amoxicillin, enrofloxacin, and rifampin are given for a long duration (more than 4 weeks) (Iannino et al. 2018).

17.4 WAY FORWARD AND CONCLUSION

Biofilms are a mechanism posed by the bacteria for their survival in the environment. Biofilms-related infections are known to cause persistent infections in human and animal. The emergence of zoonotic pathogens that can cause infection also has only recently being understood. Understanding the physiology of the pathogen is important, as it will determine the suitable therapy for the infection to avoid recurrence and failure in the treatment. Though there were many studies being conducted to assess antibiotics efficacy in the common pathogen such as *S. aureus*, *P. aeruginosa*, *E. coli*, the same knowledge is still inadequate infancy in regard to antibiotics antibiofilms efficacy against the zoonotic pathogens (Bal et al. 2017). Thus, studies need to be done to ascertain this condition to ensure the right treatment is given whenever the infections are reported in humans. This knowledge will help not only to ensure suitable antibiotics given for the recovery of the patient, the correct usage of the antibiotics will reduce exposure of pathogens to the unnecessary antibiotics, and eventually reduce the emergence of antibiotic resistant.

REFERENCES

Abdullahi, Umar Faruk, Ephraim Igwenagu, Anas Mu, Sani Aliyu, and Maryam Ibrahim Umar. 2016. "Intrigues of Biofilms: A Perspective in Veterinary Medicine." *Veterinary World* 9: 12–8. doi: 10.14202/vetworld.2016.12-18.

Alam, Munirul, Marzia Sultana, G. Balakrish Nair, A. K Siddique, Nur A. Hasan, R. Bradley Sack, David A Sack, K. U. Ahmed, A. Sadique, H. Watanabe, Christopher J. Grim, A. Huq, and Rita R. Colwell .2007. "Viable but Nonculturable Vibrio Cholerae O1 in Biofilms in the Aquatic Environment and Their Role in Cholera Transmission." *Proceedings of the National Academy of Sciences* 104 (45): 17801–6.

Allaker, Robert P. 2010. "Critical Review in Oral Biology & Medicine: The Use of Nanoparticles to Control Oral Biofilms Formation." *Journal of Dental Research* 89 (11): 1175–86. doi: 10.1177/0022034510377794.

Andersson, Dan I., and Diarmaid Hughes. 2014. "Microbiological Effects of Sublethal Levels of Antibiotics." doi: 10.1038/nrmicro3270.

Armentano, Ilaria, Carla Renata Arciola, Elena Fortunati, Davide Ferrari, Samantha Mattioli, Concetta Floriana Amoroso, Jessica Rizzo, Jose M Kenny, Marcello Imbriani, and Livia Visai. 2014. "The Interaction of Bacteria with Engineered Nanostructured Polymeric Materials: A Review." *Scientific World Journal*. 2014: 410423. doi:10.1155/2014/410423

Ashhurst-Smith, Christopher, Robert Norton, and Wendy Thoreau. 1998. "*Actinobacillus equuli* Septicemia: An Unusual Zoonotic Infection *Actinobacillus equuli* Septicemia: An Unusual Zoonotic Infection." *Journal of Clinical Microbiology* 36 (9): 2789–90.

Bal, A. M, M. Z. David, J. Garau, T. Gottlieb, T. Mazzei, F. Scaglione, P. Tattevin, and I. M Gould. 2017. "Future Trends in the Treatment of Methicillin-Resistant *Staphylococcus aureus* (MRSA) Infection: An in-Depth Review of Newer Antibiotics Active against an Enduring Pathogen." *Journal of Global Antimicrobial Resistance* 10: 295–303. doi: 10.1016/j.jgar.2017.05.019.

Bardiau, Marjorie, Kazuko Yamazaki, J. N Duprez, Bernard Taminiau, J. G Mainil, and Isabelle Ote. 2013. "Genotypic and Phenotypic Characterization of Methicillin-Resistant *Staphylococcus aureus* (MRSA) Isolated from Milk of Bovine Mastitis." *Letters in Applied Microbiology* 57 (3): 181–6. doi: 10.1111/lam.12099.

Beuron, Daniele C., Cristina S. Cortinhas, Bruno G. Botaro, Susana N. Macedo, Juliano L. Gonçalves, Maria A. V. P. Brito, and Marcos V. Santos. 2014. "Risk Factors Associated with the Antimicrobial Resistance of *Staphylococcus aureus* Isolated from Bovine Mastitis." *Pesquisa Veterinaria Brasileira* 34(10): 947–52.

Billings, Nicole, Maria Ramirez Millan, Marina Caldara, Roberto Rusconi, Yekaterina Tarasova, Roman Stocker, and Katharina Ribbeck. 2013. "The Extracellular Matrix Component Psl Provides Fast-Acting Antibiotic Defense in *Pseudomonas aeruginosa* Biofilmss." *PLoS Pathogens* 9 (8). doi: 10.1371/journal.ppat.1003526.

Bispo, Paulo J. M., Wolfgang Haas, and Michael S. Gilmore. 2015. "Biofilmss in Infections of the Eye." *Pathogens* 4: 111–36. doi: 10.3390/pathogens4010111.

Bjarnsholt, Thomas. 2013. "The Role of Bacterial Biofilmss in Chronic Infections." *Apmis* 121 (136): 1–51. doi: 10.1111/apm.12099.

Bjarnsholt, Thomas, Maria Alhede, Morten Alhede, Steffen R. Eickhardt-Sørensen, Claus Moser, Michael Ku, Peter Østrup Jensen, and Niels Høiby. 2013. "The in Vivo Biofilms." doi: 10.1016/j.tim.2013.06.002.

Boase, Sam, Andrew Foreman, Edward Cleland, Lorwai Tan, Rachel Melton-Kreft, Harshita Pant, Fen Z. Hu, Garth D. Ehrlich, and Peter John Wormald. 2013. "The Microbiome of Chronic Rhinosinusitis: Culture, Molecular Diagnostics, and Biofilms Detection." *BMC Infectious Diseases* 13 (1): 1–9. doi: 10.1186/1471-2334-13-210.

Borriello, Giorgia, Erin Werner, Frank Roe, Aana M Kim, Garth D Ehrlich, and Philip S. Stewart. 2004. "Oxygen Limitation Contributes to Antibiotic Tolerance of *Pseudomonas aeruginosa* in Biofilms." *Antimicrobial Agents and Chemotherapy* 48 (7): 2659–64. doi: 10.1128/AAC.48.7.2659.

Boukahil, Ismail, and Charles J. Czuprynski. 2015. "Characterization of Mannheimia Haemolytica Biofilms Formation in Vitro." *Veterinary Microbiology* 175 (1): 114–22. doi: 10.1016/j.vetmic.2014.11.012.

Boukahil, Ismail, and Charles J. Czuprynski. 2016. "Mannheimia Haemolytica Biofilms Formation on Bovine Respiratory Epithelial Cells." *Veterinary Microbiology* 197: 129–36. doi: 10.1016/j.vetmic.2016.11.012.

Breitschwerdt, Edward Bealmear. 2014. "Bartonellosis: One Health Perspectives for an Emerging Infectious Disease." doi: 10.1093/ilar/ilu015.

Brentel, A. S., K. Z. Kantorski, L. F. Valandro, S. B. Fúcio Bottino, R. M. Puppin-Rontani, and M. A. Bottino. 2011. "Confocal Laser Microscopic Analysis of Biofilms on Newer Fledspar Ceramic." *Operative Dentistry* 36 (1): 43–51. doi: 10.2341/10-093-LR.

Cantas, Leon, and Kaya Suer. 2014. "Review: The Important Bacterial Zoonoses in 'One Health' Concept." *Frontiers in Public Health* 2 (October): 1–8. doi: 10.3389/fpubh.2014.00144.

Catt, Cristina, and Francesca Cappitelli. 2019. "Testing Anti-Biofilms Polymeric Surfaces : Where to Start ?" *International Journal of Molecular Sciences* 20 (3794).

CDC. 2018. "Screening for Salmonella." 2018.

Chakraborty, Poushali, and Ashwani Kumar. 2019. "The Extracellular Matrix of Mycobacterial Biofilms: Could We Shorten the Treatment of Mycobacterial Infections?" *Microbial Cell* 6 (2): 105–22. doi: 10.15698/mic2019.02.667.

Cloutier, M., D. Mantovani, and F. Rosei. 2015. "Antibacterial Coatings: Challenges, Perspectives, and Opportunities." *Trends in Biotechnology*: 1–16. doi: 10.1016/j.tibtech.2015.09.002.

Confer, Anthony W. 2017. "Mannheimia Haemolytica—Zoonoses and Food Safety." CAB Reviews: Perspectives in Agriculture, Veterinary Science, Nutrition, and Natural Resources. 2017.

Costa, João Carlos Miguel, Isis de Freitas Espeschit, Fábio Alessandro Pieri, Laércio dos Anjos Benjamin, and Maria Aparecida Scatamburlo Moreira. 2012. "Increased Production of Biofilms by *Escherichia coli* in the Presence of Enrofloxacin." *Veterinary Microbiology* 160 (3–4): 488–90. doi: 10.1016/j.vetmic.2012.05.036.

Costerton, J. William, Phillip S. Stewart, and E. Peter Greenberg. 1999. "Bacterial Biofilms: A Common Cause of Persistent Infections." *Science* 284 (5418): 1318–22. doi: 10.1126/science.284.5418.1318.

Cryer, Jonathan, Ioana Schipor, Joel R. Perloff, and James N. Palmer. 2004. "Evidence of Bacterial Biofilms in Human Chronic Sinusitis." *ORL* 66: 155–8. doi: 10.1159/000079994.

Cullinan, Mary P., and Gregory J. Seymour. 2013. "Periodontal Disease and Systemic Illness: Will the Evidence Ever Be Enough?" *Periodontology 2000* 62: 271–86.

Damborg, Peter, Els M. Broens, Bruno B. Chomel, S. Guenther, and Frank Pasmans. 2015. "Bacterial Zoonoses Transmitted by Household Pets: State-of-the-Art and Future Perspectives for Targeted Research and Policy Actions." *Journal of Comparative Pathology* 155 (1): S27–40. doi: 10.1016/j.jcpa.2015.03.004.

Darouiche, Rabih O. 2004. "Treatment of Infections Associated with Surgical Implants." *N*ew England Journal of Medicine 350(14): 1422–9.

Daskalov, Hristo. 2017. "The Importance of *Aeromonas hydrophila* in Food Safety the Importance of *Aeromonas hydrophila* in Food Safety." doi: 10.1016/j.foodcont.2005.02.009.

Delcaru, Cristina, Ionela Alexandru, Paulina Podgoreanu, Mirela Grosu, Elisabeth Stavropoulos, Mariana Carmen Chifiriuc, and Veronica Lazar. 2016. "Microbial Biofilms in Urinary Tract Infections and Prostatitis: Etiology, Pathogenicity, and Combating Strategies." *Pathogens* 5 (65): 1–12. doi: 10.3390/pathogens5040065.

Desrousseaux, C., V. Sautou, S. Descamps, and O. Traore. 2013. "Modification of the Surfaces of Medical Devices to Prevent Microbial Adhesion and Biofilms Formation." 85. doi: 10.1016/j.jhin.2013.06.015.

Dogan, Belgin, S. Klaessig, M. Rishniw, R. A. Almeida, S. P. Oliver, K. Simpson, and Y. H. Schukken. 2006. "Adherent and Invasive *Escherichia coli* Are Associated with Persistent Bovine Mastitis." *Veterinary Microbiology* 116 (4): 270–82. doi: 10.1016/j.vetmic.2006.04.023.

Donlan, Rodney M., and J. William Costerton. 2002. "Biofilms: Survival Mechanisms of Clinically Relevant Microorganisms." *Clinical Microbiology Reviews* 15 (2): 167–93. doi: 10.1128/CMR.15.2.167.

Du, X., X. Huang, C. Huang, J. E. Frencken, and T. Yang. 2012. "Inhibition of Early Biofilms Formation by Glass-Ionomer Incorporated with Chlorhexidine in Vivo: A Pilot Study.": 58–64. doi: 10.1111/j.1834-7819.2011.01642.x.

Eke, Paul I., Gina Thornton-Evans, Bruce Dye, and Robert Genco. 2012. "Advances in Surveillance of Periodontitis: The Centers for Disease Control and Prevention Periodontal Disease Surveillance Project." *Journal of Periodontology* 83 (11): 1337–42. doi: 10.1902/jop.2012.110676.Advances.

El-Mahallawy, Heba S., Mahmoud Elhariri, Rehab Elhelw, and Dalia Hamza. 2017. "Zoonotic Importance of Some Pathogenic Bacteria Isolated from Small Ruminants' Milk and Hands of Dairy Workers." *Zagazig Veterinary Journal* 45 (4): 305–13. doi: 10.5281/zenodo.1162879.RESEARCH.

Fastenberg, Judd H., Wayne D. Hsueh, Ali Mustafa, Nadeem A. Akbar, and Waleed M. Abuzeid. 2016. "Biofilms in Chronic Rhinosinusitis: Pathophysiology and Therapeutic Strategies." *World Journal of Otorhinolaryngology-Head and Neck Surgery* 2 (4): 219–29. doi: 10.1016/j.wjorl.2016.03.002.

Feng, Guoping, Yifan Cheng, Shu-Yi Wang, Diana A. Borca-Tasciuc, Randy W. Worobo, and Carmen I. Moraru. 2015. "Bacterial Attachment and Biofilms Formation on Surfaces Are Reduced by Small-Diameter Nanoscale Pores: How Small Is Small Enough?" doi: 10.1038/npjbiofilms.2015.22.

Ferguson, Berrylin J., and Donna B. Stolz. 2005. "Demonstration of Biofilms in Human Bacterial Chronic Rhinosinusitis." In *Annual Meeting of the American Rhi-Nologic Society*, Los Angeles, California, September 24, 2005, 19: pp. 452–57.

Fox, L. K., R. N. Zadoks, and C. T. Gaskins. 2005. "Biofilms Production by *Staphylococcus aureus* Associated with Intramammary Infection." *Veterinary Microbiology* 107 (3–4): 295–9. doi: 10.1016/j.vetmic.2005.02.005.

Fux, Christoph A, Paul Stoodley, Luanne Hall-Stoodley, and J. William Costerton. 2003. "Bacterial Biofilms: A Diagnostic and Therapeutic Challenge." *Expert Review of Anti-Infective Therapy* 1 (14): 667–83.

Gonçalves, J. L., C. Kamphuis, C. M. M. R. Martins, J. R. Barreiro, T. Tomazi, and A. H. Gameiro. 2018. "Bovine Subclinical Mastitis Reduces Milk Yield and Economic Return." *Livestock Science* 210 (January): 25–32. doi: 10.1016/j.livsci.2018.01.016.

Gristina, Anthony G., Joanne J. Dobbins, Beverly Giammara, Jon C. Lewis, and William C. Devries. 1988. "Biomaterial-Centered Sepsis and the Total Artificial Heart." *JAMA: The Journal of the American Medical Association* 259 (1): 870–4.

Guentzel, M. N. 1996. "Escherichia, Klebsiella, Enterobacter, Serratia, Citrobacter, and Proteus. Medical Microbiology." In Medical Microbiology, 4th edition. Galveston, University of Texas Medical Branch.

Guzman-Otazo, Jessica, Lucia Gonzales-Siles, Violeta Poma, Bengtsson Johan Palme, Kaisa Thorell, Calrs Fredrik Flach, Volga Iniguez, and Asa Sjoling. 2019. "Diarrheal Bacterial Pathogens and Multi-Resistant Enterobacteria in the Choqueyapu River in La Paz, Bolivia." *PLoS One*: 14(1): e0210735. doi: 10.1371/journal.pone.0210735

Hashish, Emad, Abdallah Merwad, Shimaa Elgaml, Ali Amer, Huda Kamal, Ahmed Elsadek, Ayman Marei, and Mahmoud Sitohy. 2018. "*Mycobacterium marinum* Infection in Fish and Man: Epidemiology, Pathophysiology and Management; A review." *Veterinary Quarterly* 38 (1): 35–46. doi: 10.1080/01652176.2018.1447171.

Hassett, Daniel J., Michael T. Borchers, and Ralph J. Panos. 2014. "Chronic Obstructive Pulmonary Disease (COPD): Evaluation From Clinical, Immunological and Bacterial Pathogenesis Perspectives." *Journal of Microbiology* 52 (3): 211–26. doi: 10.1007/s12275-014-4068-2.

Hathroubi, Skander, Stephanie L. Servetas, Ian Windham, and D. Scott Merrell. 2018. "Helicobacter Pylori Biofilms Formation and Its Potential Role in Pathogenesis." *Microbiology and Molecular Biology Reviews* 82(2): 1–15.

Henriques, Mariana, Fernanda Gomes, and Maria Jos. 2016. "Bovine Mastitis Disease/ Pathogenicity: Evidence of the Potential Role of Microbial Biofilmss.": 1–7. doi: 10.1093/femspd/ftw006.

Høiby, Niels. 2014. "A Personal History of Research on Microbial Biofilms and Biofilm Infections." *Pathogens and Disease* 70(3): 205–211.

Høiby, Niels, Thomas Bjarnsholt, Claus Moser, Peter Østrup Jensen, Mette Kolpen, Tavs Qvist, Kasper Aanæs, Tanja Pressler, and Marianne Skov. 2017. "Diagnosis of Biofilms Infections in Cystic Fibrosis Patients." *Acta Pathologica Microbiologica*: 339–43. doi: 10.1111/apm.12689.

Heukelekian, H., and A. Heller. 1940. "Relation between Food Concentration and Surface for Bacterial Growth." *Journal of Bacteriology* (1933): 547–558.

Huong, Vu Thi Lan, Ngo Ha, Nguyen Tien Huy, Peter Horby, Ho Dang Trung Nghia, Vu Dinh Thiem, Xiaotong Zhu, et al. 2014. "Clinical Manifestations, and Outcomes of *Streptococcus suis* Infection in Humans." *Emerging Infectious Diseases* 20(7): 1105–14.

Iannino, Filomena, Stefania Salucci, Andrea Di Provvido, Alessandra Paolini, and Enzo Ruggieri. 2018. "Bartonella Infections in Humans Dogs and Cats." *Veterinaria Iteliana*: 63–72. doi: 10.12834/VetIt.398.1883.2.

Ista, Linnea K., Oswald Baca, and Gabriel P. Lbpez. 1996. "Attachment of Bacteria to Model Solid Surfaces: Oligo (Ethylene Glycol) Surfaces Inhibit Bacterial Attachment" *FEMS Microbiology Letters* 142: 59–63.

Izano, Era A., Matthew A. Amarante, William B. Kher, and Jeffrey B. Kaplan. 2008. "Differential Roles of Poly-N-Acetylglucosamine Surface Polysaccharide and Extracellular DNA in *Staphylococcus aureus* and *Staphylococcus epidermidis* Biofilmss." *Applied and Environmental Microbiology* 74 (2): 470–6. doi: 10.1128/AEM.02073-07.

Jamal, Muhsin, Wisal Ahmad, Saadia Andleeb, Fazal Jalil, Muhammad Imran, Muhammad Asif, Tahir Hussain, Muhammad Ali, Muhammad Rafiq, and Muhammad Atif. 2017. "Bacterial Biofilms and Associated Infections." *Journal of the Chinese Medical Association*: 5–9. doi: 10.1016/j.jcma.2017.07.012.

Jamal, Muhsin, Wisal Ahmad, Saadia Andleeb, Fazal Jalil, Muhammad Imran, Muhammad Asif Nawaz, Tahir Hussain, Muhammad Ali, Muhammad Rafiq, and Muhammad Atif Kamil. 2018. "Bacterial Biofilms and Associated Infections." *Journal of the Chinese Medical Association* 81 (1): 7–11. doi: 10.1016/j.jcma.2017.07.012.

James, Garth A., Ellen Swogger, Randall Wolcott, Elinor Pulcini, Patrick Secor, Jennifer Sestrich, John W. Costerton, and Philip S. Stewart. 2007. "Biofilms in Chronic Wounds." *Wound Repair and Regeneration* 37–44. doi: 10.1111/j.1524-475X.2007.00321.x.

Jennings, Laura K., Kelly M. Storek, Hannah E. Ledvina, Charlène Coulon, Lindsey S. Marmont, Irina Sadovskaya, Patrick R. Secor, Boo Shan Tseng, Michele Scian, Alain Filloux, Daniel J. Wozniak, P. Lynne Howell, and Matthew R. Parsek. 2015. "Pel Is a Cationic Exopolysaccharide That Cross-Links Extracellular DNA in the *Pseudomonas aeruginosa* Biofilms Matrix." *Proceedings of the National Academy of Sciences* 112 (36): 11353–8. doi: 10.1073/pnas.1503058112.

Jiao, Yang, Franklin R. Tay, Li-Na Niu, and Ji-Hua Chen. 2019. "Advancing Antimicrobial Strategies for Managing Oral Biofilm Infections." *International Journal of Oral Science*: 1–11. doi: 10.1038/s41368-019-0062-1.

Johnson, Lori, Shawn R. Horsman, Laetitia Charron-Mazenod, Amy L. Turnbull, Heidi Mulcahy, Michael G. Surette, and Shawn Lewenza. 2013. "Extracellular DNA-Induced Antimicrobial Peptide Resistance in *Salmonella enterica* Serovar Typhimurium." *BMC Microbiol* 13: 115. doi: 10.1186/1471-2180-13-115.

Jung, Chiau-Jing, Chiou-Yueh Yeh, Chia-Tung Shun, Ron-Bin Hsu, Hung-Wei Cheng, Chi-Shuan Lin, and Jean-San Chia. 2012. "Platelets Enhance Biofilms Formation and Resistance of Endocarditis-Inducing Streptococci on the Injured Heart Valve." *The Journal of Infectious Diseases* 205 (1 April): 1066–75. doi: 10.1093/infdis/jis021.

Kamaruzzaman, Nor Fadhilah, Stacy Q. Y. Chong, Kamina M. Edmondson-Brown, Winnie Ntow-Boahene, Marjorie Bardiau, and Liam Good. 2017. "Bactericidal and Anti-Biofilms Effects of Polyhexamethylene Biguanide in Models of Intracellular and Biofilms of *Staphylococcus aureus* Isolated from Bovine Mastitis." *Frontiers in Microbiology* 8 (AUG): 1–10. doi: 10.3389/fmicb.2017.01518.

Kamaruzzaman, Nor Fadhilah, Li Peng Tan, Khairun Anisa Mat Yazid, Shamsaldeen Ibrahim Saeed, Ruhil Hayati Hamdan, Siew Shean Choong, Weng Kin Wong, Alexandru Chivu, and Amanda Jane Gibson. 2018. "Targeting the Bacterial Protective Armor; Challenges and Novel Strategies in the Treatment of Microbial Biofilms." *Materials* 11 (9): 1–27. doi: 10.3390/ma11091705.

Kandulski, A., M. Selgrad, and P. Malfertheiner. 2008. "Gastric Precancerous Lesions Part 2 Helicobacter Pylori Infection: A Clinical Overview." 40: 619–26. doi: 10.1016/j.dld.2008.02.026.

Kaya, Ercan, Ilknur Dag, Armagan Incesulu, Melek Kezban Gurbuz, Mustafa Acar, and Leman Birdane. 2013. "Investigation of the Presence of Biofilms in Chronic Suppurative Otitis Media, Nonsuppurative Otitis Media, and Chronic Otitis Media with Cholesteatoma by Scanning Electron Microscopy." *The Scientific World Journal* 6. http://dx.doi.org/10.1155/2013/638715.

Kunimitsu, Mao, Gojiro Nakagami, Aya Kitamura, Yuko Mugita, Kaname Akamata, Sanae Sasaki, Chieko Hayashi, Yukie Mori, and Hiromi Sanada. 2019. "The Combination of High Bacterial Count and Positive Biofilms Formation Is Associated with the

Inflammation of Pressure Ulcers." *Chronic Wound Care Management and Research* 6: 1–8.

La, Barbara, Tourette Prosser, Doris Taylor, Barbara A. Dix, and Roy Cleeland. 1987. "Method of Evaluating Effects of Antibiotics on Bacterial Biofilms." *Antimicrobial Agents and Chemotherapy* 31(10): 1502–6.

Lagree, Katherine, Htwe H. Mon, Aaron P. Mitchell, and William A. Ducker. 2018. "Impact of Surface Topography on Biofilms Formation by *Candida albicans*.": 1–13. doi: 10.7294/W4D21VRB.All.

Lasserre, Jérôme Frédéric, Michel Christian Brecx, and Selena Toma. 2018. "Oral Microbes, Biofilms and Their Role in Periodontal and Peri-Implant Diseases Jérôme." *Materials* 11 (1802): 1–17. doi: 10.3390/ma11101802.

Lawrence, J. R., G. D. W. Swerhone, U. Kuhlicke, and T. R. Neu. 2007. "In Situ Evidence for Microdomains in the Polymer Matrix of Bacterial Microcolonies." *Canada Journal of Microbiology* 458: 450–8. doi: 10.1139/W06-146.

Li, Xiaojing, Kristin M. Kolltveit, Leif Tronstad, and Ingar Olsen. 2000. "Systemic Diseases Caused by Oral Infection." *Clinical Microbiology Reviews* 13(4): 547–58.

Lopez, Alfonso and Shannon A. Martinson. 2017. "Respiratory System, Mediastinum, and Pleurae." In *Pathologic Basis of Veterinary Disease*, pp. 471–560. St. Louis, MI.

Lopez, Analette I., Amit Kumar, Megan R. Planas, Yan Li, Thuy V. Nguyen, and Chengzhi Cai. 2011. "Colonization, Biofunctionalization of Silicone Polymers Using Poly(Amidoamine) Dendrimers and a Mannose Derivative for Prolonged Interference against Pathogen." *Biomaterials* 32 (19): 4336–46. doi: 10.1016/j.biomaterials.2011.02.056.Biofunctionalization.

Lun, Zhao-Rong, Qiao-Ping Wang, Xiao-Guang Chen, An-Xing Li, and Xing-Quan Zhu. 2007. "*Streptococcus suis*: An Emerging Zoonotic Pathogen." *Lancet Infect Dis* 7 (March 2007): 38–43. doi: 10.1016/S1473-3099(07)70001-4.

Lynch, Martin J., Simon Swift, David F. Kirke, C. William Keevil, Christine E. R. Dodd, and Paul Williams. 2002. "The Regulation of Biofilms Development by Quorum Sensing in *Aeromonas hydrophila*." *Environmental Microbiology* 4: 18–28.

Lynskey, Nicola N., Richard A. Lawrenson, and Shiranee Sriskandan. 2011. "New Understandings in *Streptococcus pyogenes*." *Current Opinion in Infectious Diseases* 24 (3): 196–202. doi: 10.1097/QCO.0b013e3283458f7e.

Mah, Thien-Fah C., and George A. O. Toole. 2001. "Mechanisms of Biofilms Resistance to Antimicrobial Agents." *Trends in Microbiology* 9 (1): 34–9. doi: 10.1016/S0966-842X(00)01913-2.

Maisch, Tim, Tetsuji Shimizu, Yang Fang Li, Julia Heinlin, Sigrid Karrer, Gregor Morfill, and Julia L. Zimmermann. 2012. "Decolonization of MRSA, *S. aureus* and *E. coli* by Cold-Atmospheric Plasma Using a Porcine Skin Model in Vitro." *PLoS One* 7 (4): 1–9. doi: 10.1371/journal.pone.0034610.

Marin, C, A. Hernandiz, and M. Lainez. 2007. "Biofilms Development Capacity of Salmonella Strains Isolated in Poultry Risk Factors and Their Resistance against Disinfectants.": 424–31. doi: 10.3382/ps.2008-00241.

Mason, Timothy, Kathy Snell, Erika Mittge, Ellie Melancon, Rebecca Montgomery, Marcie McFadden, Javier Camoriano, Michael L. Kent, Christopher M. Whipps, and Judy Peirce. 2016. "Strategies to Mitigate a Mycobacterium Marinum Outbreak in a Zebrafish Research Facility." *Zebrafish* 13: S77–87.

Mazumder, S. A., and M. Gelfand. 2019. "*Mycobacterium marinum* Infection: Background, Pathophysioloy, Epidemiology." *Medscape* 2019.

Merchan, Martha, Jana Sedlarikova, Vladimir Sedlarik, Michal Machovsky, Jitka Svobodova, and Petr Saha. 2010. "Antibacterial Polyvinyl Chloride / Antibiotic Films: The Effect of Solvent on Morphology, Antibacterial Activity, and Release Kinetics." *Journal of Applied Polymer Science* 118(4): 2369–78. doi: https://doi.org/10.1002/app.32185.

Metcalf, Daniel G., David Parsons, and Philip G. Bowler. 2016. "Clinical Safety and Effectiveness Evaluation of a New Antimicrobial Wound Dressing Designed to Manage Exudate, Infection, and Biofilms." *Internation Wound Journal*: 1–11. doi: 10.1111/iwj.12590.

Morck, D. W., J. W. Costerton, D. O. Bolingbroke, H. Ceri, N. D. Boyd, and M. E. Olson. 1990. "A Guinea Pig Model of Bovine Pneumonic Pasteurellosis." *Canadian Journal of Veterinary Research* 54(1): 139–45.

Neill, Eoghan O., Clarissa Pozzi, Patrick Houston, Davida Smyth, Hilary Humphreys, D. Ashley Robinson, and James P. O. Gara. 2007. "Association between Methicillin Susceptibility and Biofilms Regulation in *Staphylococcus aureus* Isolates from Device-Related Infections." 45 (5): 1379–88. doi: 10.1128/JCM.02280-06.

Nelson, Christina A., Shubhayu Saha, and Paul S. Mead. 2016. "Cat-Scratch Disease in the United States, 2005–2013." *Emerging Infectious Diseases* 22(10): 1741–6.

Nguyen, Dao, Amruta Joshi-Datar, Francois Lepine, Elizabeth Bauerle, Oyebode Olakanmi, Karlyn Beer, Geoffrey McKay, Richard Siehnel, James Schafhauser, Yun Wang, Bradley E. Britigan, Pradeep K. Singh. 2011. "Active Starvation Responses Mediate Antibiotic Tolerance in Biofilmss and Nutrient-Limited Bacteria." *Science* 334 (6058): 982–6. doi: 10.1126/science.1211037.

Nickel, J. C., I. Ruseska, J. B. Wright, and J. W. Costerton. 1985. "Tobramycin Resistance of Cells of *Pseudomonas aeruginosa* Growing as Biofilms on Urinary Catheter Material." *Antimicrob Agents Chemother* 27 (4): 619–24. doi: 10.1128/AAC.27.4.619.

Okshevsky, Mira, Viduthalai R. Regina, and Rikke Louise Meyer. 2015. "Extracellular DNA as a Target for Biofilms Control." *Current Opinion in Biotechnology* 33: 73–80. doi: 10.1016/j.copbio.2014.12.002.

Palmer, James. 2006. "Bacterial Biofilmss in Chronic Rhinosinusitis." *Annals of Otology, Rhinology & Laryngology* 115 (9): 35–9.

Percival, Steven L., and Ana B. Garcı. 2011. *Zoonotic Infections: The Role of Biofilmss*. doi: 10.1007/978-3-642-21289-5.

del Pozo, J. L., and R. Patel. 2007. "The Challenge of Treating Biofilms-Associated Bacterial Infections." *Clinical Pharmacology and Therapeutics* 82 (2): 204–9. doi: 10.1038/sj.clpt.6100247.

Prijck, Kristof De, Nele De Smet, and Tom Coenye. 2010. "Prevention of Candida Albicans Biofilms Formation by Covalently Bound Dimethylaminoethylmethacrylate and Polyethylenimine.": 213–21. doi: 10.1007/s11046-010-9316-3.

Primm, Todd P., Christie A. Lucero, and Joseph O. Falkinham. 2004. "Health Impacts of Environmental Mycobacteria." *Clinical Microbiology Reviews* 17 (1): 98–106. doi: 10.1128/CMR.17.1.98.

Raad, Issam, Hend Hanna, Ying Jiang, Tanya Dvorak, Ruth Reitzel, Gassan Chaiban, Robert Sherertz, and Ray Hachem. 2007. "Comparative Activities of Daptomycin, Linezolid, and Tigecycline against Catheter-Related Methicillin-Resistant Staphylococcus Bacteremic Isolates Embedded in Biofilms." 51 (5): 1656–60. doi: 10.1128/AAC.00350-06.

Shin, Bora, and Woojun Park. 2018. "Zoonotic Diseases and Phytochemical Medicines for Microbial Infections in Veterinary Science: Current State and Future Perspective." *Frontiers in Veterinary Science* 5 (July): 1–9. doi: 10.3389/fvets.2018.00166.

Singh, Rachna, Simmi Sahore, Preetinder Kaur, Alka Rani, and Pallab Ray. 2016. "Penetration Barrier Contributes to Bacterial Biofilms-Associated Resistance against Only Select Antibiotics, and Exhibits Genus-, Strain- and Antibiotic-Specific Differences." *Pathogens and Disease* 74 (6): 1–6. doi: 10.1093/femspd/ftw056.

Singla, Saloni, Kusum Harjai, and Sanjay Chhibber. 2014. "Artificial *Klebsiella Pneumoniae* Biofilms Model Mimicking in Vivo System: Altered Morphological Characteristics and Antibiotic Resistance." *Journal of Antibiotics* 67 (4): 305–9. doi: 10.1038/ja.2013.139.

Soto, Sara M. 2014. "Importance of Biofilmss in Urinary Tract Infections: New Therapeutic Approaches." *Advances in Biology*. doi: 10.1155/2014/543974.Review.

State, Ebonyi, Iroha Ifeanyichukwu, Mkpuma Nicodemus, Ejikeugwu Chika, and Ugbo Emmanuel. 2015. "Detection of Biofilms-Producing Isolates of Aeromonas Species from Drinking Water Sources in Ezza South Local Government Area of Ebonyi State, Nigeria." 3 (August): 316–23.

Stevens, Mark P., Tom J. Humphrey, and Duncan J. Maskell. 2009. "Molecular Insights into Farm Animal and Zoonotic Salmonella Infections." *Philosophical Transactions of the Royal Society* 364: 2709–23. doi: 10.1098/rstb.2009.0094.

Stewart, Allison J., Kenneth W. Hinchcliff, William J. A. Saville, Eduard Jose-Cunilleras, Joanne Hardy, Catherine W. Kohn, Stephen M. Reed, and Joseph J. Kowalski. 2002. "*Actinobacillus* sp. Bacteremia in Foals: Clinical Signs and Prognosis." *Journal of Veterinary Internal Medicine* 16: 464–71.

Stewart, Philip S. 2014. "Biophysics of Biofilms Infection.": 212–8. doi: 10.1111/2049-632X.12118.

Stewart, Philip S., Frank Roe, Joanna Rayner, James G. Elkins, Zbigniew Lewandowski, Urs A. Ochsner, and Daniel J. Hassett. 2000. "Effect of Catalase on Hydrogen Peroxide Penetration into *Pseudomonas Aeruginosa* Biofilmss Effect of Catalase on Hydrogen Peroxide Penetration into *Pseudomonas aeruginosa* Biofilmss" 66 (2): 8–11. doi: 10.1128/AEM.66.2.836-838.2000.Updated.

Sunil, Vidya, Andrew W. Harris, Bob Sine, Anne Marie, A. Lynn Noseworthy, Doug Sider, Frances B. Jamieson, Shelley White, Cassandra Johnston, and Olivia Spohn. 2018. "Investigation of a Community Cluster of Cutaneous *Mycobacterium marinum* Infection, an Emerging Zoonotic Pathogen in Aquaculture Industry, Haliburton, Kawartha, Pine Ridge District Health Unit, Ontario, Canada, July – August 2015." *Zoonoses Public Health* 66 (May): 164–8. doi: 10.1111/zph.12521.

Takeda, Seiji, Yasutomo Arashima, Kimitoshi Kato, Masahiro Ogawa, Kenji Kono, Kentaro Watanabe, and Takao Saito. 2003 "A Case of *Pasteurella haemolytica* Sepsis in a Patient with Mitral Valve Disease Who Developed a Splenic Abscess.": 764–5. doi: 10.1080/00365540310016385.

Tasneem, Umber, Nusrat Yasin, Iqbal Nisa, Faisal Shah, Ubaid Rasheed, Faiza Momin, and Sadir Zaman. 2018. "Biofilms-Producing Bacteria: A Serious Threat to Public Health in Developing Countries." *Journal of Food Science Nutrition* 1 (2): 25–31.

Tomaras, Andrew P., Caleb W. Dorsey, Richard E. Edelmann, and Luis A. Actis. 2003. "Attachment to and Biofilms Formation on Abiotic Surfaces by *Acinetobacter baumannii*: Involvement of a Novel Chaperone-Usher Pili Assembly System Printed in Great Britain.": 3473–84. doi: 10.1099/mic.0.26541-0.

Tribedi, P., and A. K. Sil. 2014. "Cell Surface Hydrophobicity: A Key Component in the Degradation of Polyethylene Succinate by *Pseudomonas* sp.": 295–303. doi: 10.1111/jam.12375.

Vestby, Lene K., Trond Møretrø, Solveig Langsrud, Even Heir, and Live L. Nesse. 2009. "Biofilms-Forming Abilities of Salmonella Are Correlated with Persistence in Fish Meal and Feed Factories." *BMC Veterinary Research* 5 (20): 1–6. doi: 10.1186/1746-6148-5-20.

Vishnupriya, S., P. X. Antony, H. K. Mukhopadhyay, R. M. Pillai, J. Thanislass, V. M. Vivek Srinivas, and R. Sumanth Kumar. 2014. "Methicillin-Resistant Staphylococci Associated with Bovine Mastitis and Their Zoonotic Importance." *Veterinary World* 7: 422–7. doi:10.14202/vetworld.2014.422-427.

Walters III, Marshall C., Frank Roe, Amandine Bugnicourt, Michael J. Franklin, and Philip S. Stewart. 2003. "Contributions of Antibiotic Penetration, Oxygen Limitation." *Antimicrobial Agents and Chemotherapy* 47 (1): 317–23. doi: 10.1128/AAC.47.1.317.

Wang, Zhejun, Ya Shen, and Markus Haapasalo. 2013. "Dental Materials with Antibiofilms Properties." *Dental Materials* 30 (2): e1–16. doi: 10.1016/j.dental.2013.12.001.

Waters, Elaine M., Sarah E. Rowe, James P. O'Gara, and Brian P. Conlon. 2016. "Convergence of *Staphylococcus aureus* Persister and Biofilms Research: Can Biofilmss Be Defined

as Communities of Adherent Persister Cells?" *PLoS Pathogens* 12 (12): 2–6. doi: 10.1371/journal.ppat.1006012.
Werner, Erin, Frank Roe, Amandine Bugnicourt, Michael J. Franklin, Arne Heydorn, Søren Molin, Betsey Pitts, and Philip S. Stewart. 2004. "Stratified Growth in *Pseudomonas aeruginosa* Biofilmss Stratified Growth in *Pseudomonas aeruginosa* Biofilmss." *Applied and Environmental Microbiology* 70 (10): 6188–96. doi: 10.1128/AEM.70.10.6188.
Wolcott, Randall D., Daniel D. Rhoads, and Scot E. Dowd. 2008. "Biofilmss and Chronic Wound Inflammation." *Journal of Wound Care* 17 (8): 333–41.
Yang, Cheng-Hong, Pai-Wei Su, Sin-Hua Moi, and Li-Yeh Chuang. 2019. "Biofilms Formation in *Acinetobacter baumannii*" *Molecules* 24(1849): 1–12.
Zhang, Ke, Fang Li, Satoshi Imazato, Lei Cheng, Huaibing Liu, Dwayne D. Arola, Yuxing Bai, and Hockin H. K. Xu. 2013. "Dual Antibacterial Agents of Nano-Silver and 12-Methacryloyloxydodecylpyridinium Bromide in Dental Adhesive to Inhibit Caries." *Journal of Biomedical Research Part B Applied Biomaterial* 101 (6): 697–704. doi: 10.1002/smll.201201811.Involvement.
Zhao, X., and P. Lacasse. 2008. "Mammary Tissue Damage during Bovine Mastitis: Causes and Control." *Journal of Animal Science* 86 (13 Suppl): 57–65. doi: 10.2527/jas.2007-0302.
Zuanazzi, David, Renata Souto, Marcelo Barbosa, Accioly Mattos, Maura Rodrigues, Bernardo Rangel, Carmelo Sansone, Ana Paula, and Vieira Colombo. 2010. "Prevalence of Potential Bacterial Respiratory Pathogens in the Oral Cavity of Hospitalized Individuals." *Archives of Oral Biology* 55: 21–8. doi: 10.1016/j.archoralbio.2009.10.005.

18 The Utilization of Essential Oils to Treat Biofilm-Associated Vaginal Infections

Lúcia G. V. Sousa, Joana Castro, and Nuno Cerca
University of Minho

CONTENTS

18.1 INTRODUCTION: VAGINAL MICROBIOTA

The human vaginal ecosystem is particularly diversified and can influence women's health and disease (Kim et al. 2009; Diop et al. 2019). Microbial species that inhabit the vaginal tract play an important role in the maintenance of a healthy vaginal microbiota, which consequently results in a functional equilibrium (Martin 2012). From this perspective, the recognition of the composition and ecology of the host vaginal microbiome is the main factor to understand how this unique environment prevents the acquisition of vaginal infections (Zhou et al. 2004; Diop et al. 2019).

Advances in culture-independent approaches, such as high-throughput 16S rRNA gene sequencing, have generated renewed knowledge in the composition and abundance of vaginal bacterial species in asymptomatic reproductive-age women, showing at least five major types of vaginal microbiota, known as community state types (CSTs). Four of these CSTs are dominated by *Lactobacillus crispatus*, *L. iners*, *L. gasseri*, and *L. jensenii*, and one does not contain a significant number of lactobacilli (20%–30% of the cases) but is composed of a diverse array of facultative and strictly anaerobic microorganisms, including *Prevotella*, *Dialiester*, *Atopobium*, *Gardnerella*, *Megasphaera*, *Peptoniphilus*, *Sneathia*, *Eggerthella*, *Aerococcus*, *Finegoldia*, and *Mobiluncus* (Ravel et al. 2011). Interestingly, these differences between CSTs appear to be driven by a combination of cultural, behavioral, genetic, and other uncharacterized underlying factors (Ma, Forney, and Ravel 2012). These findings challenged the wisdom that the occurrence of the high number of lactobacilli is synonymous with *normal* or *healthy*. However, it is important to note that all CSTs contain certain members which have the capacity to produce lactic acid suggesting that this is an important function conserved among different communities (Ravel et al. 2011). The production of lactic acid by bacteria contributes to the maintenance of a low vaginal pH (Amabebe and Anumba 2018). This acidic vaginal environment ($pH<4.5$) constitutes an efficient mechanism of protection of the vaginal epithelium since it makes the environment inhospitable to pathogenic microorganisms (Eschenbach et al. 1989; Turovskiy, Sutyak Noll, and Chikindas 2011). Other mechanisms have been studied that highlight the importance of endogenous microbiota in preventing other microorganisms from colonizing the vaginal tract (Eschenbach et al. 1989; Zhou et al. 2004). Another of such mechanisms is the ability of many *Lactobacillus* spp. have of producing antimicrobial compounds, including hydrogen peroxide (H_2O_2) (Eschenbach et al. 1989) and target-specific bacteriocins (Alpay Karaoğlu et al. 2003). However, although some studies have previously demonstrated that H_2O_2 could inhibit the colonization of pathogenic bacteria (Sgibnev and Kremleva 2015), it was also shown that under normal physiological concentrations, this antimicrobial effect was not detected, at least in 17 vaginal pathogens grown under anaerobic growth conditions (O'Hanlon, Moench, and Cone 2011). The vagina is virtually an anaerobic environment wherein dissolved oxygen levels are low. Therefore, it is unlikely that significant amounts of H_2O_2 are produced and accumulate to a toxic level to prevent the colonization of pathogens (O'Hanlon, Moench, and Cone 2011). Regarding bacteriocins, their antimicrobial activity is usually based on the permeabilization of the target membrane (Oscáriz and Pisabarro 2001). Thus, in the vagina, bacteriocins could play a significant role in fending off nonindigenous bacteria or pathogenic microorganisms (Stoyancheva et al. 2014). In addition, vaginal lactobacilli competitively block the adhesion of pathogens to vaginal epithelial cells (Ravel et al. 2011; Ojala et al. 2014; Amabebe and Anumba 2018).

Another important aspect to take into consideration is that the composition of the vaginal microbiota can vary throughout a woman's life in response to endogenous and exogenous factors, such as menstrual cycle (Eschenbach et al. 2000), pregnancy (DiGiulio et al. 2015), the use of medications (Stokholm et al. 2014), contraceptive methods (Brooks et al. 2017), sexual behaviors (Schwebke, Richey, and Weiss 1999), as well as racial groups (Ravel et al. 2011). Alterations in the microbial composition associated with a healthy vaginal microbiota can result in unbalances, and the *normal* state of health can be affected resulting in vaginal infections.

18.2 VAGINAL INFECTIONS

Vaginitis is characterized as any condition that predisposes women to vaginal symptoms such as abnormal vaginal discharge, odor, irritation, itching, or burning (Paladine and Desai 2018). The most common vaginitis, which will be discussed in this book chapter, are bacterial vaginosis (BV) and vulvovaginal candidiasis (VVC) (Abdul-Aziz et al. 2019; Konadu et al. 2019; Rosca et al. 2019). It is important to highlight that the prevalence of these two vaginal infections is high globally, with a concomitant high economic burden (Denning et al. 2018; Peebles et al. 2019). Therefore, sustainable treatment strategies are urgently needed to reduce the burden. This chapter overviews the use of essential oils (EOs) as potential alternatives for the treatment of these two infections.

18.2.1 Bacterial Vaginosis (BV)

BV is the most common vaginal infection in reproductive-aged women (Sobel 2000; Turovskiy, Sutyak Noll, and Chikindas 2011), and it affects more than 30% of the population worldwide (Livengood 2009), being more prevalent among sexually active women, and in women with a history of sexually transmitted infections (Cherpes et al. 2008). In addition, the prevalence is higher in women of African ethnicity (Turovskiy, Sutyak Noll, and Chikindas 2011). Some risk factors seem to be related to the prevalence and frequency of BV, such as vaginal douches (Ness et al. 2002), smoking (Bradshaw et al. 2005), use of intrauterine devices (Hodoglugil, Aslan, and Bertan 2000), and sexual behaviors (Verstraelen et al. 2010).

BV is also associated with a negative impact on self-esteem, sexual relationships, and quality of life (Bradshaw and Sobel 2016). Furthermore, it has been reported that BV can predispose women to several more serious complications, namely post-operative infections (Larsson et al. 2000), pelvic inflammatory disease (Ness et al. 2005), urinary tract infections (Harmanli et al. 2000), preterm delivery (Svare et al. 2006), miscarriage and pregnancy losses (Isik et al. 2016), and postpartum endometritis (Jacobsson et al. 2002). Moreover, the presence of BV increases the risk to the acquisition of *Trichomonas vaginalis* (Brotman et al. 2010), *Neisseria gonorrhoeae* and *Chlamydia trachomatis* (Wiesenfeld et al. 2003), and human immunodeficiency virus (Spear, St John, and Zariffard 2007).

18.2.1.1 Symptoms of BV

While some women with BV have symptoms, in other cases BV is asymptomatic (Gibbs 2007). Symptomatic BV classically exhibits an unpleasant, off-white, and thin discharge and a fishy odor (Sobel 2000; Hay 2014). This infection is also characterized by an increase in vaginal pH and by the presence of clue cells under a microscopic analysis (Livengood 2009). Clue cells are epithelial cells that have granular, cloudy, and rough edges and whose surfaces are coated with bacteria (Swidsinski et al. 2005).

18.2.1.2 Etiology of BV

Despite many years of research, BV etiology is still unclear (Onderdonk, Delaney, and Fichorova 2016; Jung et al. 2017; Muzny et al. 2019). It has been suggested that due to the high diversity of BV-associated vaginal microbiota, BV is not caused by the mere presence of a single bacterium but by an association of different bacterial species (Fredricks, Fiedler, and Marrazzo 2005; Oakley et al. 2008; Ceccarani et al. 2019). Unfortunately, in spite of the development of a more comprehensive picture of the vaginal microbiota during BV through the use of high-throughput 16S rRNA sequencing, the significance of these findings remains unclear, since it is not known whether these microorganisms are pathogens that cause BV or if they simply are opportunistic microorganisms that take advantage of the temporary higher pH environment and thus increase in numerical dominance (Ma, Forney, and Ravel 2012). For instance, it was shown before that *Gardnerella* spp. outcompeted 29 other bacterial species isolated from patients with BV, clearly demonstrating that not all BV-associated species have the same virulence potential (Alves et al. 2014). Interestingly, it has been reported from several studies that *Gardnerella vaginalis* is the most prevalent bacteria found in cases of BV (Fredricks, Fiedler, and Marrazzo 2005; Janulaitiene et al. 2017). However, *G. vaginalis* is also commonly found in asymptomatic or BV-negative women (Hickey and Forney 2014). This has generated interest in the question of whether genetic differences among isolates might distinguish pathogenic from commensal isolates (Castro et al. 2015; Castro, Jefferson, and Cerca 2019). Very recently, Vaneechoutte and colleagues showed that what has been referred to as *G. vaginalis* are, in fact, distinct species of the genus *Gardnerella* (Vaneechoutte et al. 2019). Furthermore, there is evidence that women with BV are colonized with multiple strains or species of *Gardnerella* spp. (Janulaitiene et al. 2017; Vodstrcil et al. 2017). Interestingly, Hill and colleagues recently showed several significant co-occurrences of different *Gardnerella* spp. in vaginal microbiota of 413 reproductive-aged Canadian women (Hill, Albert, and the VOGUE Research Group 2019). An abundance of *G. vaginalis* and *Gardnerella swidsinskii* was associated with vaginal symptoms of abnormal odor and discharge (Hill, Albert, and the VOGUE Research Group 2019). Of note, throughout this chapter we decided to use the term *Gardnerella* spp. when referring to previous studies concerning *G. vaginalis*.

18.2.2 Vulvovaginal Candidiasis (VVC)

VVC is the second more common vaginal infection worldwide and around 75% of all women experience, at least, once in their lifetime an episode of this infection (Sobel 2007; Bitew and Abebaw 2018). Although there is still some controversy behind the key factors associated with the development of VVC (Bitew and Abebaw 2018), often these are attributed to host-related risk factors and behaviors such as pregnancy (Aguin and Sobel 2015), diabetes mellitus (de Leon et al. 2002), immunosuppression (Duerr et al. 2003), antibiotics (Pirotta and Garland 2006), oral contraceptives and intrauterine devices (Cetin et al. 2007), and hygienic and clothing habits (Corsello et al. 2003). The morbidity inherently associated with VVC cases is associated with some issues such as discomfort, pain, mental distress, and altered self-esteem. Nevertheless, in some more complicated cases, and particularly if VVC is left untreated, it can result in many complications, namely pelvic inflammatory disease, infertility, ectopic pregnancy, pelvic abscess, spontaneous abortion, and menstrual disorders (Gonçalves et al. 2016). VVC can thus be classified as either uncomplicated or complicated disease, based on clinical presentation, host factors, microbiology, and response to therapy (Sobel 2007; Nyirjesy 2008). Complicated cases are normally characterized by (i) moderate to severe disease or (ii) four or more episodes per year or (iii) only budding yeast visible on microscopy analysis or (iv) adverse factors in the host such as pregnancy, diabetes, immunocompromised. Uncomplicated VVC is considered when (i) mild to moderate severity and (ii) fewer than four episodes per year happen and (iii) pseudohyphae or hyphae on microscopy and the host is non-immunocompromised (CDC 2015).

18.2.2.1 Symptoms of VVC

Some of the clinical symptoms of VVC are nonspecific for the infection and can be indicators of other vaginal infections (Gonçalves et al. 2016). The most common clinical manifestations of VVC cases are associated with the vaginal inflammation, characteristic of this infection, resulting in vaginal soreness, pruritus, irritation, burning, redness, vaginal dryness, leading to dyspareunia and external dysuria (Sobel 2016; Yano et al. 2019). A white vaginal discharge, with no odor, can also be present, but it is extremely nonspecific to the diagnosis of VVC (Sobel 2007). Vaginal pH is almost normal (<4.5) in VVC cases (Paladine and Desai 2018).

18.2.2.2 Etiology of VVC

VVC is usually caused by an overgrowth of *Candida* species, mainly *Candida albicans* (Sobel 2007), which is the dominant pathogen in cases of VVC associated with an acute inflammatory condition of the vulva and vaginal mucosa (Yano et al. 2019). However, in other cases non-*C. albicans* species can also be implicated, such as *C. glabrata*, *C. tropicalis*, *C. parapsilosis*, *C. krusei*, *C. stelloidea*, *C. dubliniensis*, and *C. lusitaniae* (Nyirjesy and Sobel 2003; Gonçalves et al. 2016). Importantly, an estimated 5% of women with VVC experience recurrent VVC, which is defined as four or more distinct episodes in a single year (Sobel 2007). Even though in some women, *Candida* species are a common commensal microorganism, under specific changes in the host vaginal environmental, *Candida* spp. can become a pathogen with the capacity to develop infection (Beigi, Meyn, et al. 2004). However, the exact mechanisms by which *Candida* spp. induce VVC are still unclear. It has been noted that the hyphal formation and phenotypic switching is an important virulence factor since filamentous forms improve mechanical strength and consequently the colonization and tissue invasion (Sobel 2007). Other virulence factors might include extracellular hydrolytic enzyme production and biofilm formation (Sobel 2007; Gonçalves et al. 2016).

18.2.3 Association between BV and VVC

Although BV and VVC are extremely common vaginal infections, information regarding the co-occurrence of both infections is still scarce. Rivers and colleagues found that, among women with BV, 33.1% were also colonized with yeast and 4.4% of the total women analyzed presented a BV/VVC mixed infection, based on the signs and symptoms described (Rivers, Adaramola, and Schwebke 2011). The study by Donders and colleagues corroborates these findings reporting that 6.3% of patients with symptomatic VVC also had BV (Donders et al. 2011). Furthermore, Wei and colleagues demonstrated the coexistence of *Candida* spp., *Gardnerella* spp., and other BV-associated bacteria on Pap smears (Wei et al. 2012). They showed a decrease in the exhibition of blastospores by *C. albicans* in coexistence with BV-associated bacteria. In addition, there is some evidence that VVC is a common side effect of BV treatment with antibiotics (Pirotta, Gunn, and Chondros 2003). Nevertheless, more studies are needed to better understand the co-occurrence of BV and VVC and their impact on treatment failure.

18.3 BIOFILM IN VAGINAL INFECTIONS

It is well known that in many infections, bacteria do not grow planktonically but assume a biofilm structure (Stewart and Costerton 2001). A biofilm can be defined as a dynamic and complex community that involves multiple interactions between single or multiple microbial species which are enveloped in a polymeric matrix and adherent to an inert or living interface (Visick et al. 2016). In the context of vaginal biofilms, biofilm development initiates during the initial attachment to the vaginal epithelial cells, cell clusters then start to be formed, and eventual dispersion of some cells will occur (Hardy et al. 2017).

18.3.1 BV-Associated Biofilms

It has been proposed that *Gardnerella* spp. have a preponderant role in the initial stages of biofilm formation (Muzny et al. 2019). Furthermore, it has been shown that *Gardnerella* spp. were able to displace beneficial *Lactobacillus* spp. (Castro et al. 2013) and had a higher initial adhesion to vaginal cells than other BV-associated bacteria (Patterson et al. 2010; Alves et al. 2014). Based on this evidence, it was postulated that *Gardnerella* spp. is the primary colonizer involved in BV biofilms development, paving the way to subsequent colonization by other species to adhere (Machado and Cerca 2015). This hypothesis has been supported by ex vivo studies. Notably, Swidsinski and colleagues were the first to demonstrate the presence of a dense biofilm in 90% on the vaginal epithelium of the subjects with BV (Swidsinski et al. 2005). They found that *Gardnerella* spp. composed of 60%–95% of the mass of the biofilm and *Atopobium vaginae* accounted for 1%–40% of the biofilm mass (Swidsinski et al. 2005). A subsequent in vivo study also showed a polymicrobial biofilm attached to the endometrial tissue in women with BV, composed of *Gardnerella* spp. and other bacteria (Swidsinski et al. 2013). More recently, Hardy and colleagues showed the presence of a polymicrobial biofilm in vaginal specimens from women with BV, in which *A. vaginae* was part of a *Gardnerella*-dominated biofilm (Hardy et al. 2015, 2016). Of note these polymicrobial biofilms could allow bacteria to persist in spite of antibiotic treatment (Swidsinski et al. 2008). Furthermore, *A. vaginae* has been found to promote an immune response from vaginal epithelial cells and contribute to the appearance of the typical symptoms (Mendling et al. 2019). There is also in vitro evidence of microbial interactions between *Gardnerella* spp.

and other bacterial species, such as *Prevotella bivia* (Castro, Machado, and Cerca 2019). *Gardnerella* and *P. bivia* are two bacteria that produce sialidase, which are a large group of enzymes, the majority of which catalyze the cleavage of terminal sialic acids from complex carbohydrates on glycoproteins or glycolipids of the vaginal epithelium. It has been proposed that the production of this virulence factor may induce the vaginal epithelial cells exfoliation and promote the invasion of *P. bivia* (Gilbert et al. 2019). Furthermore, other species were shown to induce specific gene expression changes in *Gardnerella* spp. mediated biofilms, concurrent with the typical symptoms of BV can develop (Cerca 2019).

18.3.2 VVC-Associated Biofilms

The role of *Candida* biofilms in VVC is still a matter of controversy (Xu, Qu, and Deighton 2019). Very recently, Swidsinski and colleagues demonstrated the absence of *Candida* biofilms in human vaginal tissue biopsies from patients with VVC. These authors showed that histopathological lesions were exclusively invasive and accompanied by co-invasion with *Gardnerella* spp. or *Lactobacillus* spp. (Swidsinski et al. 2019). Other efforts have been done to evaluate *Candida* biofilms using rodent models (Harriott et al. 2010; Zhang et al. 2013). Although these two studies showed that *C. albicans* forms biofilms on the vaginal mucosa, they were not able to confirm the biofilm on the vaginal epithelium. Importantly, Swidsinski and Sobel underlined that up to now, no direct microscopic evidence has been produced that *Candida* forms a biofilm on the human vaginal wall (Swidsinski and Sobel 2019). These authors suggested that until the visual proof of biofilm on the surface of intact human vaginal epithelium is available, all speculations about the existence of vaginal *Candida* biofilms will remain scientifically unsound.

Although most clinical isolates from patients with VVC remain sensitive to conventional antifungal agents, the ineffectiveness of these drugs against non-*C. albicans* strains commonly associated with recurrent VVC has been reported (Richter et al. 2005). Importantly, Xu and colleagues suggested that other strategies, other than intrinsic resistance, might be involved (Xu, Qu, and Deighton 2019). According to these authors, such possible strategies include the presence of a transient biofilm formation and the production of persister cells (a small population of "transiently resistant" cells). The possibility of the presence of such transient biofilms in patients whose infections have progressed to a later stage should not be excluded. The further clarification of the possible dynamic change of *Candida* growth in the vagina of VVC women and whether the endocytosed hyphal cell growth mode promotes persister cell production needs collaborative efforts from experimental microbiologists and clinical investigators.

18.4 CURRENT TREATMENT OF VAGINAL INFECTIONS

Since vaginal infections can result in serious complications for women's health, the treatment, cure, and management of BV and VVC are of high importance. It is also important to note that an inadequate treatment could be applied, as the result of a misdiagnosis, by the reason of the signs/symptoms characteristics of BV/VVC are similar (Nyirjesy 2008).

18.4.1 Standard BV Treatment

According to the Centers for Disease Control and Prevention (CDC), all symptomatic BV women, pregnant or not, should be treated (CDC 2015). The antibiotics commonly recommended are metronidazole and clindamycin, both available in oral and intravaginal preparations (Beigi, Austin, et al. 2004). The recommended regimens include oral metronidazole 500 mg twice daily for 7 days, intravaginal 5 g metronidazole gel once daily for 5 days, intravaginal 2% clindamycin cream once daily for 7 days. Alternative regimens propose to use, clindamycin 300 mg orally twice daily for 7 days, clindamycin ovules 100 mg intravaginally daily for 3 days, tinidazole 2 g orally daily for 2 days or 1 g orally daily for 5 days (CDC 2015). Metronidazole is a widely applied nitroimidazole used to treat infectious diseases, and it is the first-line antibiotic for treating BV (Beigi, Austin, et al. 2004; Deng et al. 2018). Clindamycin belongs to the family of macrolide antibiotics, and it is widely used due to its efficacy and a wide spectrum of action against most anaerobic bacteria as well as many aerobic Gram-positive cocci (Beigi, Austin, et al. 2004). As an alternative for BV treatment, tinidazole has been proposed when metronidazole or clindamycin is not well tolerated by the patients (Livengood et al. 2007; Dickey, Nailor, and Sobel 2009).

18.4.2 Standard VVC Treatment

In cases of VVC, several antifungals are nowadays available for treatment in a variety of formulations. Over-the-counter regimens include clotrimazole, miconazole, and tioconazole all available in intravaginal formulations. Regimens with prescription consist of butoconazole 2% vaginal cream 5 g in a single application, vaginal terconazole (0.4% cream 5 g daily for 7 days or 0.8% cream 5 g daily for 3 days and 80 mg vaginal suppository daily for 3 days) and oral fluconazole 150 mg orally in a single dose (CDC 2015). The choice of therapy will depend on the VVC status, meaning that complicated VVC will be addressed differently than uncomplicated VVC. Short-term local therapy or single-dose oral treatment with azoles is effective for treating 90% of uncomplicated VVC cases (Dovnik et al. 2015). Apparently, there is no difference in the efficacy of treatment between oral and intravaginal antifungal drugs (Watson et al. 2002). Topical azoles are remarkably safe and well-tolerated in 80%–90% of the cases (Dovnik et al. 2015). Women with complicated VVC can require more aggressive and prolonged therapy (Nyirjesy 2008). Despite that recurrent VVC responds well to short duration oral or topical therapy, it is necessary to maintain a mycological control, thus a longer duration of the initial therapy (7–14 days) is recommended before maintaining the antifungal regimen by oral fluconazole weekly for 6 months. If this therapy is not efficient, the use of topical treatments intermittently can be considered (CDC 2015).

18.5 BIOFILMS AND ANTIMICROBIAL THERAPY FAILURE

Antimicrobial therapy failure resulting in high recurrence rates of infections is one of the most important concerns among physicians and scientists. Besides specific resistance mechanisms (Khan, Miller, and Arias 2018), an important contributor to the increase in recurrent infections is associated with the biofilm phenotype, which is known to increase bacterial tolerance to antimicrobials (Rabin et al. 2015; Venkatesan, Perumal, and Doble 2015).

Different hypotheses have been proposed, aiming to clarify why biofilms are more tolerant to antimicrobials, including (i) the low or incomplete penetration of the antibiotic into the biofilm (Anderl, Franklin, and Stewart 2000; Tseng et al. 2013); (ii) the altered chemical environment throughout the biofilm that may antagonize the action of the antibiotic (Walters et al. 2003; Borriello et al. 2004); and (iii) the phenotypic state of some microorganisms on biofilm that can differentiate into an antimicrobial-resistant phenotype (Keren et al. 2004; Gaio and Cerca 2019).

Biofilms formed by *Gardnerella* spp. and *Candida* spp. in cases of BV and VVC, respectively, represent important virulence factors and mechanisms that contribute to the recurrence of infections (Muzny and Schwebke 2015).

While addressing BV biofilms, Swidsinski et al. conducted a study to observe the resistance levels to metronidazole (Swidsinski et al. 2008). It could be observed that polymicrobial biofilm was temporarily suppressed during the treatment; however, it regained its activity after the treatment (Swidsinski et al. 2008). Reasons for recurrence are unclear, but it might occur when the antibiotics fail to complete eradicate all pathogenic bacteria and, at the same time, normal lactobacilli fail to recolonize the vaginal microbiota (Sobel 2000) and due to the presence of multiple bacteria biofilm associated with the infection (Verstraelen and Swidsinski 2019). It has been highlighted that multi-species biofilms are harder to treat since antimicrobial agents directed to one of the species could facilitate non-target microorganisms to continue the infection (Thein et al. 2009). However, it is still not clear how the different BV-associated bacteria contribute to this recurrence. *A. vaginae*, which is rarely found alone but commonly co-isolated with *Gardnerella* spp. during BV, appears to be naturally resistant to metronidazole, and this might be one of the reasons why BV has high levels of recurrence (Ferris et al. 2004). Recently, it was also shown that genes related to antimicrobial resistance were upregulated in a *Gardnerella* spp. biofilm when specific BV-associated species were present (Castro, Machado, and Cerca 2019).

Similar to what has been shown for BV-associated biofilms, *Candida* biofilms also showed reduced susceptibility to antifungal agents (Mathé and Van Dijck 2013). Mechanisms of increased tolerance to antibiotics in VVC biofilms are multiple and include differential regulation of drug targets, upregulation of drug efflux pumps, persister cells in a dormant state, reduced growth rate, cell density, and matrix components that prevent the agents to reach their targets through the biofilm (Mathé and Van Dijck 2013; Muzny and Schwebke 2015). Due to this increase in both antimicrobial resistance and tolerance, reduced treatment options are now available, and this has led to an urgent need to develop novel therapies (Palmeira-de-Oliveira, Palmeira-de-Oliveira, and Martinez-de-Oliveira 2015; Machado et al. 2016).

18.6 ESSENTIAL OILS AS ALTERNATIVES IN THE TREATMENT OF VAGINAL INFECTIONS

Due to the chemical diversity of plant secondary metabolites, they are a potential source of new drugs (Savoia 2012). Among natural extracts, EOs have important properties (Bakkali et al. 2008; Seow et al. 2014). EOs derived from aromatic plants are complex mixtures of volatile substances containing volatile components, mainly terpenes, terpenoids, aromatic and aliphatic compounds (Bakkali et al. 2008; Savoia 2012). EOs have a wide spectrum of biological activities and are being studied in different fields due to their good antibacterial (Hammer, Carson, and Riley 1999b), antiviral (Garozzo et al. 2009), antioxidant (Bozin et al. 2007), and antifungal (Palmeira-de-Oliveira et al. 2009) properties. Interestingly, the use of EOs is associated with low antimicrobial resistance (Hammer, Carson, and Riley 2012). On the downside, they have a short life, are volatile and reactive in the presence of light, heat, moisture, and oxygen (Figueiredo et al. 2008).

The antimicrobial activity of EOs is somehow related to the activity of the main components, however, other works suggest that components present in small amounts in EOs can also play an important role in the antimicrobial activity due to possible synergistic effects (Herman, Tambor, and Herman 2016; Merghni et al. 2018). One of the disadvantages of using EOs is the fact that they present some variability in the compositions among different batches (Tommasi et al. 2007; Figueiredo et al. 2008) that can affect their properties and antimicrobial activities. Importantly, interactions between major and minor constituents can have a critical influence on global antimicrobial activity (Burt 2004).

During the last few years, the potential to use plant-derived substances against vaginal pathogens has been on the rise. In fact, in vaginal infections, some of these natural alternatives have been tested with promising results (Pietrella et al. 2011; Bogavac et al. 2015). Herein, we decided to present an overview of the use of EOs against microorganisms commonly associated with vaginal infections, namely VVC and BV. Several in vitro success stories have been reported, with a multitude of EOs demonstrating to have significant antimicrobial activity against BV or VVC. The biological effects of the EOs are reflected in high susceptibilities of the microorganisms to reduced concentrations of the EOs, resulting in a good capacity to inhibit the growth of planktonic cells and the formation of biofilms (Khan and Ahmad 2012; Mertas et al. 2015; Machado et al. 2017).

Other successful examples using EOs against microorganisms isolated from vaginal infections are listed in Table 18.1. Up to now, synergistic effects have been demonstrated between different EOs (Fu et al. 2007; Vieira et al. 2017), between components of the same EO (Pina-Vaz et al. 2004), and between EOs and conventional therapies, namely antibiotics (Silva et al. 2011; Knezevic et al. 2016).

Furthermore, it has been proposed that the use of EOs in combination with antimicrobial agents has the potential to increase the activity of some antimicrobial agents against a target microorganism and to combat antibiotic resistance (Owen and Laird 2018). As such, synchronous applications of EOs and antimicrobial agents have gained interest. However, up to know, no such study has been reported addressing BV. On the other hand, the application of antibiotics and EO against VVC have demonstrated evident synergistic effects, with small amounts of EO contributing to a significant decreased in the amount of antifungals necessary to inhibit the growth of *C. albicans* (Saad et al. 2010; Silva et al. 2011). Other examples are described in Table 18.2, in which in the majority of the cases, synergistic effects are reported.

The early successes obtained during in vitro testing have promoted the development of some clinical trials. Some examples are given in Table 18.3. In all examples

TABLE 18.1
Examples of the Use of Essential Oils Against Microorganisms Isolated from Cases of VVC and BV

Vaginal Infection	Microorganism	Essential Oil	Effect on Planktonic Cells	Effect on Biofilm	References
BV	*Gardnerella* spp. *Mobiluncus* spp. *Bacteroides*, *Prevotella*, fusobacteria, anerobic Gram-positive cocci	*Melaleuca alternifolia*	High capacity of EO to inhibit BV associated-bacteria	EOs showed a great capacity to inhibit biofilm formation and reduce the viability of cells	Hammer, Carson, and Riley (1999a)
BV	*Gardnerella* spp.	Thymol from thyme oil	Strains were susceptible to the EO	EO at a concentration of MIC showed efficacy to inhibit the formation of biofilm (around 30%) and reduced the biomass of a mature biofilm (~25%)	Braga et al. (2010)
VVC	*Candida albicans, C. tropicalis, C. glabrata, C. krusei*	*Cymbopogon citratus and Syzygium aromaticum*	*Candida* species were susceptible to the EOs	EOs showed a great capacity to inhibit biofilm formation and reduce the viability of cells	Khan and Ahmad (2012)
VVC	*C. albicans*	*M. alternifolia*	*C. albicans* strains were susceptible to low concentrations of the EO. Exposure to EO improved the susceptibility to fluconazole	N/A	Mertas et al. (2015)
BV	*Gardnerella* spp.	*Thymbra capitata*	The isolates were highly susceptible to the EO	High capacity of EO to reduce *Gardnerella* biofilm	Machado et al. (2017)
VVC	*C. albicans*	*Origanum vulgare*	The EO showed high activity against *C. albicans* and inhibited germ tube formation	N/A	Karaman et al. (2017)

BV: bacterial vaginosis; EO: essential oil; EOs: essential oils; N/A: not applicable; VVC: vulvovaginal candidiasis

TABLE 18.2
Examples of Essential Oils in Combination with Antimicrobial Agents Against Microorganisms Commonly Associated with VVC

Vaginal Infection	Microorganism	Antimicrobial Agent and EOs	EOs Activity	Combination	References
VVC	*Candida albicans*	Amphotericin B and fluconazole and *Thymus maroccanus* and *T. broussonettii*	Susceptibility of *Candida* to the EOs	Combinations between EOs and antifungal agents resulted in a significant decrease in the MIC values of antifungals, with the best combination observed in the combinations with fluconazole	Saad et al. (2010)
VVC	*C. albicans and C. tropicalis*	Amphotericin B and *Coriandrum sativum*	All *Candida* strains were susceptible to the EO and the germ tube formation was inhibited	Evident synergism between the oil and amphotericin B against *C. albicans*, and additive effect against *C. tropicalis*	Silva et al. (2011)
VVC	*C. albicans*	Fluconazole, econazole, ketoconazole and itraconazole, and *Melaleuca leucadendra*	EO showed activity against *C. albicans*	EO showed a synergistic activity in combination with all the antifungal agents	Zhang et al. (2019)
VVC	*C. albicans, C. glabrata, C. krusei*	Fluconazole and itraconazole and *Mentha x piperita*	The EO showed high antimicrobial activity against all the species	The combination between EO and fluconazole resulted in an additive effect against *C. albicans* and *C. glabrata* but indifferent effect against *C. krusei*. The combination of itraconazole and EO appeared to be a synergistic interaction against all tested strains	Tullio et al. (2019)

BV: bacterial vaginosis; EO: essential oil; EOs: essential oils; VVC: vulvovaginal candidiasis

TABLE 18.3
Examples of Clinical Trials Using Essential Oils in Cases of VVC and BV

Vaginal Infection	Conditions	Agents	Formulation and Application	Activity	References
VVC	2 groups (31 patients received treatment with clotrimazole, and 30 patients received the EO) diagnosed with VVC	*Zataria multiflora and clotrimazole*	*Z. multiflora* vaginal cream 0.1%, clotrimazole cream 1%, once daily for 7 days	The EO cream was effective to reduce symptoms associated with VVC, as well as the treatment with clotrimazole	Khosravi et al. (2008)
VVC and BV	221 cases of BV (122 treated with vaginal douche and 99 with metronidazole); 209 VVC cases (104 treated with vaginal douche and 105 with econazole)	Vaginal douche of thymol+eugenol versus metronidazole (BV) and econazole (VVC)	Thymol and eugenol (140 mL single dose, 1 vaginal douche/day for 1 week); econazole 150 mg suppositories 1/day for 1 week and metronidazole 500 mg suppository 1/day for 1 week	No significant differences were observed in the reduction of symptoms between the 2 groups in BV and VVC. The use of vaginal douches with active agents seemed to be as effective as the use of metronidazole or econazole	Sosto and Benvenuti (2011)
BV	2 groups of 70 women diagnosed with BV	*Z. multiflora* and metronidazole	*Z. multiflora* vaginal cream 0.1%, 50 g for 7 days and metronidazole 250 mg oral every 12 h for 7 days	The use of *Z. multiflora* cream reduced the symptoms associated with BV. No differences were found between the treatments. Side effects were higher using metronidazole	Abdali et al. (2015)
BV	2 groups of 40 women diagnosed with BV	*Myrtus communis* and metronidazole	*M. communis* 2% vaginal gel (5 g) in a base of metronidazole and metronidazole vaginal gel 0.75%, once daily for 5 days	Vaginal gel of EO in a base of metronidazole had better efficacy in the treatment of BV. The recurrence ratio was 30% in the metronidazole group and no recurrence was observed using EO in a base of metronidazole	Masoudi, Rafieian Kopaei, and Miraj (2019)

BV: bacterial vaginosis; EO: essential oil; VVC: vulvovaginal candidiasis given, the use of EOs showed capacity to eliminate symptoms associated with the infection (Abdali et al. 2015), and high effectiveness in the treatment of BV (Masoudi, Miraj, and Rafieian-Kopaei 2016). However, in many studies, recurrence rates are not improved, with some rare exceptions (Masoudi, Rafieian Kopaei, and Miraj 2019).

18.7 CONCLUSIONS

Vaginal infections are a common problem that can result in serious complications for women's health. The increasing antimicrobial tolerance and the high levels of recurrence are the main reasons why the development of novel therapeutic alternatives is of utmost importance. The use of natural products as therapeutic alternatives has been practiced since ancient times. EOs appear to be a good therapeutic alternative due to their good antibacterial, antifungal, and antioxidant properties.

The studies conducted to understand the effect of EOs against microorganisms commonly associated with BV and VVC have accentuated the potential application of this therapeutic alternative in vaginal infections. Indeed, the results described in some of these studies indicated that the use of EOs is a good option to recover from vaginal infections since EOs showed similar or even better effects than the common antimicrobial agents in relieving symptoms associated with the infections.

Future work should focus on better understanding how multi-species interaction might contribute to recurrence and EOs combinations might help overcome current treatment limitations.

ACKNOWLEDGMENT

Joana Castro and Lúcia Sousa are funded by the research project [PTDC/BIA-MIC/28271/2017], under the scope of COMPETE 2020 [POCI-01–0145-FEDER-028271].

REFERENCES

Abdali, Khadijeh, Leila Jahed, Sedigheh Amooee, Mahnaz Zarshenas, Hamidreza Tabatabaee, and Reza Bekhradi. 2015. "Comparison of the Effect of Vaginal *Zataria multiflora* Cream and Oral Metronidazole Pill on Results of Treatments for Vaginal Infections Including Trichomoniasis and Bacterial Vaginosis in Women of Reproductive Age." *BioMed Research International* 2015. Hindawi Publishing Corporation. doi:10.1155/2015/683640.

Abdul-Aziz, Maha, Mohammed A. K. Mahdy, Rashad Abdul-Ghani, Nuha A. Alhilali, Leena K. A. Al-Mujahed, Salma A. Alabsi, Fatima A. M. Al-Shawish, Noura J. M. Alsarari, Wala Bamashmos, Shahad J. H. Abdulwali, Mahdi Al Karawani, and Abdullah A. Almikhlafy 2019. "Bacterial Vaginosis, Vulvovaginal Candidiasis and Trichomonal Vaginitis among Reproductive-Aged Women Seeking Primary Healthcare in Sana'a City, Yemen." *BMC Infectious Diseases* 19 (1): 879. doi:10.1186/s12879-019-4549-3.

Aguin, Tina J., and Jack D. Sobel. 2015. "Vulvovaginal Candidiasis in Pregnancy." *Current Infectious Disease Reports*. Current Medicine Group LLC 1. doi:10.1007/s11908-015-0462-0.

Alpay Karaoğlu Şengül, Faruk Aydin, S. Sirri Kiliç, and Ali O. Kiliç. 2003. "Antimicrobial Activity and Characteristics of Bacteriocins Produced by Vaginal Lactobacilli." *Turkish Journal of Medical Sciences* 33 (1): 7–13.

Alves, Patrícia, Joana Castro, Cármen Sousa, Tatiana B. Cereija, and Nuno Cerca. 2014. "*Gardnerella vaginalis* Outcompetes 29 Other Bacterial Species Isolated from Patients With Bacterial Vaginosis, Using in an In Vitro Biofilm Formation Model." *The Journal of Infectious Diseases* 210 (4): 593–6. doi:10.1093/infdis/jiu131.

Amabebe, Emmanuel, and Dilly O. C. Anumba. 2018. "The Vaginal Microenvironment: The Physiologic Role of Lactobacilli." *Frontiers in Medicine* 5. Frontiers Media SA: 181. doi:10.3389/fmed.2018.00181.

Anderl, Jeff N., Michael J. Franklin, and Philip S. Stewart. 2000. "Role of Antibiotic Penetration Limitation in *Klebsiella pneumoniae* Biofilm Resistance to Ampicillin and Ciprofloxacin." *Antimicrobial Agents and Chemotherapy* 44 (7). American Society for Microbiology (ASM): 1818–24. doi:10.1128/aac.44.7.1818-1824.2000.

Bakkali, Fadil, Simone Averbeck, Dietrich Averbeck, and Mouhamed Idaomar. 2008. "Biological Effects of Essential Oils – A Review." *Food and Chemical Toxicology* 46 (2): 446–75. doi:10.1016/j.fct.2007.09.106.

Beigi, Richard H., Michele N. Austin, Leslie A. Meyn, Marijane A. Krohn, and Sharon L. Hillier. 2004. "Antimicrobial Resistance Associated with the Treatment of Bacterial Vaginosis." *American Journal of Obstetrics and Gynecology* 191 (4): 1124–9. doi:10.1016/j.ajog.2004.05.033.

Beigi, Richard H., Leslie A. Meyn, Donna M. Moore, Marijane A. Krohn, and Sharon L. Hillier. 2004. "Vaginal Yeast Colonization in Nonpregnant Women: A Longitudinal Study." *Obstetrics and Gynecology* 104 (5): 926–30. doi:10.1097/01.AOG.0000140687.51048.73.

Bitew, Adane, and Yeshiwork Abebaw. 2018. "Vulvovaginal Candidiasis: Species Distribution of *Candida* and Their Antifungal Susceptibility Pattern." *BMC Women's Health* 18 (1). BioMed Central Ltd. doi:10.1186/s12905-018-0607-z.

Bogavac, Mirjana, Maja Karaman, Ljiljana Janjušević, Jan Sudji, Bojan Radovanović, Z. Novaković, Jelica Simeunović, and Biljana Božin. 2015. "Alternative Treatment of Vaginal Infections—*In Vitro* Antimicrobial and Toxic Effects of *Coriandrum sativum* L. and *Thymus vulgaris* L. Essential Oils." *Journal of Applied Microbiology* 119 (3): 697–710. doi:10.1111/jam.12883.

Borriello, Giorgia, Erin Werner, Frank Roe, Aana M. Kim, Garth D. Ehrlich, and Philip S. Stewart. 2004. "Oxygen Limitation Contributes to Antibiotic Tolerance of *Pseudomonas aeruginosa* in Biofilms." *Antimicrobial Agents and Chemotherapy* 48 (7). American Society for Microbiology (ASM): 2659–64. doi:10.1128/AAC.48.7.2659-2664.2004.

Bozin, Biljana, Neda Mimica-Dukic, Isidora Samojlik, and Emilija Jovin. 2007. "Antimicrobial and Antioxidant Properties of Rosemary and Sage (*Rosmarinus officinalis* L. and *Salvia officinalis* L., Lamiaceae) Essential Oils." *Journal of Agricultural and Food Chemistry* 55 (19): 7879–85. doi:10.1021/jf0715323.

Bradshaw, Catriona S., Anna N. Morton, Suzanne M. Garland, Margaret B. Morris, Lorna M. Moss, and Christopher K. Fairley. 2005. "Higher-Risk Behavioral Practices Associated with Bacterial Vaginosis Compared with Vaginal Candidiasis." *Obstetrics and Gynecology* 106 (1). Lippincott Williams and Wilkins: 105–14. doi:10.1097/01.AOG.0000163247.78533.7b.

Bradshaw, Catriona S., and Jack D. Sobel. 2016. "Current Treatment of Bacterial Vaginosis—Limitations and Need for Innovation." *Journal of Infectious Diseases* 214 (suppl 1): S14–20. doi:10.1093/infdis/jiw159.

Braga, Pier Carlo, Monica Dal Sasso, Maria Culici, and Alessandra Spallino. 2010. "Inhibitory Activity of Thymol on Native and Mature *Gardnerella vaginalis* Biofilms: *In Vitro* Study." *Arzneimittel-Forschung/Drug Research* 60 (11): 675–81. doi:10.1055/s-0031-1296346.

Brooks, J. Paul, David J. Edwards, Diana L. Blithe, Jennifer M. Fettweis, Myrna G. Serrano, Nihar U. Sheth, Jerome F. Strauss, Gregory A. Buck, and Kimberly K. Jefferson. 2017. "Effects of Combined Oral Contraceptives, Depot Medroxyprogesterone Acetate and the Levonorgestrel-Releasing Intrauterine System on the Vaginal Microbiome." *Contraception* 95 (4): 405–13. doi:10.1016/j.contraception.2016.11.006.

Brotman, Rebecca M., Mark A. Klebanoff, Tonja R. Nansel, Kai F. Yu, William W. Andrews, Jun Zhang, and Jane R. Schwebke. 2010. "Bacterial Vaginosis Assessed by Gram Stain and Diminished Colonization Resistance to Incident Gonococcal, Chlamydial, and Trichomonal Genital Infection." *The Journal of Infectious Diseases* 202 (12): 1907–15. doi:10.1086/657320.

Burt, Sara. 2004. "Essential Oils: Their Antibacterial Properties and Potential Applications in Foods—A Review." *International Journal of Food Microbiology* 94 (3): 223–53. doi:10.1016/j.ijfoodmicro.2004.03.022.

Castro, Joana, Patrícia Alves, Cármen Sousa, Tatiana Cereija, Ângela França, Kimberly K. Jefferson, and Nuno Cerca. 2015. "Using an *In-Vitro* Biofilm Model to Assess the Virulence Potential of Bacterial Vaginosis or Non-Bacterial Vaginosis *Gardnerella vaginalis* Isolates." *Scientific Reports* 5 (June): 11640. doi:10.1038/srep11640.

Castro, Joana, Ana Henriques, António Machado, Mariana Henriques, Kimberly K. Jefferson, and Nuno Cerca. 2013. "Reciprocal Interference between *Lactobacillus* spp. and *Gardnerella vaginalis* on Initial Adherence to Epithelial Cells." *International Journal of Medical Sciences* 10 (9). Ivyspring International Publisher: 1193–8. doi:10.7150/ijms.6304.

Castro, Joana, Kimberly K. Jefferson, and Nuno Cerca. 2019. "Genetic Heterogeneity and Taxonomic Diversity among *Gardnerella* Species." *Trends in Microbiology*. Elsevier Ltd. doi:10.1016/j.tim.2019.10.002.

Castro, Joana, Daniela Machado, and Nuno Cerca. 2019. "Unveiling the Role of *Gardnerella vaginalis* in Polymicrobial Bacterial Vaginosis Biofilms: The Impact of Other Vaginal Pathogens Living as Neighbors." *ISME Journal* 13 (5). Nature Publishing Group: 1306–17. doi:10.1038/s41396-018-0337-0.

CDC. 2015. "Sexually Transmitted Diseases Treatment Guidelines, 2015." *MMWR. Recommendations and Reports: Morbidity and Mortality Weekly Report. Recommendations and Reports* 64 (RR-03): 1–137. http://www.ncbi.nlm.nih.gov/pubmed/26042815.

Ceccarani, Camilla, Claudio Foschi, Carola Parolin, Antonietta D'Antuono, Valeria Gaspari, Clarissa Consolandi, Luca Laghi, Tania Camboni, Beatrice Vitali, Marco Severgnini, and Antonella Marangoni. 2019. "Diversity of Vaginal Microbiome and Metabolome during Genital Infections." *Scientific Reports* 9 (1). Nature Publishing Group. doi:10.1038/s41598-019-50410-x.

Cerca, Nuno. 2019. "Could Targeting Neighboring Bacterial Populations Help Combat Bacterial Vaginosis?" *Future Microbiology* 14 (5): 365–8. doi:10.2217/fmb-2019-0045.

Cetin, Meryem, Sabahattin Ocak, Arif Gungoren, and Ali Ulvi Hakverdi. 2007. "Distribution of *Candida* Species in Women with Vulvovaginal Symptoms and Their Association with Different Ages and Contraceptive Methods." *Scandinavian Journal of Infectious Diseases* 39 (6–7): 584–8. doi:10.1080/00365540601148491.

Cherpes, Thomas L., Sharon L. Hillier, Leslie A. Meyn, James L. Busch, and Marijane A. Krohn. 2008. "A Delicate Balance: Risk Factors for Acquisition of Bacterial Vaginosis Include Sexual Activity, Absence of Hydrogen Peroxide-Producing Lactobacilli, Black Race, and Positive Herpes Simplex Virus Type 2 Serology." *Sexually Transmitted Diseases* 35 (1): 78–83. doi:10.1097/OLQ.0b013e318156a5d0.

Corsello, Salvatore, Arsenio Spinillo, Giuseppe Osnengo, Carlo Penna, Secondo Guaschino, Anna Beltrame, Nicola Blasi, and Antonio Festa . 2003. "An Epidemiological Survey of Vulvovaginal Candidiasis in Italy." *European Journal of Obstetrics and Gynecology and Reproductive Biology* 110 (1). Elsevier Ireland Ltd: 66–72. doi:10.1016/S0301-2115(03)00096-4.

Deng, Zhi-Luo, Cornelia Gottschick, Sabin Bhuju, Clarissa Masur, Christoph Abels, and Irene Wagner-Döbler. 2018. "Metatranscriptome Analysis of the Vaginal Microbiota Reveals

Potential Mechanisms for Protection against Metronidazole in Bacterial Vaginosis." Edited by Craig D. Ellermeier. *MSphere* 3 (3). doi:10.1128/mSphereDirect.00262-18.

Denning, David W., Matthew Kneale, Jack D. Sobel, and Riina Rautemaa-Richardson. 2018. "Global Burden of Recurrent Vulvovaginal Candidiasis: A Systematic Review." *The Lancet Infectious Diseases*. Lancet Publishing Group. doi:10.1016/S1473-3099(18)30103-8.

Dickey, Laura J., Michael D. Nailor, and Jack D. Sobel. 2009. "Guidelines for the Treatment of Bacterial Vaginosis: Focus on Tinidazole." *Therapeutics and Clinical Risk Management* 5 (3): 485–9. http://www.ncbi.nlm.nih.gov/pubmed/19707258.

DiGiulio, Daniel B., Benjamin J. Callahan, Paul J. McMurdie, Elizabeth K. Costello, Deirdre J. Lyell, Anna Robaczewska, Christine L. Sun, Daniela S. A. Goltsman, Ronald J. Wong, Gary Shaw, David K. Stevenson, Susan P. Holmes, and David A. Relman . 2015. "Temporal and Spatial Variation of the Human Microbiota during Pregnancy." *Proceedings of the National Academy of Sciences of the United States of America* 112 (35): 11060–5. doi:10.1073/pnas.1502875112.

Diop, Khoudia, Jean Charles Dufour, Anthony Levasseur, and Florence Fenollar. 2019. "Exhaustive Repertoire of Human Vaginal Microbiota." *Human Microbiome Journal*. Elsevier Ltd. doi:10.1016/j.humic.2018.11.002.

Donders, Gilbert, Gert Bellen, Jannie Ausma, Luc Verguts, Johan Vaneldere, Piet Hinoul, Marcel Borgers, and Dirk Janssens. 2011. "The Effect of Antifungal Treatment on the Vaginal Flora of Women with Vulvovaginal Yeast Infection with or without Bacterial Vaginosis." *European Journal of Clinical Microbiology and Infectious Diseases* 30 (1): 59–63. doi:10.1007/s10096-010-1052-6.

Dovnik, Andraž, Andrej Golle, Dušan Novak, Darja Arko, and Iztok Takač. 2015. "Treatment of Vulvovaginal Candidiasis: A Review of the Literature." *Acta Dermatovenerologica Alpina, Pannonica et Adriatica* 24 (1). Slovenian Medical Society: 5–7. doi:10.15570/actaapa.2015.2.

Duerr, Ann, Charles M. Heilig, Susan F. Meikle, Susan Cu-Uvin, Robert S. Klein, Anne Rompalo, and Jack D. Sobel. 2003. "Incident and Persistent Vulvovaginal Candidiasis among Human Immunodeficiency Virus-Infected Women: Risk Factors and Severity." *Obstetrics and Gynecology* 101 (3). Elsevier Inc.: 548–56. doi:10.1016/S0029-7844(02)02729-1.

Eschenbach, David A., Palmella R. Davick, Betsy L. Williams, Seymour J. Klebanoff, Karen Young-Smith, Cathy M. Critchlow, and King K. Holmes. 1989. "Prevalence of Hydrogen Peroxide-Producing *Lactobacillus* Species in Normal Women and Women with Bacterial Vaginosis." *Journal of Clinical Microbiology* 27 (2): 251–6. http://www.ncbi.nlm.nih.gov/pubmed/2915019.

Eschenbach, David A., Soe Soe Thwin, Dorothy L. Patton, Thomas M. Hooton, Ann E. Stapleton, Kathy Agnew, Carol Winter, Amalia Meier, and Walter E. Stamm. 2000. "Influence of the Normal Menstrual Cycle on Vaginal Tissue, Discharge, and Microflora." *Clinical Infectious Diseases* 30 (6): 901–7. doi:10.1086/313818.

Ferris, Michael J., Alicia Masztal, Kenneth E. Aldridge, J. Dennis Fortenberry, Paul L. Fidel, and David H. Martin. 2004. "Association of *Atopobium vaginae*, a Recently Described Metronidazole-Resistant Anaerobe, with Bacterial Vaginosis." *BMC Infectious Diseases* 4 (1): 5. doi:10.1186/1471-2334-4-5.

Figueiredo, Ana C., José G. Barroso, Luís G. Pedro, Lígia Salgueiro, Maria G. Miguel, and Maria L. Faleiro. 2008. "Portuguese *Thymbra* and *Thymus* Species Volatiles: Chemical Composition and Biological Activities." *Current Pharmaceutical Design* 14 (29): 3120–40. http://www.ncbi.nlm.nih.gov/pubmed/19075695.

Fredricks, David N., Tina L. Fiedler, and Jeanne M. Marrazzo. 2005. "Molecular Identification of Bacteria Associated with Bacterial Vaginosis." *New England Journal of Medicine* 353 (18): 1899–911. doi:10.1056/NEJMoa043802.

Fu, YuJie, YuanGang Zu, LiYan Chen, XiaoGuang Shi, Zhe Wang, Su Sun, and Thomas Efferth. 2007. "Antimicrobial Activity of Clove and Rosemary Essential Oils Alone and in Combination." *Phytotherapy Research* 21 (10): 989–94. doi:10.1002/ptr.2179.

Gaio, Vânia, and Nuno Cerca. 2019. "Cells Released from *S. epidermidis* Biofilms Present Increased Antibiotic Tolerance to Multiple Antibiotics." *PeerJ* 7 (May): e6884. doi:10.7717/peerj.6884.

Garozzo, Adriana, Rossella Timpanaro, Benedetta Bisignano, Pio M. Furneri, Giuseppe Bisignano, and Angelo Castro. 2009. "*In Vitro* Antiviral Activity of *Melaleuca alternifolia* Essential Oil." *Letters in Applied Microbiology* 49 (6): 806–8. doi:10.1111/j.1472-765X.2009.02740.x.

Gibbs, Ronald S. 2007. "Asymptomatic Bacterial Vaginosis: Is It Time to Treat?" *American Journal of Obstetrics and Gynecology*, June. doi:10.1016/j.ajog.2007.04.001.

Gilbert, Nicole M., Warren G. Lewis, Guocai Li, Dorothy K. Sojka, Jean Bernard Lubin, and Amanda L. Lewis. 2019. "*Gardnerella vaginalis* and *Prevotella bivia* Trigger Distinct and Overlapping Phenotypes in a Mouse Model of Bacterial Vaginosis." *The Journal of Infectious Diseases* 220 (7). Oxford University Press (OUP): 1099–108. doi:10.1093/infdis/jiy704.

Gonçalves, Bruna, Carina Ferreira, Carlos Tiago Alves, Mariana Henriques, Joana Azeredo, and Sónia Silva. 2016. "Vulvovaginal Candidiasis: Epidemiology, Microbiology and Risk Factors." *Critical Reviews in Microbiology*. Taylor and Francis Ltd. doi:10.3109/1040841X.2015.1091805.

Hammer, Katherine A., Christine F. Carson, and Thomas V. Riley. 1999a. "In Vitro Susceptibilities of Lactobacilli and Organisms Associated with Bacterial Vaginosis to *Melaleuca alternifolia* (Tea Tree) Oil." *Antimicrobial Agents and Chemotherapy* 43 (1): 196. http://www.ncbi.nlm.nih.gov/pubmed/10094671.

Hammer, Katherine A., Christine F. Carson, and Thomas V. Riley. 1999b. "Antimicrobial Activity of Essential Oils and Other Plant Extracts." *Journal of Applied Microbiology* 86 (6): 985–90. doi:10.1046/j.1365-2672.1999.00780.x.

Hammer, Katherine A., Christine F. Carson, and Thomas V. Riley. 2012. "Effects of *Melaleuca alternifolia* (Tea Tree) Essential Oil and the Major Monoterpene Component Terpinen-4-Ol on the Development of Single- and Multistep Antibiotic Resistance and Antimicrobial Susceptibility." *Antimicrobial Agents and Chemotherapy* 56 (2): 909–15. doi:10.1128/AAC.05741-11.

Hardy, Liselotte, Nuno Cerca, Vicky Jespers, Mario Vaneechoutte, and Tania Crucitti. 2017. "Bacterial Biofilms in the Vagina." *Research in Microbiology* 168 (9–10): 865–74. doi:10.1016/j.resmic.2017.02.001.

Hardy, Liselotte, Vicky Jespers, Said Abdellati, Irith De Baetselier, Lambert Mwambarangwe, Viateur Musengamana, Janneke Van De Wijgert, Mario Vaneechoutte, and Tania Crucitti. 2016. "A Fruitful Alliance: The Synergy between *Atopobium vaginae* and *Gardnerella vaginalis* in Bacterial Vaginosis-Associated Biofilm." *Sexually Transmitted Infections* 92 (7). BMJ Publishing Group: 487–91. doi:10.1136/sextrans-2015-052475.

Hardy, Liselotte, Vicky Jespers, Nassira Dahchour, Lambert Mwambarangwe, Viateur Musengamana, Mario Vaneechoutte, and Tania Crucitti. 2015. "Unravelling the Bacterial Vaginosis-Associated Biofilm: A Multiplex *Gardnerella vaginalis* and *Atopobium vaginae* Fluorescence *In Situ* Hybridization Assay Using Peptide Nucleic Acid Probes." Edited by A Al-Ahmad. *PLoS One* 10 (8): e0136658. doi:10.1371/journal.pone.0136658.

Harmanli, Ozgur H., Grace Y. Cheng, Paul Nyirjesy, Ashwin Chatwani, and John P. Gaughan. 2000. "Urinary Tract Infections in Women with Bacterial Vaginosis." *Obstetrics & Gynecology* 95 (5). No longer published by Elsevier: 710–2. doi:10.1016/S0029-7844(99)00632-8.

Harriott, Melphine M., Elizabeth A. Lilly, Tobias E. Rodriguez, Paul L. Fidel, and Mairi C. Noverr. 2010. "*Candida albicans* Forms Biofilms on the Vaginal Mucosa." *Microbiology* 156 (12): 3635–44. doi:10.1099/mic.0.039354-0.

Hay, Phillip. 2014. "Bacterial Vaginosis." *Medicine* 42 (7). Elsevier: 359–63. doi:10.1016/j.mpmed.2014.04.011.

Herman, Anna, Krzysztof Tambor, and Andrzej Herman. 2016. "Linalool Affects the Antimicrobial Efficacy of Essential Oils." *Current Microbiology* 72 (2): 165–72. doi:10.1007/s00284-015-0933-4.

Hickey, Roxana J., and Larry J. Forney. 2014. "*Gardnerella vaginalis* Does Not Always Cause Bacterial Vaginosis." *Journal of Infectious Diseases* 210 (10): 1682–3. doi:10.1093/infdis/jiu303.

Hill, Janet E., Arianne Y. K. Albert, and the VOGUE Research Group. 2019. "Resolution and Cooccurrence Patterns of *Gardnerella leopoldii*, *G. swidsinskii*, *G. piotii*, and *G. vaginalis* within the Vaginal Microbiome." *Infection and Immunity*. doi:10.1128/IAI.00532-19.

Hodoglugil, Nuriye Nalan Sahin, Dilek Aslan, and Munevver Bertan. 2000. "Intrauterine Device Use and Some Issues Related to Sexually Transmitted Disease Screening and Occurrence." *Contraception* 61 (6): 359–64. doi:10.1016/S0010-7824(00)00118-9.

Isik, Gözde, Şayeste Demirezen, HanifeGüler Dönmez, and MehmetSinan Beksaç. 2016. "Bacterial Vaginosis in Association with Spontaneous Abortion and Recurrent Pregnancy Losses." *Journal of Cytology* 33 (3): 135. doi:10.4103/0970-9371.188050.

Jacobsson, Bo, Peter Pernevi, Lene Chidekel, and Jens Jörgen Platz-Christensen. 2002. "Bacterial Vaginosis in Early Pregnancy May Predispose for Preterm Birth and Postpartum Endometritis." *Acta Obstetricia et Gynecologica Scandinavica* 81 (11): 1006–10. http://www.ncbi.nlm.nih.gov/pubmed/12421167.

Janulaitiene, Migle, Virginija Paliulyte, Svitrigaile Grinceviciene, Jolita Zakareviciene, Alma Vladisauskiene, Agne Marcinkute, and Milda Pleckaityte. 2017. "Prevalence and Distribution of *Gardnerella vaginalis* Subgroups in Women with and without Bacterial Vaginosis." *BMC Infectious Diseases* 17 (1). BioMed Central: 394. doi:10.1186/s12879-017-2501-y.

Jung, Hyun-Sul, Marthie M. Ehlers, Hennie Lombaard, Mathys J. Redelinghuys, and Marleen M. Kock. 2017. "Etiology of Bacterial Vaginosis and Polymicrobial Biofilm Formation." *Critical Reviews in Microbiology* 43 (6): 651–67. doi:10.1080/1040841X.2017.1291579.

Karaman, Maja, Mirjana Bogavac, Bojan Radovanović, Jan Sudji, Kristina Tešanović, and Ljiljana Janjušević. 2017. "*Origanum vulgare* Essential Oil Affects Pathogens Causing Vaginal Infections." *Journal of Applied Microbiology* 122 (5). Blackwell Publishing Ltd: 1177–85. doi:10.1111/jam.13413.

Keren, Iris, Niilo Kaldalu, Amy Spoering, Yipeng Wang, and Kim Lewis. 2004. "Persister Cells and Tolerance to Antimicrobials." *FEMS Microbiology Letters* 230 (1). Narnia: 13–8. doi:10.1016/S0378-1097(03)00856-5.

Khan, Mohd Sajjad Ahmad, and Iqbal Ahmad. 2012. "Biofilm Inhibition by *Cymbopogon citratus* and *Syzygium aromaticum* Essential Oils in the Strains of *Candida albicans*." *Journal of Ethnopharmacology* 140 (2): 416–23. doi:10.1016/j.jep.2012.01.045.

Khan, Ayesha, William R. Miller, and Cesar A. Arias. 2018. "Mechanisms of Antimicrobial Resistance among Hospital-Associated Pathogens." *Expert Review of Anti-Infective Therapy*. Taylor and Francis Ltd. doi:10.1080/14787210.2018.1456919.

Khosravi, Ali Reza, Ali Reza Eslami, Hojjatollah Shokri, and Mohammed Kashanian. 2008. "*Zataria multiflora* Cream for the Treatment of Acute Vaginal Candidiasis." *International Journal of Gynecology and Obstetrics* 101 (2). John Wiley and Sons Ltd: 201–2. doi:10.1016/j.ijgo.2007.11.010.

Kim, Tae Kyung, Susan M. Thomas, Mengfei Ho, Shobha Sharma, Claudia I. Reich, Jeremy A. Frank, Kathleen M. Yeater, Diana R. Biggs, Noriko Nakamura, Rebecca Stumpf, Steven R. Leigh, Richard I. Tapping, Steven R. Blanke, James M. Slauch, H. Rex Gaskins, Jon S. Weisbaum, Gary J. Olsen, Lois L. Hoyer, and Brenda A. Wilson . 2009. "Heterogeneity of Vaginal Microbial Communities within Individuals." *Journal of Clinical Microbiology* 47 (4): 1181–9. doi:10.1128/JCM.00854-08.

Knezevic, Petar, Verica Aleksic, Natasa Simin, Emilija Svircev, Aleksandra Petrovic, and Neda Mimica-Dukic. 2016. "Antimicrobial Activity of *Eucalyptus camaldulensis* Essential Oils and Their Interactions with Conventional Antimicrobial Agents against Multi-Drug-Resistant *Acinetobacter baumannii*." *Journal of Ethnopharmacology* 178 (February): 125–36. doi:10.1016/j.jep.2015.12.008.

Konadu, Dennis Gyasi, Alex Owusu-Ofori, Zuwera Yidana, Farrid Boadu, Louisa Fatahiya Iddrisu, Dennis Adu-Gyasi, David Dosoo, Robert Lartey Awuley, Seth Owusu-Agyei, and Kwaku Poku Asante. 2019. "Prevalence of Vulvovaginal Candidiasis, Bacterial Vaginosis, and Trichomoniasis in Pregnant Women Attending Antenatal Clinic in the Middle Belt of Ghana." *BMC Pregnancy and Childbirth* 19 (1). BioMed Central Ltd. doi:10.1186/s12884-019-2488-z.

Larsson, Per Göran, Jens-Jörgen Platz-Christensen, Knut Dalaker, Katarina Eriksson, Lars Fåhraeus, Kristine Irminger, Fritjof Jerve, Babill Stray-Pedersen, and Pal Wölner-Hanssen. 2000. "Treatment with 2% Clindamycin Vaginal Cream Prior to First Trimester Surgical Abortion to Reduce Signs of Postoperative Infection: A Prospective, Double-Blinded, Placebo-Controlled, Multicenter Study." *Acta Obstetricia et Gynecologica Scandinavica* 79 (5): 390–6. http://www.ncbi.nlm.nih.gov/pubmed/10830767.

de Leon, Ella M., Scott J. Jacober, Jack D. Sobel, and Betsy Foxman. 2002. "Prevalence and Risk Factors for Vaginal Candida Colonization in Women with Type 1 and Type 2 Diabetes." *BMC Infectious Diseases* 2 (January). doi:10.1186/1471-2334-2-1.

Livengood, Charles H. 2009. "Bacterial Vaginosis: An Overview for 2009." *Reviews in Obstetrics & Gynecology* 2 (1). MedReviews, LLC: 28–37. http://www.ncbi.nlm.nih.gov/pubmed/19399292.

Livengood, Charles H., Daron G. Ferris, Harold C. Wiesenfeld, Sharon L. Hillier, David E. Soper, Paul Nyirjesy, Jeanne Marrazzo, Ashwin Chatwani, Paul Fine, Jack Sobel, Stephanie N. Taylor, Lindsey Wood, and John J. Kanalas . 2007. "Effectiveness of Two Tinidazole Regimens in Treatment of Bacterial Vaginosis." *Obstetrics & Gynecology* 110 (2, Part 1): 302–9. doi:10.1097/01.AOG.0000275282.60506.3d.

Ma, Bing, Larry J. Forney, and Jacques Ravel. 2012. "Vaginal Microbiome: Rethinking Health and Disease." *Annual Review of Microbiology* 66 (1). Annual Reviews: 371–89. doi:10.1146/annurev-micro-092611-150157.

Machado, António, and Nuno Cerca. 2015. "Influence of Biofilm Formation by *Gardnerella vaginalis* and Other Anaerobes on Bacterial Vaginosis." *Journal of Infectious Diseases* 212 (12). Oxford University Press: 1856–61. doi:10.1093/infdis/jiv338.

Machado, Daniela, Joana Castro, Ana Palmeira-de-Oliveira, José Martinez-de-Oliveira, and Nuno Cerca. 2016. "Bacterial Vaginosis Biofilms: Challenges to Current Therapies and Emerging Solutions." *Frontiers in Microbiology*. Frontiers Media S.A. doi:10.3389/fmicb.2015.01528.

Machado, Daniela, Carlos Gaspar, Ana Palmeira-de-Oliveira, Carlos Cavaleiro, Lígia Salgueiro, José Martinez-de-Oliveira, and Nuno Cerca. 2017. "*Thymbra capitata* Essential Oil as Potential Therapeutic Agent against *Gardnerella vaginalis* Biofilm-Related Infections." *Future Microbiology* 12: 407–16. doi:10.2217/fmb-2016-0184.

Martin, David H. 2012. "The Microbiota of the Vagina and Its Influence on Women's Health and Disease." *American Journal of the Medical Sciences* 343: 2–9. Lippincott Williams and Wilkins. doi:10.1097/MAJ.0b013e31823ea228.

Masoudi, Mansoureh, Mahmoud Rafieian Kopaei, and Sepideh Miraj. 2019. "A Comparison of the Efficacy of Metronidazole Vaginal Gel and Myrtus (*Myrtus communis*) Extract Combination and Metronidazole Vaginal Gel Alone in the Treatment of Recurrent Bacterial Vaginosis." *Avicenna Journal of Phytomedicine* 7 (2): 129–36. Accessed November 28. http://www.ncbi.nlm.nih.gov/pubmed/28348968.

Masoudi, Mansoureh, Sepideh Miraj, and Mahmoud Rafieian-Kopaei. 2016. "Comparison of the Effects of *Myrtus communis L*, *Berberis vulgaris*, and Metronidazole Vaginal Gel Alone for the Treatment of Bacterial Vaginosis." *Journal of Clinical and Diagnostic Research* 10 (3). doi:10.7860/JCDR/2016/17211.7392.

Mathé, Lotte, and Patrick Van Dijck. 2013. "Recent Insights into *Candida albicans* Biofilm Resistance Mechanisms." *Current Genetics* 59 (4): 251–64. doi:10.1007/s00294-013-0400-3.

Mendling, Werner, Ana Palmeira-de-Oliveira, Stephan Biber, and Valdas Prasauskas. 2019. "An Update on the Role of *Atopobium vaginae* in Bacterial Vaginosis: What to Consider When Choosing a Treatment? A Mini Review." *Archives of Gynecology and Obstetrics*. Springer Verlag. doi:10.1007/s00404-019-05142-8.

Merghni, Abderrahmen, Emira Noumi, Ons Hadded, Neyla Dridi, Harsh Panwar, Ozgur Ceylan, Maha Mastouri, and Mejdi Snoussi. 2018. "Assessment of the Antibiofilm and Antiquorum Sensing Activities of *Eucalyptus globulus* Essential Oil and Its Main Component 1,8-Cineole against Methicillin-Resistant *Staphylococcus aureus* Strains." *Microbial Pathogenesis* 118 (May): 74–80. doi:10.1016/j.micpath.2018.03.006.

Mertas, Anna, Aleksandra Garbusińska, Ewelina Szliszka, Andrzej Jureczko, Magdalena Kowalska, and Wojciech Król. 2015. "The Influence of Tea Tree Oil (*Melaleuca alternifolia*) on Fluconazole Activity against Fluconazole-Resistant *Candida albicans* Strains." *BioMed Research International* 2015. Hindawi Publishing Corporation. doi:10.1155/2015/590470.

Muzny, Christina A., and Jane R. Schwebke. 2015. "Biofilms: An Underappreciated Mechanism of Treatment Failure and Recurrence in Vaginal Infections." *Clinical Infectious Diseases*. Oxford University Press. doi:10.1093/cid/civ353.

Muzny, Christina A., Christopher M. Taylor, W. Edward Swords, Ashutosh Tamhane, Debasish Chattopadhyay, Nuno Cerca, and Jane R. Schwebke. 2019. "An Updated Conceptual Model on the Pathogenesis of Bacterial Vaginosis." *The Journal of Infectious Diseases* 220 (9). Oxford University Press (OUP): 1399–405. doi:10.1093/infdis/jiz342.

Ness, Roberta B., Sharon L. Hillier, Holly E. Richter, David E. Soper, Carol Stamm, James McGregor, Debra C. Bass, Richard L. Sweet, and Peter Rice. 2002. "Douching in Relation to Bacterial Vaginosis, Lactobacilli, and Facultative Bacteria in the Vagina." *Obstetrics and Gynecology* 100 (4): 765. http://www.ncbi.nlm.nih.gov/pubmed/12383547.

Ness, Roberta B., Kevin E. Kip, Sharon L. Hillier, David E. Soper, Carol A. Stamm, Richard L. Sweet, Peter Rice, and Holly E. Richter. 2005. "A Cluster Analysis of Bacterial Vaginosis-Associated Microflora and Pelvic Inflammatory Disease." *American Journal of Epidemiology* 162 (6): 585–90. doi:10.1093/aje/kwi243.

Nyirjesy, Paul. 2008. "Vulvovaginal Candidiasis and Bacterial Vaginosis." *Infectious Disease Clinics of North America*. doi:10.1016/j.idc.2008.05.002.

Nyirjesy, Paul, and Jack D. Sobel. 2003. "Vulvovaginal Candidiasis." *Obstetrics and Gynecology Clinics of North America* 30 (4): 671–84. http://www.ncbi.nlm.nih.gov/pubmed/14719844.

O'Hanlon, Deirdre E., Thomas R. Moench, and Richard A. Cone. 2011. "In Vaginal Fluid, Bacteria Associated with Bacterial Vaginosis Can Be Suppressed with Lactic Acid but Not Hydrogen Peroxide." *BMC Infectious Diseases* 11 (July). doi:10.1186/1471-2334-11-200.

Oakley, Brian B., Tina L. Fiedler, Jeanne M. Marrazzo, and David N. Fredricks. 2008. "Diversity of Human Vaginal Bacterial Communities and Associations with Clinically Defined Bacterial Vaginosis." *Applied and Environmental Microbiology* 74 (15). American Society for Microbiology (ASM): 4898–909. doi:10.1128/AEM.02884-07.

Ojala, Teija, Matti Kankainen, Joana Castro, Nuno Cerca, Sanna Edelman, Benita Westerlund-Wikström, Lars Paulin, Liisa Holm, and Petri Auvinen. 2014. "Comparative Genomics of *Lactobacillus crispatus* Suggests Novel Mechanisms for the Competitive Exclusion of *Gardnerella vaginalis*." *BMC Genomics* 15 (1): 1070. doi:10.1186/1471-2164-15-1070.

Onderdonk, Andrew B., Mary L. Delaney, and Raina N. Fichorova. 2016. "The Human Microbiome during Bacterial Vaginosis." *Clinical Microbiology Reviews* 29 (2): 223–38. doi:10.1128/CMR.00075-15.

Oscáriz, Juan C., and Antonio G. Pisabarro. 2001. "Classification and Mode of Action of Membrane-Active Bacteriocins Produced by Gram-Positive Bacteria." *International Microbiology* 4 (1). Springer-Verlag GmbH Co. KG: 13–9. doi:10.1007/s101230100003.

Owen, Lucy, and Katie Laird. 2018. "Synchronous Application of Antibiotics and Essential Oils: Dual Mechanisms of Action as a Potential Solution to Antibiotic Resistance." *Critical Reviews in Microbiology* 44 (4): 414–35. doi:10.1080/1040841X.2018.1423616.

Paladine, Heather L., and Urmi A. Desai. 2018. "Vaginitis: Diagnosis and Treatment." *American Family Physician*. NLM (Medline). doi:10.2165/00003495–197204050–00004.

Palmeira-de-Oliveira, Rita, Ana Palmeira-de-Oliveira, and José Martinez-de-Oliveira. 2015. "New Strategies for Local Treatment of Vaginal Infections." *Advanced Drug Delivery Reviews* 92 (September): 105–22. doi:10.1016/j.addr.2015.06.008.

Palmeira-de-Oliveira, Ana, Lígia Salgueiro, Rita Palmeira-de-Oliveira, José Martinez-de-Oliveira, Cidália Pina-Vaz, João A. Queiroz, and Acácio G. Rodrigues. 2009. "Anti-*Candida* Activity of Essential Oils." *Mini-Reviews in Medicinal Chemistry* 9 (11): 1292–305. doi:10.2174/138955709789878150.

Patterson, Jennifer L., Annica Stull-Lane, Philippe H. Girerd, and Kimberly K. Jefferson. 2010. "Analysis of Adherence, Biofilm Formation and Cytotoxicity Suggests a Greater Virulence Potential of *Gardnerella vaginalis* Relative to Other Bacterial-Vaginosis-Associated Anaerobes." *Microbiology (Reading, England)* 156 (Pt 2). Microbiology Society: 392–9. doi:10.1099/mic.0.034280-0.

Peebles, Kathryn, Jennifer Velloza, Jennifer E. Balkus, R. Scott McClelland, and Ruanne V. Barnabas. 2019. "High Global Burden and Costs of Bacterial Vaginosis: A Systematic Review and Meta-Analysis." *Sexually Transmitted Diseases*. Lippincott Williams and Wilkins. doi:10.1097/OLQ.0000000000000972.

Pietrella, Donatella, Letizia Angiolella, Elisabetta Vavala, Anna Rachini, Francesca Mondello, Rino Ragno, Francesco Bistoni, and Anna Vecchiarelli. 2011. "Beneficial Effect of *Mentha suaveolens* Essential Oil in the Treatment of Vaginal Candidiasis Assessed by Real-Time Monitoring of Infection." *BMC Complementary and Alternative Medicine* 11 (1): 18. doi:10.1186/1472-6882-11-18.

Pina-Vaz, Cidália, Acácio Goncalves Rodrigues, Eugénia Pinto, Sofia https://microbialbiofilms.wordpress.com/current-team/hristina Tavares, Lígia Salgueiro, Carlos Cavaleiro, Maria J. Goncalves, and José Martinez-de-Oliveira. 2004. "Antifungal Activity of *Thymus* Oils and Their Major Compounds." *Journal of the European Academy of Dermatology and Venereology* 18 (1): 73–8. doi:10.1111/j.1468-3083.2004.00886.x.

Pirotta, Marie V., and Suzanne M. Garland. 2006. "Genital *Candida* Species Detected in Samples from Women in Melbourne, Australia, before and after Treatment with Antibiotics." *Journal of Clinical Microbiology* 44 (9): 3213–7. doi:10.1128/JCM.00218-06.

Pirotta, Marie V., Jane M. Gunn, and Patty Chondros. 2003. "'Not Thrush Again!' Women's Experience of Post-Antibiotic Vulvovaginitis." *The Medical Journal of Australia* 179 (1): 43–6. http://www.ncbi.nlm.nih.gov/pubmed/12831384.

Rabin, Nira, Yue Zheng, Clement Opoku-Temeng, Yixuan Du, Eric Bonsu, and Herman O. Sintim. 2015. "Biofilm Formation Mechanisms and Targets for Developing Antibiofilm Agents." *Future Medicinal Chemistry*. Future Science. doi:10.4155/fmc.15.6.

Ravel, Jacques, Pawel Gajer, Zaid Abdo, G. Maria Schneider, Sara S. K. Koenig, Stacey L. McCulle, Shara Karlebach, Reshma Gorle, Jennifer Russell, Carol O. Tacket, Rebecca

M. Brotman, Catherine C. Davis, Kevin Ault, Ligia Peralta, and Larry J. Forney . 2011. "Vaginal Microbiome of Reproductive-Age Women." *Proceedings of the National Academy of Sciences* 108 (Supplement_1): 4680–7. doi:10.1073/pnas.1002611107.

Richter, Sandra S., Rudolph P. Galask, Shawn A. Messer, Richard J. Hollis, Daniel J. Diekema, and Michael A. Pfaller. 2005. "Antifungal Susceptibilities of *Candida* Species Causing Vulvovaginitis and Epidemiology of Recurrent Cases." *Journal of Clinical Microbiology* 43 (5): 2155–62. doi:10.1128/JCM.43.5.2155-2162.2005.

Rivers, Charles A., Oluwaseun O. Adaramola, and Jane R. Schwebke. 2011. "Prevalence of Bacterial Vaginosis and Vulvovaginal Candidiasis Mixed Infection in a Southeastern American STD Clinic." *Sexually Transmitted Diseases* 38 (7): 672–4. doi:10.1097/OLQ.0b013e31820fc3b8.

Rosca, Aliona S., Joana Castro, Lúcia G. V. Sousa, and Nuno Cerca. 2019. "*Gardnerella* and Vaginal Health: The Truth Is out There." *FEMS Microbiology Reviews*. doi:10.1093/femsre/fuz027.

Saad, Asmaa, Mariam Fadli, Mohamed Bouaziz, Ahmed Benharref, Noureddine E. Mezrioui, and Lahcen Hassani. 2010. "Anticandidal Activity of the Essential Oils of *Thymus maroccanus* and *Thymus broussonetii* and Their Synergism with Amphotericin B and Fluconazol." *Phytomedicine* 17 (13). Urban und Fischer Verlag Jena: 1057–60. doi:10.1016/j.phymed.2010.03.020.

Savoia, Dianella. 2012. "Plant-Derived Antimicrobial Compounds: Alternatives to Antibiotics." *Future Microbiology* 7 (8): 979–90. doi:10.2217/fmb.12.68.

Schwebke, Jane R., Charity M. Richey, and Heidi L. Weiss. 1999. "Correlation of Behaviors with Microbiological Changes in Vaginal Flora." *The Journal of Infectious Diseases* 180 (5): 1632–6. doi:10.1086/315065.

Seow, Yi Xin, Chia Rou Yeo, Hui Ling Chung, and Hyun-Gyun Yuk. 2014. "Plant Essential Oils as Active Antimicrobial Agents." *Critical Reviews in Food Science and Nutrition* 54 (5): 625–44. doi:10.1080/10408398.2011.599504.

Sgibnev, Andrey V., and Elena A. Kremleva. 2015. "Vaginal Protection by H_2O_2-Producing Lactobacilli." *Jundishapur Journal of Microbiology* 8 (10). Kowsar Medical Publishing Company. doi:10.5812/jjm.22913.

Silva, Filomena, Susana Ferreira, Andreia Duarte, Dina I. Mendonça, and Fernanda C. Domingues. 2011. "Antifungal Activity of *Coriandrum sativum* Essential Oil, Its Mode of Action against *Candida* Species and Potential Synergism with Amphotericin B." *Phytomedicine* 19 (1): 42–7. doi:10.1016/j.phymed.2011.06.033.

Sobel, Jack D. 2000. "Bacterial Vaginosis." *Annual Review of Medicine* 51 (1): 349–56. doi:10.1146/annurev.med.51.1.349.

Sobel, Jack D. 2007. "Vulvovaginal Candidiasis." *The Lancet* 369 (9577): 1961–71. doi:10.1016/S0140-6736(07)60917-9.

Sobel, Jack D. 2016. "Recurrent Vulvovaginal Candidiasis." *American Journal of Obstetrics and Gynecology*. Mosby Inc. doi:10.1016/j.ajog.2015.06.067.

Sosto, Francesco, and Claudio Benvenuti. 2011. "Controlled Study on Thymol+Eugenol Vaginal Douche versus Econazole in Vaginal Candidiasis and Metronidazole in Bacterial Vaginosis." *Arzneimittel-Forschung/Drug Research* 61 (2): 126–31. doi:10.1055/s-0031-1296178.

Spear, Gregory T., Elizabeth St John, and M. Reza Zariffard. 2007. "Bacterial Vaginosis and Human Immunodeficiency Virus Infection." *AIDS Research and Therapy* 4 (October). BioMed Central: 25. doi:10.1186/1742-6405-4-25.

Stewart, Philip S., and J. William Costerton. 2001. "Antibiotic Resistance of Bacteria in Biofilms." *Lancet (London, England)* 358 (9276): 135–8. http://www.ncbi.nlm.nih.gov/pubmed/11463434.

Stokholm, Jakob, Susanne Schjørring, Carl E. Eskildsen, Louise Pedersen, Anne L. Bischoff, Nilofar Følsgaard, Charlotte G. Carson, Bo L. K. Chawes, Klaus Bønnelykke, Anne

Mølgaard, Bo Jacobsson, Karen A. Krogfelt, and Hans Bisgaard . 2014. "Antibiotic Use during Pregnancy Alters the Commensal Vaginal Microbiota." *Clinical Microbiology and Infection* 20 (7): 629–35. doi:10.1111/1469-0691.12411.

Stoyancheva, Galina, Marta Marzotto, Franco Dellaglio, and Sandra Torriani. 2014. "Bacteriocin Production and Gene Sequencing Analysis from Vaginal *Lactobacillus* Strains." *Archives of Microbiology* 196 (9). Springer Verlag: 645–53. doi:10.1007/s00203-014-1003-1.

Svare, Jens A., Henrik Schmidt, Bent B. Hansen, and Gunnar Lose. 2006. "Bacterial Vaginosis in a Cohort of Danish Pregnant Women: Prevalence and Relationship with Preterm Delivery, Low Birthweight and Perinatal Infections." *BJOG: An International Journal of Obstetrics & Gynaecology* 113 (12): 1419–25. doi:10.1111/j.1471-0528.2006.01087.x.

Swidsinski, Alexander, Alexander Guschin, Qionglan Tang, Yvonne Dörffel, Hans Verstraelen, Alexander Tertychnyy, Guzel Khayrullina, Xin Luo, Jack D. Sobel, and Xuefeng Jiang. 2019. "Vulvovaginal Candidiasis: Histologic Lesions Are Primarily Polymicrobial and Invasive and Do Not Contain Biofilms." *The American Journal of Obstetrics & Gynecology* 220: 91.e1–91.e8. doi:10.1016/j.ajog.2018.10.023.

Swidsinski, Alexander, Werner Mendling, Vera Loening-Baucke, Axel Ladhoff, Sonja Swidsinski, Laura P. Hale, and Herbert Lochs. 2005. "Adherent Biofilms in Bacterial Vaginosis." *Obstetrics & Gynecology* 106 (5, Part 1): 1013–23. doi:10.1097/01.AOG.0000183594.45524.d2.

Swidsinski, Alexander, Werner Mendling, Vera Loening-Baucke, Sonja Swidsinski, Yvonne Dörffel, Jürgen Scholze, Herbert Lochs, and Hans Verstraelen. 2008. "An Adherent *Gardnerella vaginalis* Biofilm Persists on the Vaginal Epithelium after Standard Therapy with Oral Metronidazole." *American Journal of Obstetrics and Gynecology* 198 (1): 97.e1–97.e6. doi:10.1016/j.ajog.2007.06.039.

Swidsinski, Alexander, and Jack Sobel. 2019. "Reply." *American Journal of Obstetrics and Gynecology*. Mosby Inc. doi:10.1016/j.ajog.2019.07.008.

Swidsinski, Alexander, Hans Verstraelen, Vera Loening-Baucke, Sonja Swidsinski, Werner Mendling, and Zaher Halwani. 2013. "Presence of a Polymicrobial Endometrial Biofilm in Patients with Bacterial Vaginosis." Edited by Adam J. Ratner. *PLoS One* 8 (1). Public Library of Science: e53997. doi:10.1371/journal.pone.0053997.

Thein, Zaw M., Chaminda J. Seneviratne, Yuthika H. Samaranayake, and Lakshman P. Samaranayake. 2009. "Community Lifestyle of *Candida* in Mixed Biofilms: A Mini Review." *Mycoses*. doi:10.1111/j.1439-0507.2009.01719.x.

Tommasi, Luca, Carmine Negro, Antonio Cerfeda, Eliana Nutricati, Vincenzo Zuccarello, Luigi De Bellis, and Antonio Miceli. 2007. "Influence of Environmental Factors on Essential Oil Variability in *Thymbra capitata* (L.) Cav. Growing Wild in Southern Puglia (Italy)." *Journal of Essential Oil Research* 19 (6): 572–80. doi:10.1080/10412905.2007.9699335.

Tseng, Boo Shan, Wei Zhang, Joe J. Harrison, Tam P. Quach, Jisun Lee Song, Jon Penterman, Pradeep K. Singh, David L. Chopp, Aaron I. Packman, and Matthew R. Parsek. 2013. "The Extracellular Matrix Protects *Pseudomonas aeruginosa* Biofilms by Limiting the Penetration of Tobramycin." *Environmental Microbiology* 15 (10). NIH Public Access: 2865–78. doi:10.1111/1462-2920.12155.

Tullio, Vivian, Janira Roana, Daniela Scalas, and Narcisa Mandras. 2019. "Evaluation of the Antifungal Activity of *Mentha x piperita* (Lamiaceae) of Pancalieri (Turin, Italy) Essential Oil and Its Synergistic Interaction with Azoles." *Molecules* 24 (17). MDPI AG. doi:10.3390/molecules24173148.

Turovskiy, Yevgeniy, Katia Sutyak Noll, and Michael L. Chikindas. 2011. "The Etiology of Bacterial Vaginosis." *Journal of Applied Microbiology* 110 (5): 1105–28. doi:10.1111/j.1365-2672.2011.04977.x.

Vaneechoutte, Mario, Alexander Guschin, Leen Van Simaey, Yannick Gansemans, Filip Van Nieuwerburgh, and Piet Cools. 2019. "Emended Description of *Gardnerella vaginalis* and Description of *Gardnerella leopoldii* Sp. Nov., *Gardnerella piotii* Sp. Nov. and *Gardnerella swidsinskii* Sp. Nov., with Delineation of 13 Genomic Species within the Genus *Gardnerella*." *International Journal of Systematic and Evolutionary Microbiology* 69 (3): 679–87. doi:10.1099/ijsem.0.003200.

Venkatesan, Nandakumar, Govindaraj Perumal, and Mukesh Doble. 2015. "Bacterial Resistance in Biofilm-Associated Bacteria." *Future Microbiology*. Future Medicine Ltd. doi:10.2217/fmb.15.69.

Verstraelen, Hans, and Alexander Swidsinski. 2019. "The Biofilm in Bacterial Vaginosis: Implications for Epidemiology, Diagnosis, and Treatment: 2018 Update." *Current Opinion in Infectious Diseases*. Lippincott Williams and Wilkins. doi:10.1097/QCO.0000000000000516.

Verstraelen, Hans, Rita Verhelst, Mario Vaneechoutte, and Marleen Temmerman. 2010. "The Epidemiology of Bacterial Vaginosis in Relation to Sexual Behavior." *BMC Infectious Diseases* 10 (March): 81. doi:10.1186/1471-2334-10-81.

Vieira, Maria, Lucinda J. Bessa, M. Rosário Martins, Sílvia Arantes, António P. S. Teixeira, Ângelo Mendes, Paulo Martins da Costa, and Anabela D. F. Belo. 2017. "Chemical Composition, Antibacterial, Antibiofilm, and Synergistic Properties of Essential Oils from *Eucalyptus globulus* Labill. and Seven Mediterranean Aromatic Plants." *Chemistry & Biodiversity* 14 (6): e1700006. doi:10.1002/cbdv.201700006.

Visick, Karen L., Mark A. Schembri, Fitnat Yildiz, and Jean Marc Ghigo. 2016. "Biofilms 2015: Multidisciplinary Approaches Shed Light into Microbial Life on Surfaces." *Journal of Bacteriology*. American Society for Microbiology. doi:10.1128/JB.00156-16.

Vodstrcil, Lenka A., Jimmy Twin, Suzanne M. Garland, Christopher K. Fairley, Jane S. Hocking, Matthew G. Law, Erica L. Plummer, Katherine A. Fethers, Eric P. F. Chow, Sepehr N. Tabrizi, and Catriona S. Bradshaw . 2017. "The Influence of Sexual Activity on the Vaginal Microbiota and *Gardnerella vaginalis* Clade Diversity in Young Women." Edited by David N Fredricks. *PLoS One* 12 (2): e0171856. doi:10.1371/journal.pone.0171856.

Walters, Marshall C., Frank Roe, Amandine Bugnicourt, Michael J. Franklin, and Philip S. Stewart. 2003. "Contributions of Antibiotic Penetration, Oxygen Limitation, and Low Metabolic Activity to Tolerance of *Pseudomonas aeruginosa* Biofilms to Ciprofloxacin and Tobramycin." *Antimicrobial Agents and Chemotherapy* 47 (1). American Society for Microbiology (ASM): 317–23. doi:10.1128/aac.47.1.317-323.2003.

Watson, Margaret C., Jeremy M. Grimshaw, Christine M. Bond, Jill Mollison, and Anne Ludbrook. 2002. "Oral versus Intravaginal Imidazole and Triazole Antifungal Agents for the Treatment of Uncomplicated Vulvovaginal Candidiasis (Thrush): A Systematic Review." *BJOG: An International Journal of Obstetrics and Gynaecology*. doi:10.1111/j.1471-0528.2002.01142.x.

Wei, Qingzhu, Bo Fu, Jianghuan Liu, Zhixiong Zhang, and Tong Zhao. 2012. "*Candida albicans* and Bacterial Vaginosis Can Coexist on Pap Smears." *Acta Cytologica* 56 (5). S. Karger AG: 515–9. doi:10.1159/000339155.

Wiesenfeld, Harold C., Sharon L. Hillier, Marijane A. Krohn, Daniel V. Landers, and Richard L. Sweet. 2003. "Bacterial Vaginosis Is a Strong Predictor of *Neisseria gonorrhoeae* and *Chlamydia trachomatis* Infection." *Clinical Infectious Diseases* 36 (5): 663–8. doi:10.1086/367658.

Xu, Boyun, Yue Qu, and Margaret Deighton. 2019. "Should We Absolutely Reject the Hypothesis That Epithelium-Based *Candida* Biofilms Contribute to the Pathogenesis of Human Vulvovaginal Candidiasis?" *American Journal of Obstetrics and Gynecology*. Mosby Inc. doi:10.1016/j.ajog.2019.07.007.

Yano, Junko, Jack D. Sobel, Paul Nyirjesy, Ryan Sobel, Valerie L. Williams, Qingzhao Yu, Mairi C. Noverr, and Paul L. Fidel. 2019. "Current Patient Perspectives of Vulvovaginal Candidiasis: Incidence, Symptoms, Management, and Post-Treatment Outcomes." *BMC Women's Health* 19 (1). BioMed Central Ltd. doi:10.1186/s12905-019-0748-8.

Zhang, Jin-E, Dan Luo, Rong Yi Chen, Yan Ping Yang, Ying Zhou, and Yi Ming Fan. 2013. "Feasibility of Histological Scoring and Colony Count for Evaluating Infective Severity in Mouse Vaginal Candidiasis." *Experimental Animals* 62 (3): 205–10. doi:10.1538/expanim.62.205.

Zhang, Jing, Huihui Wu, Dan Jiang, Yongan Yang, Wenjian Tang, and Kehan Xu. 2019. "The Antifungal Activity of Essential Oil from *Melaleuca leucadendra* (L.) L. Grown in China and Its Synergistic Effects with Conventional Antibiotics against *Candida*." *Natural Product Research* 33 (17). Taylor and Francis Ltd.: 2545–8. doi:10.1080/14786419.2018.1448979.

Zhou, Xia, Stephen J. Bent, Maria G. Schneider, Catherine C. Davis, Mohammed R. Islam, and Larry J. Forney. 2004. "Characterization of Vaginal Microbial Communities in Adult Healthy Women Using Cultivation-Independent Methods." *Microbiology* 150 (8): 2565–73. doi:10.1099/mic.0.26905-0.

19 Essential Oils
Potential Remedy in Controlling Biofilms-Associated Infections of Human Pathogenic Fungi

Mohd Sajjad Ahmad Khan
Imam Abdulrahman Bin Faisal University

Mohd Musheer Altaf
Institute of Information Management and Technology

CONTENTS

19.1 INTRODUCTION

Fungi are eukaryotic microorganisms that mainly cause life-threatening diseases in immunocompromised individuals; however, fungal infections have also been frequent in healthy individuals. Various forms of mycoses differ in their nature, causative agents, and etiology. Among the various human fungal pathogens, the species of *Candida*, *Aspergillus*, *Fusarium*, and *Trichosporum* are responsible for comparatively higher rates of morbidity and mortality (Fluckiger et al. 2006). Also, many pathogenic fungi can easily colonize to abiotic surfaces such as prostheses and catheters and may result in deep-seated infections. Especially, yeasts take up the benefit of this condition to enter blood circulation and the inner organs of patients. This is frightening, as disseminated fungal infections have an elevated mortality rate (Blankenship and Mitchell 2006). Oropharyngeal candidiasis is the repeated opportunistic fungal infection among HIV patients, and it has been anticipated that more than 90% of affected patients acquire this ailment (De Repentigny et al. 2004). Approximately, 70% of women at least experience vaginal candidiasis once in their lifetime and 20% endured from reappearance (Fidel et al. 1999; Achkar and Fries 2010). It has been reported that annual occurrence of invasive mycoses is 72–228 infections per million peoples for *Candida* spp., 30–66 infections per million peoples for *Cryptococcus neoformans*, and 12–34 infections per million peoples for *Aspergillus* spp. (Pfaller et al. 2006).

The prominent antifungal agents being used in clinical settings are categorized into four main classes: azoles, polyenes, echinocandins, and pyrimidine analogues (5-fluorocytosine). Additionally, allylamines are regularly used against superficial fungal infections (Gupta et al. 2017). However, the existing antimicrobial treatments are capable to deal with various forms of fungal diseases, but the drug toxicity and development of drug resistance have challenged the current armamentarium of antifungals. The increase in resistance against azoles and echinocandins and cross-resistance to at least 2 antifungal classes, i.e., multi-drug resistance are worrying trends, mainly in large tertiary and oncology hospitals. Many workers have witnessed increased antifungal resistance and multi-drug resistance (MDR) in *Candida* spp. (Farmakiotis and Kontoyiannis 2017). In addition, azole resistance in *Aspergillus* spp. has been described globally, and such resistant strains can produce invasive infections with high mortality rates (Verweij et al. 2016). Moreover, existing antimicrobial treatments are commonly associated with therapeutic failure due to the formation of drug-tolerant biofilms (Davies 2003; Morace et al. 2014; Van Acker et al. 2014; Muzny and Schwebke 2015).

Establishment of biofilm in pathogenic fungi is considered as a crucial virulence factor. Both yeasts and filamentous fungi can cling to biotic and abiotic surfaces, evolving into well-structured populations that are resistant to antimicrobial agents and other environmental restrictions (Muzny and Schwebke 2015). Mostly pathogenic microorganisms exist predominantly in biofilms and account for more than 80% of human infectious diseases (Dongari-Bagtzoglou 2008). Development of biofilms in yeast and filamentous fungi differs, and often polymicrobial communities are co-existing and have become gradually more important to be studied. However, *Candida* biofilms have been most studied at morphological and molecular perceptions. The extracellular matrix of biofilms is highly vital as it encompasses and shields biofilm cells from the adjoining environment. Furthermore, to attain cell-to-cell interaction, microbes exude quorum-sensing molecules that regulate their biological activities and behaviors and play a role in fungal resistance and pathogenicity (Costa-Orlandi et al. 2017).

The surge in antimicrobial resistance in fungal pathogens mostly associated with biofilm formation, coupled with the host toxicities of antifungal drugs demands for the innovation and advancement of new agents to confront antibiotic resistance. Even though it is tough to beat the pathogens, the discovery and development of compounds with pleiotropic modes or mechanisms of action distinct from the conventional drugs can help us deal with biofilm-forming fungal pathogens. Since ancient times, plant-based medicines and its derivatives such as essential oils have been used to combat different diseases including fungal infections. Plant products have been known to provide prolific supply of bioactive compounds with varied structures and functional group chirality. Many workers have explored the modes of action of essential oils and their major active compounds (Angioni et al. 2006; Bounatirou et al. 2007; Bakkali et al. 2008). Several researchers have highlighted medicinal plants for their antibacterial, antibiofilm, efflux pump inhibition, cure of respiratory and urinary tract infections, and wound healing effects (Bakkali et al. 2008; Adorjan and Buchbauer 2010). There is a pressing requirement for research into plants to obtain bioactive compound with antibiofilm mechanism of action. This chapter throws more light on such antimicrobials. The aim of this chapter is to illustrate the properties of essential oils, primarily as antifungal agents, and their role in opposing fungal biofilm growth.

19.2 COMMON FUNGAL DISEASES IN HUMANS

Human fungal pathogens are mainly represented by four classes viz. zygomycetes, ascomycetes, deuteromycetes, and basidiomycetes. The fungi of these groups can produce a variety of illnesses under two categories as (i) superficial infections affecting skin, nails, scalp, or mucous membrane and (ii) systemic; infections involving deep-seated tissues and organs. Superficial infections are quite prevailing such as tinea or ringworms involving the skin, hair, and nails. The causing agents are expert saprophytes with the ability to break down keratin like *Trichophyton* spp., *Microsporum* spp., *Epidermophyton* spp., and *Candida albicans*, whereas, systemic mycoses are produced by soil saprophytes. The causative agents are *Cryptococcus* spp., *Sporothrix* spp., *Paracoccidioides* spp., *Blastomyces* spp., and *Histoplasma* spp. Another kind of fungal infection occurs in immunocompromised patients such as AIDS, cancer, diabetes by the opportunistic fungal pathogens. For these, causative agents are *C. albicans*, *Aspergillus* spp., *Mucor* spp., *Penicillium* spp., *Fusarium* spp., and *Rhizopus* spp. (Pfaller and Diekema 2004; Chakrabarti 2005; Reedy et al. 2007).

19.2.1 CANDIDIASIS

Candidiasis is the group of secondary infections caused by *Candida* spp. extending from acute and chronic to life-threatening mycoses. Generally, healthy individuals face superficial infections such as vulvovaginal candidiasis, candiduria, onychomycosis, and oropharyngeal candidiasis, whereas, in immunocompromised populations, invasive infections are produced viz. candidemia and disseminated candidiasis infecting inner organs (Pfaller et al. 2006). Several *Candida* spp. can cause candidiasis such as *C. albicans*, *Candida dubliniensis*, *Candida glabrata*, *Candida krusei*, *Candida tropicalis*, and *Candida parapsilosis* (Hayens and Westerneng 1996; Pfaller et al. 2006). Among these, *C. albicans* is the commonest infective agent in infections of cutaneous sites, reproductive organs, oral cavity, bloodstream infections, bone marrow transplantation, and nosocomial infections (Pfaller et al. 2006). Among others, *C. glabrata* has surfaced as common pathogen due to the heightened use of immune-suppressive drugs (Malani et al. 2005). *C. krusei* is a substantial pathogen in patients with hematological cancers and transplants (Viudes et al. 2002). *C. parapsilosis* affects critically ill neonates and ICU patients and very often isolated from blood cultures of patients with insertive medical devices (Sarvikivi et al. 2005), and it is also the most common species exist on health personnel (Trofa et al. 2008). *C. tropicalis* is one of the causative agents of candidemia and frequently shows its presence in the leukemic persons with bone marrow transplantation. Occasionally, 60%–80% of neutropenic develop invasive infections of *C. tropicalis* (Wingard 1995; Pfaller and Diekema 2007). *C. dubliniensis* is mainly related to oral and systemic infections in AIDS patients (Gutierrez et al. 2002; Loreto et al. 2010).

19.2.2 Aspergillosis

Aspergillosis is the range of ailments caused by members of the genus *Aspergillus*. Majority of these invasive infections are produced by *Aspergillus fumigatus*. The second common species is *Aspergillus flavus* followed by *Aspergillus niger*, and *Aspergillus terreus* (Krishnan et al. 2009). Aspergillosis primarily includes allergic bronchopulmonary aspergillosis (ABPA), invasive aspergillosis (IA) and pulmonary aspergilloma, and its occurrence prevails mainly in immunocompromised patients (Dagenais and Keller 2009). About 1%–2% of asthmatic patients may develop ABPA, and it has been discovered that 15% of asthmatic patients may be colonized by *A. fumigatus*, whereas, 7%–35% of cystic fibrosis patients suffer ABPA (Krasnick et al. 1995; Moss 2007). Aspergilloma, a fungus ball, appears in prior established pulmonary cavities resulted from tuberculosis, sarcoidosis, or other bullous lung disorders (Ma et al. 2011). Transplant recipients are at the highest risk of invasive aspergillosis. About 2%–26% of hematopoietic stem cell transplant recipients and 1%–15% of organ transplant recipients develop *Aspergillus* infections. IA could occur in 5%–25% of acute leukemics, 5%–10% of patients with allogeneic bone marrow transplant (BMT) and 0.5%–5% of patients undergoing cytotoxic treatment of blood infections and solid-organ transplantation (Latge 1999; Singh and Paterson 2005; Fuqua Jr et al. 2010).

19.2.3 Dermatophytosis

Dermatophytosis is infections of skin, hair, scalp, and nails and is mainly caused by a group of fungi which are morphologically and physiologically similar and termed as dermatophytes such as *Microsporum* spp., *Trichophyton* spp., and *Epidermophyton* spp. These members can easily use keratin of subcutaneous layer of human skin as a nutrient source (Weitzman and Summerbell 1995). Approximately, 90% of cases of onychomycosis have been reported to be caused by these dermatophytes in the United States and Europe (Ellis et al. 1997) and prevalent in developing countries involving India (Kannan et al. 2006). In immunocompromised individual, *Trichophyton* spp. and *Microsporum* spp. can infest the dermis or subcutaneous tissue and can result in skin lesions leading to disseminated mycoses (Galhardo et al. 2004; Marconi et al. 2010).

19.3 PROBLEM OF DRUG RESISTANCE ASSOCIATED WITH BIOFILM-FORMING ABILITY OF FUNGAL PATHOGENS

In the management of fungal diseases, drug resistance is considered as an important feature and its epidemiological characteristics persist to progress (Kanafani and Perfect 2008). The risk of development of drug resistance in pathogens is always associated with drug use (Cowen 2008). *In vitro* resistance, under clinical settings, to almost all currently available antifungal drugs has been described (White et al. 1998; Sanglard et al. 2009). A lot of research has revealed various mechanisms of drug resistance in fungi at molecular or physiological levels. Such as drug import, drug modification or degradation, reduced buildup of drugs, target site modification by mutations, upregulation of target enzyme, deployment of compensatory metabolic pathways, development of biofilms (White et al. 1998). In the last decade, many genera of fungi have been associated with the ability to form biofilms (Costa-Orlandi et al. 2017). Development of biofilms by *Candida* spp. is reported to bestow a greater level of drug resistance (Cannon et al. 2007; Sanglard et al. 2009). The mechanism of drug resistance by biofilms is summarized in Table 19.1 and depicted in Figure 19.1 (Douglas 2003; Ramage et al. 2005).

TABLE 19.1
Resistance Mechanisms Associated with Biofilm Formation

Resistance Mechanisms	Effects
Cellular density	Quorum sensing
Differential regulation drug target	Alteration in target level; associated with changes in target structure that make the drug unable to bind to the target
Upregulation drug efflux pump	Antifungal is pumped out of cells and thus cannot perform its intracellular function
Persister cells	Because of the dormant state of the persistors, antifungal targets are inactive
Presence of a matrix	Specific binding of antifungals to β-1,3-glucans, a major component of the matrix, prevents antifungal agents from reaching their targets
Diverse stress responses	Possible indirect effects through the regulation of other resistance mechanisms

Source: Adapted from Van Acker, H., Van Dijck, P., Coenye, T., *Trends Microbiol.*, 22, 326–33, 2014; Sardi, J.D.C., Pitangui, N.D.S., Voltan, A.R. et al., *Virulence* 6, 642–51, 2015; Costa-Orlandi, C.B., Sardi, J.C.O., Pitangui, N.S. et al., J. *Fungi.*, 3, 22, 2017.

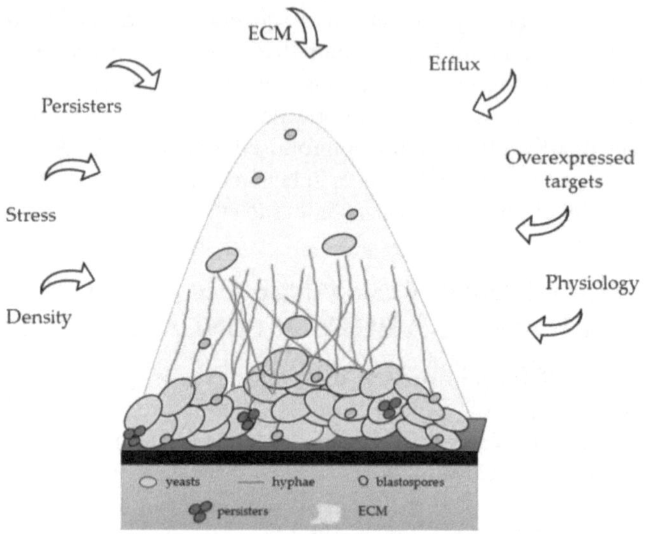

FIGURE 19.1 Scheme of the mechanisms and factors that promote fungal biofilm resistance, which are common to several fungi. (Adapted from Costa-Orlandi et al. 2017.)

19.3.1 Biofilm Formation in Pathogenic Fungi

Biofilms are the coordinated forms containing microbial populations that are fastened to some inert surfaces or tissues and skirted in a self-manufactured matrix of exopolymeric substances. Cells inside the biofilms display different phenotypes, growth rate, and gene expression compared to planktonic cells (Muzny and Schwebke 2015). This kind of growth structure is expected to facilitate microbial cells to endure in antagonistic environments, enhances their resistance to physical and chemical stresses, and stimulates metabolic collaboration (Van Acker et al. 2014). Such coordinated communities are under tight regulation of gene expression monitored through quorum sensing that, in turn, are controlled by farnesol and tyrosol molecules (Ramage et al. 2002). This cell-to-cell contact precludes and controls unnecessary overpopulation, nutritional fight and has repercussion in dissemination and launch of infection at distal site from old biofilm (Alem et al. 2006).

Development of fungal evolves through synchronized stages of early, intermediate, and maturation. It starts with adherence of a microbial cell to a surface, pursued by a cascade of differential gene expression. It has been witnessed that studies on the formation of biofilms by filamentous fungi are narrow compared to those of yeasts (Blankenship and Mitchell 2006). The phases of advancement of biofilms in filamentous fungi are described in Figure 19.2a, and it includes many stages such as (i) firstly, propagule adsorption involves interaction of spores with the surfaces followed by hyphal fragmentations or sporangia, (ii) active adhesion occurs because of adhesins which are secreted by spores during germination and other reproductive structures, (iii) formation of monolayer of a first microcolony as a result of elongation and hyphal branching and also the production of extracellular matrix, (iv) second

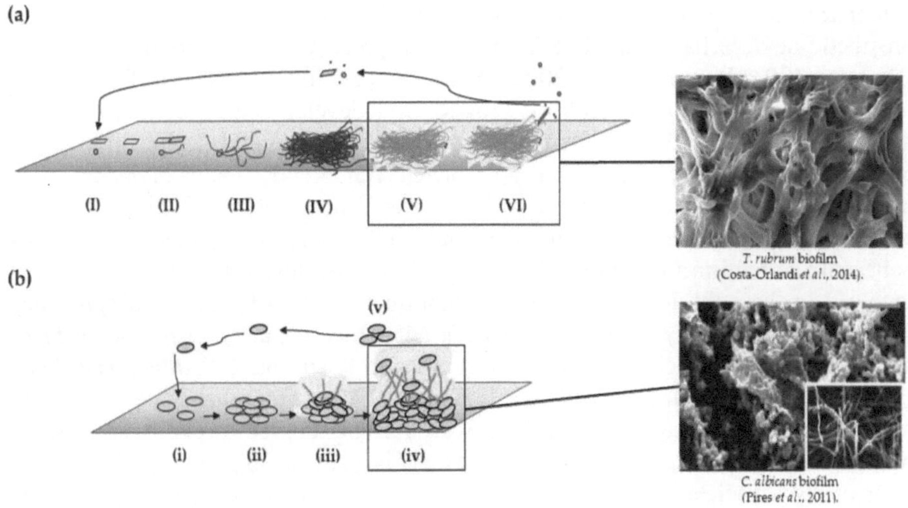

FIGURE 19.2 Models of biofilm development in filamentous fungi (a) and *C. albicans* (b). The stages of development are similar, although the morphology and number of stages are different. In the first model (a), six stages were proposed by Harding et al. [12]: (I) adsorption, (II) active attachment, (III) first formation of microcolony through germination and/or monolayer development, (IV) mycelial development, (V) biofilm maturation, and (VI) dispersion of conidia and/or arthroconidia. The second model corresponds to classical *C. albicans* biofilm development (b) which includes five stages, such as in bacteria: (i) adsorption, (ii) adhesion, (iii) microcolony formation, (iv) mature biofilm, and (v) dispersion. (Courtesy of Costa-Orlandi et al. 2017.)

microcolony formation or initial maturation occurs as dense hyphae grids form in three dimensions, casing by an extracellular matrix, and construction of water channels, (v) final maturation occurs in the form of fruiting bodies and other survivor structures contingent of the fungal strains, (vi) finally, the dispersion of conidia and/or hyphae fragments occurs to jump a newfangled cycle. Filamentous fungi are known to emit small proteins, hydrophobins that enable the adhesion of hyphae to hydrophobic surfaces paving the way for biofilm formation (Harding et al. 2009). Whereas, regarding yeasts, *C. albicans* is the best examined model of biofilm development. Establishment process comprises lesser phases of progress compared to filamentous fungi as (i) first, the yeast cells adhere to a surface, (ii) foundation of basal layers of yeast with initial expansion of hyphae and extracellular matrix, (iii) maturation of biofilms occurs comprising a substantial quantity of yeast, hyphae, pseudohyphae; construction of water channels and covering with extracellular matrix, and (iv) finally, cells are dispersed (Figure 19.2b) (Kojic and Darouiche 2004; Blankenship and Mitchell 2006).

In recent years, investigations pertaining to fungal biofilms have been escalated considerably and numerous species have exhibited the capacity to develop these populations. The perseverance of fungal infections is fostered by the aptitude of fungi to establish biofilms on a wide-ranging implanted medical device. The ability of

Candida to form biofilms on catheters, endotracheal tubes, pacemakers, and other prosthetic devices has offered to its prevailing presence in nosocomial infections (Douglas 2003; Ramage et al. 2005). Such devices, in addition to offering stage for candidal cells to form biofilm, growth, and development, create way out of host barrier fortifications for dissemination. Vinitha and Ballal (2007) showed the biofilm-forming ability of *Candida* spp. isolated from blood samples of patients with indwelling devices. Their findings signify that the virulence of *Candida* spp. to cause central venous catheter-related candidaemia in ICU patients or patients on dialysis is because of biofilm formation. Blanco et al. (2010) evaluated the virulence factors involved in the adhesion process during biofilm formation by *Candida* spp. They studied cell surface hydrophobicity (CSH), adherence capacity to plastic surfaces and buccal epithelial cells, and biofilm formation in 17 strains of *C. albicans* isolated from bronchial aspirates of critically ill patients. They found that these factors are unambiguously implicated in adhesion, with variable degree of expression. In the lights of their findings, they suggested that CSH is a key attribute in *C. albicans* and is precisely associated with adherence to plastic and epithelial surface to promote the formation of biofilm. On the other hand, De Souza et al. (2009) assessed the variations in the CSH of *Candida* spp. isolated from denture users with or without denture-associated candidiasis. The results of their study revealed that cells of *C. albicans* strains with better CSH rates displayed higher adhesion achievement in distinct host tissues than cells with lesser rates.

Villar-Vidal et al. (2011) compared the capability of biofilm assembly by blood and oral isolates of *C. albicans* and *C. dubliniensis*. In their study, biofilm formation by *C. albicans* isolates was statistically significantly greater than biofilm formation by *C. dubliniensis*. They concluded that this varying ability allows *C. albicans* and *C. dubliniensis* to preserve their oral ecological niches as commensal or pathogenic microbes and could be a key virulence factor during invasive candidiasis. Another dimorphic fungus *Paracoccidioides brasiliensis* responsible for paracoccidioidomycosis, a systemic mycosis, has been found to form biofilms in the yeast phase (Sardi et al. 2015). Formation of biofilms in *Histoplasma capsulatum*, a causative agent of histoplasmosis, was first explained by Pitangui et al. (2012). On the other hand, dermatophytes have also been known to form biofilms as proved in vitro by two of the highly predominant dermatophytes causing dermatophytomas, *Trichophyton rubrum* and *Trichophyton mentagrophytes* (Costa-Orlandi et al. 2014). Moreover, *Aspergillus* spp. are saprophytic and opportunistic fungi that also form biofilms as Aspergilloma is a fungal mass showing characteristics of biofilms (Ramage et al. 2011). Additionally, biofilms formed by several other fungi have been examined and described such as those formed by *Cryptococcus* spp., *Fusarium* spp., *Malassezia* spp., *Scedosporium* spp., and *Trichosporon* spp. (Costa-Orlandi et al. 2017). Siqueira and Lima (2013) studied *Aspergillus* spp., *Alternaria* spp., *Botrytis* spp., *Cladosporium* spp., and *Penicillium* spp. isolated from water sources to evaluate their ability to expand as biofilms under in vitro. All these isolates formed biofilm but showed different patterns of progress. A direct relationship between biomass and cell activity was not noticed, but biomass rates and morphological limitations, i.e., monolayer and exopolysaccharide substances (EPS) production, were directly related. Thus, the findings presented here emphasize the potential of fungi to develop biofilms and the growing need to develop approaches for more research in this area to combat fungal infections.

19.4 COMBATING FUNGAL INFECTIONS

To combat mycoses, several strategies have been endeavored and employed, however, to develop further newer approaches, novel targets, new compounds, and advanced delivery systems are prerequisite. In this concern, the cell membrane, cell wall, virulence factors, and putative genes have been targeted. Fungal cell wall is a structure vital to fungi but deficient in human cells and therefore could be considered as novel target. Their key macromolecular targets are mannoproteins and chitin β-glucan. Likewise, plasma membrane could also be targeted in terms of interference with the synthesis of ergosterol, phospholipids, and shingolipid. Whereas, proton ATPase, efflux pump, protein synthesis, nucleic acid metabolism, signal transduction, and cell cycle could also serve innovative antifungal drug targets (Sangamwar et al. 2008). Several newer molecular drug targets have been recognized and validated, but novel drug development is still anticipated.

19.4.1 Targeting Virulence and Biofilm

An advantage of using virulence factors including biofilm formation as drug target is that they are usually pathogen specific and not found in human host. Fortunately, in the last few years, virulence factors of fungi, molecular mechanism of biofilm formation, expression of genes regulating quorum sensing, and their inhibitors have, at least to some extent, been discovered and characterized (Gauwerky et al. 2009). This should provide new options for the development of potential antifungal therapeutics.

19.4.2 Use of Combination Therapy

To overcome the high mortality due to fungal infections resulting from biofilm-forming drug-tolerant fungal pathogens, a possible approach could be a combination of two or more different classes of antifungals, possessing different mechanisms of action. The major objective of the combination approach is to obtain improved clinical efficacy through a synergistic interaction of drugs and to reduce host toxicity (Johnson and Perfect 2010).

19.5 ROLE OF PLANT PRODUCTS IN ALTERNATIVE MEDICINE

In these views, plant products including essential oils having ethnomedicinal values against various diseases including mycoses are anticipated to bring antifungal agents with broad-spectrum and novel modes of action. Lots of anti-infective agents derived from medicinal plants have been validated by researchers (Kamboj 2000; Adorjan and Buchbauer 2010). Last decade has witnessed an ongoing resurgence of importance of conducting research and employing the herbs or phyto-products as alternative medicine in developed as well as in developing countries because phyto-drugs have been described to be inexpensive, innocuous, and without side-effects (Bansod and Rai 2008). It is also believed that phyto-compounds exhibiting target sites varying from those displayed by currently used antifungals will be especially effective against drug-resistant pathogens. Yet, the data available on phytomedicine against biofilm-forming drug-resistant fungal pathogens is not adequate for successful formulations for human use (Agarwal et al. 2010). It is esteemed that these compounds could be used alone or to potentiate existing antifungals for attaining effective therapeutic values.

19.5.1 Essential Oils

Essential oils are volatile secondary metabolites of medicinal plants normally with strong odors (Bakkali et al. 2008). Essential oils are well known since ancient times for their biological activities such as antibacterial, antifungal, antioxidant, antiviral, anti-cancerous, and anti-inflammatory, and are also used for flavoring and preserving foods and drinks (Bounatirou et al. 2007; Bakkali et al. 2008). Their bioactivity of oils differs depending upon the composition and orientation of functional groups of active compounds. The oils are acquired, by hydro or steam distillation, from varying parts of plants viz. flowers, fruits, leaves, seeds, stems, barks, and woods. Composition of essential oils and therefore its medicinal values varies broadly depending on the plant parts, age, vegetative cycle stage, geographical location, season, cultivation techniques, soil texture, environmental and agronomic conditions (Angioni et al. 2006; Yesil et al. 2007). Table 19.2 shows ethnomedicinal values of certain medicinal plants/essential oils.

19.5.1.1 Chemical Composition of Essential Oils

Majority of essential oils are principally comprised of terpenes and their oxygenated derivatives such as mono- and sesquiterpenoids, benzoids, phenylpropanoids, etc. and therefore exhibit distinct biological activities (Adorjan and Buchbauer 2010). Essential oils are complex mixtures of ingredients of different chemical groups which can contain about 20–60 elements at different concentrations. They are characterized by two or three major components at higher concentrations (20%–70%) compared to other components present in trace quantity (Bakkali et al. 2008). Chemical structures of some of the selected components of essential oils are depicted in Figure 19.3. More thorough structure of different compounds from medicinal plants can be reviewed from Harborne and Baxter (1995).

19.5.1.2 Anti-Pathogenic and Antibiofilm Activity of Essential Oils

Essential oils are expected to offer potential newer bioactive compounds with novel mode of action with various cellular and molecular targets in fungal pathogens. Many researchers have highlighted and validated antifungal activities from many essential oils against pathogenic fungi including yeasts, dermatophytes, and *Aspergillus* spp. (Rai et al. 2003; Angioni et al. 2006; Bounatirou et al. 2007; Bakkali et al. 2008; Adorjan and Buchbauer 2010; Khan and Ahmad 2011). Use of natural compounds has boosted the evaluation of anti-biofilm action of oils and their active components because as these molecules are less likely to stimulate resistant phenotypes (Khan and Ahmad 2012a,b; Sanglard et al. 2003). For example, tea tree oil (TTO) has been

TABLE 19.2
Plant Essential Oils, Their Major Active Compounds and Their Ethnobotanical Uses

Medicinal Plants (Botanical Name)	Common Name	Family	Active Compounds	Ethnomedicine Uses
Apium graveolens	Celery	Umbelliferae	Limonene, selinene, apigenin, rutaretin, apiol, carvone	Sedative and tonics, rheumatism, antispasmodic, nerve stimulant, tranquilizer
Carum copticum	Ajowan	Apiaceae	Thymol, carvacrol, cymene	Antiseptic, digestive
Cinnamomum verum	Cinnamon	Lauraceae	Cinnamaldehyde, eugenol, cinnamic aldehyde	
Citrus limon	Lemon	Rutaceae	Limonene, psoralene, bergaptene, citral	Scurvy, antioxidants
Citrus paradisi	Grape	Rutaceae	Limonene, psoralene, bergaptene, myrcene, sabinene, geraniol, linalool, citrnellal, α-pinene	Scurvy, antioxidants
Citrus sinensis	Orange	Rutaceae	Limonene, psoralene, bergaptene, linalool, geraniol	Scurvy, antioxidants
Cymbopogon citratus	Lemongrass	Poaceae	Citral, citronellol, geraniol, myrcene, luteolin	Elephantiasis, analgesic, antirheumatic, antipyretic, expectorant, anticarcinogenic, digestive, antifungal
Cymbopogon martini	Palmarosa	Poaceae	Geraniol, gernial aldehyde, citral	Rheumatism, antifungal
Eucalyptus globulus	Eucalyptus		Cineole, caryophyllene, limonene, α-pinene, eucalyptol	Expectorant, antiseptic, diarrhea
Foeniculum vulgare	Sweet fennel	Apiaceae	Anethole, transanthole, fenchane, limonene, α-pinene	Spice, carminative, anabolic, analgesic, antipyretic, antiflatulence, antimicrobial
Mentha piperita	Peppermint	Lamiaceae	Menthol, menthone, limonene, sesquiterpine, apigenin	Carminative, spasmolytic, anti-inflammatory, diuretic, mild sedative
Myristica fragrans	Nutmeg	Myristicaceae	Pinene, camphene, myristicin, eugenol, isoeugenol, sabinene	Carminative, indigestion, stomach, cough

(*Continued*)

TABLE 19.2 (*Continued*)
Plant Essential Oils, Their Major Active Compounds and Their Ethnobotanical Uses

Medicinal Plants (Botanical Name)	Common Name	Family	Active Compounds	Ethnomedicine Uses
Petroselinum crispum	Parsley	Apiaceae	Apigenin, apiol, myristicin	Diuretic, anticancer, antimicrobial
Rosmarinus officinalis	Rosemary	Lamiaceae	Borneol, cineole, camphene, α-pinene, camphor	Memory enhancer, Alzheimer disease, antioxidant
Santalum album	Sandalwood	Santalaceae	Santol, santene, sesquiterpene, santalene	Sedative, diuretic, tonic, bronchitis, gonorrhea, ulcers, headache, anti-inflammation
Syzygium aromaticum	Clove	Myrtaceae	Eugenol, caryophyllene, vanillin	Toothache, antiseptic, antispasmodic
Thymus vulgaris	Thyme	Lamiaceae	Thymol, carvacrol, linalool, ρ-cymene	Antiseptic, fungicidal
Zea mays	Corn	Poaceae	Coumarin, apigenin, maysin, ionene	Diuretic, diarrhea
Zingiber officinale	Ginger	Zingiberaceae	Zingerone, gingerols, zingiberene, cineol, citral, farnesene, bisabolene	Bronchitis, indigestion, abdominal colic

Source: From Singh, M.P., Panda, H., *Medicinal Herbs with Their Formulations*, Daya Publications, New Delhi, 2005; Daniel, M., *Medicinal Plants: Chemistry and Properties*, Oxford & IBH Publishers Co. Pvt Ltd, New Delhi, 2006; Singh, A., *Medicinal Plants of the World*, Oxford & IBH publishers Co. Pvt Ltd, New Delhi, 2006.

shown to possess antimicrobial properties and effectively active against *C. albicans* biofilms development (Carson et al. 2006). Different EOs, such as those extracted by *Coriandrum sativum* (Ramage et al. 2012) and *Ocimum americanum* (Furletti et al. 2011), exhibited attractive activities as inhibitors of biofilms in *C. albicans*. Others such as cinnamon oil, linalool, and xanthorrhizol have also shown potent biofilm inhibitory properties (Nazzaro et al. 2017). Khan and Ahmad (2012a) have described antibiofilm activity of oils of *Cymbopogon citratus* and *Syzygium aromaticum* against drug-resistant strains of *C. albicans*. Whereas, Taweechaisupapong et al. (2012) have demonstrated antibiofilm activity of lemongrass oil against *C. dubliniensis*.

An array of virulence factors viz. adhesions, melanin, calcineurin, hydrolytic enzymes, catalases, lipid signaling molecules including biofilm formation has been observed in fungal pathogenesis. These traits have been addressed under *in vitro* and *in vivo* for the development of newer antifungals targeting pathogenicity. It has been shown that some phyto-compounds such as tacrolimus and cyclosporine A impede calcineurin in *C. albicans* and *Cryptococcus neoformans* (Liu et al. 1991; Huai et al. 2002). Aureobasdin A and Khafrefungin have been shown to inhibit inositol phosphoryl ceramide synthase in *Candida* spp., *C. neoformans*, and *Aspergillus* spp.

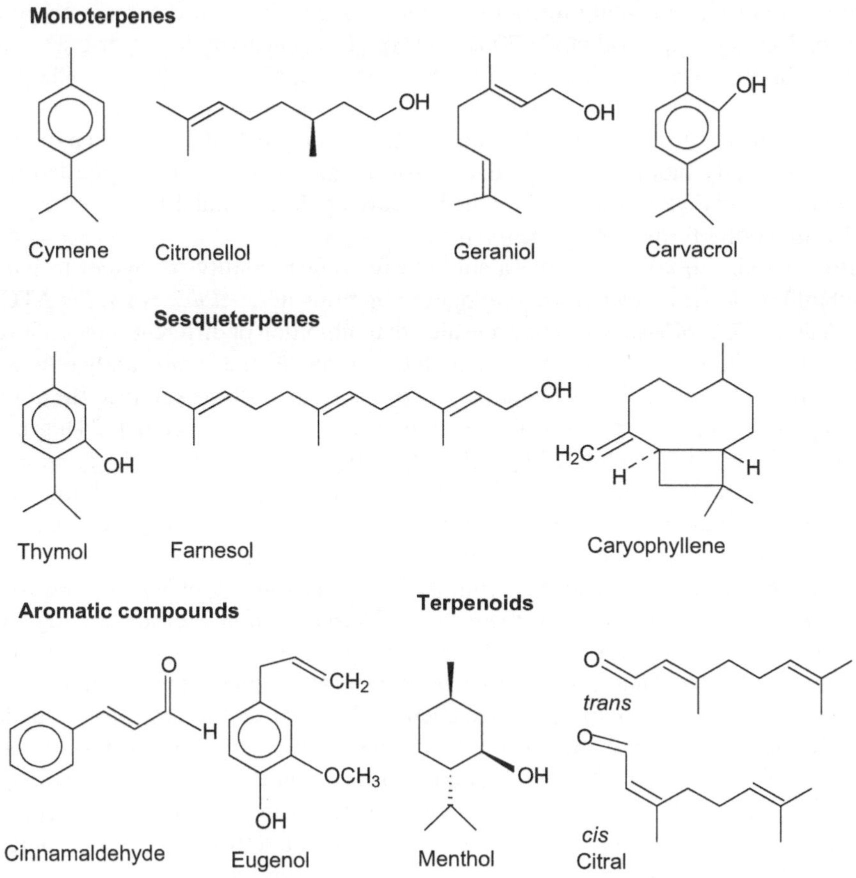

FIGURE 19.3 Chemical structures of selected components of common essential oils.

(Gauwerky et al. 2009). Hammer et al. (2000) reported inhibitory effects of TTO on germ tube formation (GTF) in *C. albicans*. They observed that GTF by *C. albicans* is influenced by the exposure of *Candida* cells to sub-inhibitory concentrations of TTO. Manoharan et al. (2017) revealed the chemical makeup, antibiofilm, and antihyphal activities of oil from cedar leaf. They observed potential antibiofilm activity of cedar leaf oil against *C. albicans*. They identified α-thujone, camphor, fenchone, fenchyl alcohol, and borneol as significant inhibitor of biofilm development in *C. albicans*. As inhibition of hyphal formation is responsible for antibiofilm effects, their transcriptomic analyses revealed camphor and fenchyl alcohol to be downregulating a few hypha-specific and biofilm-related genes (ECE1, ECE2, RBT1, and EED1).

He et al. (2007) studied the impacts of eugenol on adherent cells, biofilm formation, and cell morphogenesis of *C. albicans*. Their findings suggested that the impact of eugenol on adherent cells and following biofilm formation was reliant on the early adherence time and the amount of the eugenol-inhibiting hyphal growth of *Candida* cells. They established potential therapeutic implication of eugenol biofilm-associated candidal infections. In a scanning electron microscopic observation conducted by Agarwal et al. (2008), oils of clove, eucalyptus, ginger grass, and peppermint displayed excellent antibiofilm potential against *C. albicans*. Whereas, Manganyi et al. (2015) revealed antibiofilm potential of clove and thyme oils against *Fusarium* sp. Their observation showed that these oils precluded adherence of cells and subsequently biofilm development on soft contact lenses. They suggested that oils of clove and thyme could be exploited to develop drugs-inhibiting device associated with biofilm formation in *C. albicans*.

Braga et al. (2008) carried out a study to investigate ability of thymol to interrupt biofilm formation and destroying mature biofilms in *C. albicans strains* ATCC 3153A and ATCC MYA 2876. They revealed that inhibition of different stages of evolution of biofilm is easily interfered by thymol, a terpene, and significantly effective against mature biofilm. On the other hand, Dalleau et al. (2008) examined the antibiofilm potential of 10 terpenes against *Candida* spp. They observed that 24 h grown *C. albicans* biofilms is inhibited by >80% upon treatment of carvacrol, geraniol, or thymol. This study proved the antibiofilm potential of terpenes and highlighted the competence of carvacrol, geraniol, and thymol to signify newer drug candidates in the therapy of candidiasis related with medical devices. Furletti et al. (2011) assessed the efficiency of extracts and oils from *Allium tuberosum*, *Cymbopogon martini*, *Cymbopogon winterianus*, *C. sativum*, and *Santolina chamaecyparissus* against *Candida* spp. isolates from the oral cavity of patients with periodontal disease. *C. sativum* oil exhibited its promise for a novel natural antifungal formulation. F8–10 fractions from *C. sativum* displayed excellent antibiofilm activity as observed under scanning electron microscopy. Furthermore, chemical analysis of the F8–10 fractions showed presence of 2-hexen-1-ol and 3-hexen-1-ol and cyclodecane as major active ingredients. When they tested these compounds in standard form, a greater activity was obtained than the oil. This finding suggested that these active components could interact synergistically to enhance antibiofilm potential.

The studies presented here highlight the potential of various essential oils and their active compounds to be explored to obtain newer antibiofilm compounds to deal with the variety of human fungal pathogens (Table 19.3).

19.5.1.3 Antifungal Activity of Essential Oils in Combination with Antifungal Drugs

Combination antifungal therapy offers the greater spectrum of drug action against potential drug-resistant pathogens, increased efficiency of drugs in killing the pathogens, reduction in drug dose to lower the toxicity, and lowering in the emergence of fungal resistance (Baddley and Pappas 2005; Johnson and Perfect 2010). Combination therapy has been used successfully against fungal infections caused by yeast or filamentous fungi and is highly preferred in clinical settings as many workers have described effective *in vitro* and *in vivo* combinations of antifungals (Cuenca-Estrella 2005, Baddley and Pappas 2007, Espinel-Ingroff 2009; de Oliveira Santos et al. 2018). Combination of drugs needs a thorough assessment of the interactive behavior of drugs as synergism, antagonism, and agonism. Where, synergy is defined as combination of drugs resulting in greater potency of each drug compared alone, and if, it

TABLE 19.3
Principal Effects of Essential Oils and/or Their Components on Fungi-Targeting Biofilm and Quorum Sensing

Essential Oils/Compounds	Activity
Essential Oils	
Coriandrum sativum, Croton cajucara, Cymbopogon, Cytrus, Eucalyptus, Laurus nobilis, Litsea, Melaleuca alternifolia, Mentha, Myrtus communis, Ocimum, Piper claussenianum, Rosmarinus officinalis, Syzygium aromaticum	Inhibition of biofilm development
Citrus, Juniperus communis, Mentha piperita, Origanum, Salvia sclarea	Inhibition of quorum sensing
Active Compounds	
p-Cymene, 1-8-cineole, linalool, terpinen-4-ol terpinolene, α-terpineol, eucarobustol E, eugenol, α-terpinene, β-terpinene	Inhibition of biofilm development
Limonene, linalool, α-pinene, terpinene-4-ol	Inhibition of quorum sensing

Source: Adapted from Nazzaro, F., Fratianni, F., Coppola, R., De Feo, V., *Pharmaceuticals*, 10, 86, 2017.

is lower than antagonism. Additivity is defined as if efficacy of drugs in combination summation of all drugs' activities, whereas, indifference is no increase in potencies of drugs in combination (Baddley and Pappas 2005). Combination of drugs depends on modes of action of participating drugs. Combination of drugs acting at different target sites could produce synergistic killing of fungal pathogens, and several models have been studied on this assumption (Johnson et al. 2004; Johnson and Perfect 2007). Nonetheless, prudence is crucial as some antifungals in combination have also demonstrated antagonistic interaction. Therefore, well-monitored experimental trials are obliged to describe the most efficient antifungal treatment. Furthermore, these experiments should assess the side effects of combination approach including pharmaco-fiscal impact (Vazquez 2007).

Many newer antifungals have shown synergistic or additive interaction with conventional antifungal drugs against a range of pathogenic fungi *in vitro* and *in vivo* (Johnson and Perfect 2010; de Oliveira Santos et al. 2018). Nowadays, plant products or their compounds have been studied to exhibit synergistic effects with conventional antifungal drugs. Many researchers have found plant products to be less toxic than current antifungal drugs; therefore, their union with these drugs could overall reduce the toxicity and enhance the antifungal value (Newman and Cragg 2012). Many researchers have studied combination of plant oils, extracts, or active compounds with antifungal agents against various human pathogenic fungi. Tangarife-Castano et al. (2011) investigated some plant oils or extracts are synergistic when combined with antifungal drugs against *C. albicans.* They found excellent synergistic effects for combination of *Piper bredemeyeri* extract with itraconazole. Zaidi et al.

(2018) showed that combination of methanolic extract of leaves of *Ocimum sanctum* with fluconazole is synergistic against drug-resistant *Candida* spp. Zore et al. (2011) showed that terpenoids have excellent synergistic interaction with fluconazole against fluconazole-resistant strains of *C. albicans*. Xanthorrhizol, a sesquiterpene from *Curcuma xanthorrhiza* Roxb, has been shown synergistic in combination with ketoconazole or amphotericin B against *C. albicans*, *C. glabrata*, *Candida guilliermondii*, *C. krusei*, *C. parapsilosis* and *C. tropicalis* (Rukayadi et al. 2009). Thymol has displayed synergistic effects in combination with fluconazole against *Candida* spp. (Sharifzadeh et al. 2018). Ahmad et al. (2010) showed that eugenol and methyleugenol have great potential to be used in combination with fluconazole to enhance antifungal effects against drug-resistant strains of *C. albicans*.

In a study conducted by Khan and Ahmad (2011), *Cinnamomum verum*, *S. aromaticum* oils, and their major active ingredients, i.e., cinnamaldehyde and eugenol potentiated the activity of fluconazole against drug-resistant strains of *A. fumigatus* and *T. rubrum*, whereas, oil of *C. martini* and its active compound geraniol could not exhibit any interaction with fluconazole. This group has also reported synergistic interaction of cinnamaldehyde and eugenol with fluconazole as well as amphotericin B against biofilm-forming drug-resistant strains of *C. albicans* (Khan and Ahmad 2012b). Methanolic extract of *Buchenavia tetraphylla* has been reported to augment the efficacy of fluconazole against azole-resistant strains of *C. albicans* isolated from vaginal secretions (Cavalcanti Filho et al. 2017). They observed additive effects against 20% of strains whereas synergistic effects against 60% of strains. Giordani et al. (2004) found that essential oil of *Thymus vulgaris* thymol chemotype is synergistic with amphotericin B against *C. albicans*. In a study carried out by Hirasawa and Takada (2004), catechins exhibited synergism with many antifungals against *C. albicans*. However, pyrogallol catechin appeared to be more effective than catechol catechin. Giordani et al. (2006) found oil of *Cinnamomum cassia* to be in synergy with amphotericin B against *C. albicans*. Sim and Shin (2008) have highlighted that ligustilide and butylidene phthalide, the major active ingredients of oil of *Ligusticum chuanxiong*, have enhanced the susceptibility of *Trichophyton erinacei*, *Trichophyton mentagrophytes*, *T. rubrum*, *T. schoenleinii*, *T. tonsurans* and *T. soudanense* ketoconazole and itraconazole.

Saad et al. (2010) investigated combination of *Thymus broussonetii* and *Thymus maroccanus* with amphotericin B and fluconazole against *C. albicans*. Their studies showed that both oils were more synergistic with fluconazole than with amphotericin B. Synergistic interaction of coriander oil with amphotericin B against *C. albicans* strains was demonstrated by Silva et al. (2011). They also observed additive effects of these combinations against *C. tropicalis*. Avijgan et al. (2014) reported a synergistic effect of ethanolic extract of *Echinophora platyloba* with itraconazole or fluconazole against *C. albicans* isolated from vaginal secretions of patients with recurrent vulvovaginitis. In a study, methanolic extract of Terminalia *catappa* leaves exhibited synergism when mixed with nystatin or amphotericin B against *C. albicans*, *Candida neoformans*, *C. glabrata*, *C. apicola*, and *Trichosporon beigelii* (Chanda et al. 2013). Moraes et al. (2015) detected synergism between a water-insoluble fraction isolated from Uncaria *tomentosa* bark with terbinafine and fluconazole against drug-resistant isolates of *C. glabrata* and *C. krusei*. The study revealed that synergistic interactions led to cell damage. As analyzed through differential scanning calorimetry and infrared analysis, they suggested that intermolecular interactions between the extract components and terbinafine or fluconazole occurring outer to cell wall is responsible for this mechanism of action.

Though we have described here various in vitro studies investigating synergistic effects among potential antifungal phytocompounds and conventional antifungal agents, the mechanisms core to these synergistic effects are yet not well known. Moreover, higher cost to conduct these approaches and a lesser number of clinical cases have necessitated to wisely analyze synergism between newer phytocompounds and currently used antifungals in order to get more insights into combination therapy. Understanding the cellular action of each drug in combination is also a vital measure in inferring approaches to use this strategy in the clinics. A lack of consent in the medical clinics underlines the necessity to carry out more clinical trials using combinations of antifungals. The researches and findings referred in this chapter support more investigation of new phyto-drugs with antifungal properties to be exploited in combination with existing antifungal drugs for the development of novel antifungal approaches.

19.6 CONCLUSION

The increase in fungal infections is alarming and has resulted in high rates of morbidity and mortality worldwide. Whereas, emergence of drug resistance strains of yeasts such as *Candida* spp. and filamentous fungi such as *Aspergillus* app and *Trichophyton* spp. has challenged the available antifungal arsenal. In addition, drugs such as azoles and amphotericin B have limitations manifested by their fungistatic character and associated host toxicity, respectively. In addition, ability of fungal pathogens to form biofilm results in therapeutic disappointment and incidence of resistance. Biofilm-mediated resistance is inherited to the extracellular matrix, which conceals and shields biofilm cells from the adjacent environment. Therefore, it is vital to develop new drugs as alternative therapies that are potentially active in combating fungal infections. Plants are rich source of drug compounds and considered as safe. Their antifungal and antibiofilm activities may be beneficial in attaining competent, and cost-effective drugs for the inhibition of fungal diseases. Consistent with the current weightage on inhibiting infections, a lot of efforts are focused on developing antibiofilm coatings for medical devices such as catheters and implants. Consequently, regular testing of such materials is of great importance. Many phytocompounds are exceptionally effective in combination therapy with traditional or other phytochemicals, which can be further exploited to lead to novel drug therapies against recalcitrant fungal infections.

ACKNOWLEDGMENT

Mohd Sajjad Ahmad Khan acknowledges the research support from the Department of Scientific Research, Imam Abdulrahman Bin Faisal University, Saudi Arabia, in completing this work.

REFERENCES

Achkar, J.M., Fries, B.C. 2010. *Candida* infections of the genitourinary tract. *Clin Microbiol Rev* 23:253–373.

Adorjan, B., Buchbauer, G. 2010. Biological properties of essential oils: An updated review. *Flavour Fragr J* 25:407–26

Agarwal, V., Lal, P., Pruthi, V. 2008. Prevention of *Candida albicans* biofilm by plant oils. *Mycopathologia* 165:13–9.

Agarwal, V., Lal, P., Pruthi, V. 2010. Effect of plant oils on *Candida albicans*. *J Microbiol Immunol Infect* 43:447–51.

Ahmad, A., Khan, A., Khan, L.A., Manzoor, N. 2010. *In vitro* synergy of eugenol and methyleugenol with fluconazole against clinical *Candida* isolates. *J Med Microbiol* 59:1178–4.

Alem, M.A.S., Oteef, M.D.Y., Flowers, T.H., Douglas, L.J. 2006. Production of tyrosol by *Candida albicans* biofilms and its role in quorum sensing and biofilm development. *Eukaryot Cell* 5:1770–9.

Angioni, A., Barra, A., Coroneo, V., Dessi, S., Cabaras, P. 2006. Chemical composition, seasonal variability, and antifungal activity of *Lavandula stoechas* L. ssp. *Stoechas* essential oils from stem/leaves and flowers. *J Agric Food Chem* 54:4364–70.

Avijgan, M., Mahboubi, M., Nasab, M.M., Nia, E.A., Yousefi, H. 2014. Synergistic activity between *Echinophora platyloba* DC ethanolic extract and azole drugs against clinical isolates of *Candida albicans* from women suffering chronic recurrent vaginitis. *J de Mycol Med* 24:112–6.

Baddley, J.W., Pappas, P.G. 2005. Antifungal combination therapy: Clinical potential. *Drugs* 65:1461–80.

Baddley, J.W., Pappas, P.G. 2007. Combination antifungal therapy for the treatment of invasive yeast and mold infections. *Curr Infect Dis Rep* 9:448–56.

Bakkali, F., Averbeck, S., Averbeck, D., Idaomar, M. 2008. Biological effects of essential oils—a review. *Food Chem Toxicol* 46:446–75.

Bansod, S., Rai, M. 2008. Antifungal activity of essential oils from Indian medicinal plants against human pathogenic *Aspergillus fumigatus* and *A. niger*. *W J Med Scie* 3:81–8.

Blanco, M.T., Sacristan, B., Lucio, L., Blanco, J., Perez-Giraldo, C., Gomez Garcia, A.C. 2010. Cell surface hydrophobicity as an indicator of other virulence factors in *Candida albicans*. *Rev Iberoam Micol* 27:195–9.

Blankenship, J.R., Mitchell, A.P. 2006. How to build a biofilm: A fungal perspective. *Curr Opin Microbiol* 9:588–94.

Bounatirou, S., Smiti, S., Miguel, M.G. et al. 2007. Chemical composition, antioxidant, and antimicrobial activities of the essential oils isolated from Tunisian *Thymus capitus* Hoff. et Link. *Food Chem* 105:146–55.

Braga, P.C., Culici, M., Alfieri, M., Dal Sasso, M. 2008. Thymol inhibits *Candida albicans* biofilm formation and mature biofilm. *Int J Antimicrob Agents* 31:472–7.

Cannon, R.D., Lamping, E., Holmes, A.R. et al. 2007. *Candida albicans* drug resistance—Another way to cope with stress. *Microbiology* 153:3211–7.

Carson, C.F., Hammer, K.A., Riley, T.V. 2006. *Melaleuca alternifolia* (tea tree) oil: A review of antimicrobial and other medicinal properties. *Clin Microbiol Rev* 19:50–62.

Cavalcanti Filho, J.R., Silva, T.F., Nobre, W.Q. et al. 2017. Antimicrobial activity of *Buchenavia tetraphylla* against *Candida albicans* strains isolated from vaginal secretions. *Pharmaceut Biol* 55:1521–7.

Chakrabarti, A. 2005. Microbiology of systemic fungal infection. *J Postgrad Med* 51 (Suppl1):S16–20.

Chanda, S., Rakholiya, K., Dholakia, K., Baravalia, Y. 2013. Antimicrobial, antioxidant, and synergistic properties of two nutraceutical plants: *Terminalia catappa* L. and *Colocasia esculenta* L. *Turk J Biol* 37:81–91.

Costa-Orlandi, C.B., Sardi, J.C., Santos, C.T., Fusco-Almeida, A.M., Mendes-Giannini, M.J. 2014. In vitro characterization of *Trichophyton rubrum* and *T. mentagrophytes* biofilms. *Biofouling* 30:719–27.

Costa-Orlandi, C.B., Sardi, J.C.O., Pitangui, N.S. et al. 2017. Fungal biofilms and polymicrobial diseases. *J Fungi* 3:22.

Cowen, L.E. 2008. The evolution of fungal drug resistance: Modulating the trajectory from genotype to phenotype. *Nat Rev Microbiol* 6:187–98.

Cuenca-Estrella, M., Gomez-Lopez, A., Garcia-Effron, G. et al. 2005. Combined activity in vitro of caspofungin, amphotericin azole agents against Itraconazole-resistant clinical isolates of *Aspergillus fumigatus. Antimicrob Agents Chemother* 49:1232–5.

Dagenais, T.R.T., Keller, N.P. 2009. Pathogenesis of *Aspergillus fumigatus* in invasive aspergillosis. *Clin Microbiol Rev* 22:447–65.

Dalleau, S., Cateau, E., Bergès, T., Berjeaud, J.M., Imbert, C. 2008. *In vitro* activity of terpenes against *Candida* biofilms. *Int J Antimicrob Agents* 31:572–6.

Daniel, M. 2006. *Medicinal Plants: Chemistry and Properties*, Oxford & IBH Publishers Co. Pvt Ltd, New Delhi.

Davies, D. 2003. Understanding biofilm resistance to antibacterial agents. *Nat Rev Drug Discov* 2:114–22.

de Oliveira Santos, G.C., Vasconcelos, C.C., Lopes, A.J.O. et al. 2018. Infections and therapeutic strategies: Mechanisms of action for traditional and alternative agents. *Front Microbiol* 9:1351.

de Repentigny, L., Lewandowski, D., Jolicoeur, P. 2004. Immunopathogenesis of oropharyngeal candidiasis in human immunodeficiency virus infection. *Clin Microbiol Rev* 17:729–59.

de Souza, R.D., Mores, A.U., Cavalca, L., Rosa, R.T., Samaranayake, L.P., Rosa, E.A. 2009. Cell surface hydrophobicity of *Candida albicans* isolated from elder patients undergoing denture-related candidosis. *Gerodontology* 26:157–61.

Dongari-Bagtzoglou, A. 2008. Pathogenesis of mucosal biofilm infections: Challenges and progress. *Expert Rev Anti Infect Ther* 6(2):201–8.

Douglas, L.J. 2003. *Candida* biofilms and their role in infection. *Trends Microbiol* 11:30–6.

Ellis, D.H., Watson, A.B., Marley, J.E., Williams, T.G. 1997. Nondermatophytes in onychomycosis of the toenails. *Br J Dermatol* 136:490–3.

Espinel-Ingroff, A. 2009. Novel antifungal agents, targets, or therapeutic strategies for the treatment of invasive fungal diseases: A review of the literature (2005–2009). *Rev Iberoam Micol* 26:15–22.

Farmakiotis, D., Kontoyiannis, D.P. 2017. Epidemiology of antifungal resistance in human pathogenic yeasts: Current viewpoint and practical recommendations for management. *Int J Antimicrob Agents* 50(3):318–24.

Fidel, P.L. Jr, Vanquez, J.A., Sobel, J.D. 1999. *Candida glabrata*: A review of epidemiology, pathogenesis, and clinical disease with comparison to *Candida albicans. Clin Microbiol Rev* 12:80–96.

Fluckiger, U., Marchetti, O., Bille, J. et al. 2006. Treatment options of invasive fungal infections in adults. *Swiss Med Wkly* 136:447–63.

Fuqua, Jr T.H., Sittitavornwong, S., Knoll, M., Said-Al-Naief, N. 2010. Primary invasive oral aspergillosis: An updated literature review. *J Oral Maxillofac Surg* 68:2557–63.

Furletti, V.F., Teixeira, I.P., Obando-Pereda, G. et al. 2011. Action of *Coriandrum sativum* L. essential oil upon oral *Candida albicans* biofilm formation. *Evid Based Complement Altern Med* 2011:985832.

Galhardo, M.C., Wanke, B., Reis, R.S., Oliveira, L.A., Valle, A.C. 2004. Disseminated dermatophytosis caused by *Microsporum gypseum* in an AIDS patient: Response to terbinafine and amorolfine. *Mycoses* 47:238–41.

Gauwerky, K., Boreli, C., Korting, H.C. 2009. Targeting virulence: A new paradigm for antifungals. *Drug Disc Today* 14:214–22.

Giordani, R., Regli, P., Kaloustian, J., Mikail, C., Abou, L., Portugal, H. 2004. Antifungal effect of various essential oils against *Candida albicans*. Potentiation of antifungal action of amphotericin B by essential oil from *Thymus vulgaris*. *Phytother Res* 18:990–5.
Giordani, R., Regli, P., Kaloustian, J., Portugal, H. 2006. Potentiation of antifungal activity of amphotericin B by essential oil from *Cinnamomum cassia*. *Phytother Res* 20:58–61.
Gupta, A.K., Foley, K.A., Versteeg, S.G. 2017. New antifungal agents and new formulations against dermatophytes. *Mycopathologia* 182(1–2):127–41.
Gutierrez, J., Morales, P., Gonzalez, M.A., Quindos, G. 2002. *Candida dubliniensis*, a new fungal pathogen. *J Basic Microbiol* 42:207–27.
Hammer, K.A., Carson, C.F., Riley, T.V. 2000. *Melaleuca alternifolia* (tea tree) oil inhibits germ tube formation by *Candida albicans*. *Med Mycol* 38:355–62.
Harborne, S.B., Baxter, H. 1995. *Phytochemical Dictionary: A Handbook of Bioactive Compounds from Plants*, Taylor and Francis, London.
Harding, M.W., Marques, L.L., Howard, R.J., Olson, M.E. 2009. Can filamentous fungi form biofilms? *Trends Microbiol* 17:475–80.
Hayens, K.A., Westerneng, T.J. 1996. Rapid identification of *Candida albicans*, *C. glabrata*, *C. parapsilosis*, and *C. krusei* by species specific PCR of large subunit ribosomal DNA. *J Med Microbiol* 44:390–6.
He, M., Du, M., Fan, M., Bian, Z. 2007. *In vitro* activity of eugenol against *Candida albicans* biofilms. Mycopathologia 163:137–43.
Hirasawa, M., Takada, K. 2004. Multiple effects of green tea catechin on the antifungal activity of antimycotics against *Candida albicans*. *J Antimicrob Chemother* 53:225–9.
Huai, Q., Kim, H.Y., Liu, Y. et al. 2002. Crystal structure of calcineurin-cyclophilin -cyclosporin shows common but distinct recognition of immunophilin-drug complexes. *Proc Natl Acad Sci USA* 99: 12037–42.
Johnson, M.D., MacDougall, C., Ostrosky-Zeichner, L., Perfect, J.R., Rex, J.H. 2004. Combination antifungal therapy. *Antimicrob Agents Chemother* 48:693–715.
Johnson, M.D., Perfect, J.R. 2007. Combination antifungal therapy: What can and should we expect? *Bone Marr Transplant* 40:297–306.
Johnson, M.D., Perfect, J.R. 2010. Use of antifungal combination therapy: Agents, order, and timing. *Curr Fung Infec Rep* 4:87–95.
Kamboj, V.P. 2000. Herbal medicine. *Curr Sci* 78:35–9.
Kanafani, Z.A., Perfect, J.R. 2008. Resistance to antifungal agents: Mechanisms and clinical impact. *Clin Infect Dis* 46:120–8.
Kannan, P., Janaki, C., Selvi, G.S. 2006. Prevalence of dermatophytes and other fungal agents isolated from clinical samples. *Indian J Med Microbiol* 24:212–5.
Khan, M.S., Ahmad, I. 2012a. Biofilm inhibition by *Cymbopogon citratus* and *Syzygium aromaticum* essential oils in the strains of *Candida albicans*. *J Ethnopharmacol* 140:416–23
Khan, M.S.A., Ahmad, I. 2011. Antifungal activity of essential oils and their synergy with fluconazole against drug-resistant strains of *Aspergillus fumigatus* and *Trichophyton rubrum*. *Appl Microbiol Biotechnol* 90:1083–94
Khan, M.S.A., Ahmad, I. 2012b. Antibiofilm activity of certain phytocompounds and their synergy with fluconazole against *Candida albicans* biofilms. *J Antimicrob Chemother* 67:618–21.
Kojic, E.M., Darouiche, R.O. 2004. *Candida* infections of medical devices. *Clin Microbiol Rev* 17:255–67.
Krasnick, J., Greenberger, P.A., Roberts, M., Patterson, R. 1995. Allergic bronchopulmonary aspergillosis: Serologic update for 1995. *J Clin Lab Immunol* 46:137–42.
Krishnan, S., Manavathu, E.K., Chandrasekar, P.H. 2009. *Aspergillus flavus*: An emerging non-fumigatus *Aspergillus* species of significance. *Mycoses* 52:206–22.

Latge, J.P. 1999. *Aspergillus fumigatus* and aspergillosis. *Clin Microbiol Rev* 12:310–50.

Liu, J., Farmer, J.D. Jr, Lane, W.S., Friedman, J., Weissman, I., Schreiber, S.L. 1991. Calcineurin is a common target of cyclophilin-cyclosporin A and FKBP-FK506 complexes. *Cell* 66:807–15.

Loreto, E.S., Scheid, L.A., Nogueira, C.W., Zeni, G., Santurio, J.M., Alves, S.H. 2010. *Candida dubliniensis*: Epidemiology and phenotypic methods for identification. *Mycopathologia* 169:431–43.

Ma, J.E., Yun, E.Y., Kim, Y.E. et al. 2011. Endobronchial aspergilloma: Report of 10 cases and literature review. *Yonsei Med J* 52:787–92.

Malani, A., Hmoud, J., Chiu, L., Carver, P.L., Bielaczyc, A., Kauffman, C.A. 2005. *Candida glabrata* fungemia: Experience in a tertiary care center. *Clin Infect Dis* 41:975–81.

Manganyi, M.C., Regnier, T., Olivier, E.I. 2015. Antimicrobial activities of selected essential oils against Fusarium oxysporum isolates and their biofilms. *S Afr J Bot* 99:115–21.

Manoharan, R.K., Lee, J.H., Lee, J. 2017. Antibiofilm and antihyphal activities of cedar leaf essential oil, camphor, and fenchone derivatives against *Candida albicans*. *Front Microbiol* 8:1476.

Marconi, V.C., Kradin, R., Marty, F.M., Hospenthal, D.R., Kotton, C.N. 2010. Disseminated dermatophytosis in a patient with hereditary hemochromatosis and hepatic cirrhosis: Case report and review of the literature. *Med Mycol* 48:518–27.

Morace, G., Perdoni, F., Borghi, E. 2014. Antifungal drug resistance in *Candida* species. *J Glob Antimicrob Resist* 2:254–9.

Moraes, R.C., Lana, A.J.D., Kaiser, S. et al. 2015. Antifungal activity of *Uncaria tomentosa* (Willd.) D.C. against resistant non-albicans *Candida* isolates. *Ind Crops Prod* 69:7–14.

Moss, R.B. 2007. Allergic bronchopulmonary aspergillosis. *Clin Rev Aller Immunol* 23:87–104.

Muzny, C.A., Schwebke, J.R. 2015. Biofilms: An underappreciated mechanism of treatment failure and recurrence in vaginal infections. *Clin Infect Dis* 61:601–6.

Nazzaro, F., Fratianni, F., Coppola, R., De Feo, V. 2017. Essential oils and antifungal activity. *Pharmaceuticals* 10:86.

Newman, D.J., Cragg, G.M. 2012. Natural products as sources of new drugs over the 30 years from 1981 to 2010. *J Nat Prod* 75:311–35.

Pfaller, M.A., Diekema, D.J. 2004. Rare and emerging opportunistic fungal pathogens: Concern for resistance beyond *Candida albicans* and *Aspergillus fumigatus*. *J Clin Microbiol* 42:4419–31.

Pfaller, M.A., and Diekema, D.J. 2007. Epidemiology of invasive candidiasis: A persistent public health problem. *Clin Microbiol Rev* 20:133–63.

Pfaller, M.A., Pappas, P.G., Wingard, J.R. 2006. Invasive fungal pathogens: Current epidemiological trends. *Clin Infect Dis* 43(Suppl 1):S3–14.

Pitangui, N.S., Sardi, J.C., Silva, J.F. et al. 2012. Adhesion of *Histoplasma capsulatum* to pneumocytes and biofilm formation on an abiotic surface. *Biofouling* 28:711–8.

Rai, M.K., Acharya, D., Wadegaonkar, P. 2003. Plant derived-antimycotics: Potential of Asteraceous plants, In: *Plant-Derived Antimycotics: Current Trends and Future prospects*, pp. 165–185, Haworth press, New York, London, Oxford.

Ramage, G., Milligan, S., Lappin, D.F. et al. 2012. Antifungal, cytotoxic, and immunomodulatory properties of tea tree oil and its derivative components: Potential role in management of oral candidosis in cancer patients. *Front Microbiol* 3:220.

Ramage, G., Rajendran, R., Gutierrez-Correa, M., Jones, B., Williams, C. 2011. *Aspergillus* biofilms: Clinical and industrial significance. *FEMS Microbiol Lett* 324:89–97.

Ramage, G., Saville, S.P., Thomas, D.P., Lopez-Ribot, J.L. 2005. *Candida* biofilms: An update. *Eukaryotic Cell* 4:633–38.

Ramage, G., Saville, S.P., Wickes, B.L., Lopez-Ribot, J.L. 2002. Inhibition of *Candida albicans* biofilm formation by farnesol, a quorum-sensing molecule. *Appl Environ Microbiol* 68:5459–63.

Reedy, J.L., Bastidas, R.J., Heitman, J. 2007. The virulence of human pathogenic fungi: Notes from the south of France. *Cell Host Microbe* 2:77–83.
Rukayadi, Y., Lee, K., Lee, M.S., Yong, D., Hwang, J.K. 2009. Synergistic anticandidal activity of xanthorrhizol in combination with ketoconazole or amphotericin B. *FEMS Yeast Res* 9:1302–11.
Saad, A., Fadli, M., Bouazizb, M., Benharref, A., Mezrioui, N.E., Hassani, L. 2010. Anticandidal activity of the essential oils of *Thymus maroccanus* and *Thymus broussonetii* and their synergism with amphotericin B and fluconazole. *Phytomedicine* 17:1057–60.
Sangamwar, A.T., Deshpande, U.D., Pekamwar, S.S. 2008. Antifungals: Need to search for a new molecular target. *Ind J Pharmaceut Sci* 70:423–30.
Sanglard, D., Coste, A., Ferrari, S. 2009. Antifungal drug resistance mechanism in fungal pathogens from the perspectives of transcriptional gene regulation. *FEMS Yeast Res* 9:1029–50.
Sanglard, D., Ischer, F., Parkinson, T., Falconer, D., Bille, J. 2003. *Candida albicans* mutations in the ergosterol biosynthetic pathway and resistance to several antifungal agents. *Antimicrob Agents Chemother* 47:2404–12.
Sardi, J.D.C., Pitangui, N.D.S., Voltan, A.R. et al. 2015. In vitro *Paracoccidioides brasiliensis* biofilm and gene expression of adhesins and hydrolytic enzymes. *Virulence* 6:642–51.
Sarvikivi, E., Lyytikainen, O., Soll, D.R. et al. 2005. Emergence of fluconazole resistance in a *Candida parapsilosis* strain that caused infections in a neonatal intensive care unit. *J Clin Microbiol* 43:2729–35.
Sharifzadeh, A., Khosravi, A.R., Shokri, H., Shirzadi, H. 2018. Potential effect of 2-isopropyl-5-methylphenol (thymol) alone and in combination with fluconazole against clinical isolates of *Candida albicans, C. glabrata* and *C. krusei. J de Mycol Med* 28:294–9.
Silva, F., Ferreiraa, S., Duartea, A., Mendonc, D.I., Dominguesa, F.C. 2011. Antifungal activity of *Coriandrum sativum* essential oil, its mode of action against *Candida* species and potential synergism with amphotericin B. *Phytomedicine* 19:42–7.
Sim, Y., Shin, S. 2008. Combinatorial anti-*Trichophyton* effects of *Ligusticum chuanxiong* essential oil components with antibiotics. *Arch Pharm Res* 31:497–502.
Singh, A. 2006. *Medicinal Plants of the World*, Oxford & IBH publishers Co. Pvt Ltd, New Delhi.
Singh, M.P., Panda, H. 2005. *Medicinal Herbs with Their Formulations*, Daya Publications, New Delhi
Singh, N., Paterson, D.L. 2005. *Aspergillus* infections in transplant recipients. *Clin Microbiol Rev* 18:44–69.
Siqueira, V.M., Lima, N. 2013. Biofilm formation by filamentous fungi recovered from a water system. *J Mycol* 2013: Article ID 152941, 9.
Tangarife-Castano, V., Correa-Royero, C., Zapata-Londono, B., Duran, C., Stanshenko, E., Mesa-Arango, A.C. 2011. Anti-*Candida albicans* activity, cytotoxicity and interaction with antifungal drugs of essential oils and extracts from aromatic and medicinal plants. *Infectio* 15:160–7.
Taweechaisupapong, S., Ngaonee, P., Patsuk, P., Pitiphat, W., Khunkitti, W. 2012. Antibiofilm activity and post antifungal effect of lemongrass oil on clinical *Candida dubliniensis* isolate. *S Afr J Biot* 78:37–43.
Trofa, D., Gacser, A., Nosanchuk, J.D. 2008. *Candida parapsilosis*, an Emerging Fungal Pathogen. *Clin Microbiol Rev* 21:606–25.
Van Acker, H., Van Dijck, P., Coenye, T. 2014. Molecular mechanisms of antimicrobial tolerance and resistance in bacterial and fungal biofilms. *Trends Microbiol* 22:326–33.
Vazquez, J.A. 2007. Combination antifungal therapy: The new frontier. *Fut Microbiol* 2:115–39.
Verweij, P.E., Chowdhary, A., Melchers, W.J.G., Meis, J.F. 2016. Azole resistance in *Aspergillus fumigatus*: Can we retain the clinical use of mold-active antifungal azoles? *Clin Infect Dis* 62(3):362–8.

Villar-Vidal, M., Marcos-Arias, C., Eraso, E., Quindos, G. 2011. Variation in biofilm formation among blood and oral isolates of *Candida albicans* and *Candida dubliniensis*. *Enferm Infecc Microbiol Clin* 29:660–5.

Vinitha, M., Ballal, M. 2007. Biofilm as virulence marker in *Candida* isolated from blood. *W J Med Sci* 2:46–8.

Viudes, A., Peman, J., Canton, E., Ubeda, P., Lopez-Ribot, J.L., Gobernado, M. 2002. Candidemia at a tertiary-care hospital: Epidemiology, treatment, clinical outcome and risk factors for death. *Eur J Clin Microbiol Infect Dis* 21:767–4.

Weitzman, I., Summerbell, R.C. 1995. The Dermatophytes. *Clin Microbiol Rev* 8:240–59.

White, T.C., Marr, K.A., Bowden, R.A. 1998. Clinical, cellular, and molecular factors that contribute to antifungal drug resistance. *Clin Microbiol Rev* 11:382–402.

Wingard, J.R. 1995. Importance of *Candida* species other than *C. albicans* as pathogens in oncology patients. *Clin Infect Dis* 20:115–25.

Yesil Celiktas, O., Hames Kocabas, E.E., Bedir, E., Vardar Sukan, F., Ozek, T., Baser, K.H.C. 2007. Antimicrobial activities of methanol extracts and essential oils of *Rosmarinus officinalis*, depending on location and seasonal variations. *Food Chem* 100:553–9.

Zaidi, K.U., Shah, F., Parmar, R., Thawani, V. 2018. Anticandidal synergistic activity of *Ocimum sanctum* and fluconazole of azole resistance strains of clinical isolates. *J de Mycol Med* 28:289–93.

Zore, G.B., Thakre, A.D., Jadhav, S., Karuppayil, S.M. 2011. Terpenoids inhibit *Candida albicans* growth by affecting membrane integrity and arrest of cell cycle. *Phytomedicine* 18:1181–90.

20 Biomedical Applications of Microbial Biofilm

Muhammad Shahid and Muhammad Tjammal Rehman
University of Agriculture

Fozia Anjum
Government College University

Hina Fatima
University of Agriculture

CONTENTS

20.1 INTRODUCTION

A group of bacteria, yeast, or fungi can grow in the form of heterogeneous colonies both on the living and non-living surface with the help of extracellular polymeric substances (EPSs). These EPSs make the single cell resistant to many hostile factors, i.e., antibacterial agents, nutrient deficiency, and host immune system (Whitchurch et al. 2002). The freely floating bacteria living in the form of biofilm have different phenotypic and genotypic properties which are playing a role in making them highly resistant against antibiotics. Biofilm has resistance due to the following reasons:

- Multi-drug efflux pump induction in biofilm
- Availability of persisters
- Antibiotics are unable to enter biofilm (Lewis 2005).

Bacteria in the form of biofilm are fruitful in many ways to human beings such as these are involved in the recycling of nutrients, wastewater treatment, removal of pollutants from environment, decomposition of complex organic substances and those that are present in large intestine helping in digestion process (Gilbert et al. 2002).

Biofilm producing different substances are causing difficulties in the field of medicine to manage them. The reason is that these biofilms are highly impossible to remove by antibiotics and the immune system of host body. Biofilms have 100–1,000 times more susceptibility than their planktonic form against available antibiotics. Biofilm formation leads to many diseases in humans and plants, some of which are discussed below with their treatment (Thomas et al. 2012).

Bacteria in the form of biofilm lead to many chronic infections which can be defined as tissue damaged and long-lasting inflammation (Bjarnsholt et al. 2009). These are the infections that

> Highly resistant to antibiotics as well as response of host's adaptive and innate immune system
>
> **(Høiby et al. 2010).**

Bacterial biofilm grows and leads to many chronic diseases. Bacteria grow in the form of biofilms on many natural surfaces like endocarditis (valves of heart), leading to bronchopneumonia in patients with cystic fibrosis (CF) (Bjarnsholt et al. 2009), teeth, otitis media (in the middle ear), chronic wounds, stents, chronic rhinosinusitis (CRS), chronic osteomyelitis, infection in prosthetic joints I.V (intra venous) catchers (Del & Robin 2009).

The bacterial biofilm produces a bio-polymer matrix known as EPS that helps them to grow in the form of colony. The matrix consists of DNA, polysaccharide, and proteins secreted by microbes. It is observed that there may be species of one or more than one microbe living together in the form of socio-microbiological way (Whitchurch et al. 2002). The matrix helps in the stability of biofilm against antibiotics (Klausen et al. 2003).

20.2 CHRONIC RHINOSINUSITIS (CRS)

Inflammation in the mucosal lining of the paranasal sinuses as well as nasal cavity persisting more than a period of 3 months is mentioned as CRS. Symptoms include pain in fascial muscles, nasal obstruction leading to no or reduced sensation of smell (Figure 20.1) (Kartush et al. 2019). It is the major cause of chronic diseases all over the world affecting 15% of the total population (Khawar et al. 2016). It has many phenotypes often explained into two forms without polyps and with polyps. It is attributed to be the end of many inflammatory pathways as well as secondary in many cases, i.e., fungal etiology and CF.

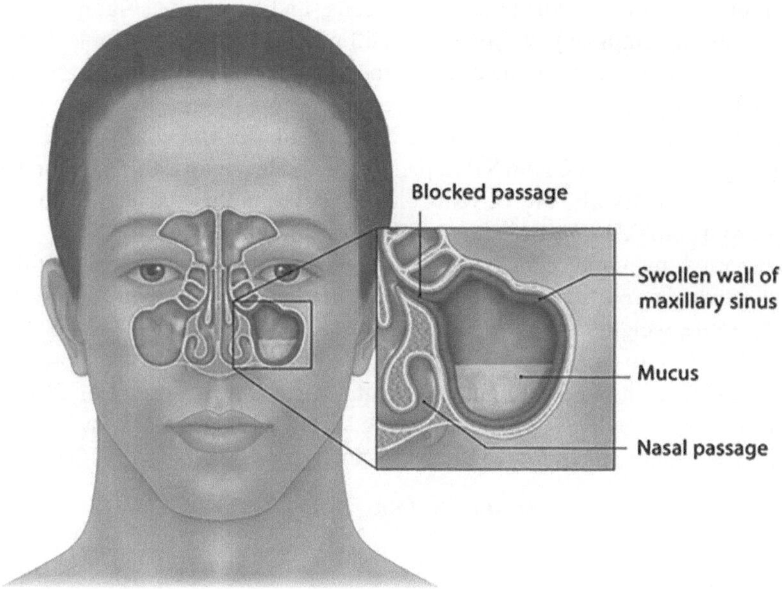

FIGURE 20.1 Symptoms of chronic rhinosinusitis swollen wall of maxillary sinus, increased mucus production, and blockage of nasal passage. (From Kartush, A. G., et al., *Am. J. Rhinol. Allergy*, 33, 2, 203–211, 2019. With permission.)

CRS is divided into two forms (CRSsNP and CRSwNP) due to the sign and symptoms diagnosed using endoscopy.

- CRSsNP with minimal asthma death rate known as inflammatory non-type 2 disease associated with cytokines or IL-17 in patients
- CRSwNP attributed as type 2 which includes IL-4, IL-5, IL-13 cytokines, IgE, and eosinophils. It is attributed as higher level of serum IgE and eosinophil in the patients' blood (Walker et al. 2019).

The highly sensitive form of rhinosinusitis is CRSwNP type 2 because it cannot be treated easily using antibiotics as well as surgery. This is treated by using biological agents in clinical trials on patients with asthma. It can be controlled by using anti IL4, IL5, IL13, GATA 3 DNA-zyme, and anti-IgE but here the mechanism of IgE is explained in detail (Rivero & Liang 2017).

20.2.1 Anti-IgE Antibodies

Omalizumab is a monoclonal antibody used against IgE because it can interact and bind with the Fc receptor of the IgE present on the surface of different inflammatory cells in the blood thereby involving in the blockage of IgE inflammation cascade pathway. In this way, it also reduces the increased level of free IgE in blood. It also controls the mediators by triggering its secondary downstream effect on the

Fc receptor present on dendritic cell, mast cells, and basophils (Figure 20.2). This effect is induced completely within a time of 3 months (16 weeks) (Table 20.1) (Kim et al. 2017). This medicine is helpful for the treatment of patients diagnosed with CRSwNP having asthma and increased level of eosinophil, IgE, and toxins produced by *Staphylococcus aureus* in nasal polyp (Baba et al. 2014). Omalizumab is a drug approved by FDA against CRSwNP in patients with laborious asthma. It can be used only for those patients who are 12 years old or older than this age having no effect of corticosteroids inhaled (Kartush et al. 2019).

It is reported that Omalizumab has a bad effect on the patients because it involves in the anaphylactic reactions. It is taken as intravenous within a period of 8–16 days according to the weight of the patient. it has 0.2% risk of anaphylaxis in patients due to which it is given to patients within clinic as well as the frequency and dose level are estimated by checking body weight and serum IgE count (Chipps et al. 2012). This drug has improved the lives of patients having severe asthma and CRSswNP which is checked by using computed tomography and polyp score (Tsetsos et al. 2018).

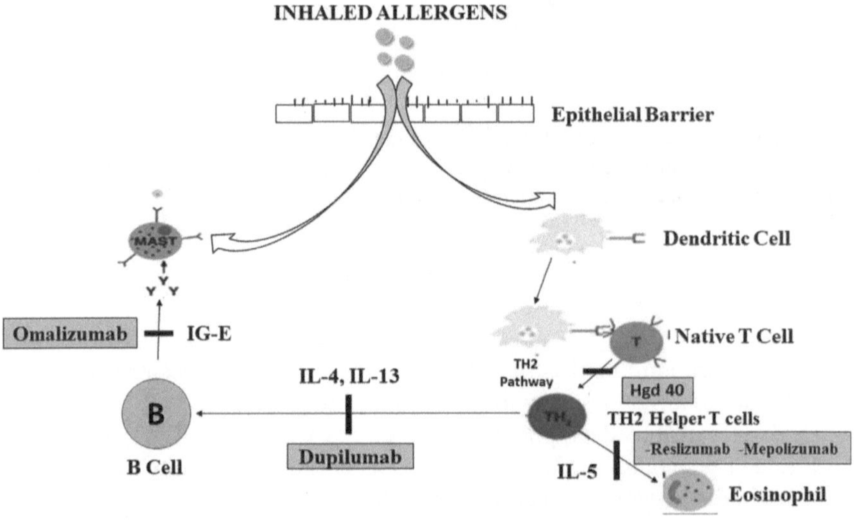

FIGURE 20.2 The immunologic response to inhaled allergens, demonstrating the points at which relevant biologic agents inhibit the process. (From Kartush, A. G., et al., *Am. J. Rhinol. Allergy*, 33, 2, 203–211, 2019. With permission.)

TABLE 20.1
Biologic Agents Discussed with Mechanism, Route of Administration

Generic Name	Mechanism	Route of Administration	Reference
Omalizumab	1. Fixes to free IgE 2. Down orders FceRI 3. Lessened mediator release	Intravenous injection each 8–16 days Dosing and frequency level is resolute by serum total IgE level and body weight	Kartush et al. (2019)

20.3 DENTAL CARIES

Bacteria form biofilm on teeth indirectly leading to dental caries (Rosan & Lamont 2000). *Streptococci mutant*, *Actinomycetes* and *Lactobacillus* are the Gram-positive bacteria that are involved in the formation of biofilm on teeth, while the Gram-negative bacteria that are involved in dental caries are *Actinobacillus*, *Fusobacterium*, *Porphyromonas gingivalis*, and *Prevotella* (Jenkinson & Lamont 2005). The cause of dental diseases and dental caries is the subgingival and supragingival plaque of bacterial biofilm. These bacterial biofilms increased persistence to antibiotics lead to their massive appearance and development. Due to the increased resistance of biofilm against previously used antibiotics in dental caries, researchers are looking forward to a new range of compounds having more efficacy and antimicrobial activity against biofilm (Projan & Youngman 2002). The methods used to prevent dental caries include:

- Prevention and
- Disruption of biofilm.

It can not only be done by using bacteriostatic antibiotics only rather there is a need for antiadhesive agents for the prevention of bacteria to adhere to natural surfaces (Sbordone & Bortolaia 2003).

20.3.1 Natural Compounds as Modulators of Bacterial Biofilms

More than a period of 1,000 years, plant extracts are used by humans against many diseases (Gyawali & Ibrahim 2014). Plant extract contains many biochemically active compounds that have activity to inhibit the growth of biofilm, affects the quorum sensing, and reduces the over-expression of efflux pump. The effect of following discussed plant extracts on biofilm formation is elaborated in Table 20.2.

TABLE 20.2
Inhibitory Effects of Plant Extracts on Biofilm Forming Dental Caries

Plants	Extracts	Strains	Inhibition Effect	References
Salvadora persica	• Miswak • Methanol • Ethanol • Chloroform • Acetone • Aqueous	• *Mutans streptococci* • *Lactobacilli* • *Staphylococcus aureus* • *Streptococcus mutans*	• In vivo growth • Growth • Biofilm • Quorum sensing	Almas & Al-Zeid (2004) Al-Sohaibani & Murugan (2012)
Juglans regia	• Water	• *Streptococcus mutans*	• Biofilm	Faraz et al. (2012)
VitisVinifera	• Phenol	• *Streptococcus mutans*	• Growth • Adherence • Glucosyltransferase	Furiga et al. (2008)

20.3.1.1 *Salvadora persica*

Salvodora persica a species of Salvadora family also known as Arak is usually used as toothbrush from centuries in the Middle East. Extracts of this plant are used as a cleansing agent for teeth (Kouidhi et al. 2015). It is studied that *S. persica* has anti-cariogenic and anti-periodontal activities against *Streptococcus salivarius*, *Streptococcus mutans*, *S. aureus*, and *Actinomycetes* (Almas & Al-Zeid 2004). The extract includes bioactive compounds, namely 3-benzyloxy-1-nitro-butan-2-ol, 1,3-cyclohexane dicarbohydrazide, and benzyl-6,9,12-octadecatrienoate which have greater ability to inhibit the growth of biofilm as well as it has effect on the quorum-sensing regulators of bacterial communities (Al-Sohaibani & Murugan 2012). Moreover, extracts from miswak are also studied having activity against the acid-producing bacteria, thus preventing the dental caries to happen and inhibit plaque formation (Amoian et al. 2010).

20.3.1.2 *Juglans regia*

Juglans regia is known as Persian walnut tree domestically found in central Asia where it develops into complete plant without cultivation or semi-cultivation and covers an area from South Eastern Europe to Iran. The bioactive compounds found in its bark are quercetin-3-a-L-arabinoside, folic acid, b-sitosterol, ascorbic acid, regiolone, and gallic acid (Figure 20.3) (Kouidhi et al. 2015). These plant extracts have antioxidant as well as antimicrobial activities. *J. regia* also exist in Pakistan, and its antibacterial activities are studied showing that it inhibits the growth of *S. mutans* preventing from dental caries as compared to previously used antibiotics (Faraz et al. 2012).

20.3.1.3 *Vitis vinifera*

Vitis vinifera is also known as common grapevine belongs to Vitis family inborn to the Mediterranean area (Southern Europe to Western Asia) and now grown all around the world (Nassiri-Asl & Hosseinzadeh 2009). Its fruits are used all around the world as a source of dietary supplements while leaves and seeds as a medicine. Its seeds and leaves contain bioactive compounds that directly involved in the inhibition of bacterial biofilm in oral dental caries and teeth diseases (Wu 2009). The biochemically active compounds in this plant are trans-resveratrol, flavanols, hydroxy cinnamic acids, and gallic acid that have the anti-inhibitory effect against biofilm (Wen et al. 2003). It is reported that the extract from grapevine have the potential to inhibit the *S. mutans* biofilm formation in dental caries as well as lessened the activity of

(a) (b) (c)

FIGURE 20.3 (a–c) *Salvadora persica*, *Juglans regia*, and *Vitis vinifera*, respectively. (From Kouidhi, B., et al., *Microb. Pathog.*, 80, 39–49, 2015. With permission.)

Medical devices having antibacterial coverings are proved very effective because these kill the bacteria as soon as it comes in its contact without allowing the formation of biofilm. This phenomenon is used in orthodontic, bone cements, and orthopedic (Jaeblon 2010). The bioactive compounds used in modified medical devices include biosurfactants, lactoferrin, antibody-releasing surfaces, and quorum sensing inhibitors (Rendueles & Ghigo 2012).

20.5 CONCLUSION

Technologies under investigation for the control and prevention of biofilm are using enzymes directly targeting matrix, bactericidal molecules, and antiadhesion coverings. Bioactive molecules from different plant extracts have been proved to inhibit and destroy the bacterial biofilm. Medical devices are modified by covering with antimicrobial agents preventing biofilm formation and do not allow adhesion of bacteria to these surfaces. These novel antibiofilm technologies could eventually lead to antibiofilm therapies that are superior to the previously available antibiotic treatment.

ACKNOWLEDGMENT

The authors of this book chapter gratefully acknowledge the funding support from Higher Education Commission (HEC), Government of Pakistan for this support under the project NRPU#4927.

REFERENCES

Almas, Khalid, and Zuhair Al-Zeid. "The immediate antimicrobial effect of a toothbrush and miswak on cariogenic bacteria: a clinical study." *The Journal of Contemporary Dental Practice* 5, no. 1 (2004): 105–114.

Al-Sohaibani, Saleh, and Kasi Murugan. "Antibiofilm activity of *Salvadora persica* on cariogenic isolates of *Streptococcus mutans*: in vitro and molecular docking studies." *Biofouling* 28, no. 1 (2012): 29–38.

Amoian, Babak, Ali Akbar Moghadamnia, Soozan Barzi, Sepideh Sheykholeslami, and Afsaneh Rangiani. "*Salvadora persica* extract chewing gum and gingival health: improvement of gingival and probe-bleeding index." *Complementary Therapies in Clinical Practice* 16, no. 3 (2010): 121–123.

Baba, Shintaro, Kenji Kondo, Makiko Toma-Hirano, Kaori Kanaya, Keigo Suzukawa, Munetaka Ushio, MahoSuzukawa, Ken Ohta, and Tatsuya Yamasoba. "Local increase in IgE and class switch recombination to IgE in nasal polyps in chronic rhinosinusitis." *Clinical & Experimental Allergy* 44, no. 5 (2014): 701–712.

Banerjee, Indrani, Ravindra C. Pangule, and Ravi S. Kane. "Antifouling coatings: recent developments in the design of surfaces that prevent fouling by proteins, bacteria, and marine organisms." *Advanced Materials* 23, no. 6 (2011): 690–718.

Bixler, Gregory D., and Bharat Bhushan. "Biofouling: lessons from nature." *Philosophical Transactions of the Royal Society A: Mathematical, Physical, and Engineering Sciences* 370, no. 1967 (2012): 2381–2417.

Bjarnsholt, Thomas, Peter Østrup Jensen, Mark J. Fiandaca, Jette Pedersen, Christine Rønne Hansen, Claus Bøgelund Andersen, Tacjana Pressler, Michael Givskov, and Niels Høiby. "*Pseudomonas aeruginosa* biofilms in the respiratory tract of cystic fibrosis patients." *Pediatric Pulmonology* 44, no. 6 (2009): 547–558.

Chaieb, Kamel, Bochra Kouidhi, Rihab BEN Slama, Kais Fdhila, Tarek Zmantar, and Amina Bakhrouf. "Cytotoxicity, antibacterial, antioxidant, and antibiofilm properties of tunisian *Juglans regia* bark extract." *Journal of Herbs, Spices & Medicinal Plants* 19, no. 2 (2013): 168–179.

Chipps, Bradley E., Maria Figliomeni, and Sheldon Spector. "Omalizumab: An update on efficacy and safety in moderate-to-severe allergic asthma." In *Allergy & Asthma Proceedings*, vol. 33, no. 5. 2012.

Del Pozo, Jose L., and Robin Patel. "Infection associated with prosthetic joints." *New England Journal of Medicine* 361, no. 8 (2009): 787–794.

Dongari-Bagtzoglou, Anna. "Pathogenesis of mucosal biofilm infections: challenges and progress." *Expert Review of Anti-Infective Therapy* 6, no. 2 (2008): 201–208.

Faraz, Naveed, Z. Islam, and Rehana Rehman. "Sehrish Antibiofilm-forming activity of naturally occurring compound." *Biomedica* 28, no. 2 (2012): 171–175.

Furiga, Aurélie, Aline Lonvaud-Funel, Georges Dorignac, and Cecile Badet. "In vitro antibacterial and antiadherence effects of natural polyphenolic compounds on oral bacteria." *Journal of Applied Microbiology* 105, no. 5 (2008): 1470–1476.

Gilbert, Peter, Tomas Maira-Litran, Andrew J. McBain, Alexander H. Rickard, and Fraser W. Whyte. "The physiology and collective recalcitrance of microbial biofilm communities." *Advances in Microbial Physiology* 46 (2002): 202–256.

Gyawali, Rabin, and Salam A. Ibrahim. "Natural products as antimicrobial agents." *Food Control* 46 (2014): 412–429.

Hasan, Jafar, Russell J. Crawford, and Elena P. Ivanova. "Antibacterial surfaces: the quest for a new generation of biomaterials." *Trends in Biotechnology* 31, no. 5 (2013): 295–304.

Høiby, Niels, Thomas Bjarnsholt, Michael Givskov, Søren Molin, and Oana Ciofu. "Antibiotic resistance of bacterial biofilms." *International Journal of Antimicrobial Agents* 35, no. 4 (2010): 322–332.

Jaeblon, Todd. "Polymethylmethacrylate: properties and contemporary uses in orthopedics." *JAAOS-Journal of the American Academy of Orthopaedic Surgeons* 18, no. 5 (2010): 297–305.

Jenkinson, Howard F., and Richard J. Lamont. "Oral microbial communities in sickness and in health." *Trends in Microbiology* 13, no. 12 (2005): 589–595.

Kartush, Alison G., Jane K. Schumacher, Rachna Shah, and Monica O. Patadia. "Biologic agents for the treatment of chronic rhinosinusitis with nasal polyps." *American Journal of Rhinology & Allergy* 33, no. 2 (2019): 203–211.

Khawar, Muhammad Babar, Muddasir Hassan Abbasi, and Nadeem Sheikh. "IL-32: a novel pluripotent inflammatory interleukin, towards gastric inflammation, gastric cancer, and chronic rhinosinusitis." *Mediators of Inflammation* 2016 (2016).

Kim, Harold, Anne K. Ellis, David Fischer, Mary Noseworthy, Ron Olivenstein, Kenneth R. Chapman, and Jason Lee. "Asthma biomarkers in the age of biologics." *Allergy, Asthma & Clinical Immunology* 13, no. 1 (2017): 48.

Klausen, Mikkel, Arne Heydorn, Paula Ragas, Lotte Lambertsen, Anders Aaes-Jørgensen, Søren Molin, and Tim Tolker-Nielsen. "Biofilm formation by *Pseudomonas aeruginosa* wild type, flagella and type IV pili mutants." *Molecular Microbiology* 48, no. 6 (2003): 1511–1524.

Kouidhi, Bochra, Yasir Mohammed A. Al Qurashi, and Kamel Chaieb. "Drug resistance of bacterial dental biofilm and the potential use of natural compounds as alternative for prevention and treatment." *Microbial Pathogenesis* 80 (2015): 39–49.

Lee, Ashlynn L. Z., Victor W. L. Ng, Weixin Wang, James L. Hedrick, and Yi Yan Yang. "Block copolymer mixtures as antimicrobial hydrogels for biofilm eradication." *Biomaterials* 34, no. 38 (2013): 10278–10286.

Lewis, Kim. "Persister cells and the riddle of biofilm survival." *Biochemistry (Moscow)* 70, no. 2 (2005): 267–274.

glucotransferase B according to concentration of dose (Furiga et al. 2008). When its extracts were separated and tested by using phenol, it gives better results against the glucotransferase inhibition showing a maximum of 60% inhibition at a minimum concentration (Thimothe et al. 2007).

20.4 DEVICE-RELATED BIOFILMS

It is reported that bacteria also form biofilm on the inert materials which include artificially synthesized polymers, internally used medical devices. The individual cell from biofilm can cause disease and transmit it to other parts of the body (Dongari-Bagtzoglou 2008). It is stated that the infections caused by planktonic bacterial cells are easy to handle and cure as compared to those caused by biofilm which leads to death all over the world. Due to this, scientists are looking forward to the devices that have antimicrobial activities to prevent patients from lethal effects (Zarb et al. 2012).

20.4.1 Approaches to Biofilm Control and Treat Device-Related Biofilm

Biofilm formation on the biomedical devices occurs due to contamination while implantation in patients' body, and more sophisticatedly these bacteria form biofilm on these devices due to their surface and structure functionalities. Nowadays, scientists are focusing to make devices having antimicrobial activities and anti-inhibitory effects according to their surface functionalities and structural composition. For this purpose, different ways are used which include:

- Addition of antibacterial to the surfaces of health devices (Shintani 2004)
- Antiadhesive surface changes (Neoh & Kang 2011)
- Covering devices through polymer substances (Li 2013)
- Surface manufacturing with biochemical moieties (Pavlukhina et al. 2012)
- Antifouling coverings (Banerjee et al. 2011)
- Covering, control, adsorption, or lamination of biomolecules (Lee et al. 2013).

The attachment of microbes to the medical devices happened due to the protein layer that adsorbed on them followed by immobilization of bacterial cells with the help of oligo ethylene glycol or poly ethylene glycol and production of antifouling surfaces due to zwitterion species (Hasan et al. 2013).

Structurally modified medical devices contain sulfonate unit that do not allow the interaction and attachment of *Candida albicans*, *Pseudomonas aeruginosa*, *Staphylococcus epidermidis*, *Candida tropicalis*, and *Escherichia coli* (Roosjen et al. 2004). More characteristics associated with the attachment of bacteria to inert surfaces are hydrophobicity of surface, charge, and roughness (Bixler & Bhushan 2012). Natural antibacterial drugs are also tested for disease-causing *Listeria monocytogenes*, spoilage organisms, and *E. coli*, and the glass having poly vinyl-N-alkyl pyridinium bromide is the most effective agent that kills airborne bacteria (Rojas-Graü et al. 2009).

Li, Li, Soeren Molin, Liang Yang, and Sokol Ndoni. "Sodium dodecyl sulfate (SDS)-loaded nanoporous polymer as antibiofilm surface coating material." *International Journal of Molecular Sciences* 14, no. 2 (2013): 3050–3064.

Nassiri-Asl, Marjan, and Hossein Hosseinzadeh. "Review of the pharmacological effects of *Vitis vinifera* (Grape) and its bioactive compounds." *Phytotherapy Research: An International Journal Devoted to Pharmacological and Toxicological Evaluation of Natural Product Derivatives* 23, no. 9 (2009): 1197–1204.

Neoh, Koon Gee, and En-Tang Kang. "Combating bacterial colonization on metals via polymer coatings: relevance to marine and medical applications." *ACS Applied Materials & Interfaces* 3, no. 8 (2011): 2808–2819.

Pavlukhina, Svetlana V., Jeffrey B. Kaplan, Li Xu, Wei Chang, Xiaojun Yu, Srinivasa Madhyastha, Nandadeva Yakandawala, Almagul Mentbayeva, Babar Khan, and Svetlana A. Sukhishvili. "Noneluting enzymatic antibiofilm coatings." *ACS Applied Materials & Interfaces* 4, no. 9 (2012): 4708–4716.

Projan, Steven J., and Philip J. Youngman. "Antimicrobials: new solutions badly needed." *Current Opinion in Microbiology* 5, no. 5 (2002): 463–465.

Rendueles, Olaya, and Jean-Marc Ghigo. "Multi-species biofilms: how to avoid unfriendly neighbors." *FEMS Microbiology Reviews* 36, no. 5 (2012): 972–989.

Rivero, Alexander, and Jonathan Liang. "Anti-IgE and anti-IL5 biologic therapy in the treatment of nasal polyposis: a systematic review and meta-analysis." *Annals of Otology, Rhinology & Laryngology* 126, no. 11 (2017): 739–747.

Rojas-Graü, María Alejandra, Robert Soliva-Fortuny, and Olga Martín-Belloso. "Edible coatings to incorporate active ingredients to fresh-cut fruits: a review." *Trends in Food Science & Technology* 20, no. 10 (2009): 438–447.

Roosjen, Astrid, Henny C. van der Mei, Henk J. Busscher, and Willem Norde. "Microbial adhesion to poly (ethylene oxide) brushes: influence of polymer chain length and temperature." *Langmuir* 20, no. 25 (2004): 10949–10955.

Rosan, Burton, and Richard J. Lamont. "Dental plaque formation." *Microbes and Infection* 2, no. 13 (2000): 1599–1607.

Sbordone, Ludovico, and Claudia Bortolaia. "Oral microbial biofilms and plaque-related diseases: microbial communities and their role in the shift from oral health to disease." *Clinical Oral Investigations* 7, no. 4 (2003): 181–188.

Shintani, Hideharu. "Modification of medical device surface to attain anti-infection." *Trends Biomater Artif Organs* 18, no. 1 (2004): 1–8.

Thimothe, Joanne, Illeme A. Bonsi, Olga I. Padilla-Zakour, and Hyun Koo. "Chemical characterization of red wine grape (*Vitis vinifera* and *Vitis interspecific* hybrids) and pomace phenolic extracts and their biological activity against *Streptococcus mutans*." *Journal of Agricultural and Food Chemistry* 55, no. 25 (2007): 10200–10207.

Thomas, John, Sara Linton, Linda Corum, Will Slone, Tyler Okel, and Steven L. Percival. "The affect of pH and bacterial phenotypic state on antibiotic efficacy." *International Wound Journal* 9, no. 4 (2012): 428–435.

Tsetsos, Nikolaos, John K. Goudakos, Dimitrios Daskalakis, Iordanis Konstantinidis, and Konstantinos Markou. "Monoclonal antibodies for the treatment of chronic rhinosinusitis with nasal polyposis: a systematic review." *Rhinology* 56, no. 1 (2018): 11–21.

Walker, Abigail, Carl Philpott, and Claire Hopkins. "What is the most appropriate treatment for chronic rhinosinusitis?" *Postgraduate Medical Journal* 95, no. 1127 (2019): 493–496.

Wen, Aimin, Pascal Delaquis, Kareen Stanich, and Peter Toivonen. "Antilisterial activity of selected phenolic acids." *Food Microbiology* 20, no. 3 (2003): 305–311.

Whitchurch, Cynthia B., Tim Tolker-Nielsen, Paula C. Ragas, and John S. Mattick. "Extracellular DNA required for bacterial biofilm formation." *Science* 295, no. 5559 (2002): 1487–1487.

Wu, Christine D. "Grape products and oral health." *The Journal of Nutrition* 139, no. 9 (2009): 1818S–1823S.

Zarb, Peter, B. Coignard, J. Griskeviciene, A. Muller, V. Vankerckhoven, K. Weist, Mathijs Michiel Goossens Vaerenberg, S., Hopkins, S., Catry, B., Dominique, M.. "The European Centre for Disease Prevention and Control (ECDC) pilot point prevalence survey of healthcare-associated infections and antimicrobial use." *Eurosurveillance* 17, no. 46 (2012): 20316.

Index

Note: **Bold** page numbers refer to tables and *italic* page numbers refer to figures